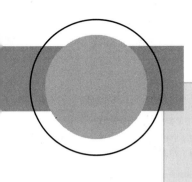

Introduction to Wireless and Mobile Systems
Second Edition

Dharma Prakash Agrawal
Qing-An Zeng

Department of Electrical & Computer Engineering and Computer Science
University of Cincinnati

THOMSON

Australia   Canada   Mexico   Singapore   Spain   United Kingdom   United States

Introduction to Wireless and Mobile Systems, 2e
by Dharma Prakash Agrawal and Qing-An Zeng

Associate Vice-President and Editorial Director:
Evelyn Veitch

Publisher:
Bill Stenquist

Sales and Marketing Manager:
John More

Developmental Editor:
Kamilah Reid Burrell

Permissions Coordinator:
Vicki Gould

Production Services:
RPK Editorial Services

Copy Editor:
Harlan James

Proofreader:
Patricia Daly

Indexer:
Dharma Agrawal
Qing-An Zeng
RPK Editorial Services

Production Manager:
Renate McCloy

Creative Director:
Angela Cluer

Interior Design:
Carmela Pereira

Cover Design:
Andrew Adams

Compositor:
PreTEX, Inc.

Printer:
Transcontinental Printing

North America
Nelson
1120 Birchmount Road
Toronto, Ontario M1K 5G4
Canada

Asia
Thomson Learning
5 Shenton Way #01-01
UIC Building
Singapore 068808

Australia/New Zealand
Thomson Learning
102 Dodds Street
Southbank, Victoria
Australia 3006

Europe/Middle East/Africa
Thomson Learning
High Holborn House
50/51 Bedford Row
London WC1R 4LR
United Kingdom

Latin America
Thomson Learning
Seneca, 53
Colonia Polanco
11560 Mexico D.F.
Mexico

Spain
Paraninfo
Calle/Magallanes, 25
28015 Madrid, Spain

In memory of my parents Shri Saryoo Prasad Agrawal and Shrimati Chandrakanta Bai Agrawal who raised me affectionately and made me learn how to excel from a small unknown village.

Dharma Prakash Agrawal

To my wife Min, and to our children Yao and Andrew.

Qing-An Zeng

Contents

Preface to the Second Edition xv

Preface to the First Edition xvii

Acknowledgments xxi

● ● ● ● ● ● ● ● ● ● ● ● ● ● ● ●
1 Introduction 1

1.1   History of Cellular Systems 1
1.2   Characteristics of Cellular Systems 8
1.3   Fundamentals of Cellular Systems 12
1.4   Cellular System Infrastructure 17
1.5   Satellite Systems 21
1.6   Network Protocols 22
1.7   Ad Hoc and Sensor Networks 22
1.8   Wireless MANs, LANs, and PANs 24
1.9   Outline of the Book 25
1.10 References 26
1.11 Problems 27

● ● ● ● ● ● ● ● ● ● ● ● ● ● ● ●
2 Probability, Statistics, and Traffic Theories 29

2.1   Introduction 29
2.2   Basic Probability and Statistics Theories 29
  2.2.1   Random Variables 29
  2.2.2   Cumulative Distribution Function 30
  2.2.3   Probability Density Function 31
  2.2.4   Expected Value, nth Moment, nth Central Moment, and Variance 31
  2.2.5   Some Important Distributions 33
  2.2.6   Multiple Random Variables 35
2.3   Traffic Theory 39
  2.3.1   Poisson Arrival Model 39

2.4 Basic Queuing Systems 40
    2.4.1 What Is Queuing Theory? 40
    2.4.2 Basic Queuing Theory 40
    2.4.3 Kendall's Notation 41
    2.4.4 Little's Law 41
    2.4.5 Markov Process 42
    2.4.6 Birth–Death Process 42
    2.4.7 $M/M/1/\infty$ Queuing System 43
    2.4.8 $M/M/S/\infty$ Queuing System 45
    2.4.9 $M/G/1/\infty$ Queuing System 47

2.5 Summary 52

2.6 References 53

2.7 Problems 53

3 Mobile Radio Propagation 57

3.1 Introduction 57

3.2 Types of Radio Waves 57

3.3 Propagation Mechanisms 59

3.4 Free Space Propagation 60

3.5 Land Propagation 61

3.6 Path Loss 62

3.7 Slow Fading 65

3.8 Fast Fading 67
    3.8.1 Statistical Characteristics of Envelope 67
    3.8.2 Characteristics of Instantaneous Amplitude 70

3.9 Doppler Effect 71

3.10 Delay Spread 72

3.11 Intersymbol Interference 73

3.12 Coherence Bandwidth 74

3.13 Cochannel Interference 75

3.14 Summary 76

3.15 References 76

3.16 Problems 76

4 Channel Coding and Error Control 79

4.1 Introduction 79

4.2 Linear Block Codes 80

4.3 Cyclic Codes 85

4.4   Cyclic Redundancy Check (CRC) 86

4.5   Convolutional Codes 87

4.6   Interleaver 90

4.7   Turbo Codes 91

4.8   ARQ Techniques 92

        4.8.1   Stop-And-Wait ARQ Scheme 93
        4.8.2   Go-Back-N ARQ Scheme 94
        4.8.3   Selective-Repeat ARQ Scheme 96

4.9   Summary 97

4.10   References 97

4.11   Problems 98

5 Cellular Concept 102

5.1   Introduction 102

5.2   Cell Area 102

5.3   Signal Strength and Cell Parameters 104

5.4   Capacity of a Cell 108

5.5   Frequency Reuse 110

5.6   How to Form a Cluster? 112

5.7   Cochannel Interference 114

5.8   Cell Splitting 116

5.9   Cell Sectoring 117

5.10   Summary 119

5.11   References 120

5.12   Problems 120

6 Multiple Radio Access 125

6.1   Introduction 125

6.2   Multiple Radio Access Protocols 126

6.3   Contention-Based Protocols 127

        6.3.1   Pure ALOHA 128
        6.3.2   Slotted ALOHA 130
        6.3.3   CSMA 131
        6.3.4   CSMA/CD 134
        6.3.5   CSMA/CA 136

6.4   Summary 139

6.5   References 139

6.6   Problems 141

● ● ● ● ● ● ● ● ● ● ● ● ● ● ● ●

7 Multiple Division Techniques 143

7.1   Introduction 143

7.2   Concepts and Models for Multiple Divisions 143
    7.2.1   FDMA 144
    7.2.2   TDMA 146
    7.2.3   CDMA 148
    7.2.4   OFDM 155
    7.2.5   SDMA 156
    7.2.6   Comparison of Multiple Division Techniques 158

7.3   Modulation Techniques 159
    7.3.1   AM 159
    7.3.2   FM 160
    7.3.3   FSK 161
    7.3.4   PSK 161
    7.3.5   QPSK 162
    7.3.6 $\pi/4$QPSK 163
    7.3.7   QAM 164
    7.3.8   16QAM 165

7.4   Summary 165

7.5   References 166

7.6   Problems 166

● ● ● ● ● ● ● ● ● ● ● ● ● ● ● ●

8 Channel Allocation 169

8.1   Introduction 169

8.2   Static Allocation versus Dynamic Allocation 170

8.3   Fixed Channel Allocation (FCA) 171
    8.3.1   Simple Borrowing Schemes 172
    8.3.2   Complex Borrowing Schemes 172

8.4   Dynamic Channel Allocation (DCA) 174
    8.4.1   Centralized Dynamic Channel Allocation Schemes 174
    8.4.2   Distributed Dynamic Channel Allocation Schemes 175

8.5   Hybrid Channel Allocation (HCA) 176
    8.5.1   Hybrid Channel Allocation Schemes 176
    8.5.2   Flexible Channel Allocation Schemes 177

8.6   Allocation in Specialized System Structure 177
    8.6.1   Channel Allocation in One-Dimensional Systems 177
    8.6.2   Reuse Partitioning–Based Channel Allocation 178
    8.6.3   Overlapped Cells–Based Channel Allocation 179

8.7   System Modeling 181
    8.7.1   Basic Modeling 181

8.7.2 Modeling for Channel Reservation 183

8.8   Summary 184

8.9   References 185

8.10 Problems 185

● ● ● ● ● ● ● ● ● ● ● ● ● ● ● ● ●

9 Mobile Communication Systems 190

9.1   Introduction 190

9.2   Cellular System Infrastructure 190

9.3   Registration 193

9.4   Handoff Parameters and Underlying Support 195
      9.4.1 Parameters Influencing Handoff 195
      9.4.2 Handoff Underlying Support 196

9.5   Roaming Support 198
      9.5.1 Home Agents, Foreign Agents, and Mobile IP 200
      9.5.2 Rerouting in Backbone Routers 203

9.6   Multicasting 204

9.7   Security and Privacy 207
      9.7.1 Encryption Techniques 207
      9.7.2 Authentication 209
      9.7.3 Wireless System Security 212

9.8   Firewalls and System Security 215

9.9   Summary 216

9.10 References 216

9.11 Problems 218

● ● ● ● ● ● ● ● ● ● ● ● ● ● ● ● ●

10 Existing Wireless Systems 221

10.1 Introduction 221

10.2 AMPS 221
      10.2.1 Characteristics of AMPS 222
      10.2.2 Operation of AMPS 223
      10.2.3 General Working of AMPS Phone System 225

10.3 IS-41 226
      10.3.1 Introduction 226
      10.3.2 Support Operations 228

10.4 GSM 229
      10.4.1 Frequency Bands and Channels 230
      10.4.2 Frames in GSM 232
      10.4.3 Identity Numbers used by a GSM System 232
      10.4.4 Interfaces, Planes, and Layers of GSM 235

10.4.5 Handoff 237
10.4.6 Short Message Service (SMS) 238

10.5 PCS 239
10.5.1 Chronology of PCS Development 240
10.5.2 Bellcore View of PCS 241

10.6 IS-95 243
10.6.1 Power Control 247

10.7 IMT–2000 249
10.7.1 International Spectrum Allocation 250
10.7.2 Services Provided by Third-Generation Cellular Systems 250
10.7.3 Harmonized 3G Systems 251
10.7.4 Multimedia Messaging Service (MMS) 252
10.7.5 Universal Mobile Telecommunications System (UMTS) 253

10.8 Summary 258

10.9 References 259

10.10 Problems 259

● ● ● ● ● ● ● ● ● ● ● ● ● ● ●
11 Satellite Systems 261

11.1 Introduction 261

11.2 Types of Satellite Systems 261

11.3 Characteristics of Satellite Systems 267

11.4 Satellite System Infrastructure 268

11.5 Call Setup 271

11.6 Global Positioning System 273
11.6.1 Limitations of GPS 277
11.6.2 Beneficiaries of GPS 279

11.7 A-GPS and E 911 280

11.8 Summary 281

11.9 References 281

11.10 Problems 282

● ● ● ● ● ● ● ● ● ● ● ● ● ● ●
12 Network Protocols 284

12.1 Introduction 284

12.2 OSI Model 284
12.2.1 Layer 1: Physical Layer 285
12.2.2 Layer 2: Data Link Layer 286
12.2.3 Layer 3: Network Layer 286
12.2.4 Layer 4: Transport Layer 286
12.2.5 Layer 5: Session Layer 287

12.2.6 Layer 6: Presentation Layer 287
12.2.7 Layer 7: Application Layer 287

12.3   TCP/IP Protocol 288
12.3.1 Physical and Data Link Layers 288
12.3.2 Network Layer 289
12.3.3 TCP 290
12.3.4 Application Layer 291
12.3.5 Routing using Bellman-Ford Algorithm 291

12.4   TCP over Wireless 292
12.4.1 Need for TCP over Wireless 292
12.4.2 Limitations of Wired Version of TCP 292
12.4.3 Solutions for Wireless Environment 292

12.5   Internet Protocol Version 6 (IPv6) 296
12.5.1 Transition from IPv4 to IPv6 296
12.5.2 IPv6 Header Format 297
12.5.3 Features of IPv6 297
12.5.4 Differences between IPv6 and IPv4 298

12.6   Summary 299

12.7   References 299

12.8   Problems 300

13 Ad Hoc and Sensor Networks 303

13.1   Introduction 303

13.2   Characteristics of MANETs 304

13.3   Applications 305

13.4   Routing 306
13.4.1 Need for Routing 307
13.4.2 Routing Classification 308

13.5   Table-Driven Routing Protocols 308
13.5.1 Destination-Sequenced Distance-Vector Routing 309
13.5.2 Cluster Head Gateway Switch Routing 309
13.5.3 Wireless Routing Protocol 311

13.6   Source-Initiated On-Demand Routing 311
13.6.1 Ad Hoc On-demand Distance Vector Routing 312
13.6.2 Dynamic Source Routing 313
13.6.3 Temporarily Ordered Routing Algorithm 315
13.6.4 Associativity-Based Routing 318
13.6.5 Signal Stability-Based Routing 319

13.7   Hybrid Protocols 320
13.7.1 Zone Routing 320
13.7.2 Fisheye State Routing 321

13.7.3 Landmark Routing (LANMAR) for MANET with Group Mobility 321

13.7.4 Location-Aided Routing 321

13.7.5 Distance Routing Effect Algorithm for Mobility 323

13.7.6 Relative Distance Microdiscovery Ad Hoc Routing 323

13.7.7 Power Aware Routing 324

13.7.8 Multipath Routing Protocols 324

13.8 Wireless Sensor Networks 334

13.8.1 Case Study 336

13.8.2 DARPA Efforts toward Wireless Sensor Networks 338

13.9 Fixed Wireless Sensor Networks 338

13.9.1 Classification of Sensor Networks 339

13.9.2 Fundamentals of MAC Protocol for Wireless Sensor Networks 340

13.9.3 Flat Routing in Sensor Networks 341

13.10 Summary 349

13.11 References 349

13.12 Problems 353

14 Wireless MANs, LANs, and PANs 358

14.1 Introduction 358

14.2 Wireless Metropolitan Area Networks (WMANs) 359

14.2.1 IEEE 802.16 359

14.2.2 Ricochet 367

14.3 Wireless Local Area Networks (WLANs) 369

14.3.1 IEEE 802.11 369

14.3.2 ETSI HiperLAN 372

14.3.3 HomeRF 374

14.4 Wireless Personal Area Networks (WPANs) 377

14.4.1 Introduction 377

14.4.2 IEEE 802.15.1 (Bluetooth) 378

14.4.3 IEEE 802.15.3 385

14.4.4 IEEE 802.15.4 387

14.5 Summary 392

14.6 References 392

14.7 Problems 394

15 Recent Advances 397

15.1 Introduction 397

15.2 Ultra-Wideband Technology 398

15.2.1 UWB System Characteristics 399

15.2.2 UWB Signal Propagation 400

15.2.3 Current Status and Applications of UWB Technology 400

15.2.4 Difference Between UWB and Spread Spectrum Techniques 401

15.2.5 UWB Technology Advantages 401

15.2.6 UWB Technology Drawbacks 402

15.2.7 Challenges for UWB Technology 402

15.2.8 Future Directions 403

15.3 Multimedia Services Requirements 403

15.3.1 Media Codecs 404

15.3.2 File Formats 404

15.3.3 HTTP 404

15.3.4 Media Control Protocols 405

15.3.5 SIP 405

15.3.6 Multimedia Messaging Service 405

15.3.7 Multimedia Transmission in MANETs 406

15.4 Push-to-Talk (PTT) Technology 408

15.4.1 PTT Network Technology 408

15.4.2 PTT in iDEN Cellular Networks 409

15.4.3 PTT in Non-iDEN Cellular Networks: PoC 409

15.4.4 Limitations of Current Services 410

15.5 Mobility and Resource Management for Integrated Systems 411

15.5.1 Mobility Management 411

15.5.2 Resource Management 412

15.5.3 Recent Advances in Resource Management 414

15.6 Enhancement for IEEE 802.11 WLANs 415

15.6.1 Issues in MAC Protocols 417

15.7 Multicast in Wireless Networks 419

15.7.1 Recent Advances in Multicast over Mobile IP 419

15.7.2 Reliable Wireless Multicast Protocols 421

15.7.3 Future Directions 423

15.8 Directional and Smart Antennas 423

15.8.1 Types of Antennas 424

15.8.2 Smart Antennas and Beamforming 424

15.8.3 Smart Antennas and SDMA 425

15.9 Design Issues in Sensor Networks 427

15.9.1 Sensor Databases 428

15.9.2 Collaborative Information Processing 428

15.9.3 Operating System Design 429

15.9.4 Multipath Routing in Sensor Networks 429

15.9.5 Service Differentiation 431

15.10 Bluetooth Networks 432

15.10.1 Interference on Bluetooth Networks 432

15.10.2 Bluetooth Dynamic Slot Assignment 433

15.10.3 BlueStar: Enabling Efficient Integration between Bluetooth WLANs and WPANs 434

15.10.4 Traffic Engineering over Bluetooth MANETs 435

15.10.5 Distributed Topology Construction 436

15.11 Low-Power Design 437

15.12 XML 438

15.12.1 HTML versus Markup Language 438

15.12.2 WML: XML Application for Wireless Handheld Devices 439

15.13 Threats and Security Issues 440

15.13.1 Security Threats to Wireless Networks 440

15.13.2 Why Existing Wired Solutions Are Not Applicable to Wireless Networks 441

15.13.3 Current Approaches 441

15.13.4 Intrusion Detection in MANETs 442

15.14 Summary 445

15.15 References 446

15.16 Problems 455

A Erlang B Table 459

B Simulation Projects 465

Acronyms 467

Index 479

Preface to the Second Edition

We are very pleased to see an overwhelming acceptance of our book by the worldwide wireless community. In response to recent changes in this technological field, it is our honor to present a new edition of our book within two years of the first printing. The draft version of the second edition was sent to six reviewers by the publisher. Many thanks for their constructive criticism. We have made special efforts to incorporate their useful suggestions and we hope the readers will find this edition comprehensive, easy to understand, and up to date. The task has been very challenging, and we hope our efforts are reflected in this edition by making it easy to appreciate the advances in this exciting technology.

In this edition, we have retained all the chapters and their sequence as used in the original edition. The major changes could be summarized as follows: addition of explanations and motivation for many of the concepts, numerical examples wherever possible, additional problems at the end of each chapter, and an introduction of some new concepts to reflect the state of the art. Specifically, we have emphasized the importance of the probability theory in the wireless and mobile systems area. We have also added the generalized Nakagami distribution to show the usefulness of the CRC scheme. We have explicitly illustrated how to form a cluster of given size for FDMA/TMDA systems. We have included derivations of pure and slotted ALOHA and ARQ; we have added CSMA/CD protocol and augmented security schemes. We have added two new multiple access concepts of OFDM and SDMA. A description of SMS has been added and the explanation of the Bellman-Ford algorithm has been given to calculate the shortest path between any two nodes. We have reorganized sections on routing in ad hoc networks and added multipath routing and explicitly identified WiFi as 802.11b. We have changed Chapter 14 to Wireless MAN, LAN, and PAN by adding a MAN portion and organizing the contents for enhanced clarity. We have incorporated many new topics in Chapter 15 such as Multimedia Transmission in Multimedia, PTT Technology, WiMax, Scheduling in Piconets, and Use of Directional Antenna.

Putting together this second edition has not been an easy job. Help from numerous individuals has made this formidable task both manageable and enjoyable. Professor Anup Kumar of University of Louisville and Professor Hassan Peyravi of Kent State University were the first ones to provide feedback on our first edition. Professor Ramesh C. Joshi, Indian Institute of Technology—Roorkee gave very useful comments on the draft of this second edition. Ashok Roy and Wei Li helped in reorganizing Chapter 14, while Anurag Gupta and Kumar Anand helped redoing part of Chapter 13. Many students in our research group provided comments on the contents of Chapter 15 including Torsha Banerjee, Carlos Cordeiro, Chittabrata Ghosh, Hrishikesh Gossain, Neha Jain, and Dhananjay Lal. Many thanks are also due to Weiqun Chen, Yunli Chen, Hang Chen, Hongmei Deng, Aditya Gupta, Vivek Jain, Xiaodong Li, Anindo Mukherjee, Wei Shen, Demin Wang, Haitang Wang, and Qi Zhang for reading different versions of our book and providing many helpful hints.

Our sincere thanks goes to our publisher, Mr. Bill Stenquest, for asking us to prepare a second edition within such a short time following the first edition. We would also like to thank Kamilah Reid Burrell, Development Editor, Thomson Engineering and Rose Kernan, Production Editor, for converting our electronic version of the text, figures, tables, and index into the final form.

We are very grateful to our families for their encouragement and countless hours of patience and endurance during the course of this revision.

DHARMA PRAKASH AGRAWAL
QING-AN ZENG

Preface to the First Edition

Wireless systems have been around for quite some time and their obvious use in garage-door openers and cordless phones has gone unnoticed until recently. Their unique capability of maintaining the same contact number even if the user moves from one location to another has made them increasingly popular. Wireless telephones are not only convenient but are also providing flexibility and versatility. The introduction of affordably priced wireless and mobile telephones has made them attractive for the general population worldwide. Thus, the number of wireless phone subscribers as well as service providers has proliferated.

Wireless and mobile communications have been useful in areas such as commerce, education, and defense. According to the nature of a particular application, they can be used in home-based and industrial systems or in commercial and military environments. In a home-based system, a central access point communicates with various appliances and controls them using a localized wireless node. This kind of system enables close coordination among appliances in the home (or industry) and achieves control over the home (or industry) access point using voice or a short message. To facilitate this, a consortium of companies is working on the Bluetooth project. There are many novel applications of such a wireless system—for example, a bracelet worn by a subscriber can constantly monitor body parameters and take action if needed (like informing the family physician about a health problem). However, the design and implementation of such a system brings with it a lot of important issues, such as standardization and infrastructure for Internet access, audio/video editing, and distributed decision-making software.

In a commercial system, the common issues are the range of the system, number of distribution infrastructure access points, number of users for each access point, and so on. For instance, we need to have several access points uniformly distributed in each floor of a factory so that users have continuous access to them. But this gives rise to problems such as appropriate coordination of channels between access points and the channel bandwidth requirements. Any loss of information (voice or data packet) in wireless switching is unacceptable, hence care should be taken to ensure the reliable transmission and reception of information.

Wireless systems, such as the traditional infrastructure system, satellite system, or the more recent ad hoc networks formed by mobile users find tremendous use in defense applications. Ad hoc networks involve information transfer in the peer-to-peer mode but we have to deal with the problem of power consumption for a wide coverage area. Other problems involve channel allocation based on address, traffic types (voice, video, data, or audio), mobility pattern, and routing techniques, etc.

The wireless technology has also influenced instructional infrastructure at many institutions. Carnegie Mellon University has taken the lead in creating a campus-wide wireless network. Steps have also been taken at the University of Cincinnati by installing wireless access points at several selected building and by requiring all incoming engineering undergraduate students to have laptops with wireless capability. Similar phenomena can be observed across the country at different organizations. Within Engineering and Computer and Information Science disciplines, communication technology recently has advanced at an unparalleled speed. In particular, combinations of wireless communication and computer technologies have revolutionized the world of telecommunications. To fully explore and utilize this new technology, universities need to offer new courses and train students in the field so that they could continue their graduate work in this area. However, the students in Computer Science and Engineering (CSE) and Electrical Engineering (EE) are at best exposed to data communication aspects, while wireless communication systems remain untouched, as it is relatively difficult to learn about wireless technology without having substantial background in communications technology. On the other hand, EE students learn about the radio frequency (RF) communication aspect only, and the topic of data communication and computing system issues and their correlation in nomadic seamless computing remains untouched.

Although there are many books related to wireless and mobile communications, these books can be roughly classified into two groups. The first group focuses on readers in the RF communication field, and the other covers only the general knowledge of data communication and is designed for sales agents and managers. The books in the first group require a detailed background in RF communication and signal processing and, therefore, are not suitable for students in CSE. Many recent texts emphasize microwave radar and sensor systems. However, books in the second group do not provide any depth in the data communication aspects of wireless technology. Many institutions do offer courses in the wireless and mobile networking area, primarily for graduate students, and then only as special topics. Most of these courses are EE types with many prerequisites as EE courses. Thus, most undergraduate seniors in CSE are deprived of exposure to wireless and mobile communications. In addition, most existing books are tailored toward RF communications and antenna design aspects of the technology, making them difficult to use for CSE students.

Dharma Agrawal envisioned the need for this book when he spent his sabbatical five years ago with AT&T Laboratory. After joining the University of Cincinnati in the autumn of 1998, he started offering an introductory-level course in the wireless and mobile systems area for upper-level undergraduate and entering graduate students. Agrawal primarily used an old textbook, self-prepared notes, and some recent papers. Qing-An Zeng joined the University of Cincinnati in 1999 and started helping organize the course. He noticed the need to develop class notes so that the CSE students, with limited communications background, could understand the subject matter. This led to the foundation of this textbook. The designed course complements the RF communications background of EE students.

Creating such a unique instructional curriculum requires a great deal of efforts. Planning such a text is a relatively difficult task because of the diverse background requirements. The limitations of most existing books and courses affect the wireless industries in the United States. Companies must train newly hired college graduates for a long time before they can get into the wireless industry. To the best of our knowledge, such an organized course has not been taught anywhere in the United

States or the world. Teaching the introductory course strictly from research papers is difficult for the professor, which in turn causes students to learn the material inefficiently. Preparing systematic notes in this emerging area will enhance training, increase the availability of well-educated personal, shorten the new employee training period within industries, encourage students to do graduate work in this area, and allow nations to continue to advance the research in this technological field.

This book explains how wireless systems work, how mobility is supported, how infrastructure underlies such systems, and what interactions are needed among different functional components. It is not our intention to cover various existing wireless technologies, the chronological history behind their development, or the work being carried out, but to make EE and CSE students understand how a cell phone starts working as soon as you get out of an airplane. We have selected chapter topics that focus on qualitative descriptions and realistic explanations of relationships between wireless systems and performance parameters. The chapters are organized as follows:

Chapter 1:	Introduction
Chapter 2:	Probability, Statistics, and Traffic Theories
Chapter 3:	Mobile Radio Propagation
Chapter 4:	Channel Coding
Chapter 5:	Cellular Concept
Chapter 6:	Multiple Radio Access
Chapter 7:	Multiple Division Techniques
Chapter 8:	Channel Allocation
Chapter 9:	Mobile Communication Systems
Chapter 10:	Existing Wireless Systems
Chapter 11:	Satellite Systems
Chapter 12:	Network Protocols
Chapter 13:	Ad Hoc and Sensor Networks
Chapter 14:	Wireless MANs, LANs and PANs
Chapter 15:	Recent Advances

Mathematical formulations are needed in engineering and computer science work, and we include some of the important concepts so that students can appreciate their usefulness in numerous wireless and mobile systems. In all these applications, both security and privacy issues are important. Both ad hoc and sensor networks are finding increasing use in military and commercial applications, so detailed discussions are included. The introduction of the Bluetooth standard allows easy replacement of connector cables with wireless devices and is discussed in detail. Recent advances are covered in the last chapter with emphasis on the research work being carried out in wireless and mobile computing area, even though a comprehensive discussion is beyond the scope of this book. In the questions at the end of each chapter, special efforts have been made to explore potential uses of the various technologies. Depending on availability of time (especially for undergraduates), students should be encouraged to use one of the simulators (ns, OPNET, or other stable simulators) to get a feel for the overall system complexity. A list of possible group simulation projects is included as an Appendix B. The authors have tried such projects for several years and have found them highly effective in training students. Many undergraduates have also used them as their follow-up, year-long capstone design project.

This book is written both for academic institutions and for working professionals. It can be used as a textbook for a one-semester or a one-quarter course. The book also can be used for training current or new employees of wireless companies and could be adopted for short-term training courses. The chapters are organized to provide a great deal of flexibility; emphasis can be given on different chapters, depending on the scope of the course and the instructor's own interests or emphasis. The following are some suggestions for undergraduate students:

- For a one-quarter system, Chapter 15 can be skipped and the project could be optional for extra credit. Chapters 2, 10, 11, 13, and 14 can be covered in brief. Chapter 7 on modulation techniques could be skipped as well.

- For a one-semester system, Chapter 15 can be skipped. Chapters 2 and 10 can be covered briefly, or Chapter 2 could be used for self-study and a simplified version of the project could be assigned.

In this textbook, we have tried to provide an overview of the basic principles behind wireless technology and its associated support infrastructure. We hope that we have been able to achieve our goal of helping students and others working in this area to have a basic knowledge about this exciting technology. Our efforts will not go to waste if we are able to accomplish this to some extent.

DHARMA PRAKASH AGRAWAL
QING-AN ZENG

Acknowledgments

This project would have not been possible without help from numerous individuals. Therefore, the authors would like to acknowledge the time and effort put in by all past and present members of our Research Center for Distributed and Mobile Computing at the University of Cincinnati. Special sincere thanks are due to Ranganath Duggirala, Dilip M. Kutty, Ashok L. Roy, Arati Manjeshwar, Carlos D. Cordeiro, Dhananjay Lal, Wei Li, Yunli Chen, Hrishikesh Gossain, Siddesh Kamat, Sonali Bhargava, Hang Chen, Neha Jain, and Ramnath Duggirala for collecting material for some chapters. We would also like to thank (the names in alphabetical order) Sachin Abhyankar, Nitin Auluck, Shruti Chugh, Hongmei Deng, Sagar Dharia, Sarjoun Doumit, Rahul Gupta, Abinash Mahapatra, Rajani Poorsarla, Rishi Toshniwal, Sasidhar Vogety, Jingao Wang, Qihe Wang, Jun Yin, and Qi Zhang for proofreading numerous versions of this manuscript and their direct and indirect contributions to this book.

We are extremely grateful to our families for their patience and support, especially during the late night and weekend spent writing chapters near different production milestones. Thanks are also due to our wives for their patience and dedication.

We are very grateful to Ms. Christine Sheckels, Sales Consultant for persuading and convincing us to communicate with Thomson for the possible publication of our book, to our Publisher, Mr. Bill Stenquist, and to Ms. Rose Kernan, Production Editor, for their help in publishing this book so quickly.

The authors welcome any comments and suggestions for improvements or changes that could be incorporated in forthcoming editions of this book. Please contact them at <dpa@ececs.uc.edu> and <qzeng@ececs.uc.edu>.

DHARMA PRAKASH AGRAWAL
QING-AN ZENG

About the Authors

Dr. Dharma P. Agrawal is the Ohio Board of Regents Distinguished Professor of Computer Science and Engineering and the founding director for the Center for Distributed and Mobile Computing in the Department of ECECS, University of Cincinnati, OH. His current research interests include energy efficient routing, information retrieval, and secured communication in ad hoc and sensor networks, effective handoff handling and multicasting in integrated wireless networks, interference analysis in piconets and routing in scatternet and use of smart directional antennas for enhanced QoS. He has published over 400 papers in different journals and conferences. He is an editor for the *Journal of Parallel and Distributed Systems, International Journal of High Speed Computing* and founding Editorial Board Member for three new journals: *International Journal on Distributed Sensor Networks*, Taylor and Francis Journal, Philadelphia, 2005, *International Journal of Ad Hoc and Ubiquitous Computing (IJAHUC)*, Interscience Publishers, 2004, and *International Journal of Ad Hoc & Sensor Wireless Networks*, Old City Publishing, 2004. He has served as an editor of the IEEE *Computer Magazine*, and the *IEEE Transactions on Computers*. He has been the Program Chair and General Chair for numerous international conferences and meetings. He has received numerous certificates and awards from the IEEE Computer Society. He was awarded a "*Third Millennium Medal*," by the IEEE for his outstanding contributions. He has also delivered keynote speech for five international conferences. He also has four patents in wireless networking area. He is a Fellow of the IEEE, ACM, AAAS and WIF.

Dr. Qing-An Zeng is with the Department of Electrical and Computer Engineering and Computer Science at the University of Cincinnati since November 1999 and is currently an Assistant Professor of Computer Science and Computer Engineering. In 1997, he joined NEC Corporation, Japan, where he has been engaged in the research and development of the third generation mobile communication systems. Dr. Zeng has published a number of papers in the areas of Performance Modeling and Analysis for Wireless and Mobile Networks, Handoffs, Channel Allocation, Ad Hoc Network, Sensor Network, QoS Issues, and Smart Antenna System. His current research interests include design and analysis of wireless and mobile networks, handoffs, performance modeling, channel allocation, ad hoc and sensor networks, QoS issues, smart antenna, and queuing theory. He received his M.S. and Ph.D. degrees in electrical engineering from Shizuoka University, Japan, in 1994 and 1997, respectively. Dr. Zeng is a member of the IEEE.

Introduction

● ● ● ● ● ● ● ● ● ● ● ● ● ● ●

1.1 History of Cellular Systems

Long-distance communication began with the introduction of telegraphs and simple coded pulses, which were used to transmit short messages. Since then, numerous advances have rendered reliable transfer of information both easier and quicker. There is a long history of how the field has evolved and how telephony has introduced a convenient way of conversing by transmitting audio signals. Hardware connections and electronic switches have made transfer of digital data feasible. The use of the Internet has added another dimension to the wireline communication field, and both voice and data are being processed extensively. In parallel to wireline communication, radio transmission has progressed substantially. Feasibility of wireless transmission has brought drastic changes in the way people live and communicate. New innovations in radio communication have brought about the use of this technology in new application areas [1.1]. A chronological evolution of radio communication is given in Table 1.1, with specific events that occurred in different years clearly marked [1.2]. Table 1.2 lists how, for different applications, radio frequency (RF) bands have been allocated [1.3].

Wireless systems have been around for quite some time, and their obvious use in garage-door openers and cordless telephones has gone unnoticed until recently. The introduction of affordably priced wireless telephones has made them attractive for the general population. Their main usefulness is their capability to maintain the same contact number even if the user moves from one location to another, and this is illustrated in Figure 1.1. Wireless systems have evolved over time, and the chronological development of first-generation (1G) and second-generation (2G) cellular systems (known as mobile systems outside North America) is given in Tables 1.3 and 1.4, respectively.

Table 1.1: ▶
History of Radio and Cellular Systems
Credit: From *Mobile Communications Engineering: Theory and Applications*, 2nd Ed., by W.C.Y. Lee. Copyright 1997 The McGraw Hill Companies.

Year	Event and Characteristics
1860	Maxwell's equation relating electric and magnetic fields
1880	Hertz—Initial demonstration of practical radio communication
1897	Marconi—Radio transmission to a tugboat over an 18-mile path
1921	Detroit Police Department—Police car radio dispatch (2 MHz frequency band)
1933	FCC (Federal Communications Commission)—Authorized four channels in the 30 to 40 MHz range
1938	FCC—Ruled for regular service
1946	Bell Telephone Laboratories—152 MHz (simplex)
1956	FCC—450 MHz (simplex)
1959	Bell Telephone Laboratories—Suggested 32 MHz band for high capacity mobile radio communication
1964	FCC—152 MHz (full duplex)
1964	Bell Telephone Laboratories—Active research at 800 MHz
1969	FCC—450 MHz (full duplex)
1974	FCC—40 MHz bandwidth allocation in the 800 to 900 MHz range
1981	FCC—Release of cellular land mobile phone service in the 40 MHz bandwidth in the 800 to 900 MHz range for commercial operation
1981	AT&T and RCC (radio common carrier) reach an agreement to split 40 MHz spectrum into two 20 MHz bands. Band A belongs to nonwireline operators (RCC), and band B belongs to wireline operators (telephone companies). Each market has two operators
1982	AT&T is divested, and seven RBOCs (regional Bell operating companies) are formed to manage the cellular operations
1982	MFJ (modified final judgment) is issued by the U.S. Department of Justice. All the operators were prohibited to (1) operate long-distance business, (2) provide information services, and (3) do manufacturing business
1983	Ameritech system in operation in Chicago
1984	Most RBOC markets in operation
1986	FCC allocates 5 MHz in extended band
1987	FCC makes lottery on the small metropolitan service area and all rural service area licenses
1988	TDMA (time division multiple access) voted as a digital cellular standard in North America
1992	GSM (global system for mobile communications) operable in Germany D2 system
1993	CDMA (code division multiple access) voted as another digital cellular standard in North America
1994	American TDMA operable in Seattle, Washington
1994	PDC (personal digital cellular) operable in Tokyo, Japan
1994	Two of six broadband PCS (personal communication services) license bands in auction
1995	CDMA operable in Hong Kong
1996	U.S. Congress passes Telecommunication Reform Act Bill
996	The auction money for six broadband PCS licensed bands (120 MHz) almost reaches 20 billion U.S. dollars
1997	Broadband CDMA considered as one of the third-generation mobile communication technologies for UMTS (universal mobile telecommunication systems) during the UMTS workshop conference held in Korea
1999	ITU (International Telecommunication Union) decides the next generation mobile communication systems (e.g., W-CDMA (wideband-CDMA), cdma2000, TD-SCDMA (time division synchronous CDMA))
2001	W-CDMA commercial service beginning from October in Japan
2002	FCC approves additional frequency band for Ultra-Wideband (UWB)

Table 1.2: ▶
Selected U.S. Radio Frequency Allocations and Applications (*continued on next page*)

Application		Radio Frequency Allocation (MHz)
Amateur Radio		144–148, 216–220, 222–225, 420–450, 902–928, 1240–1300, 2300–2305, 2390–2450, 3300–3500, 5650–5925
Aeronautical	Aviation	74.8–75.2, 108–137, 328.6–335.4, 960–1215, 1427–1525, 2200–2290, 2310–2320, 2345–2390
	Commercial Aviation Air-Ground Systems	849–851, 894–896
	General Aviation Air-Ground Radio Telephone	454–455, 459–460
Broadcast	Auxiliary	161.625–161.775, 450–451, 455–456, 941.5–944, 2450–2500, 6425–6525
	Cable TV Relay	6425–6525
	FM Radio	88–99, 100–108
	National Weather Service	162.0125–173.2
	Television	76–88, 174–216
Cellular Telephone		824–849, 869–894
Emergency	Position Indicating Radio Beacons	235–267, 406–406.1
	Locator Transmitters	117.975–121.9375, 235–267
Fixed Microwave	Private and Public	928–929, 932–935, 941–960, 1850–1990, 2110–2220, 2450–2690, 3700–4200, 5925–6875
General Wireless Communications Service		4500–4660*, 4660–4685*
Industrial, Scientific, and Medical (ISM)	Devices/Unlicensed Communication Devices	902–928, 2450–2500, 5725–5875
Location and Monitoring Services		902–928
Maritime Mobile Services		156.2475–157.45, 161.575–161.625, 161.775–162.0125
Medical Telemetry		174–216, 460–668
Microwave Ovens		2400–2483.5
Mobile Radio	Private	72–73, 74.6–74.8, 75.2–76, 150.05–156.2475, 157.1875–161.575, 162.0125–173.4, 220–222, 421–430, 451–454, 456–459, 460–512, 746–824, 851–869, 896–901, 935–940
	Specialized	806–821, 851–866, 896–901, 935–940
Multiple Address Systems		928–929, 932–935, 944–960
Multipoint Distribution & Instructional TV Fixed		2150–2160, 2500–2690

* Frequency allocations pending

Table 1.2: ▶
Selected U.S. Radio Frequency Allocations and Applications (Continued)

Application		Radio Frequency Allocation (MHz)
Paging & Radiotelephone Service		152–152.255, 152.495–152.855, 157.755–158.115, 454–455, 459–460, 929–930, 931–932
Personal Communications	Broadband	1850–1990
	Narrowband	901–902, 930–931, 940–941
	Unlicensed	1910–1930, 2390–2400
Personal Radio Services	Family Radio	462.5375–462.7375, 467.5375–467.7375
	General Mobile Radio	462.5375–462.7375, 467.5375–467.7375
	Low Power Radio	216–220
	Radio Control	72–73, 75.2–76
	General Wireless Communications Service Interactive Video & Data	218–219
Radar		216–220, 235–267, 406.1–450, 902–928, 960–1215, 1215–2290, 2320–2345, 2360–2390, 2700–3100, 3100–3700, 5000–5470, 5600–5925, 8500–10000
Rural Radiotelephone		152–159, 454–460
RFID (Radio Frequency Identification)	Dedicated Short Range Communications and Radio ID Devices	902–928, 2400–2500, 5850–5925*
Satellite Services	Digital Audio Radio	2320–2345
	Geostationary Fixed	3700–4200
	Geostationary Mobile	1525–1559, 1626.5–1660.5
	Global Positioning Systems (GPS)	1215–1240, 1350–1400, 1559–1610
	Non-Geostationary Mobile	1610–1626.5, 2483–2500, 5091–5250*, 6700–7075*
	Non-Geostationary Mobile Non-Voice	137–138, 148–150.05, 400.15–401, 455–456, 459–460
Unlicensed National Information Infrastructure		5000–5350, 5650–5850
Vehicle Recovery Systems		162.0125–173.2
Wireless Communications Service		746–764, 776–794*, 2305–2320, 2345–2360

* Frequency allocations pending

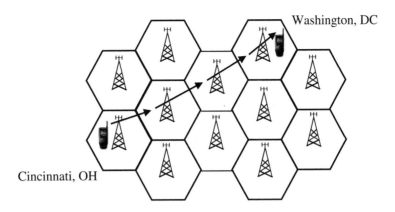

Figure 1.1
Maintaining the telephone number in a wireless and mobile system.

Washington, DC

Cincinnati, OH

Table 1.3: ▶
First-Generation Wireless Systems and Services

Year	Events
1970s	Developments of radio and computer technologies for 800/900 MHz mobile communication
1976	WARC (world administrative radio conference) allocates spectrum for cellular radio
1979	NTT (Nippon telephone & telegraph) introduces the first cellular system in Japan
1981	NMT (Nordic mobile telephone) 900 system introduced by Ericsson Radio System AB and deployed in Scandinavia
1984	AMPS (advanced mobile phone service) introduced by AT&T in North America

Table 1.4: ▶
Second-Generation Wireless Systems and Services

Year	Events
1982	CEPT (Conference European des Post of Telecommunications) establishes GSM (global special mobile) to define future Pan-European cellular radio standards
1990	Interim Standard IS-54 (USDC: United States digital cellular) adopted by TIA (Telecommunications Industry Association)
1990	Interim Standard IS-19B (NAMPS: narrowband AMPS) adopted by TIA
1991	Japanese PDC system standardized by the MPT (Ministry of Posts and Telecommunications)
1992	Phase I GSM system is operational
1993	Interim Standard IS-95 (CDMA) adopted by TIA
1994	Interim Standard IS-136 adopted by TIA
1995	PCS Licenses issued in North America
1996	Phase II GSM is operational
1997	North American PCS deploys GSM, IS-54, IS-95
1999	IS-54: used in North America; IS-95: used in North America, Hong Kong, Israel, Japan, South Korea, and China; GSM: used in 110 countries

The first-generation wireless systems were primarily developed for voice communication using frequency division multiplexing. To have efficient use of communication channels, time division multiplexing was used in the second generation systems so that data can be also processed. The third generation systems evolved due to the need for transmitting integrated voice, data, and multimedia traffic. The channel capacity is still limited, and attempts are being made to compress the amount of information without compromising the quality of received signals.

The second-generation wireless systems have been designed for both indoor and vehicular environments with an emphasis on voice communication. An increased acceptance of mobile communication networks for conventional services has led to demands for high bandwidth wireless multimedia services. These ever growing demands require a new generation of high-speed mobile infrastructure networks that can provide the capacity needed for high traffic volumes as well as flexibility in communication bandwidth or services. There is a need for frequent Internet access and multimedia data transfer, both of which may also involve the use of satellite communication. Thus, the third-generation (3G) systems (IMT-2000: International Mobile Telecommunications 2000) need to support real-time data communication while maintaining compatibility with second-generation systems. There are two schools of thought on the third-generation systems. In the United States, people are inclined to use cdma2000 as the basic technology, while in Europe and Japan, W-CDMA is being considered as the future scheme. In principle, both these schemes are similar, but there are differences in their implementations. These are basically design issues, and anticipated characteristics are identified in Table 1.5. There are subtle differences between wireless and mobile systems—for example, a system could be immobile but wireless, or a system could be mobile but not wireless. For the purpose of this text, we do not differentiate between the two and use these terms interchangeably.

Table 1.5: ▶
Third-Generation Wireless Systems and Services

IMT-2000	- Fulfill one's dream of anywhere, anytime communication
Key Features	- High degree of commonality of design worldwide - Compatibility of services within IMT-2000 and with the fixed networks - High quality - Small terminal for worldwide use - Worldwide roaming capability - Capability for multimedia applications, and a wide range of services and terminals
Important Component	- 2 Mbps for fixed environment - 384 kbps for indoor/outdoor and pedestrian environment - 144 kbps for vehicular environment
Standardization Work	- In progress
Scheduled Service	- Started in October 2001 in Japan (W-CDMA)

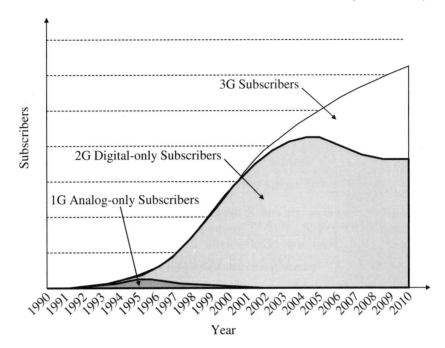

Figure 1.2
Subscriber growth
for wireless phones.

Wireless telephones are not only convenient but are also providing flexibility and versatility. Thus, there has been a growing number of wireless phone service providers as well as subscribers. Past numbers and future projections are given in Figure 1.2. It is expected that third-generation wireless systems will have many subsystems, with different requirements, characteristics, and coverage areas (Figure 1.3). The term *cell* basically represents the area that can be covered by a transmitting station, usually called a base station (BS), and pico, micro, macro, and so on primarily

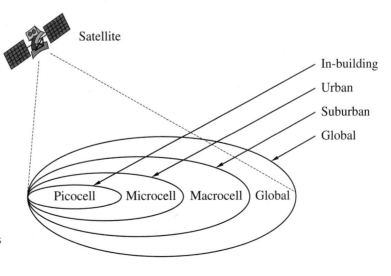

Figure 1.3
Coverage aspect
of third-generation
wireless communications
systems.

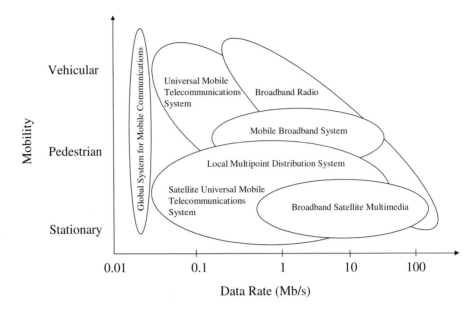

Figure 1.4
Transmission capacity as a function of mobility in some radio access systems.

indicate the relative size of the area that can be covered. The transmission capacity as a function of support for mobility in different radio access systems is illustrated in Figure 1.4. To cater to the different needs, different wireless technologies have been developed and are discussed next.

Different size cells are primarily needed due to the fact that in some areas, such as downtown or a big office complex, a large number of wireless telephone users may be present and served by a smaller size cell. This enables having a larger number of channels allocated to each cell, which is assumed to be the same or independent of the cell size. The idea is to maintain the same number of channels per customer and try to have a similar quality of service in all areas.

● ● ● ● ● ● ● ● ● ● ● ● ● ● ● ● · · ·
1.2 Characteristics of Cellular Systems

The network characteristics largely depend on the type of applications being explored, and a brief account is given in Table 1.6 [1.1]. One major partition of requirements is based on whether it is being envisioned for the home-based or industrial system versus the commercial and private environment (Table 1.7). In a house, a central access point (AP) is expected to communicate with various appliances and control them using localized wireless mode. This would not only enable close coordination among appliances, but also enable control from a remote location to the house AP using voice or a short message. A similar mechanism could be used to control devices in an industrial floor as well. To provide such wireless control, a consortium of companies are pursuing the Bluetooth project. For example, a system like this could support a bracelet, which would constantly monitor various body

Table 1.6: ▶
Wireless Technologies and Associated Characteristics
Credit: L. Malladi and D.P. Agrawal, "Current and Future Applications of Mobile and Wireless Networks," *Communications of the ACM*, 45:10, October 2002, pp. 144-146. (c) 2004 ACM, Inc. Reprinted by permission.

Technology	Services or Features	Coverage Area	Limitations	Examples
Cellular	Voice and data through handheld phones	Continuous coverage limited to metropolitan regions	Available bandwidth is very low for most data intensive applications	Cellular phones, personal digital assistant
Wireless local area network (LAN)	Traditional LAN extended with wireless interface	Used only in local environments	Limited range	NCR's wavelan, Motorola's ALTAIR, Proxim's range LAN, Telesystem's ARLAN
GPS	Helps to determine the three-dimensional position, velocity, and time	Anyplace on the surface of earth	It is still not affordable by everyone	GNSS, NAVSTAR, GLONASS
Satellite-based PCS	Applications mainly for voice paging and messaging	Almost anyplace on earth	It is costly	Iridium, Teledesic
Ricochet	High-speed, secure mobile access to the desktop (data) from outside the office	Some major cities, airports, and some university areas	Has a transmission limitation. Environmental conditions affect quality of service	MicroCellular Data Network (MCDN)
Home networking	To connect different PCs in the house to share files and devices such as printers	Anywhere in the house	Limited to a home	Netgear Phoneline 10X, Intel AnyPoint Phoneline Home Network, 3Com Home Connect Home Network Phoneline
Ad hoc networks	Group of people come together for a short time to share data	Equal to that of local area network, but without fixed infrastructure	Limited range	Defense applications
WPAN (Bluetooth)	All digital devices can be connected without any cable	Private ad hoc groupings away from fixed network infrastructures	Range is limited due to the short-range radio link used	Home devices
Sensor networks	A large number of tiny sensors with wireless capabilities	Relatively small terrain	Very limited range	Defense and civilian applications

Table 1.7: ▶
Characteristics of Wireless and Mobile Systems

Public Sphere	Traffic information system, personal security, disaster information system.
Business Sphere	Mobile videophone, video conferencing, database e-mail.
Private Sphere	Information services, music on demand portable TV, interactive TV, interactive games, video on demand, electronic newspapers and books, shopping, home schooling system, information service for pagers, news, weather forecasts, financial information.

functions/parameters and take corrective action (like informing a family physician about a health problem). Substantial efforts are needed to make such a system fully operational. To design such a generic system with plug-and-play capability requires standardization and necessary infrastructure for Internet access, audio/video editing, and distributed decision-making software. Wireless communication has become very popular in major fields such as commerce, medicine, education, and military defense. A simple example is when doctors are diagnosing a patient and can receive advice from medical specialists located in any part of the world (Figure 1.5).

In a commercial environment there are many issues involved, like the range of the system, the number of distribution infrastructures (APs) that are installed, and the number of users for each AP. For a department store, each floor may have one AP, while in a factory there is a need for several uniformly spaced APs per floor so that users are connected to an AP at all times. Thus channel bandwidth requirements and coordination of channels between APs govern the complexity. Communication can be either by a voice or a data packet, or a combination of both. The corresponding

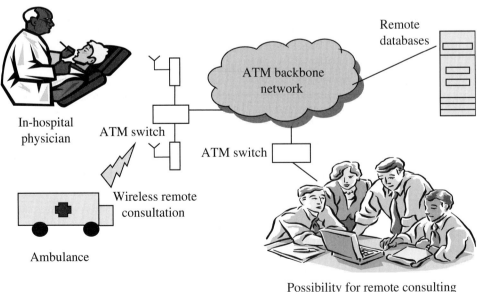

Figure 1.5
An example of medical and health application.

data loss is unacceptable in connection-oriented as well as connectionless wireless switching schemes; therefore correctness of transmitted and received data is important in all such applications. A new high-speed technology (WiMAX [Worldwide Interoperability for Microwave Access]) is being introduced to cover larger areas, possibly large metropolitan areas.

In a defense application, effective communication could be achieved using an infrastructure system or could be supported by a decentralized ad hoc network formed with close-by mobile users or mobile stations (MSs). It may also involve satellite systems. In ad hoc networks, information transfer is achieved in peer-to-peer mode, and there is a tradeoff between coverage area and power consumption. Other issues include channel allocation based on address and type of traffic (voice, video, audio, or data), utilization, routing techniques, and mobility pattern (e.g., moving speed, moving direction, etc.). It is also not clear how to optimize power usage, routing table size, and sustainability of path during each transmission session and diversity for unicasting and multicasting. Issues like handling of congestion, overloading of resources, adaptations of protocols, and queue length need to be considered carefully.

In all these systems, security, both in terms of authentication and encryption, is critical. This is fairly expensive in terms of hardware and software resources, and it affects channel capacity and information contents. Often, many levels of security may be useful and desirable. In all these systems, mobility is an integral factor and can be characterized by personal, terminal, and service mobilities. The effect of handoff needs to be viewed in various layers, and changing of radio resources needs to be minimized as much as possible. In order to minimize handoff and switching, the use of a macrocellular infrastructure (a larger coverage area per cell) has been advocated, and multilevel overlapped schemes have also been proposed to service users with different mobility patterns. In actual practice, however, a typical user on an average utilizes a mobile phone one minute per day. A tradeoff between cost versus performance encourages the use of smaller-size cells. The idea is to have a large number of small cells, with each cell effectively covering users located in that area.

A wireless system is expected to provide "anytime anywhere" type of service, and this characteristic has made it a very attractive technology. This kind of feature is essential for military and defense areas as well as to a limited class of potentially life-threatening applications like nuclear power, aviation, and medical emergencies. Different wireless features and their potential application areas are summarized in Table 1.8. However, for most day-to-day operations, the "anytime anywhere" feature may not be needed. Therefore, a "many time" or "many where" attribute may be adequate for Internet access, wherein you wait for resources to pass by; or you wait until you are close to a resource access point to have wireless or Internet access [1.4]. Also, there is no need to wait for completion of a transaction or data transfer completely for a MS as long as the remaining part could be made available (automatically routed) to an AP that the unit will be reaching along the path within the synchronized time constraints. In addition, emphasis should be on a scalable communication paradigm to reach multiple destinations and to support a query in a distributed fashion. Transfer of data at the right time is also guided by associated cost. Therefore, efficient design of a protocol is a challenge, as users may not always be connected ubiquitously.

Table 1.8: ▶
Potential Applications of Different Services

Wireless Features	Electronic Mail	WMAN/WLAN (Wireless MAN/LAN)	GPS	Satellite-Based PCS
Application Areas	- Field Service - Sales Force - Transportation Industry - Vending - Public Safety - Stock Trading - Airline Activities - Bill Paying - Field Audit	- Retail - Warehouses - Manufacturing - Students - Telediagnostics - Hospitality - General Office - Health Care	- Surveying - Car Rental Agency - Robin Toll Collection - Sports	- Iridium - Teledesic

● ● ● ● ● ● ● ● ● ● ● ● ● ● ● ● ● ●

1.3 Fundamentals of Cellular Systems

As discussed earlier, there are many ways of providing wireless and mobile communications, and each has relative advantages and disadvantages. For example, a cordless telephone used at home also employs wireless technology, except that it has a transmitter with a small amount of power and hence has a very limited coverage area. In fact, such range makes all users use more or less the same frequency range without much interferences among users. The same principle of frequency interference avoidance is used in cellular systems with a much more powerful transmitting station, or base station (BS). All users in the cell are served by the BS. Under ideal radio environments, the shape of the cell can be circular around the microwave transmitting tower. The radius of the circle is equal to the reachable range of the transmitted signal. It means that if the BS is located at the center of the cell, the cell area and periphery are determined by the signal strength within the region, which in turn depends on many factors, such as the contour of the terrain; height of the transmitting antenna; presence of hills, valleys, and tall buildings; and atmospheric conditions. Therefore, the actual shape of the cell, indicating a true coverage area, may be of a zigzag shape. However, for all practical purposes, the cell is approximated by a hexagon (Figure 1.6).

The hexagon is a good approximation of a circular region. Moreover, it allows a larger region to be divided into nonoverlapping hexagonal subregions of equal size, with each one representing a cell area. The square is another alternative shape that can be used to represent the cell area. The triangle is another less frequently used coverage area. Octagons and decagons do represent shapes closer to a circular area as compared to a hexagon. However, as explained in Chapter 5, they are not used to model a cell as it is not possible to divide a larger area into non-overlapping subareas

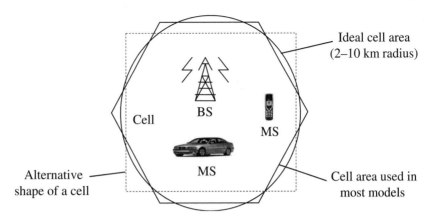

Figure 1.6
Illustration of a cell with a BS and MSs.

of the same shape. One practical example of a hex-based building block is that of hives made by bees; hives are three-dimensional hexagons in nature.

In each cell area, multiple users or wireless subscribers are served by a single BS. If the coverage area is to be increased, then additional BSs are placed to take care of the added area. Moreover, only a limited amount of bandwidth is allocated for the wireless service. Therefore, to increase the effectiveness of the overall system, some kind of multiplexing technique needs to be employed. Four basic multiplexing techniques that are employed are primarily known as frequency division multiple access (FDMA), time division multiple access (TDMA), code division multiple access (CDMA), and orthogonal frequency division multiplexing (OFDM). A new technique of space division multiple access (SDMA) is also being explored using specialized microwave antennas. In FDMA, the allocated frequency band is divided into a number of subbands, called channels, and one channel is allocated by the BS to each user (as illustrated in Figures 1.7, 1.8, and 1.9). FDMA is used in all first-generation cellular systems.

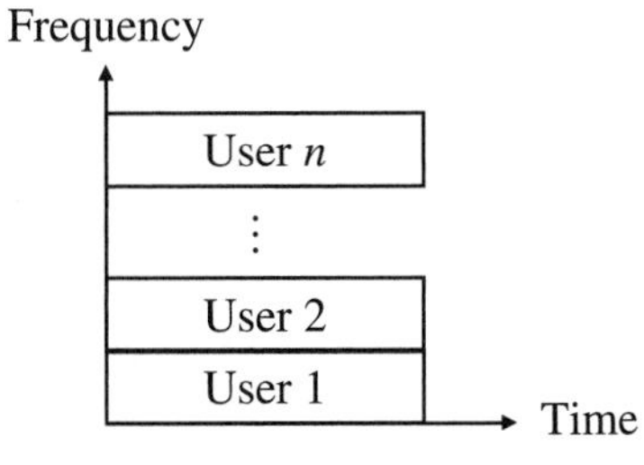

Figure 1.7
Frequency division multiple access (FDMA).

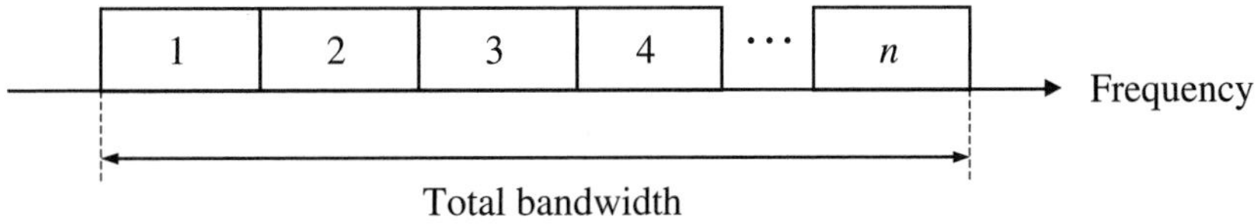

Figure 1.8
FDMA bandwidth structure.

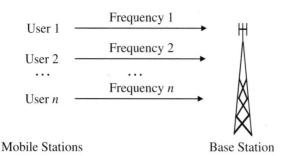

Figure 1.9
Illustration of FDMA
channel allocation.

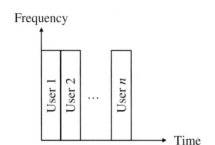

Figure 1.10
Time division multiple
access (TDMA).

In TDMA, one channel is used by several users, with BS assigning time slots for different users, and each user is served in a round-robin method. This fixed time slot scheme is shown in Figures 1.10, 1.11, and 1.12. Most second-generation cellular systems are based on TDMA.

The third and most promising CDMA technique utilizes a wider frequency band for each user. As the transmission frequency is distributed over the allocated spectrum, this technique is also known as spread spectrum. This scheme (Figure 1.13) is totally different from FDMA or TDMA. In this technique, one unique code is assigned by the BS to each user and distinct codes are used for different users. This

Figure 1.11
TDMA frame structure.

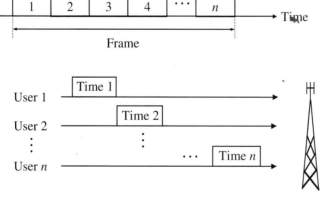

Figure 1.12
TDMA frame illustration
by multiple users.

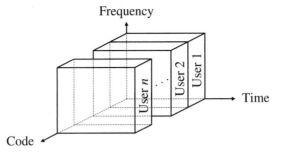

Figure 1.13
Code division multiple
access (CDMA).

code is employed by a user to mix with each bit of information before it is transmitted. The same code (or key) is used to decode these encoded bits, and any variation of the code interprets the received information simply as noise. This is illustrated for a 10-bit codeword in Figure 1.14. The orthogonality of the codes (described in Chapter 7 in more detail) enables transmission of data from multiple subscribers simultaneously using the full frequency band assigned for a BS. Each receiver is provided the corresponding code so that it can decode the data it is expected to receive. The number of users being serviced simultaneously is determined by the number of possible orthogonal codes that could be generated. The encoding step in the transmitter and the corresponding decoding at the receiver make the system design robust but complex. Some second-generation and most third-generation cellular systems

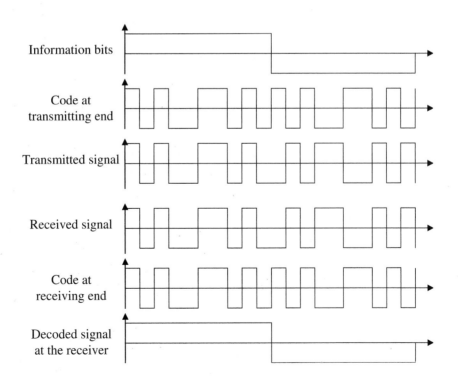

Figure 1.14
Transmitted and
received code in a
CDMA system.

Table 1.9: ▶
Frequency Range Used in Different Systems (an Example)

Systems	BS Transmitting Range/ MS Receiving Range	BS Receiving Range/ MS Transmitting Range	RF Channel
FDMA (AMPS)	870–890 MHz	825–845 MHz	0.03 MHz
TDMA (GSM 900)	935–960 MHz	890–915 MHz	0.20 MHz
TDMA (GSM 1800)	1805–1880 MHz	1710–1785 MHz	0.20 MHz
CDMA (IS-95)	869–894 MHz	824–849 MHz	1.25 MHz

employ CDMA. The frequency ranges used by FDMA, TDMA and CDMA in the United States, are shown in Table 1.9.

One of the newest and upcoming modulation techniques, known as OFDM, has recently been introduced, allowing parallel data transmission using multiple frequency channels. In radio communications, reflection and diffractions cause the transmitted signal to arrive at the receiver traversing different path lengths. Since there are many objects such as buildings, automobiles, trees, etc., which can serve as obstacles, the radio signals are affected and scattered throughout the area. Thus, in general, multipath signals arrive at the receiver with intersymbol interference (ISI). Therefore, it is relatively harder to extract the original signal. One approach to decrease the ISI is to use multicarrier transmission techniques, which requires converting high-speed data stream to slow transmission of parallel bit streams and employing several channels. Therefore, OFDM provides super quality signals with decreased ISI. OFDM is different from FDMA systems. In FDMA, the total bandwidth is divided into non-overlapping frequency subbands, which are used to eliminate the interference between adjacent channels and do not contribute to enhance the bandwidth utilization. In OFDM, the chosen subcarrier frequencies are spaced apart by the inverse of the symbol time, and the spectrum of each subchannel may overlap to fully utilize the available bandwidth. Figure 1.15 illustrates the two different multicarrier techniques.

There are several variants and combinations of FDMA, TDMA, and CDMA schemes based on specific systems. A detailed comparison is beyond the scope of this book. However, one noted exception is the frequency hopping, which can be defined as a combination of FDMA and TDMA in terms of the frequency use and time multiplexing. Basically, one user employs one channel for a pre-specified time period and then changes to another channel for transmission. This kind of frequency hopping is illustrated in Figure 1.16. The receiver can tune into the transmitter provided that it also knows the frequency hopping sequence. Of course, the sequence is repeated after all channels to be used in the sequence have been exhausted. For multiple users, different frequency hopping sequences can be used for transmitting information as long as, at any given time, one channel is used by only one user. The frequency hopping technique was primarily introduced for defense purposes wherein messages could still be transmitted even if strong enemy signals were present at one particular frequency band and is widely known as the "jamming" effect.

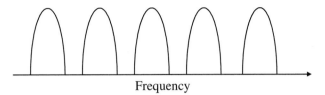

(a) Conventional multicarrier modulation used in FDMA

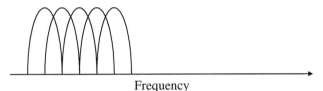

Figure 1.15
Two different
multicarrier techniques. (b) Orthogonal multicarrier modulation used in OFDM

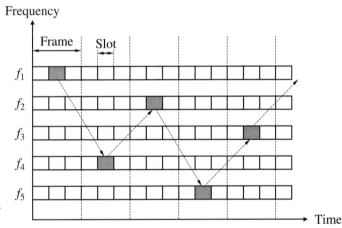

Figure 1.16
Illustration of frequency
hopping.

1.4 Cellular System Infrastructure

Early wireless systems had a high-power transmitter, covering the entire service area. This required a huge amount of power and was not suitable for many practical reasons. The cellular system replaced a large zone with a number of smaller hexagonal cells with a single BS covering a fraction of the area. Evolution of such a cellular system is shown in Figures 1.17 and 1.18, with all wireless receivers located in a cell being served by a BS.

Wireless devices need to be supported for different types of services. The wireless device could be a wireless telephone, personal digital assistant (PDA), Palm

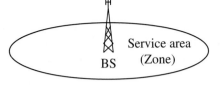

Figure 1.17
Early wireless system:
large zone.

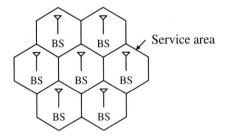

Figure 1.18
Cellular system:
small zone.

Pilot™, laptop with wireless card, or Web-enabled phone. For simplicity, it could be called a MS. The only underlying requirement is to maintain connectivity with the world while moving, irrespective of the technology used to obtain ubiquitous access. In a cellular structure, a MS needs to communicate with the BS of the cell where the MS is currently located (Figure 1.6), and the BS acts as a gateway to the rest of the world. Therefore, to provide a link, the MS needs to be in the area of one of the cells (and hence a BS) so that mobility of the MS can be supported. Several BSs are connected through hard-wires and are controlled by a BS controller (BSC), which in turn is connected to a mobile switching center (MSC). Several MSCs are interconnected to a PSTN (public switched telephone network) and the ATM (asynchronous transfer mode) backbone. To provide a better perspective of wireless communication technology, simplified system infrastructure for a cellular system is shown in Figure 1.19.

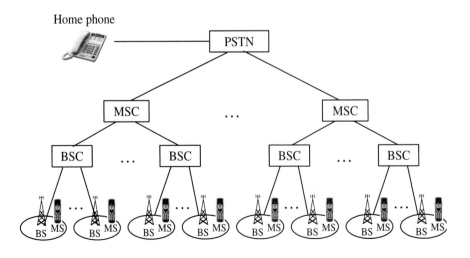

Figure 1.19
Cellular system
infrastructure.

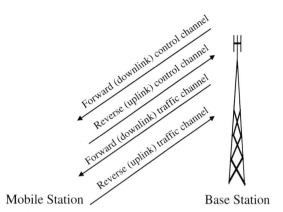

Figure 1.20
Four simplex channels between BS and MS in a cell.

Mobile Station   Base Station

A BS consists of a base transceiver system (BTS) and a BSC. Both tower and antenna are a part of the BTS, while all associated electronics are contained in the BSC. The home location register (HLR) and visitor location register (VLR) are two sets of pointers that support mobility and enable the use of the same telephone numbers worldwide. HLR is located at the MSC where the MS is registered and where the initial home location for billing and access information is maintained. In simple words, any incoming call, based on the called number, is directed to HLR of the home MSC and then HLR redirects the call to the MSC (and the BS) where the MS is currently located. VLR basically contains information about all visiting MSs in that particular MSC area.

In any cellular (mobile) scheme, four simplex channels are needed to exchange synchronization and data between BS and MS, and such a simplified arrangement is shown in Figure 1.20. The control links are used to exchange control messages (such as authentication, subscriber information, call parameter negotiations) between the BS and MS, while traffic (or information) channels are used to transfer actual data between the two. The channels from BS to MS are known as *forward channels* (called downlinks outside the United States), and the term *reverse channels* (uplinks) is used for communication from MS to BS. Control information needs to be exchanged before actual data information transfer can take place. Simplified handshake steps for call setup using control channels are illustrated in Figure 1.21.

The control channels are used for a short duration for exchanging control information between the BS and each MS needing any service. Therefore, all MSs use just a few control channels to achieve this and hence have to compete for such access in shared mode. On the other hand, traffic channels are exclusively allocated to each MS by the BS, and a large number of channels are used for the traffic. For this reason, handing of control and traffic channels must be considered in different ways, and more details on control channel access are provided in Chapter 6. Various alternative techniques for traffic channel assignments are covered in Chapter 7. The total number of channels that could be allocated for both control and traffic channels is influenced by the cell design and is discussed in Chapter 5.

There are many issues involved in wireless communication, and extensive signal processing is required before any signals are transmitted. The major steps are shown

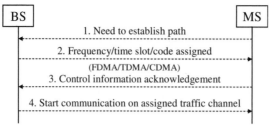

(a) Steps for a call setup from MS to BS

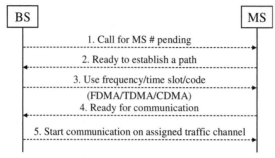

Figure 1.21
Handshake steps
for a call setup
between MS and
BS using control
channels.

(b) Steps for a call setup from BS to MS

in Figure 1.22. Many of the signal processing operations are beyond the scope of this book, and we will concentrate primarily on the system aspect of wireless data communication.

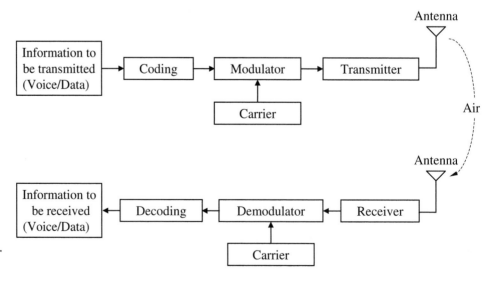

Figure 1.22
A simplified wireless
communication
system representation.

● ● ● ● ● ● ● ● ● ● ● ● ● ● ● ● ●●●●

1.5 **Satellite Systems**

Satellite systems have been in use for several decades. Satellites, which are far away from the surface of the earth, can cover a wider area, with several satellite beams being controlled and operated by one satellite [1.5]. Large areas can be covered due to the rotation of satellites around the earth. The information transmitted using satellites should be correctly received from one of the earth stations (ESs). Thus only "line of sight (LOS)" communication is possible. There is a long history of the development of satellite systems from a communications point of view, and important events are shown in Table 1.10. Possible application areas are outlined in Table 1.11. A more detailed discussion on satellite systems is given in Chapter 11.

Table 1.10: ▶
History of Satellite Systems

1945	Arthur C. Clarke publishes an essay titled "Extra Terrestrial Relays"
1957	First satellite, SPUTNIK
1960	First reflecting communication satellite, ECHO
1963	First geostationary satellite, SYNCOM
1965	First commercial geostationary satellite, "Early Bird" (INTEKSAT I): 240 duplex telephone channels or 1 TV channel, 1.5 years lifetime
1976	Three MARISAT satellites for maritime communication
1982	First mobile satellite telephone system, INMARSAT-A
1988	First satellite system for mobile phones and data communication, INMARSAT-C
1993	First digital satellite telephone system
1998	Global satellite systems for small mobile phones

Table 1.11: ▶
Application Areas of Satellite Systems

Traditionally	- Weather satellites
	- Radio and TV broadcast satellites
	- Military satellites
	- Satellites for navigation and localization (e.g., GPS)
Telecommunication	- Global telephone connections
	- Backbone for global networks
	- Connections for communication in remote places or underdeveloped areas
	- Global mobile communication

● ● ● ● ● ● ● ● ● ● ● ● ● ●

1.6 Network Protocols

Protocols are a basic set of rules that are followed to provide systematic signaling steps for information exchange. Such interfaces for smooth transfer in networks are covered in Chapter 12. Most systems evolve over a period of time. We explain early signaling systems and compare them with current systems. Separate signaling approaches are taken for narrowband and broadband transmissions and are based on some simple concepts. We introduce the concepts of OSI (Open Systems Interconnection), TCP/IP (Transmission Control Protocol/Internet Protocol), IPv4 (Internet Protocol Version 4), and IPv6 (Internet Protocol Version 6) protocols in Chapter 12.

● ● ● ● ● ● ● ● ● ● ● ● ● ●

1.7 Ad Hoc and Sensor Networks

An ad hoc (also written ad-hoc or adhoc) network is a local network with wireless or temporary plug-in connection, in which mobile or portable devices are part of the network only while they are in close proximity. Future military applications for ad hoc networks, which include a group of soldiers in close proximity sharing information on their notebook computers using RF signals, and numerous commercial applications are now being explored.

A mobile ad hoc network (MANET) is an autonomous system of mobile nodes, mobile hosts (MHs), or MSs (also serving as routers) connected by wireless links, the union of which forms a network modeled in the form of an arbitrary communication graph. The routers are free to move at any speed in any direction and organize themselves randomly. Thus, the network's wireless topology may dynamically change in an unpredictable manner. There is no fixed infrastructure, and information is forwarded in peer-to-peer mode using multihop routing. According to D. B. Johnson and D. A. Maltz [1.6], "an ad hoc network is a collection of wireless MHs forming a temporary network without the aid of any centralized administration or standard support services regularly available on the wide area network to which the hosts may normally be connected."

MANETs are basically peer-to-peer multihop mobile wireless networks where information packets are transmitted in a store-and-forward method from source to destination, via intermediate nodes, as shown in Figure 1.23. As the nodes move, the resulting change in network topology must be made known to the other nodes so that prior topology information can be updated. Such a network may operate in a stand-alone fashion, or with just a few selected routers communicating with an infrastructure network.

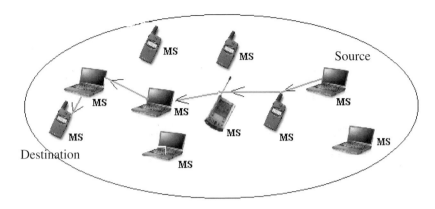

Figure 1.23
Illustration of a MANET.

MANET consists of mobile platforms, known as nodes (MSs), which are free to move around arbitrarily. Very small device-based nodes may be located inside airplanes, ships, trucks, cars, and perhaps within the human body. The system may operate in isolation or may have gateways to a fixed network. When it is communicating with hosts in a wired network, it is typically envisioned to operate as a "stub" network connected to a fixed internetwork. Stub networks carry traffic originating at and/or destined for internal nodes but do not permit exogenous traffic to "transit" through the stub network.

Each node is equipped with a wireless transmitter and a receiver with appropriate antenna, which may be omnidirectional, highly directional (point to point) [1.7], possibly steerable, or some combination thereof. At a given point in time, depending on the nodes' positions and their transmitter and receiver coverage patterns, transmission power levels, and cochannel interference levels, a wireless connectivity in the form of a random, multihop graph or ad hoc network exists between the nodes. This MANET topology may change with time as the nodes move or adjust their transmission and reception parameters.

Sensor networks [1.8][1.9][1.10][1.11] are the newest members of one special class of wireless ad hoc networks wherein a large number of tiny immobile sensors are planted on an ad hoc basis to sense and transmit some physical characteristics of the environment. An associated BS collects the information gathered by the sensors on a data-centric basis. Although tiny sensors are yet to be produced on a large scale, people are exploring their usefulness in many application areas. One such example sensing the cloud of smoke is shown in Figure 1.24, with sensor nodes being deployed in the area of interest. One of the most quoted examples is the battlefield surveillance of enemy territory, wherein a large number of sensors are dropped from an airplane so that activities on the ground can be detected and communicated. Other potential commercial fields include machinery prognosis, biosensing [1.12], and environmental monitoring [1.13].

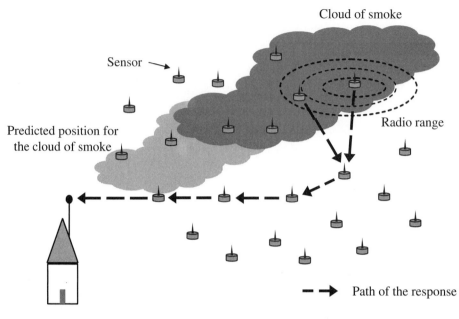

Cloud of smoke

Sensor

Radio range

Predicted position for
the cloud of smoke

Path of the response

Figure 1.24
An example of
a wireless sensor
network.

Data collection and
monitoring agency

1.8 Wireless MANs, LANs, and PANs

Wireless and mobile networking is finding extensive applications in different facets
of our life. Cellular telephones comprise a significant portion of household and business voice services, and wireless pagers have made inroads into major commercial
sectors. Plans are also underway to enable efficient transfer of data using wireless
devices. It is also anticipated that wireless multimedia support is forthcoming. Wireless devices are also influencing both office operations and the home environment. A
citywide access is now feasible using a wireless MAN (WMAN) and is being named
as WiMAX. A special class of wireless local area and personal area networks (wireless LANs [WLANs] or Wireless PANs [WPANs]) can cover smaller areas with low
power transmission (especially in the ISM [industrial, scientific, and medical] band)
and have become increasingly important for both office and home. Noteworthy techniques include the use of the IEEE 802.11 (IEEE stands for Institute of Electronics
and Electrical Engineering) [1.14], Bluetooth network [1.15][1.16], HomeRF [1.17],
and HiperLAN [1.18][1.19]. Characteristics of these networks are given in Table 1.12,
and more details are covered in Chapter 14.

Table 1.12: ▶
Noteworthy Wireless LANs and PANs Techniques

Type of Network	Range of Node	Primary Function	Deployed Locations
IEEE 802.11	30 meters	A standard for wireless nodes	Any peer-to-peer connection
HiperLAN	30 meters	High-speed indoor connectivity	Airports, warehouses
Ad Hoc Networks	≥ 500 meters	Mobile, wireless, similar to wired connectivity	Battlefields, disaster locations
Sensor Networks	2 meters	Monitor inhospitable or inaccessible terrain cheaply	Nuclear & chemical plants, ocean, etc.
HomeRF	30 meters	Share resources, connect devices	Homes
Ricochet	30 meters	High-speed wireless Internet access (128 Kbps)	Airports, office
Bluetooth Networks	10 meters	Avoid wire clutter, provide low mobility	Offices

● ● ● ● ● ● ● ● ● ● ● ● ● ● ⋯⋯

1.9 Outline of the Book

We introduce probability, statistics, queuing, and traffic theories in Chapter 2 and wireless and mobile radio propagation in Chapter 3. We discuss ways of coding channels in Chapter 4 and cellular concepts in Chapter 5. Multiple radio access techniques, multiple division techniques and modulation techniques, and different channel allocation techniques are covered in Chapters 6, 7, and 8, respectively. Design of mobile communication systems is included in Chapter 9. A summary of existing systems is included in Chapter 10. Satellite systems are becoming increasingly important because of their effective support for GPS capabilities. They are discussed in Chapter 11. Concepts of several network protocols are introduced in Chapter 12. Ad hoc and sensor networks have also become increasingly important and are discussed in Chapter 13. Wireless MANs, LANs, and PANs are described in Chapter 14. Many recent advances in technologies have emerged, and we provide a brief overview in Chapter 15. An adequate number of problems are provided to reinforce the ideas covered in the text as well as to test the knowledge gained in specific subject matter. Each chapter is also followed by important relevant references.

● ● ● ● ● ● ● ● ● ● ● ● ● ● ● ●

1.10 **References**

[1.1] R. Malladi and D. P. Agrawal, "Applications of Mobile and Wireless Networks: Current and Future," *Communications of the ACM*, Vol. 45, No. 10, pp. 144–146, October 2002.

[1.2] W. C. Y. Lee, *Mobile Communications Engineering: Theory and Applications*, 2nd edition, McGraw-Hill, 1997.

[1.3] *http://www.rfm.com/corp/new868dat/fccchart.pdf*.

[1.4] D. P. Agrawal, "Future Directions in Mobile Computing and Networking Systems," *Report on NSF Sponsored Workshop held at the University of Cincinnati*, June 13-14, 1999, *Mobile Computing and Communications Review*, October 1999, Vol. 3, No. 4, pp. 13–18, also available at *http://www.ececs.uc.edu/~dpa/tc-ds-article.pdf*.

[1.5] R. Bajaj, S. L. Ranaweera, and D. P. Agrawal, "GPS: Location Technology," *IEEE Computer*, pp. 115–117, April 2002.

[1.6] D. B. Johnson and D. A. Maltz. "The Dynamic Source Routing Protocol in Ad Hoc Networks," *Mobile Computing*, T. Imielinski and H. Korth, eds., Culwer, pp. 152–181, 1996. *http://www.ics.uci.edu/atm/adhoc/papercollection/johnson-dsr.pdf*.

[1.7] S. Jain and D. P. Agrawal, "Community Wireless Networks for Sparsely Populated Areas," *IEEE Computer*, Vol. 36, No. 8, pp. 90–92, August 2003.

[1.8] D. Estrin, et al., "Next Century Challenges: Scalable Coordination in Sensor Networks," *ACM Mobicom*, 1999.

[1.9] J. M. Kahn, etc., "Next Century Challenges: Mobile Networking for Smart Dust," *ACM Mobicom*, 1999.

[1.10] "The Ultra Low Power Wireless Sensors Project," *http://www-mtl.mit.edu*.

[1.11] A. Manjeshwar, "Energy Efficient Routing Protocols with Comprehensive Information Retrieval for Wireless Sensor Networks," *M.S. Degree Thesis*, University of Cincinnati, Cincinnati, May 2001.

[1.12] L. A. Roy and D. P. Agrawal, "Wearable Networks: Present and Future," *IEEE Computer*, Vol. 36, No. 11, pp. 31–39, November 2003.

[1.13] D. P. Agrawal, M. Lu, T. C. Keener, M. Dong, and V. Kumar, "Exploiting the Use of Wireless Sensor Networks for Environmental Monitoring," *Journal of the Environmental Management*, pp. 35–41, August 2004.

[1.14] "Wireless WAN Medium Access Control (MAC) and Physical Layer (PHY) Specification: Higher Speed Physical Layer (PHY) Extension in the 2.4 GHz Band," *IEEE*, 1999.

[1.15] The Bluetooth Special Interest Group, "Baseband Specifications," *http://www.bluetooth.com*.

[1.16] J. Haartsen, "The Bluetooth Radio System," *IEEE Personal Communications*, pp. 28–36, February 2000.

[1.17] K. Negus, A. Stephens, and J. Lansford, "HomeRF: Wireless Networking for the Connected Home," *IEEE Personal Communications*, pp. 20–27, February 2000.

[1.18] M. Johnson, "HiperLAN/2—The Broadband Radio Transmission Technology Operating in the 5 GHz Frequency Band," *http://www.hiperlan2.com/site/specific/whitepaper.exe*.

[1.19] L. Taylor, "HIPERLAN Type 1—Technology Overview," *http://www.hiperlan.com/hiper_white.pdf*.

1.11 Problems

P1.1. Why do you need wireless services when adequate wired infrastructure exists in most parts of the United States?

P1.2. What are the challenges for wireless networking?

P1.3. What are the unconventional applications of wireless networks?

P1.4. What are the household applications that use wireless schemes?

P1.5. How many cellular service providers are present in your area? Which of the multiple access techniques is supported by each system? What are the cell size and transmitting power level? What is the number of subscribers in your area?

P1.6. How is an ad hoc network different from a cellular network?

P1.7. List some prospective application areas for sensor networks.

P1.8. Look at your favorite Web site and find what is meant by "Web-in-the-sky."

P1.9. What are the advantages of different wireless service providers in an area? Explain clearly.

P1.10. Can a network be wireless, but not mobile? Explain your answer carefully.

P1.11. What are the limitations if a network is mobile with no wireless support?

P1.12. Why is "anytime anywhere" access not required for all applications? Explain clearly.

P1.13. What are the pros and cons of having different-size cells for wireless networking?

P1.14. Why do you have difficulty in using your cell phone inside an elevator?

P1.15. What phenomenon do you observe when a cell phone is used while traveling a long metallic bridge?

P1.16. How do you compare a cell phone with a satellite phone?

P1.17. In an airplane in flight, what happens if you use
(a) A walkie-talkie?
(b) A satellite phone?
(c) A cell phone?

P1.18. What are the similarities between frequency hopping and TDMA?

P1.19. If a total of 33 MHz of bandwidth is allocated to a particular cellular telephone system that uses two 25 kHz simplex channels to provide full duplex voice channels, compute the number of simultaneous calls that can be supported per cell if a system uses
(a) FDMA
(b) TDMA with 8-way time multiplexing
Assume that additional bandwidth is reserved for the control channels.

P1.20. Many types of sensors are commercially available. Looking at different Web sites, can you prepare their cost-size-performance tradeoff?

Probability, Statistics, and Traffic Theories

• • • • • • • • • • • • • ••••

2.1 Introduction

Many factors influence the performance of a wireless and mobile networking system, such as what is the density of MSs in a cell, what is the distribution of moving speed and direction of MSs, how frequently the calls are made, how many MSs simultaneously make calls, how long they use the call connection, how are the positions of MSs with respect to each other and the BS, what is the type of traffic (real-time or non–real-time) in the cell, how is the traffic in adjacent cells, and how frequently the handoff from one cell to another cell occurs. It is useful to qualify and quantify some of these parameters, which could indicate the overall effectiveness of the system under given constraints. It is important to understand the basics of the traffic patterns and the underlying probabilistic, statistical, and traffic theories . This chapter provides a brief overview of simple concepts widely employed in correlating performance with different system parameters. We start with basic theories of probability and statistics.

• • • • • • • • • • • • • ••••

2.2 Basic Probability and Statistics Theories

2.2.1 Random Variables

A random variable is a function defined by the characteristics of an arbitrary random phenomenon. If S is the sample space associated with an experiment E, then a random variable X is a function that assigns a real number $X(s)$ to each element s that belongs to S. Random variables can be divided into two types: discrete and continuous random variables. If a random variable is a continuous variable , then an associated probability density function (pdf) is defined. A discrete random variable has either an associated probability distribution or probability mass function (pmf), which reflects the behavioral characteristics of the variable at discrete times.

Discrete Random Variables

One of the widely quoted examples in real life is that of throwing a coin and finding out whether you get the head or the tail. Another practical example is to try a six-sided die and defining a probability that a particular number may appear next. Representing such a finite or countable infinite number of possible values by a random variable is an example of a discrete random variable.

For a discrete random variable X, the pmf $p(k)$ of X is the probability that the random variable X is equal to k and is defined by the following function:

$$p(k) = P(X = k), \quad \text{for} \quad k = 0, 1, 2, \ldots, \tag{2.1}$$

It must satisfy the following conditions:

1. $0 \le p(k) \le 1$, for every k,
2. $\sum p(k) = 1$, for all k.

Continuous Random Variables

If a random variable can take an infinite number of values, it is called a continuous random variable. One such example of continuous random variable is a daily temperature. Continuous random variables have probability density functions instead of probability mass functions. For a continuous random variable X, the pdf $f_X(x)$ is a nonnegative valued function defined on the whole set of real numbers $(-\infty, \infty)$ such that for any subset $S \subset (-\infty, \infty)$,

$$P(X \subset S) = \int_S f_X(x)\, dx, \tag{2.2}$$

where x is simply a variable in the integral. It must satisfy the following conditions:

1. $f_X(x) \ge 0$, for all x,
2. $\int_{-\infty}^{\infty} f_X(x)\, dx = 1$.

2.2.2 Cumulative Distribution Function

For all discrete (or continuous) random variables, a cumulative distribution function (CDF) is represented by $P(k)$ (or $F_X(x)$), indicating the probability that the random variable X is less than or equal to k (or x), for every value k (or x). Formally, the CDF is defined to be

$$P(k) = P(X \le k), \quad \text{for all } k \tag{2.3}$$

or

$$F_X(x) = P(X \le x), \quad \text{for } -\infty < x < \infty. \tag{2.4}$$

For a discrete random variable, the CDF is found by summing the probabilities as follows:

$$P(k) = \sum_{\text{all } k} P(X = k). \tag{2.5}$$

For a continuous random variable, the CDF is the integral of its pdf, i.e.,

$$F_X(x) = \int_{-\infty}^{x} f_X(x)\, dx. \tag{2.6}$$

Since $F_X(x) = P(X \leq x)$, we have

$$F X (a \leq x \leq b) = \int_{a}^{b} f_X(x)\, dx$$
$$= F_X(b) - F_X(a)$$
$$= P (a \leq X \leq b). \tag{2.7}$$

2.2.3   Probability Density Function

The pdf of a continuous random variable is a function that can be integrated to obtain the probability that the random variable takes a value in a given interval.

Formally, the pdf $f_X(x)$ of a continuous random variable X is the derivative of the CDF $F_X(x)$:

$$f_X(x) = \frac{d F_X(x)}{dx}. \tag{2.8}$$

2.2.4   Expected Value, *n*th Moment, *n*th Central Moment, and Variance

The expected value (or population mean value) of a random variable represents its average or central value. It is a useful value (a number) to summarize the variable's distribution. The variance (population) of a random variable is a nonnegative number that gives an idea of how widely spread the values of the random variable are likely to be; the larger the variance, the more scattered are the observations on average. From wireless system point of view, this can indicate how calls are generated by the subscribers in different parts of a cell in a wireless system and computing the average number of calls, would show the number of busy channels in a cell. Also, new calls from subscribers are initiated at different times, and hence the calling event from subscribers can be represented by discrete random variables, rather than a continuous random variable. In addition, the call holding time (the conversation period of a subscriber) is variable, and the percentage of time a channel is busy depends on the weighted function of the call rate and the call duration. On the other hand, interference between adjacent channels used by different subscribers depends on how long each channel is used and how long is the overlapped period during which multiple channels are used. This requires calculating associated moment functions to represent the traffic characteristics. Therefore, we need to quantify these variables and understand their impact on the system performance.

Discrete Random Variable

■ **Expected value or mean value:**

$$E[X] = \sum_{\text{all } k} kP(X = k) \tag{2.9}$$

The expected value of the function $g(X)$ of a discrete random variable X is the mean of another random variable Y that assumes the values of $g(X)$ according to the probability distribution of X. Denoted by $E[g(X)]$, it is given by

$$E[g(X)] = \sum_{\text{all } k} g(k)P(X = k). \tag{2.10}$$

■ **nth moment:**

$$E[X^n] = \sum_{\text{all } k} k^n P(X = k) \tag{2.11}$$

The first moment of X is simply the expected value of X.

■ **nth central moment:**
The central moment is the moment about the mean value; that is,

$$E\left[(X - E[X])^n\right] = \sum_{\text{all } k} (k - E[X])^n P(X = k). \tag{2.12}$$

The first central moment is equal to 0.

■ **Variance or the second central moment:**

$$\sigma^2 = \text{Var}(X) = E\left[(X - E[X])^2\right] = E\left[X^2\right] - (E[X])^2, \tag{2.13}$$

where σ is called the standard deviation.

Continuous Random Variable

■ **Expected value or mean value**:

$$E[X] = \int_{-\infty}^{\infty} x f_X(x)\, dx \tag{2.14}$$

The expected value of the function $g(X)$ of a continuous random variable X is the mean of another random variable Y that assumes the values of $g(X)$ according to the probability distribution of X. Denoted by $E[g(X)]$, it is given by

$$E[g(X)] = \int_{-\infty}^{\infty} g(x) f_X(x)\, dx. \tag{2.15}$$

■ *n*th **moment**:

$$E[X^n] = \int_{-\infty}^{\infty} x^n f_X(x)\, dx \tag{2.16}$$

■ *n*th **central moment**:

$$E\left[\left(X - E[X]\right)^n\right] = \int_{-\infty}^{\infty} \left(x - E[X]\right)^n f_X(x)\, dx \tag{2.17}$$

■ **Variance or the second central moment**:

$$\sigma^2 = \mathbf{Var}\,(X) = E\left[\left(X - E[X]\right)^2\right] = E\left[X^2\right] - \left(E[X]\right)^2 \tag{2.18}$$

2.2.5 Some Important Distributions

As discussed earlier, it is important to capture the nature of the calls, and many models have been used to represent the call arrival distribution and the service time distribution within each cell of a wireless system as well as user's mobility pattern. Therefore, we need to consider how the occurrence of a generic event could be characterized by different types of distributions.

Discrete Random Variable

■ **Poisson distribution**:

A Poisson random variable is a measure of the number of events that occur in a certain time interval. The probability distribution of having k events is

$$P(X = k) = \frac{\lambda^k e^{-\lambda}}{k!}, \qquad k = 0, 1, 2, \dots, \text{ and } \lambda > 0. \tag{2.19}$$

The Poisson distribution has expected value $E[X] = \lambda$ and variance $\mathrm{Var}(X) = \lambda$.

■ **Geometric distribution**:

A geometric random variable indicates the number of trials required to obtain the first success. The probability distribution of random variable X is given by

$$P(X = k) = p\,(1 - p)^{k-1}, \qquad k = 0, 1, 2, \dots, \tag{2.20}$$

where p is a success probability. The geometric distribution has expected value $E[X] = 1/(1 - p)$ and variance $\mathrm{Var}(X) = p/(1 - p)^2$.

■ **Binomial distribution**:

A binomial random variable represents the presence of k, and only k, out of n items and is the number of successes in a series of trials. The probability distribution of random variable X is

$$P(X = k) = \binom{n}{k} p^k (1-p)^{n-k}, \tag{2.21}$$

where $k = 0, 1, 2, \ldots, n$, $n = 0, 1, 2, \ldots$, p is a success probability, and

$$\binom{n}{k} = \frac{n!}{k!(n-k)!}.$$

The binomial distribution has expected value $E[X] = np$ and variance $\mathrm{Var}(X) = np(1-p)$.

The Poisson distribution can sometimes be used to approximate the binomial distribution with parameters n and p. When the number of observations n is large, and the success probability p is small, the binomial distribution approaches the Poisson distribution with the parameter given by $\lambda = np$. This is useful since the computations involved in calculating Poisson probabilities are substantially simpler than the binomial distributions.

The geometric distribution is related to the binomial distribution in that both are based on independent trials in which the probability of success is constant and equal to p. However, a geometric random variable is the number of trials until the first success, whereas a binomial random variable is the number of successes in n trials.

Continuous Random Variable

■ **Normal distribution**:

A normal random variable should be capable of assuming any real value, though this requirement is often waived in actual practice. The pdf of random variable X is given by

$$f_X(x) = \frac{1}{\sqrt{2\pi}\sigma} e^{-\frac{(x-\mu)^2}{2\sigma^2}}, \quad \text{for } -\infty < x < \infty, \tag{2.22}$$

and the CDF can be obtained by

$$F_X(x) = \frac{1}{\sqrt{2\pi}\sigma} \int_{-\infty}^{x} e^{-\frac{(y-\mu)^2}{2\sigma^2}} \, dy, \tag{2.23}$$

where μ is the expected value and σ^2 is the variance of random variable X. Usually, we denote $X \sim N(\mu, \sigma^2)$ indicating X as a normal random variable with expected value μ and variance σ^2. The case where $\mu = 0$ and $\sigma = 1$ is called the standard normal distribution.

■ **Uniform distribution**:

The values of a uniform random variable are uniformly distributed over an interval. A continuous random variable X is said to follow a uniform distribution with parameters a and b if its pdf is constant within a finite interval $[a, b]$, and zero outside this interval (with a less than or equal to b). The probability density distribution of random variable X is

$$f_X(x) = \begin{cases} \frac{1}{b-a}, & \text{for } a \leq x \leq b, \\ 0, & \text{otherwise} \end{cases} \tag{2.24}$$

and the CDF is

$$F_X(x) = \begin{cases} 0, & \text{for } x < a, \\ \frac{x-a}{b-a}, & \text{for } a \leq x \leq b, \\ 1, & \text{for } b < x. \end{cases} \tag{2.25}$$

The uniform distribution has expected value $E[X] = (a + b)/2$ and variance $\text{Var}(X) = (b - a)^2/12$.

■ **Exponential distribution**:

The exponential distribution is a very commonly used distribution in engineering. Due to its simplicity, it has been widely employed even in cases where it may not be applicable. The pdf of random variable X is given by

$$f_X(x) = \begin{cases} 0, & x < 0, \\ \lambda e^{-\lambda x}, & \text{for } 0 \leq x < \infty, \end{cases} \tag{2.26}$$

and the CDF is

$$F_X(x) = \begin{cases} 0, & x < 0, \\ 1 - e^{-\lambda x}, & \text{for } 0 \leq x < \infty, \end{cases} \tag{2.27}$$

where λ is the rate. The exponential distribution has expected value $E[X] = 1/\lambda$ and variance $\text{Var}(X) = 1/\lambda^2$.

2.2.6   Multiple Random Variables

In some cases, the result of one random experiment is dictated by the values of several random variables, where these values may also affect each other. For example, different users initiate calls at different rates and for different time period. If each user is characterized by a random variable, then the overall characteristics of a typical user may be represented by a global random variable. Similarly, interference depends on the traffic in adjacent cells. Therefore, to determine interference level, it may be desirable to determine call rates in many cells and computation may be quite involved. A joint pmf of the discrete random variables $X_1, X_2, \ldots, X_n$ is given by

$$p(x_1, x_2, \ldots, x_n) = P(X_1 = x_1, X_2 = x_2, \ldots, X_n = x_n) \tag{2.28}$$

and represents the probability that $X_1 = x_1, X_2 = x_2, \ldots, X_n = x_n$.

In the continuous case, the joint distribution function

$$F_{X_1 X_2 \ldots X_n}(x_1, x_2, \ldots, x_n) = P(X_1 \leq x_1, X_2 \leq x_2, \ldots, X_n \leq x_n) \tag{2.29}$$

represents the probability that $X_1 \leq x_1$, $X_2 \leq x_2, \ldots, X_n \leq x_n$. The joint pdf is given by

$$f_{X_1 X_2 \ldots X_n}(x_1, x_2, \ldots, x_n) = \frac{\partial^n F_{X_1 X_2 \ldots X_n}(x_1, x_2, \ldots, x_n)}{\partial x_1 \partial x_2 \ldots \partial x_n}. \tag{2.30}$$

Conditional Probability

A conditional probability is the probability that $X_1 = x_1$ when given $X_2 = x_2, \ldots,$ $X_n = x_n$. Therefore, for discrete random variables, we have

$$P(X_1 = x_1 \mid X_2 = x_2, \ldots, X_n = x_n) = \frac{P(X_1 = x_1, X_2 = x_2, \ldots, X_n = x_n)}{P(X_2 = x_2, \ldots, X_n = x_n)}. \tag{2.31}$$

For continuous random variables, we have

$$P(X_1 \leq x_1 \mid X_2 \leq x_2, \ldots, X_n \leq x_n) = \frac{P(X_1 \leq x_1, X_2 \leq x_2, \ldots, X_n \leq x_n)}{P(X_2 \leq x_2, \ldots, X_n \leq x_n)}. \tag{2.32}$$

Bayes's Theorem

A theorem concerning conditional probabilities of the form $P(X|Y)$ (read as: the probability of X, given Y) is

$$P(X \mid Y) = \frac{P(Y \mid X)\,P(X)}{P(Y)}, \tag{2.33}$$

where $P(Y)$ and $P(X)$ are the unconditional (or a priori) probabilities of Y and X, respectively. This is a fundamental theorem of probability theory, but its use in statistics is a subject of some controversy (Bayesian statistics). For further discussion, see [2.1], [2.2]. This is useful when we want to compute probability of additional traffic, given the current traffic condition.

Independence

Two events are independent if one may occur irrespective of the other. That is, the occurrence or nonoccurrence of one does not alter the likehood of occurrence of nonoccurrence of the other. More importantly, for example, if the occurrence of event X does not change the probability of event Y, we have

$$P(Y \mid X) = P(Y), \quad \text{when } P(X) > 0. \tag{2.34}$$

In this case, we say that the events X and Y are independent. Moreover, the multiplication rule becomes

$$P(XY) = P(X)P(Y \mid X)$$
$$= P(X)P(Y).$$

This, in turn, implies, when $P(Y) > 0$, that

$$P(X \mid Y) = \frac{P(XY)}{P(Y)}$$

$$= \frac{P(X)P(Y)}{P(Y)}$$

$$= P(X).$$

If the random variables $X_1, X_2, \ldots, X_n$ (e.g., indicating call rates in respective cells) are independent of each other, we obtain pmf for discrete random variable case as

$$p(x_1, x_2, \ldots, x_n) = P(X_1 = x_1) P(X_2 = x_2) \ldots P(X_n = x_n), \quad (2.35)$$

or for the continuous random variable case we have

$$F_{X_1 X_2 \ldots X_n}(x_1, x_2, \ldots, x_n) = F_{X_1}(x_1) F_{X_2}(x_2) \ldots F_{X_n}(x_n). \quad (2.36)$$

Important Property

- **Sum property of the expected value**:
  The expected value of a sum of random variables $X_1, X_2, \ldots, X_n$ is

$$E\left[\sum_{i=1}^{n} a_i X_i\right] = \sum_{i=1}^{n} a_i E[X_i], \quad (2.37)$$

  where a_i are arbitrary constants.

- **Product property of the expected value**:
  If the random variables $X_1, X_2, \ldots, X_n$ are stochastically independent, then the expected value of the product of the random variables $X_1, X_2, \ldots, X_n$ is

$$E\left[\prod_{i=1}^{n} X_i\right] = \prod_{i=1}^{n} E[X_i]. \quad (2.38)$$

- **Sum property of the variance**:
  The variance of a sum of random variables $X_1, X_2, \ldots, X_n$ is

$$\text{Var}\left[\sum_{i=1}^{n} a_i X_i\right] = \sum_{i=1}^{n} a_i^2 \text{Var}(X_i) + 2 \sum_{i=1}^{n-1} \sum_{j=i+1}^{n} a_i a_j \text{Cov}[X_i, X_j], \quad (2.39)$$

  where $\text{Cov}[X_i, X_j]$ is the covariance of random variables X_i and X_j and

$$\text{Cov}[X_i, X_j] = E\left[(X_i - E[X_i])(X_j - E[X_j])\right]$$

$$= E[X_i X_j] - E[X_i] E[X_j]. \quad (2.40)$$

If random variables X_i and X_j are two independent random variables (uncorrelated), i.e., $\text{Cov}[X_i, X_j] = 0$, for all $i \neq j$ we have

$$\text{Var}\left(\sum_{i=1}^{n} a_i X_i\right) = \sum_{i=1}^{n} a_i^2 \text{Var}(X_i). \quad (2.41)$$

Distribution of Sum

We assume that X and Y are continuous random variables with joint pdf $f_{XY}(x, y)$. If $Z = \phi(X, Y)$, the distribution of Z may be written as

$$F_Z(z) = P(Z \leq z) = \iint_{\phi_Z} f_{XY}(x, y) \, dx \, dy, \qquad (2.42)$$

where ϕ_Z is a subset of Z.

For a special case, $Z = X + Y$, we have

$$F_Z(z) = \iint_{\phi_Z} f_{XY}(x, y) \, dx \, dy = \int_{-\infty}^{\infty} \int_{-\infty}^{\infty} f_{XY}(x, y) \, dx \, dy. \qquad (2.43)$$

Making a variable substitution $y = t - x$, we have

$$F_Z(z) = \int_{-\infty}^{z} \int_{-\infty}^{\infty} f_{XY}(x, t - x) \, dx \, dt = \int_{-\infty}^{z} f_Z(t) \, dt. \qquad (2.44)$$

Thus the pdf of Z is given by

$$f_Z(z) = \int_{-\infty}^{\infty} f_{XY}(x, z - x) \, dx, \qquad \text{for } -\infty \leq z < \infty. \qquad (2.45)$$

If X and Y are independent random variables, then $f_{XY}(x, y) = f_X(x) f_Y(y)$, and we have

$$f_Z(z) = \int_{-\infty}^{\infty} f_X(x) f_Y(z - x) \, dx, \qquad \text{for } -\infty \leq z < \infty. \qquad (2.46)$$

Further, if both X and Y are nonnegative random variables, then

$$f_Z(z) = \int_{0}^{z} f_X(x) f_Y(z - x) \, dx, \qquad \text{for } -\infty \leq z < \infty. \qquad (2.47)$$

Thus, the pdf of the sum of two nonnegative independent random variables is the convolution of their individual pdfs, $f_X(x)$ and $f_Y(y)$.

Central Limit Theorem

The central limit theorem states that whenever a random sample $(X_1, X_2, \ldots, X_n)$ of size n is taken from any distribution with expected value $E[X_i] = \mu$ and variance $\text{Var}(X_i) = \sigma^2$, where $i = 1, 2, \ldots, n$, then their arithmetic mean is defined by

$$S_n = \frac{1}{n} \sum_{i=1}^{n} X_i. \qquad (2.48)$$

The sample mean can be approximated by normal distribution with $E[S_n] = \mu$ and variance $\text{Var}(S_n) = \sigma^2/n$. The larger the value of the sample size n, the better the approximation to the normal.

This is very useful when inference between signals needs to be considered. For example, it allows us (if the sample size is fairly large) to use hypothetical tests that assume normality even if the data do not appear to be normal. This is because the tests use the sample mean and the central limit theorem enables us to approximate with normal distribution.

• • • • • • • • • • • • • • • • •

2.3 Traffic Theory

2.3.1 Poisson Arrival Model

A Poisson process is a sequence of events randomly spaced in time. For example, customers arriving at a bank and geiger counter clicks are similar to packets arriving to a buffer. Similarly, in wireless networks, different users initiate their calls at different time, and the sequence of calls being initiated in a cell is usually identified as a Poisson process. The rate λ of a Poisson process is the average number of events per unit time (over a long time).

Properties of a Poisson Process

For a time interval $[0, t)$, the probability of n arrivals in t units of time is

$$P_n(t) = \frac{(\lambda t)^n}{n!} e^{-\lambda t}, \qquad \text{for } n = 0, 1, 2, \dots . \tag{2.49}$$

For two disjoint (nonoverlapping) intervals, (t_1, t_2) and (t_3, t_4), ($i.e.$, $t_1 < t_2 < t_3 < t_4$), the number of arrivals in (t_1, t_2) is considered independent of the number of arrivals in (t_3, t_4). For example, in wireless networks, the number of calls initiated between time (t_1, t_2) may be independent of calls during (t_3, t_4).

Interarrival Times of a Poisson Process

We pick an arbitrary starting point t in time. Let T_1 be the time until the next arrival. We have

$$P(T_1 > t) = P_0(t) = e^{-\lambda t}. \tag{2.50}$$

Thus, the distribution function of T_1 is given by

$$F_{T_1}(t) = P(T_1 \le t) = 1 - e^{-\lambda t}, \tag{2.51}$$

and the pdf of T_1 is

$$f_{T_1}(t) = \lambda e^{-\lambda t}. \tag{2.52}$$

Therefore, T_1 has an exponential distribution with mean rate λ.

Let T_2 be the time between the first and second call arrivals. We can show that

$$P(T_2 > T_1 + t \mid T_1 = \Delta) = e^{-\lambda t}, \qquad \text{for } \Delta, t > 0. \tag{2.53}$$

Thus, the distribution function of T_2 is given by

$$F_{T_2}(t) = P(T_2 \le T_1 + t \mid T_1 = \Delta) = 1 - e^{-\lambda t}, \tag{2.54}$$

and the pdf of T_2 is

$$f_{T_2}(t) = \lambda e^{-\lambda t}. \tag{2.55}$$

Similarly, we define T_3 as the time between the second and third arrivals, T_4 as the time between the third and fourth arrivals, and so on. The random variables T_1, T_2, $T_3 \dots$ are called the interarrival times of the Poisson process. We can observe that the interarrival times, $T_1, T_2, T_3, \dots$, are independent of each other and each has the same exponential distribution with mean arrival rate λ.

Memoryless Property

The importance of the Poisson process is based on the fact that it is the only continuous random variable to exhibit the memoryless property. For any nonnegative real numbers δ and t, we have

$$P(X > \delta + t \mid X > \delta) = P(X > t). \qquad (2.56)$$

If we interpret X as a lifetime, then the probability that the lifetime X exceeds $\delta + t$ given that X exceeds δ is the probability that the lifetime exceeds t. In the wireless area, it means that a new call is initiated independent on previous history calls made by the user.

Merging Property

If we merge n Poisson processes with distributions for the interarrival times $1 - e^{-\lambda_i t}$, where $i = 1, 2, \ldots, n$, into one single process, then the result is a Poisson process for which the interarrival times have the distribution $1 - e^{-\lambda t}$ with $\lambda = \lambda_1 + \lambda_2 + \cdots + \lambda_n$. In wireless networks, a cell may consist of different types of users such as one group for voice calls by pedestrians, another group for voice calls from fast moving car phones, another group primarily transmitting data, and so on. Thus, each group can be represented by a different Poisson process.

Splitting Property

If a Poisson process with interarrival time distribution $1 - e^{-\lambda t}$ is split into n processes so that the probability that the arriving job is assigned to the ith process is P_i, where $i = 1, 2, \ldots, n$, then the ith subprocess has an interarrival time distribution of $1 - e^{-P_i \lambda t}$ (i.e., n Poisson processes have been created).

● ● ● ● ● ● ● ● ● ● ● ● ● ● ● ● ● ●

2.4 Basic Queuing Systems

2.4.1 What Is Queuing Theory?

Queuing theory is the study of queues (sometimes called waiting lines). Most people are familiar with the concept of queues; they exist all around us in our daily lives. Queuing theory can be used to describe real world queues, or more abstract queues, which are often found in many branches of communications and computer science, such as operating systems. This section deals with basic mathematical formulations needed in queueing theory.

2.4.2 Basic Queuing Theory

Queuing theory has a wide range of applications, including its extensive use in wireless networks for indicating new call requests in a cell and allocation of channels to these cells. It can be divided into three main sections of traffic flow, scheduling, and employee allocation. Examples in these areas are certainly not the only applications where queuing theory can be put to good use; other examples are included to illustrate the usefulness of queuing theory.

2.4.3 Kendall's Notation

D. G. Kendall in 1951 [2.3] proposed a standard notation for classifying queuing systems into different types. The systems are described by the notation

$$A/B/C/D/E$$

where

A	Distribution of interarrival times of customers
B	Distribution of service times
C	Number of servers
D	Maximum number of customers in system
E	Calling population size

and A and B can take any of following distribution types:

M	Exponential distribution (Markovian)
D	Degenerate (or Deterministic) distribution
E_k	Erlang distribution (k = shape parameter)
G	General distribution (arbitrary distribution)
H_k	Hyperexponential with parameter k

Notes: If G is used for A, it is sometimes written as GI. C is normally taken to be either 1, or a variable, such as n, s, or m. D is usually infinite or a variable. If D or E is assumed to be infinite for modeling purposes, they can be omitted from the notation (which they frequently are). If E is included, D must be included to eliminate confusion between the two, but an infinity symbol is allowed for D.

2.4.4 Little's Law

Assuming a queuing environment operating in a steady state in which all initial transients have vanished, the key parameters characterizing the system are as follows:

- λ—the mean steady-state customer arrival rate

- N—the average number of customers in the system (both in the buffer and in service)

- T—the mean time spent by each customer in the system (time spent in the queue plus the service time)

It is intuitive to guess

$$N = \lambda T. \tag{2.57}$$

This indeed is the content of Little's theorem, which holds very generally for a very wide range of service disciplines and arrival statistics. In the next section, we study the different states of a system and transitions from one state to another.

2.4.5 Markov Process

A Markov process is one in which the next state of the process depends only on the present state, irrespective of any previous states taken by the process. This means that knowledge of the current state and the transition probabilities from this state allows us to predict the possible next state independent of any past state. A Markov chain is a discrete state Markov process.

2.4.6 Birth–Death Process

This is a special type of Markov process often used to model a population (or the number of jobs in a queue). If, at some time, the population has n entities (n jobs in the queue), then birth of another entity (arrival of another job) causes the state to change to $n + 1$. On the other hand, a death (a job is removed from the queue for service) would cause the state to change to $n - 1$. Thus we see that in any state, transitions can be made only to one of the two neighboring states. Figure 2.1 shows a state transition diagram of the continuous birth–death process. Similar arguments can be given to the number of calls in a cell of a wireless network. If a cell has n calls being serviced by n channels, then given the probability of a new call being initiated or a call being completed, the transition to servicing $(n + 1)$ calls or $(n - 1)$ calls can be represented with appropriate transition probabilities. The number $0, 1, 2, \ldots$ represent the number of channels kept busy in servicing various users.

In state n, we have

$$\lambda_{n-1} P(n - 1) + \mu_{n+1} P(n + 1) = (\lambda_n + \mu_n) P(n),$$

where $P(i)$ is the steady-state probability of state i, λ_i ($i = 0, 1, 2, \ldots$) is the average arrival rate, and μ_i ($i = 0, 1, 2, \ldots$) is the average service rate. A similar state equation can be written for states $1, 2, 3, \ldots$. For state 0, we have

$$\lambda_0 P(0) = \mu_1 P(1).$$

Writing these state equations may be viewed as a simple process of balancing the incoming and outgoing arrows from a particular state. It should be noted that

Figure 2.1
The state transition diagram of the continuous birth–death process.

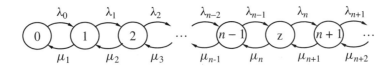

$P(0)$, $P(1)$, $P(2)$... $P(n)$,... are all steady-state probabilities, and the equations also represent steady-state transition.

Solving the set of equations obtained (one equation per state), we derive the relation between $P(n)$ and $P(0)$. Thus, we get

$$P(n) = \frac{\lambda_0\lambda_1\dots\lambda_{n-1}}{\mu_1\mu_2\dots\mu_n}P(0).$$

2.4.7 M/M/1/∞ Queuing System

Here we deal with the simplest queuing model. This is called the $M/M/1/\infty$ queue or $M/M/1$ queue shown in Figure 2.2. When a customer arrives in the system, it will be served if the server is free. Otherwise, the customer is queued. In an $M/M/1$ queuing system, customers arrive according to a Poisson distribution and compete for the service in a first-in-first-out (FIFO) or first-come-first-served (FCFS) manner. The service times are independent identically distributed (IID) random variables, the common distribution being exponential. In practice, the $M/M/1$ queuing system is useful because many complex systems can be abstracted as a composition of a simple $M/M/1$ queuing system. Theoretically, the $M/M/1$ queuing system has an accurate mathematical solution in terms of the mean arrival rate λ and the mean service rate μ. Next, we give an analytical approach to the $M/M/1$ queuing system.

Based on the preceding assumptions, $M/M/1$ queuing systems consist of a birth–death process. Let i ($i = 0, 1, 2, \dots$) be the number of customers in the system and let $P(i)$ be the steady-state probability of the system having i customers. For wireless networks, the M in $M/M/1$ represents the interarrival and service times of calls in a cell, and 1 indicates the single channel available in the cell. The state of the Markov model indicates the number of calls in progress within a cell. Therefore, the state transition diagram of system is as shown in Figure 2.3. From the state transition diagram, the equilibrium state equations are given by

$$\begin{cases} \lambda P(0) = \mu P(1), & i = 0, \\ (\lambda + \mu)\,P(i) = \lambda P\,(i-1) + \mu P\,(i+1), & i \geq 1. \end{cases} \quad (2.58)$$

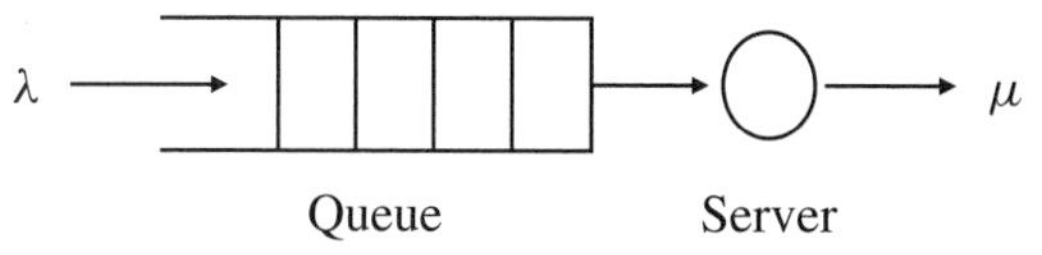

Figure 2.2
The M/M/1/∞ queuing model.

Figure 2.3
The state transition diagram of the M/M/1/∞ queuing system.

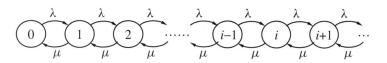

Thus, we have

$$
\begin{cases}
P(1) = \frac{\lambda}{\mu} P(0) = \rho P(0), \\
P(2) = \frac{\lambda}{\mu} P(1) = \left(\frac{\lambda}{\mu}\right)^2 P(0) = \rho^2 P(0), \\
\cdots \\
P(i) = \frac{\lambda}{\mu} P(i-1) = \left(\frac{\lambda}{\mu}\right)^i P(0) = \rho^i P(0), \\
\cdots
\end{cases}
\tag{2.59}
$$

where $\rho = \frac{\lambda}{\mu}$ and is called *traffic intensity.*

The normalized condition is given by

$$
\sum_{i=0}^{\infty} P(i) = 1.
\tag{2.60}
$$

From the preceding equations, we have

$$
\sum_{i=0}^{\infty} \rho^i P(0) = \frac{P(0)}{1-\rho}.
\tag{2.61}
$$

Thus,

$$
P(0) = 1 - \rho.
\tag{2.62}
$$

We know that $P(0)$ is the probability of the server being free. Since $P(0) > 0$, the necessary condition of a system being in a steady state is $\rho = \frac{\lambda}{\mu} < 1$. That is, the arrival rate cannot be more than service rate; otherwise the queue length will increase to infinity and jobs will experience infinite waiting time. Therefore, $\rho = 1 - P(0)$ is the probability of the server being busy. From equation (2.59), we have

$$
P(i) = \rho^i (1 - \rho).
\tag{2.63}
$$

We know that equation (2.63) is a geometric distribution.

According to the probabilities $P(i)$s, the average number of customers in the system is

$$
L_s = \sum_{i=0}^{\infty} i P(i)
$$

$$
= \rho (1-\rho) \sum_{i=1}^{\infty} i \rho^{i-1}
$$

$$
= \rho (1-\rho) \left(\frac{\rho}{1-\rho}\right)'
$$

$$
= \frac{\rho}{1-\rho}
$$

$$
= \frac{\lambda}{\mu - \lambda}.
\tag{2.64}
$$

Using Little's law, the average dwell time of a customer in the cell of a wireless system is given by

$$W_s = \frac{L_s}{\lambda}$$

$$= \frac{1}{\mu\,(1-\rho)}$$

$$= \frac{1}{\mu - \lambda}. \tag{2.65}$$

The average queue length is

$$L_q = \sum_{i=1}^{\infty} (i-1)\,P(i)$$

$$= \frac{\rho^2}{1-\rho}$$

$$= \frac{\lambda^2}{\mu\,(\mu - \lambda)}. \tag{2.66}$$

The average waiting time of customers is given by

$$W_q = \frac{L_q}{\lambda}$$

$$= \frac{\rho^2}{\lambda\,(1-\rho)}$$

$$= \frac{\lambda}{\mu\,(\mu - \lambda)}. \tag{2.67}$$

2.4.8   *M/M/S/∞* Queuing System

We consider a queuing system with arrival rate λ as before, but we assume that there are multiple servers S (≥ 1) each one with service rate μ, and they all share a common queue (see Figure 2.4). Let i ($i = 0, 1, 2, \dots$) be the number of customers in the system and let $P(i)$ be the steady-state probability of the system having i customers. Therefore, the state transition diagram of this system is shown in Figure 2.5.

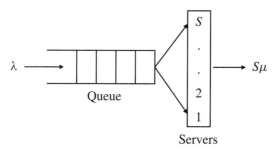

Figure 2.4
The *M/M/S/∞* queuing model.

Figure 2.5
The state transition
diagram of the
M/M/S/∞ queuing
system.

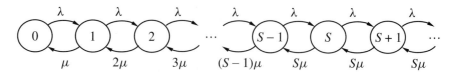

From the state transition diagram, the equilibrium state equations are given by

$$\begin{cases} \lambda P(0) = \mu P(1), & i = 0, \\ (\lambda + i\mu) P(i) = \lambda P(i-1) + (i+1)\mu P(i+1), & 1 \leq i < S, \\ (\lambda + S\mu) P(i) = \lambda P(i-1) + S\mu P(i+1), & S \leq i. \end{cases} \tag{2.68}$$

Thus, we have

$$\begin{cases} P(i) = \frac{\alpha^i}{i!} P(0), & i < S, \\ P(i) = \frac{\alpha^S}{S!} \left(\frac{\alpha}{S}\right)^{i-s} P(0), & S \leq i, \end{cases} \tag{2.69}$$

where $\alpha = \lambda/\mu$.

According to the normalized condition

$$\sum_{i=0}^{\infty} P(i) = \left[\sum_{i=0}^{S-1} \frac{\alpha^i}{i!} + \frac{\alpha^s}{S!} \sum_{i=0}^{\infty} \left(\frac{\alpha}{S}\right)^i \right] P(0) = 1, \tag{2.70}$$

we have

$$P(0) = \left[\sum_{i=0}^{S-1} \frac{\alpha^i}{i!} + \frac{\alpha^s}{S!} \sum_{i=0}^{\infty} \left(\frac{\alpha}{S}\right)^i \right]^{-1}. \tag{2.71}$$

If $\alpha < S$, we have

$$\sum_{i=0}^{\infty} \left(\frac{\alpha}{S}\right)^i = \frac{S}{S - \alpha}. \tag{2.72}$$

Thus,

$$P(0) = \left[\sum_{i=0}^{s-1} \frac{\alpha^i}{i!} + \frac{\alpha^S}{S!} \frac{S}{S - \alpha} \right]$$

$$= \left[\sum_{i=0}^{s-1} \frac{\alpha^i}{i!} + \frac{\alpha^S}{S!} \frac{1}{1 - \rho} \right], \tag{2.73}$$

where $\rho\, (= \alpha/S = \lambda/(S\mu))$ is called utilization factor. Note that for the queue to be stable we should have $\rho < 1$.

According to the probabilities $P(i)$s, the average number of customers in the system is

$$L_s = \sum_{i=0}^{\infty} i P(i)$$

$$= \alpha + \frac{\rho \alpha^S P(0)}{S! (1 - \rho)^2}. \tag{2.74}$$

Using Little's formula, the average dwell time of a customer in the system is given by

$$W_s = \frac{L_s}{\lambda}$$

$$= \frac{1}{\mu} + \frac{\alpha^S P(0)}{S\mu \cdot S! \, (1-\rho)^2}. \tag{2.75}$$

The average queue length is

$$L_q = \sum_{i=s}^{\infty} (i-S) P(i)$$

$$= \frac{\alpha^{S+1} P(0)}{(S-1)(S-\alpha)^2}. \tag{2.76}$$

The average waiting time of customers is given by

$$W_q = \frac{Lq}{\lambda}$$

$$= \frac{\alpha^S P(0)}{S\mu \cdot S! \, (1-\rho)^2}. \tag{2.77}$$

2.4.9   *M/G/1/∞* Queuing System

We consider a single server queuing system whose arrival process is Poisson with mean arrival rate λ. The service times are independent and identically distributed with distribution function F_B and pdf f_B. Jobs are scheduled for service in the order of their arrival—that is, the scheduling discipline is FCFS. As a special case of the $M/G/1$ queuing system, if we let F_B be the exponential distribution with mean rate μ, then we obtain the $M/M/1$ queuing systems. If service times are assumed to be constant, then we get the $M/D/1$ queuing system.

Let $N(t)$ denote the number of jobs in the system (those in the queue plus any in service) at time t. If $N(t) \geq 1$, then a job is in service, and since the general service time distribution need not be memoryless, besides $N(t)$, we also require knowledge of time spent by the job in service in order to predict the future behavior of the system. It follows that the stochastic process $\{N(t), t \geq 0\}$ is not a Markov chain.

To simplify the state description, we take a snapshot of the system at times of departure of jobs. These epochs of departure, called regeneration points, are used to specify the index set of a new stochastic process. Let t_n $(n = 1, 2, \ldots)$ be the time of departure (immediately following service) of the nth job and X_n be the number of jobs in the system at time t_n, so that

$$X_n = N(t_n), \qquad \text{for } n = 1, 2, \ldots . \tag{2.78}$$

The stochastic process $\{X_n, n = 1, 2, \ldots\}$ can be shown to be a discrete Markov chain, known as the imbedded Markov chain of the continuous stochastic process $\{N(t), t \geq 0\}$.

The method of the imbedded Markov chain allows us to simplify the analysis since it converts a non-Markovian problem into a Markovian one. We can then use the limiting distribution of the imbedded Markov chain as a measure of the original process $N(t)$, for it can be shown [2.4] that the limiting distribution of the number of jobs $N(t)$ observed at an arbitrary point in time is identical to the distribution of the number of jobs observed at the departure epochs; that is,

$$\lim_{t \to \infty} P\left[N(t) = k\right] = \lim_{t \to \infty} P\left(X_n = k\right). \tag{2.79}$$

For $n = 1, 2, \ldots$, let Y_n be the number of jobs arriving during the service time of nth job. Now the number of jobs immediately following the departure instant of $(n+1)$st job can be written as

$$X_{n+1} = \begin{cases} X_n - 1 + Y_n, & X_n > 0, \\ Y_{n+1}, & X_n = 0. \end{cases} \tag{2.80}$$

In other words, the number of jobs immediately following the departure of the $(n+1)$st job depends on whether the $(n+1)$st job was in the queue when the nth job departed. If $X_n = 0$, the next job to arrive is the $(n+1)$st. During its service time Y_{n+1} jobs arrive, then the $(n+1)$st job departs at time t_{n+1}, leaving Y_{n+1} jobs behind. If $X_n > 0$, then the number of jobs left behind by the $(n+1)$st job equals $X_n - 1 + Y_{n+1}$. Since Y_{n+1} is independent of $X_1, X_2, \ldots, X_n$, it follows that given the value of X_n, we need not know the values of $X_1, X_2, \ldots, X_{n-1}$ in order to determine the probabilistic behavior of X_{n+1}. Thus, $\{X_n, n = 1, 2, \ldots\}$ is a Markov chain.

The transition probabilities of the Markov chain are obtained using equation (2.80):

$$p_{ij} = P\left(X_{n+1} = j \mid X_n = i\right)$$
$$= \begin{cases} P\left(Y_{n+1} = j - i + 1\right), & i \neq 0, j \geq i - 1, \\ P\left(Y_{n+1} = j\right), & i = 0, j \geq 0. \end{cases} \tag{2.81}$$

Since all jobs are statistically identical, we expect that the Y_n's are identically distributed with pmf $P\left(Y_{n+1} = j\right) = a_j$ so that

$$\sum_{j=1}^{\infty} a_j = 1. \tag{2.82}$$

Then, the (infinite-dimensional) transition probability matrix of $\{X_n\}$ is given by

$$P = \begin{bmatrix} a_0 & a_1 & a_2 & a_3 & \cdots \\ a_0 & a_1 & a_2 & a_3 & \cdots \\ 0 & a_0 & a_1 & a_2 & \cdots \\ 0 & 0 & a_0 & a_1 & \cdots \\ 0 & 0 & 0 & a_0 & \cdots \\ \vdots & \vdots & \vdots & \vdots & \ddots \end{bmatrix}. \tag{2.83}$$

Let the limiting probability of being in state j be denoted by v_j, so that

$$v_j = \lim_{n \to \infty} P(X_n = j). \tag{2.84}$$

Using the preceding equations, we get

$$v_j = v_0 a_j + \sum_{i=1}^{j+1} v_i a_{j-i+1}. \tag{2.85}$$

If we define the generating function

$$G(z) = \sum_{j=0}^{\infty} v_j z^j, \tag{2.86}$$

then

$$\sum_{j=0}^{\infty} v_j z^j = \sum_{j=0}^{\infty} v_0 a_j z^j + \sum_{j=0}^{\infty} \sum_{i=1}^{j+1} v_i a_{j-i+1} z^j, \tag{2.87}$$

$$G(z) = v_0 \sum_{j=0}^{\infty} a_j z^j + \sum_{i=1}^{\infty} \sum_{j=i-1}^{\infty} v_i a_{j-i+1} z^j$$

$$= v_0 \sum_{j=0}^{\infty} a_j z^j + \sum_{i=1}^{\infty} \sum_{k=0}^{\infty} v_i a_k z^{k+i-1}$$

$$= v_0 \sum_{j=0}^{\infty} a_j z^j + \frac{1}{z} \left[\sum_{i=1}^{\infty} v_i z^i \sum_{k=0}^{\infty} a_k z^k \right]. \tag{2.88}$$

Defining

$$G_A(z) = \sum_{j=0}^{\infty} a_j z^j, \tag{2.89}$$

we have

$$G(z) = v_0 G_A(z) + \frac{1}{z} [G(z) - v_0] G_A(z) \tag{2.90}$$

or

$$G(z) = \frac{(z-1) v_0 G_A(z)}{z - G_A(z)}. \tag{2.91}$$

Since $G(1) = 1 = G_A(1)$, we can use L'Hôpital's rule to obtain

$$G(1) = \lim_{z \to 1} v_0 \frac{(z-1) G'_A(z) + G(z)}{1 - G'_A(z)}$$

$$= \frac{v_0}{1 - G'_A(1)}, \tag{2.92}$$

provided $G'_A(1)$ is finite and less than unity. (Note that $G'_A(1) = E[Y]$). If we let $\rho = G'_A(1)$, it follows that

$$v_0 = 1 - \rho, \tag{2.93}$$

and since v_0 is the probability that the server is idle, ρ is the server utilization in the limit. Moreover, we have that

$$G(z) = \frac{(1-\rho)(z-1)G_A(z)}{z - G_A(z)}. \tag{2.94}$$

Thus, if given the generating function $G_A(z)$, $G(z)$ can be computed from which the steady-state average number of jobs in the system can be computed by using

$$E[N] = \lim_{n \to \infty} E[X_n] = G'(1). \tag{2.95}$$

In order to evaluate $G_A(z)$, we first compute

$$a_j = P(Y_{n+1} = j). \tag{2.96}$$

This is the probability that exactly j jobs arrive during the service time of the $(n+1)$st job. Let the random variable B denote job service time. Now the conditional pmf of Y_{n+1} is obtained as

$$P(Y_{n+1} = j \mid B = t) = \frac{(\lambda t)^j}{j!} e^{-\lambda t}, \tag{2.97}$$

by the Poisson assumption. Using the theorem of total probability, we get

$$a_j = \int_0^\infty P(Y_{n+1} = j \mid B = t) f_B(t)\, dt$$

$$= \int_0^\infty \frac{(\lambda t)^j}{j!} e^{-\lambda t} f_B(t)\, dt. \tag{2.98}$$

Therefore, we have

$$G_A(z) = \sum_{j=0}^\infty a_j z^j$$

$$= \sum_{j=0}^\infty \int_0^\infty \frac{(\lambda t z)^j}{j!} e^{-\lambda t} f_B(t)\, dt$$

$$= \int_0^\infty \left[\sum_{j=0}^\infty \frac{(\lambda t z)^j}{j!} \right] e^{-\lambda t} f_B(t)\, dt$$

$$= \int_0^\infty e^{\lambda t z} e^{-\lambda t} f_B(t)\, dt$$

$$= \int_0^\infty e^{-\lambda t (1-z)} f_B(t)\, dt$$

$$= L_B [\lambda (1 - z)], \tag{2.99}$$

where $L_B [\lambda (1 - z)]$ is the Laplace transform of the service time distribution evaluated at $s = \lambda (1 - z)$. Note that

$$\rho = G'_A(1)$$

$$= \left. \frac{dL_B [\lambda (1 - z)]}{dz} \right|_{z=1}$$

$$= \left. \frac{dL_B}{ds} \right|_{s=0 \cdot (-\lambda)} \tag{2.100}$$

by the chain rule, then

$$\rho = \lambda E [B] = \frac{\lambda}{\mu} \tag{2.101}$$

by the moment-generating property of the Laplace transform. Here, the reciprocal of the service rate μ of the server equals the average service time $E [B]$.

Substituting (2.99) and (2.94), we get the well-known Pollaczek-Khinchin (P-K) transform equation

$$G(z) = \frac{(1 - \rho) (z - 1) L_B [\lambda (1 - z)]}{z - L_B [\lambda (1 - z)]}. \tag{2.102}$$

The average number of jobs in the system, in the steady-state, is determined by taking the derivation with respect to z and then taking the limit $z \to 1$

$$E [N] = \lim_{n \to \infty} E [X_n]$$

$$= \sum_{j=0}^{\infty} j v_j$$

$$= \lim_{z \to 1} G'(z)$$

$$= \rho + \frac{\lambda^2 E [B^2]}{2 (1 - \rho)}. \tag{2.103}$$

The average dwell time of customers in the system is given by

$$W_s = \frac{E [N]}{\lambda} = \frac{1}{\mu} + \frac{\lambda E [B^2]}{2 (1 - \rho)}. \tag{2.104}$$

We also discuss the average waiting time of customers in the queue. We know that

$$E[N] = \lim_{n \to \infty} E[Xn]$$

$$= \sum_{j=0}^{\infty} j\nu_j$$

$$= \sum_{j=0}^{\infty} j \int_0^\infty \int_0^\infty e^{-\lambda(t+x)} \frac{[\lambda(x+t)]^j}{j!} dW(t)\, dF_B(t)$$

$$= \int_0^\infty \int_0^\infty \lambda(t+x)\, dW(t)\, dF_B(t)$$

$$= \lambda \{W_q + E[B]\}$$

$$= \lambda W_q + \rho, \tag{2.105}$$

where $W(t)$ is the distribution of waiting time of customers in the queue and W_q is the mean value of $W(t)$.

Comparing (2.103) and (2.105), we have

$$W_q = \frac{\lambda E[B^2]}{2(1-\rho)}. \tag{2.106}$$

Thus, the average queue length is

$$L_q = \frac{\lambda^2 E[B^2]}{2(1-\rho)}. \tag{2.107}$$

● ● ● ● ● ● ● ● ● ● ● ● ● ● ●
2.5 Summary

This chapter summarizes important concepts of the probability theory that are useful in characterizing traffic in wireless networks. These concepts are also helpful in using the Markov chain model in representing an instantaneous state of the system in terms of the number of busy channels, number of calls pending in the queue, and their impact on queuing delays. Such information is employed in representing wireless system in terms of various performance parameters, which are discussed in later chapters. We need to know how a radio signal can reach users anywhere in the service area of a BS, this is covered in Chapter 3.

2.6 References

[2.1] W. T. Eadie et al., *Statistical Methods in Experimental Physics*, North Holland, 1971. (Amsterdam, London).

[2.2] D. S. Sivia, *Data Analysis: A Bayesian Tutorial*, Oxford University Press, Oxford, 1996.

[2.3] D. G. Kendall, "Stochastic Processes Occurring in the Theory of Queues and Their Analysis by the Method of the Imbedded Markov Chain," *Ann. Math. Stat.*, Vol. 24, pp. 19–53, 1953.

[2.4] L. Kleinrock, *Queuing System*, Vol. I, John Wiley & Sons, New York, 1975.

2.7 Problems

P2.1. A random number generator produces numbers between 1 and 99. If the current value of the random variable is 45, then what is the probability that the next randomly generated value for the same random variable will also be 45. Explain clearly.

P2.2. A random digit generator on a computer is activated three times consecutively to simulate a random three-digit number.
 (a) How many random three-digit numbers are possible?
 (b) How many numbers will begin with the digit 2?
 (c) How many numbers will end with the digit 9?
 (d) How many numbers will begin with the digit 2 and end with the digit 9?
 (e) What is the probability that a randomly formed number ends with 9 given that it begins with a 2?

P2.3. A snapshot of the traffic pattern in a cell with 10 users of a wireless system is given as follows:

User Number	1	2	3	4	5	6	7	8	9	10
Call Initiation Time	0	2	0	3	1	7	4	2	5	1
Call Holding Time	5	7	4	8	6	2	1	4	3	2

 (a) Assuming the call setup/connection and call disconnection time to be zero, what is the average duration of a call?
 (b) What is the minimum number of channels required to support this sequence of calls?
 (c) Show the allocation of channels to different users for part (b) of this problem.
 (d) Given the number of channels obtained in part (b), what fraction of time are the channels utilized?

P2.4. A department survey found that four of ten graduate students use CDMA cell phone service. If three graduate students are selected at random, what is the probability that the three graduate students use CDMA cell phones?

P2.5. There are three red balls and seven white balls in box A, and six red balls and four white balls in box B. After throwing a die, if the number on the die is 1 or 6, then pick a ball from box A. Otherwise, if any other number appears (i.e., 2, 3, 4, or 5), then pick a ball from box B. The selected ball must be put back before proceeding further. Answer the following:

(a) What is the probability that the selected ball is red?

(b) What is the probability a white ball is picked up in two successive selections?

P2.6. Consider an experiment consisting of tossing two dice. Let X, Y, and Z be the numbers shown on the first die, the second die, and total of both dice, respectively. Find $P(X \leq 1, Z \leq 2)$ and $P(X \leq 1) P(Z \leq 2)$ to show that X and Y are not independent.

P2.7. The following table shows the density of the random variable X.

x	1	2	3	4	5	6	7	8
$p(x)$	0.03	0.01	0.04	0.3	0.3	0.1	0.07	?

(a) Find $p(8)$.

(b) Find the table for F CDF.

(c) Find $P(3 \leq X \leq 5)$.

(d) Find $P(X \leq 4)$ and $P(X < 4)$. Are the probabilities the same?

(e) Find $F(-3)$ and $F(10)$.

P2.8. The density for X is given in the table of Problem 7.

(a) Find $E[X]$.

(b) Find $E[X^2]$.

(c) Find Var$[X]$.

(d) Find the standard deviation for X.

P2.9. Find the probability when

(a) $k = 2$ and $\lambda = 0.01$ for Poisson distribution.

(b) $p = 0.01$ and $k = 2$ for geometric distribution.

(c) Repeat (b) when binomial distribution is used and $n = 10$.

P2.10. Find the distribution function of the maximum of a finite set of independent random variables $\{X_1, X_2, \ldots, X_n\}$, where X_i has distribution function F_{X_i}. What is this distribution when X_i is exponential with a mean of $1/\mu_i$?

P2.11. The number of calls that arrive under a particular time in a cell has been established to be a Poisson distribution. The average number of calls arriving in a cell in 1 millisecond is 5. What is the probability that 8 calls arrive in a cell in a given milisecond?

P2.12. Given that the number of arrivals of data packet in the receiver follows a Poisson distribution on which arrival rate is 10 arrivals/sec. What is the probability of the number of arrivals is more than 8 but less than 11 during a time of interval of 2 seconds?

P2.13. In a wireless office environment, all calls are made between 8 AM and 5 PM over the period of 24 hours. Assuming the number of calls to be uniformly distributed between 8 AM and 5 PM, find the pdf of the number of calls over the 24 hour period. Also, determine the CDF and the variance of the call distribution.

P2.14. A gambler has a regular coin and a two-headed coin in his packet. The probability of selecting the two-head coin is given as $p = 2/3$. He selects a coin and flips it $n = 2$ times and obtains heads both times. What is the probability that the two-headed coin will be picked both times?

P2.15. A Poisson process exhibits a memoryless property and is of great importance in traffic analysis. Prove that this property is exhibited by all Poisson processes. Explain clearly every step of your analytical proof.

P2.16. What should be a relationship between call arrival rate and service rate when a cellular system is in a steady state? Explain clearly.

P2.17. Consider a cellular system with an infinite number of channels. In such a system, all arriving calls begin receiving service immediately. The average call holding time is $1/n\mu$ when there are n calls in the system. Draw a state transition diagram for this system and develop expressions for the following:
  (a) Steady-state probability p_n of n calls in the system.
  (b) Steady-state probability p_0 of no calls in the system.
  (c) Average number of calls in the system, L_s.
  (d) Average dwell time, W_s.
  (e) Average queue length, L_q.

P2.18. Consider a cellular system in which each cell has only one channel (single server) and an infinite buffer for storing the calls. In this cellular system, call arrival rates are discouraged—that is, the call rate is only $\lambda/(n+1)$ when there are n calls in the system. The interarrival times of calls are exponential distributed. The call-holding times are exponentially distributed with mean rate μ. Develop expressions for the following:
  (a) Steady-state probability p_n of n calls in the system.
  (b) Steady-state probability p_0 of no calls in the system.
  (c) Average number of calls in the system, L_s.
  (d) Average dwell time, W_s.
  (e) Average queue length, L_q.

P2.19. In a transition diagram of $M/M/5$ model, write the state transition equations and find a relation for the system to be in each state.

P2.20. In the $M/M/1/\infty$ queuing system, suppose λ and μ are doubled. How are L_s and W_s changed?

Mobile Radio Propagation

3.1 Introduction

For a wireless and mobile system design, it is very important to understand the distinguishing features of mobile radio propagation. In this chapter, we discuss some of these characteristics. A wireless mobile channel is modeled as a time-varying communication path between two stations, such as from/to one terminal to/from another terminal. The first terminal is the fixed antenna at a BS, while a moving MS or a subscriber represents the second station. This becomes a multipath propagation channel with fast fading. The mobile radio propagation properties introduce new challenges for isotropic directed antennas, choice of appropriate carrier frequency, and transmission techniques under the condition of fast fading. Propagation in a multipath channels depends on the actual environment, such as the antenna height, the profile of buildings, the trees, the roads, and the terrain. In this chapter, we describe mobile radio propagation using appropriate statistical techniques.

3.2 Types of Radio Waves

There are several kinds of radio waves, such as ground, space, and sky waves, as shown in Figure 3.1. As the name indicates, the ground wave propagates along the surface of the earth, and the sky wave propagates in the space but can return to earth by reflection either in the troposphere or in the ionosphere. Different wavelengths are reflected to dissimilar extent in troposphere and ionosphere.

Based on the attributes of these waves, we can partition the spectrum. Classification of the radio spectrum is based on propagation properties and the system aspects. Table 3.1 shows the radio frequency bands used for radio transmission. For cellular systems, we are primarily concerned with the ground and space waves and we discuss the propagation properties, path losses, and other characteristics in these areas.

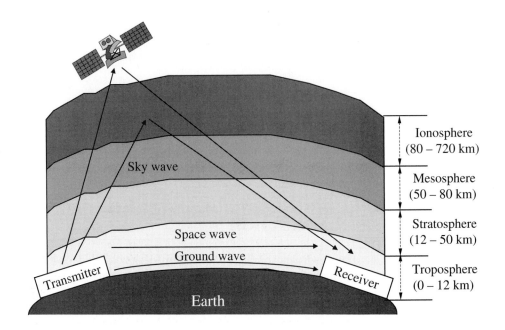

Figure 3.1
Propagation of
different types of
radio waves.

Table 3.1: ▶
Radio Frequency Bands

Classification Band	Initials	Frequency Range	Propagation Mode
Extremely low	ELF	<300 Hz ~3 kHz	Ground wave
Infra low	ILF	300 Hz ~3 kHz	Ground wave
Very low	VLF	3 kHz ~30 kHz	Ground wave
Low	LF	30 kHz ~300 kHz	Ground wave
Medium	MF	300 kHz ~3 MHz	Ground/sky wave
High	HF	3 MHz ~30 MHz	Sky wave
Very high	VHF	30 MHz ~300 MHz	Space wave
Ultra high	UHF	300 MHz ~3 GHz	Space wave
Super high	SHF	3 GHz ~30 GHz	Space wave
Extremely high	EHF	30 GHz ~300 GHz	Space wave
Tremendously high	THF	300GHz ~3000 GHz	Space wave

3.3 **Propagation Mechanisms**

Propagation in free space and without any obstacle is the most ideal situation. When the radio waves reach close to an obstacle, the following propagation effects do occur to the waves:

1. **Reflection**: Propagating wave impinges on an object that is larger as compared to its wavelength (for example, the surface of the earth, tall buildings, large walls).

2. **Diffraction**: Radio path between a transmitter and a receiver is obstructed by a surface with sharp irregular edges (for example, waves bend around the obstacle, even when line of sight (LOS) does not exist).

3. **Scattering**: When objects are smaller than the wavelength of the propagating wave (for example, foliage, street signs, lamp posts), incoming signal is scattered into several weaker outgoing signals.

Small-scale propagation characterizes signal variation over a short transmitter-receiver distance. Large-scale propagation characterizes signal variation over a larger transmitter-receiver distance. Diffraction and scattering result in small-scale fading effects, while reflection results in a large-scale fading.

A typical propagation effect of mobile radio is shown in Figure 3.2. Here, h_b is the height of antenna from the earth's surface at the BS, h_m is the height of antenna from the earth's surface at the MS, and d is the distance between the BS and the MS. The radio signals can penetrate simple walls, to some extent. However, a large street structure or hill is difficult to pass through. In those cases, diffracted and reflected radio waves enable the signals to reach these locations that are not directly in line with the direct path and help cover the neighborhood areas. The disadvantage is that MSs may receive multiple copies of the same signals with appropriate delays corresponding to the traversed paths. The advances in signal processing take care of this problem by selecting the best quality of the received signal and filtering out the rest of the weaker copies or combining the multiple signals after compensating for their arrival phases. All these operations are done in the hand-set of the MS and are transparent to the user.

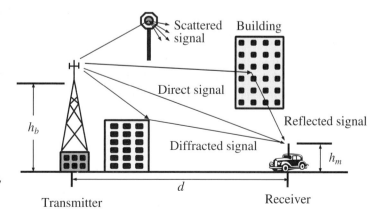

Figure 3.2
Reflection, diffraction, and scattering of radio signals.

● ● ● ● ● ● ● ● ● ● ● ● ● ● ● ● ●

3.4 Free Space Propagation

Free space is an ideal propagation medium. Consider an isotropic point source fed by a transmitter of P_t watts. At an arbitrary, large distance d from the source, the radiated power is uniformly distributed over the surface area of a sphere. Thus, the received signal power P_r at distance d is given by

$$P_r = \frac{A_e G_t P_t}{4\pi d^2},$$ (3.1)

where A_e is effective area covered by the transmitter and G_t is the transmitting antenna gain.

The relationship between an effective aperture and the receiving antenna gain G_r, derived in [3.1], can be given by

$$G_r = \frac{4\pi A_e}{\lambda^2},$$ (3.2)

where λ is the wavelength of the electromagnetic wave. By substituting A_e of equation (3.2) into equation (3.1), we obtain

$$P_r = \frac{G_r G_t P_t}{\left(\frac{4\pi d}{\lambda}\right)^2}.$$ (3.3)

Free space path loss L_f is defined as

$$L_f = \frac{P_t}{P_r}$$

$$= \frac{1}{G_r G_t}\left(\frac{4\pi d}{\lambda}\right)^2.$$ (3.4)

Basically, L_f indicates the amount of power lost in the space. A larger loss implies the use of higher transmitting power level, as the received signal strength must be at some minimal power level for correct reception at the receiving end.

When $G_t = G_r = 1$, free space path loss is given by

$$L_f = \left(\frac{4\pi d}{\lambda}\right)^2$$

$$= \left(\frac{4\pi f_c d}{c}\right)^2,$$ (3.5)

where c is essentially the speed of light ($= 2.998 \times 10^8$ m/s) and f_c is carrier frequency.

Free space path loss in decibels can be written as

$$L_f \text{ (dB)} = 32.45 + 20\log_{10} f_c \text{ (MHz)} + 20\log_{10} d \text{ (km)}.$$ (3.6)

Figure 3.3 shows the free space path loss characteristics as a function of the transmitting frequency and the distance of the receiver from the transmitter. It is clear from the figure that the signal strength reduced with the distance and the path loss also increases with the carrier frequency.

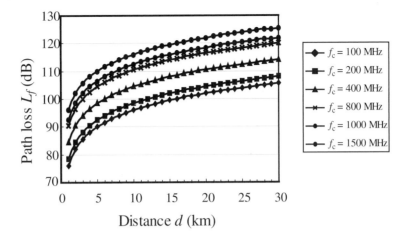

Figure 3.3
Free space path
loss.

● ● ● ● ● ● ● ● ● ● ● ● ● ● ●

3.5 **Land Propagation**

A land mobile radio channel is characterized by communication from/to a fixed station to/from a MS; it becomes a multipath propagation channel with fading. What this means is that the signal reaches the destination using many different paths, because of diffraction and reflection from various objects along the path of propagation. The signal strength and quality of received radio waves vary accordingly, as well as the time to reach the destination changes. This implies that the wave propagation in the multipath channel depends on the actual environment, including factors such as the antenna height, the profile of the buildings, roads, and the terrain. Therefore, we need to describe the behavior of mobile radio channels using a good and relevant statistical mechanism.

The received signal power P_r is expressed as

$$P_r = \frac{G_t G_r P_t}{L}, \tag{3.7}$$

where L represents the propagation loss in the channel. Wave propagation in a mobile radio channel is characterized by three aspects: path loss, slow fading (shadowing), and fast fading. Therefore, L can be expressed as

$$L = L_P L_S L_F, \tag{3.8}$$

where L_P, L_S, and L_F represent the path loss, slow fading loss, and fast fading loss, respectively (see Figure 3.4). Slow fading is long-term fading and fast fading is short-term fading and their empirical relationships are discussed later.

The path loss L_P is the average propagation loss over a wide area. It is determined by the macroscopic parameters, such as the distance between the transmitter and receiver, the carrier frequency, and the land profile. The slow fading loss L_S represents variation of the propagation loss in a local area (several tens of meters).

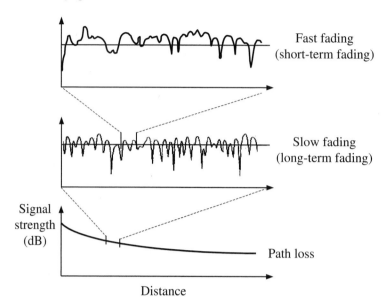

Figure 3.4
Schematic diagram
of propagation loss.

Slow fading is caused by the variation in propagation conditions due to buildings, roads, and other obstacles in a relatively small area. Slow fading is an overall average fading over some distances traveled by a MS (e.g., MS moving a couple of blocks). Fast fading loss L_F is due to the motion of the MS that consists of many diffracted waves, representing the microscopic aspect of the channel as shown in Figure 3.4. Fast fading is a fast changing fading (e.g., every step taken by a moving MS). In the following section, we discuss propagation loss (i.e., path loss of the signal).

● ● ● ● ● ● ● ● ● ● ● ● ● ● ●

3.6 Path Loss

The simplest formula for path loss of land propagation is

$$L_P = Ad^\alpha, \tag{3.9}$$

where A and α are propagation constants and d is the distance between the transmitter and the receiver. Usually, α takes a value of $3 \sim 4$ in a typical urban area. To predict propagation constants, many nomographs are obtained through propagation measurements [3.2]–[3.5]. The Okumura curves are well known for their practical use [3.2], and based on them, Hata [3.3] presented an empirical formula for prediction of path loss. The results are reproduced here for the completeness of the text.

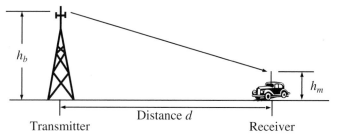

Figure 3.5
Radio propagation.

1. Urban Area

$$L_{PU} \text{ (dB)} = 69.55 + 26.16 \log_{10} f_c \text{ (MHz)} - 13.82 \log_{10} h_b \text{ (m)} - \alpha \left[h_m \text{ (m)} \right]$$

$$+ \left[44.9 - 6.55 \log_{10} h_b \text{ (m)} \right] \log_{10} d \text{ (km)}, \tag{3.10}$$

where L_{PU} (dB) $= -10 \log_{10} L_{PU}$, f_c is carrier frequency (150 MHz $\sim$ 1500 MHz), h_b is the effective BS antenna height (30 m $\sim$ 200 m), h_m is the MS antenna height (1 m $\sim$ 10 m), d is the distance (1 m $\sim$ 20 km), and $\alpha(h_m)$ is a correction factor for the mobile antenna height. Figure 3.5 illustrates the basic concept that is used as a basis for calculation of $\alpha(h_m)$ in equation (3.10). The values of $\alpha(h_m)$ are calculated differently for different environments, as shown by the following equations, and can be summarized as follows:

(a) **Large Cities**

$$\alpha \left[h_m \text{ (m)} \right] = \left[1.1 \log_{10} f_c \text{ (MHz)} - 0.7 \right] h_m \text{ (m)} - \left[1.56 \log_{10} f_c \text{ (MHz)} - 0.8 \right] \tag{3.11}$$

(b) **Medium and Small Cities**

$$\alpha \left[h_m \text{ (m)} \right] = \begin{cases} 8.29 \left[\log_{10} 1.54 h_m \text{ (m)} \right]^2 - 1.1, & f_c < 300 \text{ MHz} \\ 3.2 \left[\log_{10} 11.75 h_m \text{ (m)} \right]^2 - 4.97, & f_c > 300 \text{ MHz} \end{cases} \tag{3.12}$$

2. Suburban Area

$$L_{PS} \text{ (dB)} = L_{PU} \text{ (dB)} - 2 \left[\log_{10} \frac{f_c \text{ (MHz)}}{28} \right]^2 - 5.4 \tag{3.13}$$

3. Open Area

$$L_{PO} \text{ (dB)} = L_{PU} \text{ (dB)} - 4.78 \left[\log_{10} f_c \text{ (MHz)} \right]^2 - 18.33 \log_{10} f_c \text{ (MHz)} - 40.94 \tag{3.14}$$

Path loss characteristics in urban areas for large-, small-, and medium-size cities are shown in Figures 3.6 and 3.7. Path loss characteristics for suburban and open areas are shown in Figures 3.8 and 3.9, respectively. For a large city, the path loss is the same as that for small- and medium-size cities under $h_b = 50$ m and $h_m = 1.65$ m.

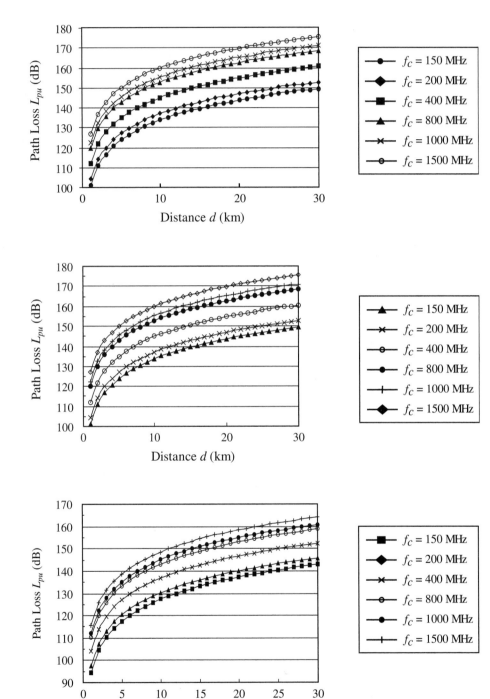

Figure 3.6
Path loss (urban: large city).

Figure 3.7
Path loss (urban area: medium and small cities).

Figure 3.8
Path loss (suburban area).

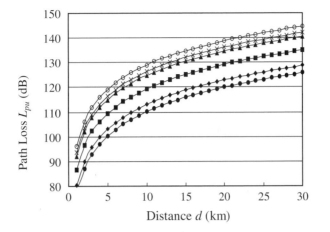

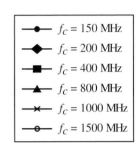

Figure 3.9
Path loss (open
area).

$\bullet \; \bullet \; \bullet \; \bullet \; \bullet \; \bullet \; \bullet \; \bullet \; \bullet \; \bullet \; \bullet \; \cdot \; \cdot \; \cdot \; \cdot$

3.7 Slow Fading

Slow fading is caused by the long term spatial and temporal varitions over distances large enough to produce gross variation in the overall path between the transmitter and receiver [3.10]. Long-term variation in the mean level is known as slow fading [3.6]. Slow fading is also called log-normal fading or shadowing , because its amplitude has a log-normal pdf.

In slow fading, the local mean value $r_m\,(d)$ at location d is defined as follows:

$$r_m\,(d) = \frac{1}{2d_w} \int_{d-d_w}^{d+d_w} r\,(x)\,dx, \tag{3.15}$$

where $r\,(x)$ is the received signal at position x and d_w is window size.

The received signal $r\,(x)$ can be expressed as the product of two parts. One is $r_s\,(x)$ which is affected by slow fading, and another is $r_f\,(x)$, which is affected by fast fading. Thus,

$$r\,(x) = r_s\,(x)\,r_f\,(x). \tag{3.16}$$

Substituting (3.16) into (3.15), we have

$$r_m\,(d) = \frac{1}{2d_w} \int_{d-d_w}^{d+d_w} r_s\,(x)\,r_f\,(x)\,dx. \tag{3.17}$$

When $x = d$, $r_s\,(d)$ is assumed as an actual local mean received signal level. Thus,

$$r_m\,(d) = r_s\,(d).$$

Therefore, based on the statistical values of the received signal, the window size d_w needs to satisfy the following condition:

$$\frac{1}{2d_w} \int_{d-d_w}^{d+d_w} r_f\,(x)\,dx \longrightarrow 1. \tag{3.18}$$

In general, the window size d_w varies from several tens to several hundreds of wavelengths. If d_w is too short, the statistical characteristics cannot represent the slow fading phenomenon. If d_w is too large, the statistical characteristics of slow fading will be lost again.

We can see that equation (3.16) is a function of the location. Since the distance can be represented as a function of speed and time (i.e., $x = vt$), equation (3.16) can be rewritten as follows:

$$r(t) = r_s(t)\, r_f(t).$$

Many experiments have indicated that slow fading obeys the log-normal distribution. In this case, the pdf of the received signal level is given in decibels by

$$p(M) = \frac{1}{\sqrt{2\pi}\,\sigma} e^{-\frac{(M-\overline{M})^2}{2\sigma^2}}, \tag{3.19}$$

where M is the true received signal level m in decibels (dB) (i.e., $M = 10\log_{10} m$), $\overline{M}$ is the area average signal level (i.e., the mean of M), and σ is the standard deviation in decibels.

When we express the pdf of the received signal level in terms of mW, it is given by

$$p(m) = \frac{1}{\sqrt{2\pi}\,m\sigma_o} e^{-\frac{\left(\log_{10}\frac{m}{\overline{m}}\right)^2}{2\sigma_o^2}}, \tag{3.20}$$

where $\overline{m}$ is the long-term average received signal level and $\sigma_o = \frac{\log_{10}\sigma}{10}$.

The average of M is defined over a distance that is long enough for average microscopic variation (several wavelengths). The variance takes values of $4 \sim 12$ dB depending on the propagation environment. Figure 3.10 shows a pdf of a log-normal distribution.

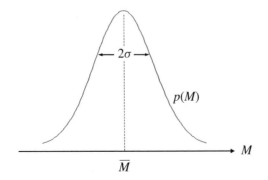

Figure 3.10
The pdf of log-normal distribution.

• • • • • • • • • • • • • • • •

3.8 Fast Fading

Fast fading is due to scattering of the signal by object near transmitter. The effects of fast fading are discussed below as they must be compensated for by adequate signal processing operations.

3.8.1   Statistical Characteristics of Envelope

Figure 3.4 illustrates the fading characteristics of a mobile radio signal. The rapid fluctuations in the spatial and temporal characteristics caused by local multipath are known as fast fading (short-term fading due to fast spatial variations). Distances of about half a wavelength results in fast fading. For VHF (very high frequency) and UHF (ultra high frequency), a vehicle traveling at 50 km (30.49 miles) per hour can pass through several fades in a second. Therefore, the mobile radio signal, as shown in Figure 3.4, consists of a short-term (fast fading) signal superimposed on a local mean value (this remains constant over a small area but varies slowly as the receiver moves). As noted previously, fading rate is low, but not zero, for a stationary handset [3.10]. Next, we consider two cases where the receiver is far from or close to the transmitter.

Receiver Far from the Transmitter

In this case, we assume that there are no direct radio waves between the transmitter and the receiver, the probability distribution of signal amplitude of every path is a Gaussian distribution and their phase distribution has a uniform distribution within $(0, 2\pi)$ radians. Therefore, the probability distribution of the envelope for the composite signals is a Rayleigh distribution and its pdf is given by

$$p\,(r) = \frac{r}{\sigma^2} e^{-\frac{r^2}{2\sigma^2}}, \qquad r > 0, \tag{3.21}$$

where r is the envelope of fading signal and σ is the standard deviation. Figure 3.11 shows the pdf of Rayleigh distribution.

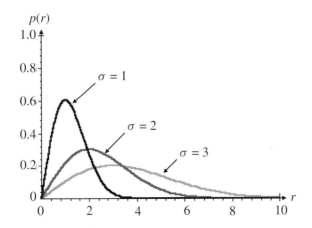

Figure 3.11
The pdf of Rayleigh distribution when σ = 1, 2, and 3.

The pdf of the phase distribution for the composite signals is given by

$$p(\theta) = \frac{1}{2\pi}, \qquad 0 \le \theta \le 2\pi. \tag{3.22}$$

Thus, the mean (first moment) of the fading signal is

$$E[r] = \int_0^\infty r p(r) \, dr$$

$$= \sqrt{\frac{\pi}{2}} \sigma \tag{3.23}$$

and the power (second moment) of the fading signal is

$$E\left[r^2\right] = \int_0^\infty r^2 p(r) \, dr$$

$$= 2\sigma^2. \tag{3.24}$$

The cumulative probability distribution (CDF) of composite signals is

$$P(r \le x) = \int_0^x p(r) \, dr$$

$$= 1 - e^{-\frac{x^2}{2\sigma^2}}. \tag{3.25}$$

Using equation (3.25), we can define that the middle value r_m of envelope signal within the sample range is satisfied by

$$P(r \le r_m) = 0.5. \tag{3.26}$$

Thus, we have $r_m = 1.777\sigma$.

Receiver Close to the Transmitter

In this case, the direct radio wave is stronger as compared to other radio waves between the receiver and transmitter. As in the previous case, we assume that the probability distribution of signal amplitudes of all paths is a Gaussian distribution and their phase has a uniform distribution within $(0, 2\pi)$. In addition, a stronger specular or direct component is considered. Therefore, the probability distribution of envelope of composite signals is a Rician distribution and its pdf is given by

$$p(r) = \frac{r}{\sigma^2} e^{-\frac{(r^2+\beta^2)}{2\sigma^2}} I_0\left(\frac{\beta r}{\sigma^2}\right), \tag{3.27}$$

where r is the envelope of fading signal, σ is standard deviation, β is the amplitude of direct signal, and $I_0(x)$ is the zero-order modified Bessel function of the first kind; that is,

$$I_0(x) = \frac{1}{2\pi} \int_0^{2\pi} e^{x \cos\theta} \, d\theta$$

$$\approx \frac{e^x}{\sqrt{2\pi x}}. \tag{3.28}$$

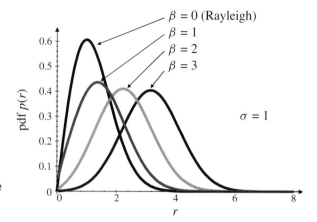

Figure 3.12
The pdf of the envelope of composite signals according to Rician distribution.

When β is very large—that is, the direct signal is very strong ($r \approx \sigma$)—equation (3.27) can be approximated by a Gaussian distribution. When β is very small—that is, there is no direct signal (the standard deviation $\sigma \approx 0$)—equation (3.27) can be approximated by a Rayleigh distribution. Figure 3.12 shows the pdf of the envelope of composite signals according to Rician distribution.

Generalized Model

Nakagami distribution (Nakagami-m distribution) is a generalized fading channel model introduced by Nakagami in the 1940s [3.4]. For the Nakagami distribution, the pdf of received signal envelope is given by

$$p(r) = \frac{2r^{2m-1}}{\Gamma(m)} \left(\frac{m}{\Omega}\right)^m e^{-\frac{mr^2}{\Omega}} \qquad \text{for } r \geq 0, \tag{3.29}$$

where $\Gamma(m)$ is the Gamma function, $\Omega = E\left[r^2\right]$ is the average power which is the second moment of the fading signal, and $m = \frac{\Omega^2}{E\left[(r^2-\Omega)^2\right]}$ is called fading factor with $m \geq 0.5$.

When $m = 1$, the Nakagami distribution becomes Rayleigh distribution. When m approaches infinity, the distribution becomes an impulse, which means there is no fading.

The advantage of Nakagami distribution is that it is easier to use than the Rician distribution which contains a Bessel function. The Rician distribution can be closely approximated by using the following relation between the Rician factor $K \left(= \frac{\beta}{2\sigma}\right)$ and the Nakagami fading factor m [3.7], i.e.,

$$m = \frac{(K+1)^2}{2K+1} \tag{3.30}$$

or

$$K = \frac{\sqrt{m^2 - m}}{m - \sqrt{m^2 - m}}, \qquad m > 1. \tag{3.31}$$

The channel for indoor and outdoor wireless and mobile communication systems can often be better modeled by a Nakagami distribution than Rician distribution. Rayleigh distribution is useful for modeling wireless and mobile communication systems where there exists no LOS.

3.8.2   Characteristics of Instantaneous Amplitude

The instantaneous amplitude of the received signal can be presented by the level crossing rate, the fading rate, the depth of fading, and the fading duration.

Level Crossing Rate

The level crossing rate $N(R_s)$ at a specified signal level (called threshold) R_s is defined as the average number of times per second that the signal envelope crosses the level in a positive going direction [3.8] [3.9] [3.10]. $N(R_s)$ is given by

$$N(R_s) = \frac{\sqrt{\pi}}{\sigma} R_s f_m e^{-\frac{R_s^2}{2\sigma^2}}, \tag{3.32}$$

where f_m is maximum Doppler frequency and is given by

$$f_m = \frac{v}{\lambda}, \tag{3.33}$$

where v is moving speed of mobile user and λ is the carrier wavelength. We introduce the Doppler effect in the next section.

Since $2\sigma^2$ is equal to mean square value, $\sqrt{2}\sigma$ is the root mean square (rms) value. The level crossing rate for a vertical monopole antenna can then be given by

$$N(R_s) = \sqrt{2\pi} f_m \rho e^{-\rho^2}, \tag{3.34}$$

where $\rho \left(= \frac{R_s}{\sqrt{2}\sigma} \right)$ is the ratio between the specified level and the rms amplitude of the fading envelope.

For example, for a Rayleigh fading signal, compute the positive going level crossing rate for $\rho = \frac{1}{\sqrt{2}}$ (i.e., at a level 3 dB below the rms level), when the maximum Doppler frequency is 100 Hz.

Using equation (3.34), the number of positive going level crossing is

$$N(R_s) = \sqrt{2\pi} \cdot 100 \cdot \frac{1}{\sqrt{2}} e^{-\frac{1}{2}}$$

$$= 107.5 \text{ crossing per second.}$$

Fading Rate

Fading rate is defined as the number of times that the signal envelope crosses the middle value r_m in a positive going direction per unit time. Usually, the fading rate is related to carrier wavelength, the velocity of mobile user, and the number of multipaths. Based on extensive experience, the average fading rate is

$$N(r_m) = \frac{2v}{\lambda}. \tag{3.35}$$

Depth of Fading

Depth of fading is defined as the ratio between the mean square value and the minimum value of the fading signal. Since the depth of fading is a random variable, the average depth of fading is used and is defined as a difference of the middle value and the amplitude value of the fading signal when $P(r \le r_{10}) = 10\%$.

Fading Duration

Fading duration is defined as the duration for which the signal is below a given threshold R_s. Since it is a random variable, we use an average fading duration to describe the fading duration. Therefore, we have

$$\tau(R_s) = \frac{P(r \le R_s)}{N(R_s)}$$

$$= \frac{e^{\rho^2} - 1}{\sqrt{2\pi} f_m \rho}. \tag{3.36}$$

● ● ● ● ● ● ● ● ● ● ● ● ● ⋯ ⋯

3.9 Doppler Effect

In a wireless and mobile system, the location of the BS is fixed while the MSs are mobile. Therefore, as the receiver is moving with respect to the wave source, the frequency of the received signal will not be the same as the source (see Figure 3.13). Here, V_1, V_2, V_3, and V_4 are different moving speeds of the receiver. When they are moving toward each other, the frequency of the received signal is higher than that of the source. When they are moving away from each other, the received frequency decreases.

Thus, the frequency f_r of the received signal is

$$f_r = f_c - f_d, \tag{3.37}$$

where f_c is the frequency of source carrier and f_d is the Doppler frequency or Doppler shift.

Doppler frequency or Doppler shift is

$$f_d = \frac{v}{\lambda} \cos\theta, \tag{3.38}$$

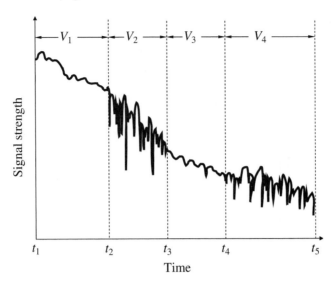

Figure 3.13
Moving speed effect.

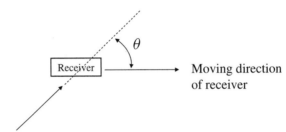

Figure 3.14
Relation of moving
speed and moving
direction.

where v is the moving speed, λ is the wavelength of carrier, and θ is as shown in Figure 3.14. $v\cos\theta$ represents the velocity component of the receiver in the direction of the sender.

● ● ● ● ● ● ● ● ● ● ● ● ● ● ●

3.10 Delay Spread

In many cases, when a signal propagates from a transmitter to a receiver, the signal suffers one or more reflections so that the path becomes indirect. This forces radio signals to follow different paths. Figure 3.15 shows the received signal due to the different multipath. Since each path has a different path length, the time of arrival for each path is different. The smearing or spreading out effect of the signal is called "delay spread." In a digital communication system, the delay spread causes intersymbol interference, thereby limiting the maximum symbol rate of a digital multipath channel. If we assume that the pdf of the delay t is $p(t)$, the average delay spread is defined as

$$\tau_m = \int_0^\infty tp(t)\,dt. \tag{3.39}$$

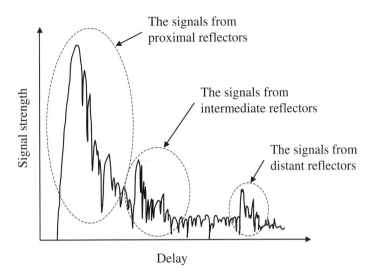

Figure 3.15
The delay spread of a signal.

Thus, the delay spread is defined as

$$\tau_d = \sqrt{\int_0^\infty (t - \tau_m)^2 \, p\,(t)\,dt}.\tag{3.40}$$

The following are well-known representative delay functions:

- Exponential:

$$p\,(t) = \frac{1}{\tau_m} e^{-\frac{t}{\tau_m}}.\tag{3.41}$$

- Uniform:

$$p\,(t) = \begin{cases} \dfrac{1}{2\tau_m}, & 0 \le t \le 2\tau_m, \\ 0, & \text{elsewhere.} \end{cases}\tag{3.42}$$

The delay spread usually takes a value of around 3 microseconds for a city area and up to 10 microseconds in hilly terrains.

• • • • • • • • • • • • • • • •

3.11 Intersymbol Interference

Intersymbol interference (ISI) is caused by time-delayed multipath signals. ISI also has an impact on the burst error rate of the channel. Such an effect is illustrated in Figure 3.16, where the second multipath signal could be delayed so much that a part could be received during the next symbol interval.

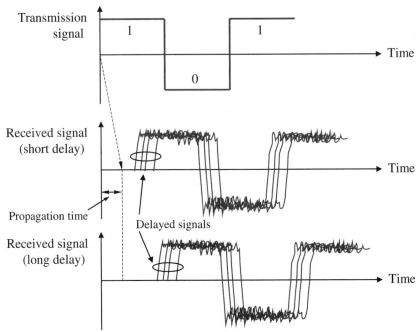

Figure 3.16
ISI caused by multipath
signals.

In a time-dispersive medium, the transmission rate R for a digital transmission is limited by the delay spread. If a low bit-error-rate (BER) performance is desired, then

$$R < \frac{1}{2\tau_d}. \tag{3.43}$$

In a real situation, R is determined based on the required BER, which may be limited by the delay spread.

3.12 Coherence Bandwidth

The coherent bandwidth is a statistical measure of the range of frequencies over which the channel can be considered "flat" (i.e., a channel which passes all spectral components with approximately equal gain and linear phase). The coherence bandwidth B_c represents the correlation between two fading signal envelopes at frequencies f_1 and f_2 and is a function of delay spread τ_d. When the correlation coefficient for two fading signal envelopes at frequencies f_1 and f_2 is equal to 0.5, the coherence bandwidth is approximated by

$$B_c \approx \frac{1}{2\pi \tau_d}. \tag{3.44}$$

Two frequencies that are larger than the coherence bandwidth fade independently. This concept is also useful for diversity reception, wherein multiple copies of the same message are sent using different frequencies.

The coherence bandwidth for two fading amplitudes of two received signals is

$$\triangle f = \mid f_1 - f_2 \mid > B_c = \frac{1}{2\pi \tau_d}. \qquad (3.45)$$

The coherence bandwidth for two random phases of two received signals is

$$\triangle f = \mid f_1 - f_2 \mid < E\left[B_c\right] = \frac{1}{4\pi \tau_d}, \qquad (3.46)$$

where $E\left[B_c\right]$ is the average value of the coherence bandwidth B_c.

If the bandwidth of transmitted signal is lower than the channel coherent bandwidth, only the gain and phase of the signal are changed, nonlinear transformation could not occur. However, if the bandwidth of transmitted signal is larger than the channel coherent bandwidth, part of the transmitted signal is truncated, which means nonlinearity is present and the signal could be severely influenced. This situation is called frequency-selective fading.

● ● ● ● ● ● ● ● ● ● ● ● ● ● ● ● ●

3.13 Cochannel Interference

In a cellular system, the key concept is the reuse of frequencies; that is, the same frequency is assigned to different cells. The frequency allocation is done in such a way that the probability P_{co} of cochannel interference between cells using the same frequency is less than a given value. It is defined as the probability that the desired signal level r_d drops below a value proportional to the interfering undesired signal level r_u; as

$$P_{co} = P\left(r_d \leq \beta r_u\right), \qquad (3.47)$$

where β is defined as the protection ratio.

We assume that desired and undesired interfering signals are independent of each other. We denote their pdfs as $p_1(r_1)$ and $p_2(r_2)$, respectively. Then P_{co} is given by

$$P_{co} = \int_0^\infty P\left(r_1 = x\right) P\left(r_2 \geq \frac{x}{\beta}\right) dx$$

$$= \int_0^\infty p_1(r_1) \int_{\frac{r_1}{\beta}}^\infty p_2(r_2)\, dr_2 dr_1. \qquad (3.48)$$

Ways for minimizing cochannel interference are discussed in Chapter 5.

●●●●●●●●●●●●●●●●

3.14 Summary

This chapter provides a brief description of how electromagnetic waves are propagated through open space. It outlines major causes that influence the propagation of these waves, and shows how these can be mathematically modeled or expressed. Attenuation of the signal due to path loss and other fading effects have been discussed. Modern wireless systems also experience other phenomena, such as perceived change in signal frequency at the receiver. Such effects have also been elaborated. In the next chapter, we consider how to minimize the effect of distortion by channel coding and other redundancy techniques, and their impact on the overall performance.

●●●●●●●●●●●●●●●●

3.15 References

[3.1] J. D. Kraus, *Antennas*, McGraw-Hill, New York, 1988.

[3.2] Y. Okumura et al., "Field Strength and Its Variability in UHF and VHF Land-Mobile Radio Service," *Review of Electrical Communication Laboratory*, 16 (1968).

[3.3] M. Hata, "Empirical Formula for Propagation Loss in Land Mobile Radio Services," *IEEE Transactions on Vehicular Technology*, VT-29, August 1980.

[3.4] J. G. Proakis, *Digital Communications*, 4th edition. New York: McGraw-Hill, 2001.

[3.5] W. C. Jakes, ed., *Microwave Mobile Communications*, John Wiley & Sons, New York, 1974.

[3.6] W. C. Y. Lee, *Mobile Communications Design Fundamentals* (Second Edition), John Wiley & Sons, New York, 1993.

[3.7] G. L. Stüber, *Principles of Mobile Communication*, Second Edition, Kluwer Academic Publishers, 2002.

[3.8] W. C. Y. Lee, *Mobile Communications Engineering*, McGraw-Hill, New York, 1982.

[3.9] A. Mehrotra, *Cellular Radio Performance Engineering*, Artech House, Boston, 1994.

[3.10] V. Garg and J. Wilkes, *Wireless and Personal Communications Systems*, Prentice-Hall, Englewood Cliffs, NJ, 1996

●●●●●●●●●●●●●●●●

3.16 Problems

P3.1. A wireless receiver with an effective diameter of 250 cm is receiving signals at 20 GHz from a transmitter that transmits at a power of 30 mW and a gain of 30 dB.

(a) What is the gain of the receiver antenna?

(b) What is the received power if the receiver is 5 km away from the transmitter?

P3.2. Consider an antenna transmitting a power of 5 W at 900 MHz. Calculate the received power at a distance of 2 km if propagation is taking place in free space.

P3.3. In a cellular system, diffraction, reflection, and direct path take a different amount of time for the signal to reach a MS. How do you differentiate and use these signals? Explain clearly. Compute the level crossing rate with respect to the rms level for a vertical monopole antenna, assuming the Rayleigh faded isotropic scattering case. The receiver speed is 20 km/hr, and the transmission occurs at 800 MHz.

P3.4. The transmission power is 40 W, under a free space propagation model,

   (a) What is the transmission power in unit of dBm?

   (b) The receiver is in a distance of 1000 m, what is the received power, assuming that the carrier frequency $f_c = 900$ MHz and $G_t = G_r = 0$ dB?

   (c) Express the free space path loss in dB.

P3.5. A receiver is tuned to 1 GHz transmission and receives signals with Doppler frequencies ranging from 10 Hz to 50 Hz when moving at a speed of 80 km/hr. What is the fading rate?

P3.6. What does a small delay spread indicate about the characteristics of a fading channel? If the delay spread is 1 microsecond, will two different frequencies that are 5 MHz apart experience correlated fading?

P3.7. Consider an antenna transmitting at 900 MHz. The receiver is traveling at a speed of 40 km/h. Calculate its Doppler shift.

P3.8. Repeat Problem P3.6. Calculate the average fading duration if $\rho = 0.1$.

P3.9. Describe the consequence of the Doppler effect on the receiver in an isotropic scattering environment. Based on your description, speculate on the meaning of the term "Doppler spread."

   (a) Is the term "Doppler spread" more appropriate in describing the channel than "Doppler shift" in a scattering environment? Why?

   (b) Observe the inverse relationship that exists between "coherence bandwidth" and "delay spread" in a wireless channel. Attempt to similarly define a term "coherence time" that has an inverse relationship with the "Doppler spread." What information does this term give about the channel?

P3.10. How can you compensate for the impact of the Doppler effect in a cellular system? Explain.

P3.11. How is radio propagation on land different from that in free space?

P3.12. What is the differences between fast fading and slow fading?

P3.13. Path loss, fading, and delay spread are the three most important radio propagation issues. Explain why those issues are important in a cellular system.

P3.14. A BS has a 900 MHz transmitter and a vehicle is moving at the speed of 50 mph. Compute the received carrier frequency if the vehicle is moving

   (a) Directly toward the BS.

   (b) Directly away from the BS.

   (c) In a direction that is 60 degrees to the direction of arrival of the transmitted signal.

P3.15. What is diversity reception? How can it be used to combat multipath?

P3.16. What is the role or usage of reflected and diffracted radio signals in a cellular system? Explain with suitable examples.

P3.17. What is intersymbol interference (ISI)? Does it affect the transmission rate of a digital channel? Explain clearly.

P3.18. A MS is not in the direct line of sight of a BS transmitting station. How is the signal received? Explain.

P3.19. Consider two random variables X and Y that are independent and Gaussian with identical variances. One is of zero mean and the other is of mean μ. Prove that the density function of $Z = \sqrt{X + Y}$ is Rician distributed.

P3.20. What causes intersymbol interference and how can you reduce intersymbol interference in the wireless communication system?

CHAPTER 4

Channel Coding and Error Control

• • • • • • • • • • • • • • • • •

4.1 Introduction

Why do we need channel coding and error control for radio communication? It is well known that severe transmission conditions are present in terrestrial mobile radio communications due to multipath fading and very low signal-to-noise ratio (S/N) and in satellite communications due to limited transmitting power in forward channels (downlink). In cellular wireless systems, messages go through the noise medium between BS and MS, and reflection, diffraction, and scattering cause deterioration to the quality of the signals. Therefore, anything that can be done to enhance correct reception of radio signals is always welcome. Channel coding adds redundancy information to the original information at the transmitter side, following some logical relation with the original information. After transmission, the receiver receives the encoded data, possibly with some degree of degradation. At the receiver side, the original information can possibly be extracted from received data based on the logical relationship between original information and redundancy information. Introduction of redundancy causes channel coding to consume more bandwidth during the transmission. However, it offers benefits of recovering from higher bit error rates. In other words, channel coding allows signal transmission power and useful bandwidth as a higher degree of redundancy can tolerate a larger number of errors. However, in cellular systems, the traffic consists of compressed data (e.g., audio and video signals in digital form) and is very sensitive to transmission errors. Therefore, channel coding can be defined as the process of coding discrete digital information in a form suitable for transmission, with an emphasis on enhanced reliability. Channel coding is applied to ensure adequacy of transmission quality (bit error rate (BER) or frame error rate (FER)) and its use in a wireless communications system is shown in Figure 4.1.

Typically, a code may have a larger or at least an equal error detecting capability than the error correction. However, from a wireless communications point of view, if an error can only be detected and not corrected, then the transmission is not successful and techniques such as retransmission (covered later in Section 4.8) need to be employed. Therefore, we primarily concentrate on error control. Here, we

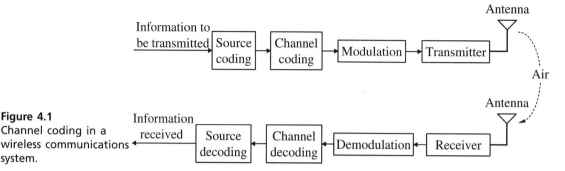

Figure 4.1
Channel coding in a wireless communications system.

discuss the three most commonly used codes: linear block codes (e.g., Hamming codes, BCH (Bose Chaudhuri Hocquenghem) codes, and Reed-Solomon codes), convolutional codes, and Turbo codes [4.1][4.2][4.3][4.5] [4.6][4.7].

4.2 Linear Block Codes

In the linear block code [4.7][4.8], the information sequence is always a multiple of a preselected length k. If not, several zeros will be generally padded at the end of the information sequence to be a multiple of k. Each k information bits are encoded into n bits in a linear block code (n, k). For example, for code $(8, 6)$ we have $n = 8$ bits in a block, $k = 6$ message bits, and $n - k = 2$ parity bits. Since the encoded sequence includes the entire information sequence, it is linear. Furthermore, the encoding is processed block by block, so the code is called linear block code.

If we assume (n, k) linear block code, there are 2^n possible combinations of different values. However, linear block codes are based on k-information bits; only 2^k possible combinations are allowed. The 2^k possibilities from a subset of the 2^n possible bit patterns are called valid codewords and hence represent the information bits.

Let the uncoded k information bits be represented by the **m** vector:

$$\mathbf{m} = (m_1, m_2, \ldots, m_k) \tag{4.1}$$

and let the corresponding codeword be represented by the n-bit **c** vector:

$$\mathbf{c} = (c_1, c_2, \ldots, c_k, c_{k+1}, \ldots, c_{n-1}, c_n) . \tag{4.2}$$

Each parity bit consists of a weighted modulo 2 sum of the data bits represented by $\oplus$ symbol. For example,

$$\begin{cases} c_1 = m_1 \\ c_2 = m_2 \\ \ldots \\ c_k = m_k \\ c_{k+1} = m_1 p_{1(k+1)} \oplus m_2 p_{2(k+1)} \cdots \oplus m_k p_{k(k+1)} \\ \ldots \\ c_n = m_1 p_{1n} \oplus m_2 p_{2n} \cdots \oplus m_k p_{kn}, \end{cases} \tag{4.3}$$

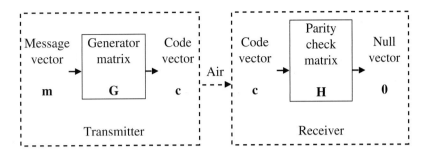

Figure 4.2
Operations of the generator matrix and the parity check matrix.

where p_{ij} $(i = 1, 2, \ldots, k; j = k+1, k+2, \ldots, n)$ is the binary weight of the particular data bit. The idea is to add parity to the information bits at the transmitting side using the generation matrix and use the parity check matrix to take care of possible errors during transmission. The operation of the generator matrix and the parity check matrix is shown in Figure 4.2.

Turning now to matrix notation, we can represent the code vector **c** as a matrix operation on the uncoded message vector **m**:

$$\mathbf{c} = \mathbf{mG}, \tag{4.4}$$

where **G** is defined as the generator matrix.

The generator matrix **G** must have dimensions k by n and is made up by concatenating the identity matrix $\mathbf{I}_k$ (k by k matrix) and the parity matrix **P** (k by $n - k$ matrix):

$$\mathbf{G} = [\mathbf{I}_k | \mathbf{P}]_{k \times n} \tag{4.5}$$

or

$$\mathbf{G} = \begin{bmatrix} 1 & 0 & 0 & \cdots & 0 & p_{11} & p_{12} & \cdots & p_{1(n-k)} \\ 0 & 1 & 0 & \cdots & 0 & p_{21} & p_{22} & \cdots & p_{2(n-k)} \\ \cdots & \cdots & \cdots & \cdots & \cdots & \cdots & \cdots & \cdots & \cdots \\ 0 & 0 & 0 & \cdots & 1 & p_{k1} & p_{k2} & \cdots & p_{k(n-k)} \end{bmatrix}. \tag{4.6}$$

The parity matrix **P** (k by $n - k$ matrix) is given by

$$\mathbf{P} = \begin{bmatrix} p_{11} & p_{12} & \cdots & p_{1(n-k)} \\ p_{21} & p_{22} & \cdots & p_{2(n-k)} \\ \cdots & \cdots & \cdots & \cdots \\ p_{k1} & p_{k2} & \cdots & p_{k(n-k)} \end{bmatrix}$$

$$= \begin{bmatrix} \mathbf{p}^1 \\ \mathbf{p}^2 \\ \cdots \\ \mathbf{p}^k \end{bmatrix}, \tag{4.7}$$

where $\mathbf{p}^i$ is the remainder of $\left[\frac{x^{n-k+i-1}}{g(x)} \right]$ for $i = 1, 2, \ldots k$, and $g(x)$ is the generator polynomial and is written as $\mathbf{p}^i = \text{rem} \left[\frac{x^{n-k+i-1}}{g(x)} \right]$. All arithmetic is performed using modulo 2 operation. The following example shows how to find linear block code generator matrix **G** given code generator polynomial $g(x)$.

Example 4.1

Find linear block encoder **G** if code generator polynomial $g(x) = 1 + x + x^3$ for a $(7, 4)$ code.

Since we have total number of bits $n = 7$, number of information bits $k = 4$, and number of parity bits $n - k = 3$, we can compute

$$\mathbf{p}^1 = \text{rem} \left[\frac{x^3}{1 + x + x^3} \right] = 1 + x \rightarrow [110], \tag{4.8}$$

$$\mathbf{p}^2 = \text{rem} \left[\frac{x^4}{1 + x + x^3} \right] = x + x^2 \rightarrow [011], \tag{4.9}$$

$$\mathbf{p}^3 = \text{rem} \left[\frac{x^5}{1 + x + x^3} \right] = 1 + x + x^2 \rightarrow [111], \tag{4.10}$$

and

$$\mathbf{p}^4 = \text{rem} \left[\frac{x^6}{1 + x + x^3} \right] = 1 + x^2 \rightarrow [101]. \tag{4.11}$$

Thus, the generator matrix is

$$G = \begin{bmatrix} 1 & 0 & 0 & 0 & 1 & 1 & 0 \\ 0 & 1 & 0 & 0 & 0 & 1 & 1 \\ 0 & 0 & 1 & 0 & 1 & 1 & 1 \\ 0 & 0 & 0 & 1 & 1 & 0 & 1 \end{bmatrix}. \tag{4.12}$$

For convenience, the code vector is expressed as

$$\mathbf{c} = [\mathbf{m} \mid \mathbf{c}_p], \tag{4.13}$$

where

$$\mathbf{c}_p = \mathbf{mP} \tag{4.14}$$

is an $(n - k)$-bit parity check vector. This binary matrix multiplication follows the usual rule with mod-2 addition, instead of conventional addition. Hence, the jth element of $\mathbf{c}_p$ can be obtained by equation (4.3).

If we define a matrix $\mathbf{H}^T$ as

$$\mathbf{H}^T = \begin{bmatrix} \mathbf{P} \\ \mathbf{I}_{n-k} \end{bmatrix} \tag{4.15}$$

and a received code vector **x** is given as

$$\mathbf{x} = \mathbf{c} \oplus \mathbf{e}, \tag{4.16}$$

where **e** is an error vector, the matrix $\mathbf{H}^T$ has the property

$$\mathbf{cH}^T = [\mathbf{m} \mid \mathbf{c}_p] \begin{bmatrix} \mathbf{P} \\ \mathbf{I}_{n-k} \end{bmatrix}$$

$$= \mathbf{mP} \oplus \mathbf{c}_p$$

$$= \mathbf{c}_p \oplus \mathbf{c}_p$$

$$= \mathbf{0}. \tag{4.17}$$

The transpose of the matrix $\mathbf{H}^T$ is

$$\mathbf{H} = \begin{bmatrix} \mathbf{P}^T \mathbf{I}_{n-k} \end{bmatrix}, \tag{4.18}$$

where $\mathbf{I}_{n-k}$ is a $n-k$ by $n-k$ unit matrix and $\mathbf{P}^T$ is the transpose of parity matrix **P**. **H** is called the parity check matrix. We can calculate a vector called the syndrome as

$$\mathbf{s} = \mathbf{xH}^T$$

$$= (\mathbf{c} \oplus \mathbf{e})\,\mathbf{H}^T$$

$$= \mathbf{cH}^T \oplus \mathbf{eH}^T$$

$$= \mathbf{eH}^T. \tag{4.19}$$

The vector **s** has $(n-k)$ dimensions. If there are no errors ($\mathbf{e} = \mathbf{0}$), the applying equation (4.19) to the vector **s** gives a null vector in the received vector **x**. Thus, we can decide that there are errors if $\mathbf{s} \neq \mathbf{0}$. An example of linear block code is shown as follows:

Consider a $(7, 4)$ linear block code, given by **G** as

$$\mathbf{G} = \begin{bmatrix} 1 & 0 & 0 & 0 & 1 & 1 & 1 \\ 0 & 1 & 0 & 0 & 1 & 1 & 0 \\ 0 & 0 & 1 & 0 & 1 & 0 & 1 \\ 0 & 0 & 0 & 1 & 0 & 1 & 1 \end{bmatrix}.$$

Then,

$$\mathbf{H} = \begin{bmatrix} 1 & 1 & 1 & 0 & 1 & 0 & 0 \\ 1 & 1 & 0 & 1 & 0 & 1 & 0 \\ 1 & 0 & 1 & 1 & 0 & 0 & 1 \end{bmatrix}.$$

For $\mathbf{m} = [1\ \ 0\ \ 1\ \ 1]$ and $\mathbf{c} = \mathbf{mG} = [1\ \ 0\ \ 1\ \ 1\ \ 0\ \ 0\ \ 1]$. If there is no error, the received vector $\mathbf{x} = \mathbf{c}$, and $\mathbf{s} = \mathbf{cH}^T = [0\ \ 0\ \ 0]$. Let **c** suffer an error in the transmission such that the received vector

$$\mathbf{x} = \mathbf{c} \oplus \mathbf{e}$$

$$= [1\ \ 0\ \ 1\ \ 1\ \ 0\ \ 0\ \ 1] \oplus [0\ \ 0\ \ 1\ \ 0\ \ 0\ \ 0\ \ 0]$$

$$= [1\ \ 0\ \ 0\ \ 1\ \ 0\ \ 0\ \ 1].$$

Then,

$$\mathbf{s} = \mathbf{x}\mathbf{H}^T$$

$$= \begin{bmatrix} 1 & 0 & 0 & 1 & 0 & 0 & 1 \end{bmatrix} \begin{bmatrix} 1 & 1 & 1 \\ 1 & 1 & 0 \\ 1 & 0 & 1 \\ 0 & 1 & 1 \\ 1 & 0 & 0 \\ 0 & 1 & 0 \\ 0 & 0 & 1 \end{bmatrix}$$

$$= \begin{bmatrix} 1 & 0 & 1 \end{bmatrix}$$

$$= \left(\mathbf{e}\mathbf{H}^T \right).$$

This basically indicates the error position, giving the corrected vector as $\begin{bmatrix} 1 & 0 & 1 & 1 & 0 & 0 & 1 \end{bmatrix}$.

There are only 2^{n-k} different syndromes generated by the 2^n possible n-bit error vectors including the no-error cases. Therefore, a given syndrome does not uniquely determine $\mathbf{e}$ and this implies that applying error vectors to different message vectors could lead to the same syndrome. This means that given $\mathbf{s}$, it is not possible to uniquely map back to single code $\mathbf{c}$. This also implies that we can uniquely map just $(2^{n-k} - 1)$ patterns of $\mathbf{s}$ with one or more errors and the remaining patterns are not correctable because of associated ambiguity. Therefore, we should design the decoder to correct $(2^{n-k} - 1)$ error patterns that can be corrected. These are also most likely errors as those patterns are generated due to fewest errors, since single errors are more probable than double errors, and so forth. This strategy, known as maximum-likehood decoding, is optimum in the sense that it minimizes the Hamming distance [4.8] between the codeword vector and received vector.

The generator matrix $\mathbf{G}$ is used in the encoding operation at the transmitter. On the other hand, the parity check matrix $\mathbf{H}$ is used in the decoding operation at the receiver. If the ith element of $\mathbf{e}$ equals $\mathbf{0}$, the corresponding element of the received vector $\mathbf{x}$ is the same as that of transmitted code vector $\mathbf{c}$. On the other hand, if the ith element of $\mathbf{e}$ equals $\mathbf{1}$, the corresponding element of the received vector $\mathbf{x}$ is different from that of the code vector $\mathbf{c}$, in which case an error has occurred in the ith location.

The receiver has the task of decoding the code vector $\mathbf{c}$ from the received vector $\mathbf{x}$. The algorithm commonly used to perform this decoding operation starts with the computation of a $1 \times (n - k)$ vector called the error syndrome vector or simply the syndrome. The importance of the syndrome lies in the fact that it depends only on the error pattern and a unique mapping to correct information bits is possible for a limited number of errors. For other types of errors, another error control technique of ARQ (discussed in Section 4.8) can be used as long as the presence of errors can be detected. We now consider a few simple coding schemes.

● ● ● ● ● ● ● ● ● ● ● ● ● ●
4.3 Cyclic Codes

Cyclic codes [4.7][4.8] are a subclass of linear block codes with acyclic structure that leads to more practical implementation. An advantage of cyclic codes over most other types of codes is that they are relatively easy to encode and decode. Thus, block codes used for forward error correction (FEC) systems are most cyclic codes wherein encoding or decoding is performed with a shift register. A mathematical expression in a polynomial form can be used because shift of a code generates another code. The codeword with n bits can be expressed as

$$c(x) = c_1 x^{n-1} + c_2 x^{n-2} + \cdots + c_n, \tag{4.20}$$

where the coefficients c_i $(i = 1, 2, \ldots, n)$ take the value either 0 or 1.

The codeword can be expressed by the data polynomial $m(x)$ and the check polynomial $c_p(x)$. Thus, we have

$$c(x) = m(x) x^{n-k} + c_p(x), \tag{4.21}$$

where the check polynomial $c_p(x)$ is the remainder from dividing $m(x) x^{n-k}$ by the generator polynomial $g(x)$, that is,

$$c_p(x) = \text{rem} \left[\frac{m(x) x^{n-k}}{g(x)} \right]. \tag{4.22}$$

Denoting the error polynomial by $e(x)$, the received signal polynomial or syndrome $s(x)$ becomes

$$s(x) = \text{rem} \left[\frac{c(x) + e(x)}{g(x)} \right]. \tag{4.23}$$

If there is no error, we have $s(x) = 0$. A (n, k) code can easily be generated with a $n - k$ linear feedback shift register. The syndrome $s(x)$ can be obtained by the same feedback shift register.

The following is an example of cyclic code.

Consider the (7, 4) cyclic code. For $m(x) = 1 + x + x^2 + 0 \cdot x^3$ and $g(x) = 1 + x + x^3$. The check polynomial is

$$c_p(x) = \text{rem} \left[\frac{x^5 + x^4 + x^3}{x^3 + x + 1} \right] = x.$$

Then, the codewords can be found as

$$c(x) = m(x) x^{n-k} + c_p(x) = x + x^3 + x^4 + x^5.$$

A similar concept is used at the message frame level and is considered in the next section.

● ● ● ● ● ● ● ● ● ● ● ● ● ● ● ● ● ●

4.4 Cyclic Redundancy Check (CRC)

Cyclic redundancy code (CRC) is an error-checking code that is widely used in data communications systems and other serial data transmission systems. Using this technique, the transmitter appends an extra n-bit sequence to every frame. The additional bit sequence is called frame check sequence (FCS). The FCS holds redundant information about the frame that helps the receivers detect errors in the frame.

CRC is based on polynomial manipulations using modulo arithmetic. The algorithm treats blocks of input bits as coefficient sets for polynomials. For example, binary 10100 implies the polynomial: $1 \cdot x^4 + 0 \cdot x^3 + 1 \cdot x^2 + 0 \cdot x^1 + 0 \cdot x^0$. This is the message polynomial. A second polynomial with constant coefficients is called the generator polynomial. This is divided into the message polynomial, giving quotient and remainder. The coefficients of the remainder form the bits of the final CRC. We define the following parameters as:

Q — k bits long frame to be transmitted

F — FCS of $n-k$ bits, which would be added to Q

J — The result after cascading Q and F

P — The CRC-generating polynomial

In the CRC algorithm, J should be exactly divisible by P. We calculate J as:

$$J = Q \cdot x^{n-k} + F. \tag{4.24}$$

This ensures that Q (which is k bits long) shifts to the left by $n-k$ bits and F (of length $n-k$) is appended to it.

Dividing $Q \cdot x^{n-k}$ by P, we have

$$\frac{Q \cdot x^{n-k}}{P} = Q + \frac{R}{P}, \tag{4.25}$$

where R is a reminder of equation (4.25). Thus, we have

$$J = Q \cdot x^{n-k} + R. \tag{4.26}$$

This value of J would yield a zero remainder for J/P. We leave the verification as an exercise. (Hint: Remember that $A + A = 0$ for modulo 2 operations.)

A list of the most commonly used CRC polynomials is as follows.

CRC-12	$x^{12} + x^{11} + x^3 + x^2 + x + 1$
CRC-16	$x^{16} + x^{15} + x^2 + 1$
CRC-CCITT	$x^{16} + x^{12} + x^5 + 1$
CRC-32	$x^{32} + x^{26} + x^{23} + x^{22} + x^{16} + x^{12} + x^{11} + x^{10} + x^8 + x^7 + x^5 + x^4 + x^2 + x + 1$

CRC-16 and CRC-CCITT transmit 8 bits and generate 16-bit FCS. CRC-32 provides more protection by generating 32-bit FCS. Few Department of Defense (DoD) applications use CRC-32, whereas most user applications in Europe and the United States use either CRC-16 or CRC-CCITT.

4.5 Convolutional Codes

Convolutional codes [4.4][4.8] are among the most widely used channel codes in practical communication systems (e.g., Global System for Mobile Communications (GSM) and Interim Standard-95 (IS-95)). These codes are developed with a separate strong mathematical structure and are primarily used for real-time error correction. The encoded bits depend not only on the current input data bits but also on past input bits. The main decoding strategy for convolutional codes is based on the widely used Viterbi algorithm [4.5][4.6]. The constraint length K for a convolutional code is defined as

$$K = M + 1, \tag{4.27}$$

where M is the maximum number of stages (memory size) in any shift register. The shift registers store the state information of the convolutional encoder and the constraint length relates the number of bits on which the output depends. The code rate r for a convolutional code is defined as

$$r = \frac{k}{n}, \tag{4.28}$$

where k is the number of parallel input information bits and n is the number of parallel output encoded bits at one time interval.

A convolutional code encoder with $n = 2$ and $k = 1$ or the code rate $r = 1/2$ is shown in Figure 4.3. The encoder outputs two bits for every one input bit. The output bits are determined from the input bit and the two previous input bits stored in the shift registers (D_1 and D_2). Usually, the convolutional encoder can be represented in several different but equivalent ways, such as the tree diagram and the trellis diagram. The state information of a convolutional encoder is maintained by the shift registers and can be represented by a state diagram. Figure 4.4 shows the state diagram of the encoder in Figure 4.3.

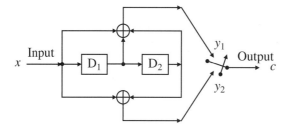

Figure 4.3
Convolutional code
encoder.

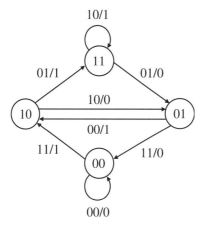

Figure 4.4
State diagram.

Each new input information bit causes a transition from one state to another. The path information between the states $(D_1 D_2)$ represents output data bits $(y_1 y_2)$ and corresponding input data bit (x). It is customary to begin convolutional encoding from an initial state of all zeros.

Based on the input sequence, the encoder state is going to change, and these state transitions will depend on the input bits and current state. Changes due to the all possible input sets could be represented by a tree diagram, and Figure 4.5 shows the tree diagram of the encoder in Figure 4.3. The branch due to input data bit "$x = 0$" is shown in upward direction in the tree diagram; likewise, for input data bit "$x = 1$" the branch direction is downward. The corresponding output bits "$y_1 y_2$" are shown along the branches of the tree. An input data sequence defines a specific path through the tree diagram from left to right. For example, the input data sequence $x = \{10011\dots\}$ produces the output encoded sequence $c = \{11\ 10\ 11\ 11\ 01\dots\}$.

Another way of jointly representing the state and the tree diagrams is to indicate all possible state transitions for each input. Such a combined information is called a trellis diagram. Figure 4.6 shows the trellis diagram for the encoder of Figure 4.3. For example, for input data sequence $x = \{10011\dots\}$ the state transition lines are represented by bold lines in Figure 4.6.

Usually, there are two typical decoding methods: namely, decoding based on hard-decision or soft-decision algorithms. A hard-decision decoding uses single-bit quantization on the received channel values. A soft-decision decoding uses multi-bit

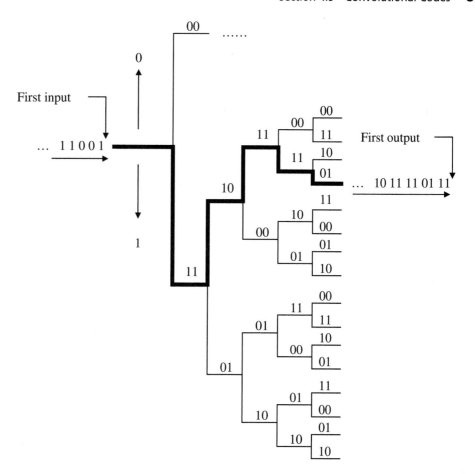

Figure 4.5
Tree diagram.

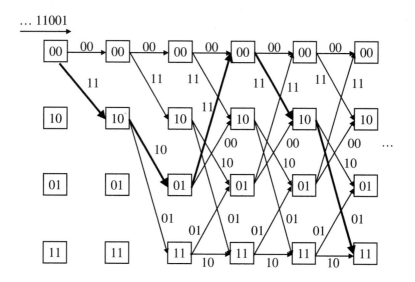

Figure 4.6
Trellis diagram.

quantization on the received channel values. For an ideal soft-decision decoding (i.e., infinite-bit quantization), the received channel values are directly used in the channel decoder.

● ● ● ● ● ● ● ● ● ● ● ● ● ● ● ● ⋅ ⋅ ⋅ ⋅

4.6 Interleaver

Interleaving is heavily used in the wireless communication. The basic objective is to protect the transmitted data from burst errors. There are many different interleavers, such as block interleaver, random interleaver, circular interleaver, semirandom interleaver, odd-even interleaver, and optimal (near-optimal) interleaver. Each one has its advantages and drawbacks in the context of noise. In this section, we only consider the block interleaver because it is the most commonly used interleaver in wireless communication systems. The basic idea is to write data in row-wise from top to bottom and left to right and read out column-wise from left to right and top to bottom. The concept of the interleaver is shown in Figure 4.7.

For example, the input data sequence {$a1$, $a2$, $a3$, $a4$, $a5$, $a6$, $a7$, $a8$, $a9$, ..., $a16$} produces the output interleaved data sequence {$a1$, $a5$, $a9$, $a13$, $a2$, $a6$, $a10$, $a14$, $a3$,..., $a12$, $a16$}. These data are transmitted over the air. At the receiving end, de-interleaving is done and the original output data sequence {$a1$, $a2$, $a3$, $a4$, $a5$, $a6$, $a7$, $a8$, $a9$, ..., $a16$} is obtained. Figure 4.8 shows an example in which

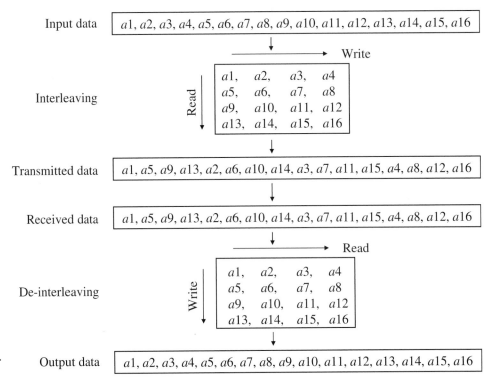

Figure 4.7
Concept of interleaver.

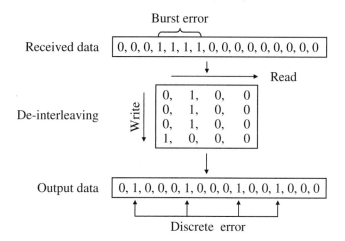

Figure 4.8
An example of an interleaver.

there are four burst error bits {0001111000000000} in the received data sequence. After interleaving, the error is dispersed and the output data sequence becomes {0100010001001000}. We can see that the burst error of length 4 is transformed into multiple individual errors. The error-correcting codes generally are capable of correcting individual errors, but not a burst error. However, in the wireless and mobile channel environment, the burst error occurs frequently. In order to correct the burst error, interleaving is needed to disperse the burst error into multiple individual errors which can be handled by the error-correcting code. Furthermore, interleaving does not have error-correcting capability. Therefore, interleaving is always used in conjunction with an error-correcting code. In other words, interleaving does not introduce any redundancy into the information sequence, so it does not add to extra bandwidth requirement.

The disadvantage of interleaving is additional delay as the sequence needs to be processed block by block. Therefore, small memory size interleaving is preferred in delay-sensitive applications.

● ● ● ● ● ● ● ● ● ● ● ● ● ● ● ● ● ● ...

4.7 Turbo Codes

Turbo codes are the most recently developed codes and are extremely powerful. This major breakthrough in channel coding theory occurred when C. Berrou et al. first developed it in 1993 [4.9] and exhibited a performance closer to the theoretical Shannon limit than any other code so far. The fundamental turbo code encoder is built using two identical recursive systematic convolutional (RSC) codes with parallel concatenation. Figure 4.9 shows an example of a turbo code encoder. The first RSC encoder uses the bit stream as it comes, whereas an interleaver precedes the second one. The two encoders introduce redundancy for the same block of bits but are different as there exists low correlation among the input bits streams due to interleaving.

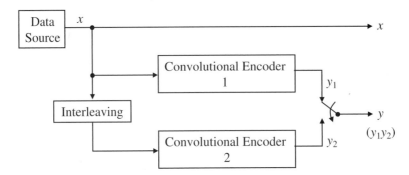

Figure 4.9
Turbo code encoder.

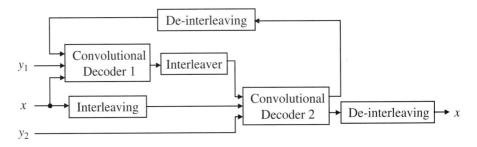

Figure 4.10
Turbo code decoder.

The interleaver randomizes the information sequence of the second encoder to make the inputs of the two encoders uncorrelated. Since there are two encoded sequences, the turbo decoder consists of two RSC decoders corresponding to the two RSC encoded sequences respectively. The decoding begins by decoding one of them to get the first estimate of the information sequence. Based on the estimate from the first RSC decoder, the second RSC decoder gets the more precise estimate of the information sequence. In order to improve the correctness of the estimate, the estimate from second RSC decoder feeds back to the first RSC decoder continuously. The repeating procedure just likes the working principle of "turbo" engine, so it is called the "turbo code." Since the estimation of the information bits is used during the decoding, the decoder must use a soft-decision input to produce some kind of soft output. Figure 4.10 shows a turbo code decoder. When coding technique can detect but cannot correct errors, then retransmission of information becomes essential and is discussed next.

● ● ● ● ● ● ● ● ● ● ● ● ● ● ● ●

4.8 ARQ Techniques

Automatic repeat request (ARQ) [4.2] is one of the error-handling mechanisms used in data communication. The concept of ARQ is illustrated in Figure 4.11. When the receiver detects bit errors in a packet (that cannot be corrected by underlying error-detecting code, if used), it simply drops the packet and the sender needs to transmit it again. ACK (acknowledgment) and NAK (negative acknowledgment) are explicit

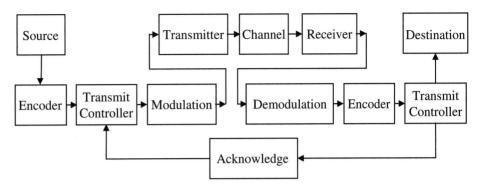

Figure 4.11
Concept of ARQ.

feedback sent by the receiver. There are three kinds of ARQ schemes: Stop-And-Wait ARQ (SAW ARQ), Go-Back-N ARQ (GBN ARQ), and Selective-Repeat ARQ (SR ARQ).

4.8.1   Stop-And-Wait ARQ Scheme

The simplest ARQ scheme is called the SAW (Stop-And-Wait) ARQ scheme. In this scheme, the sender sends one data packet each time. The receiver receives that data packet and checks if the data packet has been received correctly. If the packet is not corrupted, the receiver sends an ACK packet; otherwise, the receiver responds with a NAK packet. The process is illustrated in Figure 4.12 and is described as follows:

1. The sender transmits packet 1 and then waits for an ACK packet from the receiver.
2. The receiver receives packet 1 without error and transmits an ACK packet.
3. The sender receives the ACK packet and then proceeds to transmit packet 2.
4. Packet 2 also arrives at the receiver without error and the sender successfully gets an ACK packet from the receiver.
5. Packet 3 is sent by the sender but undergoes errors in transmission.
6. The receiver receives packet 3, but it finds the packet is corrupted. Then it sends back a NAK packet to the sender.

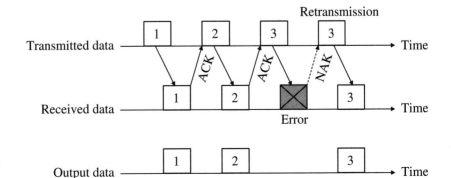

Figure 4.12
Stop-And-Wait ARQ
scheme.

7. Upon receiving the NAK, the sender retransmits packet 3.

8. A similar sequence is followed for the rest of the packets.

The throughput for the SAW ARQ scheme is given by [4.2][4.10]:

$$S_{SAW} = \frac{1}{T_{SAW}} \left(\frac{k}{n} \right), \tag{4.29}$$

where n is the number of bits in a block, k is the number of information bits in a block, D is the round-trip propagation delay time, R_b is the bit rate, P_b is the BER of the channel, and T_{SAW} is the average transmission time in terms of a block duration and it is given by

$$T_{SAW} = \left(1 + \frac{DR_b}{n} \right) P_{ACK} + 2 \left(1 + \frac{DR_b}{n} \right) P_{ACK} (1 - P_{ACK}) +$$

$$3 \left(1 + \frac{DR_b}{n} \right) P_{ACK} (1 - P_{ACK})^2 + \cdots$$

$$= \left(1 + \frac{DR_b}{n} \right) P_{ACK} \sum_{i=1}^{\infty} i (1 - P_{ACK})^{i-1} = \left(1 + \frac{DR_b}{n} \right) P_{ACK} \frac{1}{[1 - (1 - P_{ACK})]^2}$$

$$= \frac{1 + \frac{DR_b}{n}}{P_{ACK}}, \tag{4.30}$$

where P_{ACK} is the probability to return an ACK in the transceiver side and is given by

$$P_{ACK} \approx (1 - P_b)^n . \tag{4.31}$$

Therefore, the throughput for the SAW ARQ scheme is given by

$$S_{SAW} = \frac{1}{T_{SAW}} \left(\frac{k}{n} \right)$$

$$= \frac{(1 - P_b)^n}{1 + \frac{DR_b}{n}} \left(\frac{k}{n} \right) . \tag{4.32}$$

4.8.2 Go-Back-N ARQ Scheme

As we have seen in the previous section, the SAW ARQ exhibits poor utilization of the wireless communication channel since the sender does not send the next packet until it receives an ACK from the receiver. In the GBN ARQ scheme (Figure 4.13), the sender is allowed to transmit N packets without waiting for acknowledgment of prior packets. When a packet is corrupted during the transmission and it receives a NAK from the receiver, the sender has to retransmit all the packets that have been sent after that corrupted packet. As shown in Figure 4.13, when packet 3 is corrupted, packets 3, 4, and 5 all have to be retransmitted.

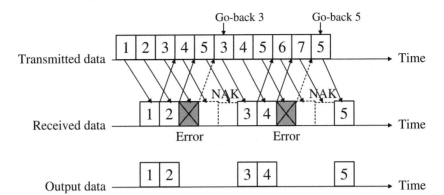

Figure 4.13
Go-Back-N ARQ
scheme.

In this scheme, all the packets that have been sent but have not been acknowledged are buffered by the sender. Since the receiver only accepts the correct and in-order packets, only a buffer of one packet size is needed at the receiver. The GBN ARQ scheme is better suited for the environment where burst errors are most probable during packet transmission.

Similar to the SAW ARQ scheme, the throughput for the GBN ARQ scheme is given by [4.2][4.10]:

$$S_{\text{GBN}} = \frac{1}{T_{\text{GBN}}} \left(\frac{k}{n} \right), \tag{4.33}$$

where T_{GBN} is the average transmission time in terms of a single block duration and is given by

$$
\begin{aligned}
T_{\text{GBN}} &= 1 \cdot P_{\text{ACK}} + (N+1) \cdot P_{\text{ACK}} (1 - P_{\text{ACK}}) + (2N+1) \cdot P_{\text{ACK}} (1 - P_{\text{ACK}})^2 + \\
&\quad (3N+1) \cdot P_{\text{ACK}} (1 - P_{\text{ACK}})^2 + \cdots \\
&= P_{\text{ACK}} + P_{\text{ACK}} \left[(1 - P_{\text{ACK}}) + (1 - P_{\text{ACK}})^2 + (1 - P_{\text{ACK}})^3 + \cdots \right] + \\
&\quad P_{\text{ACK}} \left[N (1 - P_{\text{ACK}}) + 2N (1 - P_{\text{ACK}})^2 + 3N (1 - P_{\text{ACK}})^3 + \cdots \right] \\
&= P_{\text{ACK}} + P_{\text{ACK}} \left[\frac{1 - P_{\text{ACK}}}{1 - (1 - P_{\text{ACK}})} + N \frac{1 - P_{\text{ACK}}}{[1 - (1 - P_{\text{ACK}})]^2} \right] \\
&= 1 + \frac{N (1 - P_{\text{ACK}})}{P_{\text{ACK}}}, \tag{4.34}
\end{aligned}
$$

where P_{ACK} is the probability to return an ACK in the transceiver side and is given by

$$P_{\text{ACK}} \approx (1 - P_b)^n. \tag{4.35}$$

Therefore, the throughput for the GBN ARQ scheme is given by

$$S_{GBN} = \frac{1}{T_{GBN}} \left(\frac{k}{n} \right)$$

$$= \frac{(1 - P_b)^n}{(1 - P_b)^n + N \left[1 - (1 - P_b)^n \right]} \left(\frac{k}{n} \right). \tag{4.36}$$

4.8.3 Selective-Repeat ARQ Scheme

In the GBN ARQ scheme, it is obvious that a single packet error can cause the sender to retransmit several packets, most of which may be unnecessary. The selective-repeat protocol provides improvement with respect to this issue. The receiver acknowledges all correctly received packets, and when the sender does not receive any ACK packet from a receiver—that is, some specific packet suspected to be lost or corrupted—it retransmits only that packet. Thus, it avoids unnecessary retransmissions. This scheme is shown in Figure 4.14. The sender continuously sends the packets and only retransmits the corrupted ones (packets 3 and 7 in Figure 4.14). Since the receiver may have out-of-order packets, it needs a large memory to buffer and reorder these packets before passing them to the upper layer.

Implementation of the SR ARQ protocol is more complex than that of the other two protocols. But it provides the best efficiency. If the probability of packet corruption or loss is high for communication channels, the SR ARQ scheme offers unique advantage by retransmitting only the corrupted packet.

In practice, all of the aforementioned three ARQ schemes should be implemented using a set of timers. This is because both the data packets and the ACK/NAK packets may be lost during transmission. If the sender cannot receive a response from the receiver in a certain prespecified time, it must retransmit those unACK'ed packets.

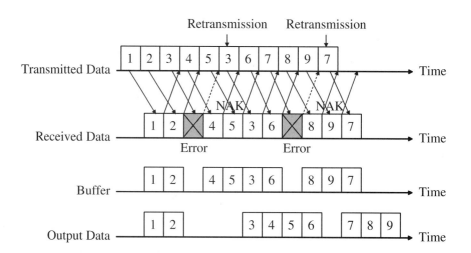

Figure 4.14
Selective-Repeat
ARQ scheme.

Similar to the GBN ARQ scheme, the throughput for the SR ARQ scheme is given by [4.2][4.10]:

$$S_{SR} = \frac{1}{T_{SR}} \left(\frac{k}{n} \right)$$

$$= (1 - P_b)^n \left(\frac{k}{n} \right), \tag{4.37}$$

where T_{SR} is the average transmission time in terms of a block duration and is given as

$$T_{SR} = 1 \cdot P_{ACK} + 2 \cdot P_{ACK} (1 - P_{ACK}) + 3 \cdot P_{ACK} (1 - P_{ACK})^2 + \cdots$$

$$= P_{ACK} \sum_{i=1}^{\infty} i \, (1 - P_{ACK})^{i-1}$$

$$= P_{ACK} \frac{1}{[1 - (1 - P_{ACK})]^2}$$

$$= \frac{1}{P_{ACK}}, \tag{4.38}$$

where P_{ACK} is the probability to return an ACK in the transceiver side and is given by

$$P_{ACK} \approx (1 - P_b)^n . \tag{4.39}$$

Therefore, the throughput for the SR ARQ scheme is given by

$$S_{SR} = \frac{1}{T_{SR}} \left(\frac{k}{n} \right)$$

$$= (1 - P_b)^n \left(\frac{k}{n} \right). \tag{4.40}$$

● ● ● ● ● ● ● ● ● ● ● ● ● ● ●

4.9 Summary

It is extremely important to control errors in wireless transmission. One way to reduce carrier-to-interference ratio is to increase the transmitting power or have appropriate use of the frequency spectrum. It may be noted that increasing transmitting power may also enhance the interference and may not be the best solution. Once these have been done, further enhancement is possible in two different ways: channel coding and retransmission, which have been discussed in this chapter. Channel coding reduces the information contents while ARQ requires retransmission. Both techniques do have associated overhead, and their impact on error reduction has been examined carefully. Both techniques can be used individually, or channel coding can follow ARQ to provide enhanced error-correcting capability. Effective use of these and other techniques in a wireless cellular design is covered in the next chapter.

● ● ● ● ● ● ● ● ● ● ● ● ● ● ● ● ●

4.10 **References**

[4.1] S. Lin, *An Introduction to Error-Correcting Codes*, Prentice Hall, Englewood Cliffs, NJ, 1970.

[4.2] S. Lin and D. J. Costello, Jr., *Error Control Coding: Fundamentals and Applications*, Prentice Hall, Englewood Cliffs, NJ, 1983.

[4.3] V. Pless, *Introduction to the Theory of Error-Correcting Codes*, John Wiley & Sons, New York, 1982.

[4.4] S. Haykin, *Communication Systems*, 4th ed., John Wiley & Sons, New York, 2001.

[4.5] A. J. Viterbi, "Error Bounds for Convolutional Codes and an Asymptotically Optimum Decoding Algorithm." *IEEE Transactions on Information Theory*, Vol. IT-13, pp. 260–269, April 1967.

[4.6] G. D. Forney, "The Viterbi Algorithm," *Proceeding of the IEEE*, Vol. 61, No. 3, pp. 268–278, March 1973.

[4.7] B. P. Lathi, *Modern Digital and Analog Communication Systems*, Holt, Rinehart and Winston, New York, 1983.

[4.8] J. G. Proakis, *Digital Communications*, 3rd ed., McGraw-Hill, New York, 1995.

[4.9] C. Berrou, A. Glavieux, and P. Thitimajshima, "Near Shannon Limit Error-Correcting Coding and Decoding: Turbo-Codes," *Proceeding of the IEEE International Conference on Communications*, Geneva, Switzerland, May 1993.

[4.10] S. Sampei, *Applications of Digital Wireless Technologies to Global Wireless Communications*, Prentice Hall, Upper Saddle River, NJ, 1997.

● ● ● ● ● ● ● ● ● ● ● ● ● ● ● ● ●

4.11 **Problems**

P4.1. Explain why channel coding reduces the bandwidth efficiency of the link.

P4.2. Can channel coding be considered as a post-detection technique?

P4.3. What is the main idea behind channel coding? Does it improve the performance of mobile communication?

P4.4. If the code generator polynomial is $g(x) = 1 + x^2$ for a (5, 3) code, find the linear block code generator matrix **G**.

P4.5. The following matrix represents a generator matrix for a (7, 4) block code.

$$G = \begin{bmatrix} 1 & 0 & 0 & 0 & 1 & 1 & 0 \\ 0 & 1 & 0 & 0 & 0 & 1 & 1 \\ 0 & 0 & 1 & 0 & 1 & 1 & 1 \\ 0 & 0 & 0 & 1 & 1 & 0 & 1 \end{bmatrix}$$

What is the corresponding parity check matrix **H**?

P4.6. Find the linear block code generator matrix **G**, if the code generator polynomial is $g(x) = 1 + x^2 + x^3$ for a $(7, 4)$ code.

P4.7. Repeat Problem P4.6 if $g(x) = 1 + x^3$ for a $(7, 4)$ code.

P4.8. Consider the rate $r = 1/2$ in the convolutional encoder shown below. Find the encoder output $(Y_1\ Y_2)$ produced by the message sequence 10111.... Assume that the initial state is zero.

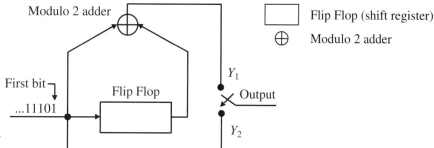

Figure 4.15
Figure for Problem P4.8.

P4.9. Find the state diagram for Problem P4.8.

P4.10. The following figure shows the encoder for a $1/2$ rate convolutional code. Determine the encoder output produced by the message sequence 1011.... Assume that the initial state of the encoder is zero.

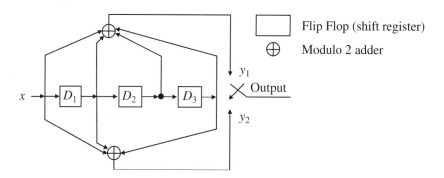

Figure 4.16
Figure for
Problem P4.10.

P4.11. Consider a SAW ARQ system between 2 nodes A (transmitting node) and B (receiving node). Assume that data frames are all of the same length and require T seconds for transmission. Acknowledgment frames require R seconds for transmission and there is a propagation delay P on the link (in both directions). One in every 3 frames that A sends is in error at B. B responds to this with a NAK and this erroneous frame is received correctly in the (first) retransmission by A. Assume that nodes send new data packets and acknowledgments as fast as possible, subject to the rules of stop and wait. What is the rate of information transfer in frames/second from A to B?

P4.12. Compare and contrast GBN ARQ and SR ARQ schemes.

P4.13. Consider the block diagram of a typical digital transmission system. Speculate where one would use source coding or channel coding. Differentiate between them. Would they increase or decrease the original message size? (Hint: We want to transmit most efficiently, i.e., message size should be the smallest possible, but enough redundancy should be added to correct small errors so that retransmission is avoided as far as possible.)

Figure 4.17
Figure for
Problem P4.13.

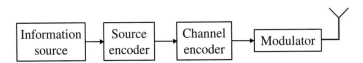

P4.14. Can you interleave an interleaved signal? What do you gain with such a system?

P4.15. Why do you need both error correction capability and ARQ in a cellular system? Explain clearly.

P4.16. In a two-stage coding system, the first stage provides (7, 4) coding while the second stage supports (11, 7) coding. Is it better to have such a two-stage coding scheme as compared to single-stage (11, 4) complex coding? Explain your answer in terms of the algorithmic complexity and error-correcting capabilities.

P4.17. Under which scenarios would cyclic codes be preferred over interleaving and vice versa?

P4.18. Polynomial $1 + x^7$ can be factored into three irreducible polynomials $(1 + x)(1 + x + x^3)(1 + x^2 + x^3)$ with $(1 + x + x^3)$ and $(1 + x^2 + x^3)$ as primitive polynomials. Using $1 + x + x^3$ as generator polynomial calculate the (7, 4) cyclic code word for given message sequence 1010.

P4.19. Repeat Problem P4.18 with $1 + x^2 + x^3$ as generator polynomial and compare the results.

P4.20. Develop the encoder and syndrome calculator with $1 + x^2 + x^3$ as generator polynomial in Problem P4.18.

P4.21. What is an RSC code? Why are these codes called systematic?

P4.22. Describe briefly syndrome decoding and incomplete decoding?

P4.23. Prove that the average transmission time in terms of block duration, T_{SR}, for Selective-Repeat ARQ is given by:

$$T_{SR} = 1.P_{ACK} + 2.P_{ACK}(1 - P_{ACK}) + 3.P_{ACK}(1 - P_{ACK})^2 + \cdots,$$

where P_{ACK} is the probability to return a ACK in the transceiver side. Also, solve the above equation for $P_{ACK} = 0.5$.

P4.24. In Stop-and-Wait ARQ, let the probability of the transmitting side receiving an ACK after exactly one loss of ACK be $P = 0.021$. Find the average transmission time in terms of a block duration if:

D = round trip propagation delay

R_b = bit rate

n = number of bits in a block

(Hint: The probability for the considered case $= P_{ACK}(1 - P_{ACK})$)

P4.25. Compare a block with a convolutional interleaver.

Cellular Concept

5.1 Introduction

The rationale behind cellular systems was given in Chapter 1, where cells were shown to constitute the design of the heart of such systems. A cell is formally defined as an area wherein the use of radio communication resources by the MS is controlled by a BS. The size and shape of the cell and the amount of resources allocated to each cell dictate the performance of the system to a large extent, given the number of users, average frequency of calls being made, average duration of call time, and so on. In this chapter, we study many parameters associated with the cell and their corresponding correlation to the cellular concept.

5.2 Cell Area

In a cellular system, the most important factor is the size and the shape of a cell. A cell is the radio area covered by a transmitting station or a BS. All MSs in that area are connected and serviced by the BS. Therefore, ideally, the area covered by a BS can be represented by a circular cell, with a radius R from the center of the BS [Figure 5.1(a)]. There are many factors that cause reflections and refractions of the signals, including elevation of the terrain, presence of a hill or valley or a tall building, and presence of particles in the air. The actual shape of the cell is determined by the received signal strength in the surrounding area. Therefore, the coverage area may be a little distorted [Figure 5.1(b)]. An appropriate model of a cell is needed before a cellular system can be analyzed and evaluated.

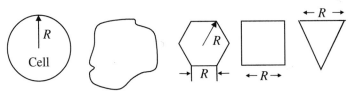

Figure 5.1
Shape of the cell coverage area.

(a) Ideal cell   (b) Actual cell   (c) Different cell models

There are many possible models that can be used, to represent a cell boundary and the most popular alternatives of hexagon, square, and equilateral triangle are shown in Figure 5.1(c). In most modeling and simulation, hexagons are used as a hexagon is closer to a circle and multiple hexagons can be arranged next to each other, without having any overlapping area and without leaving any uncovered space in between. In other words, hexagons can fit just like tiles on the floor and an arrangement of such multiple hexagons (cells) could cover a larger area over the surface of earth. The second most popular cell type is a rectangular shape, which can also function similar to a hexagon model. The size and capacity of the cell per unit area and the impact of the shape of a cell on service characteristics are shown in Table 5.1. It is clear that if the cell area is increased, the number of channels per unit area is reduced for the same number of channels and is good for less populated areas, with fewer cell phone subscribers. On the other hand, if the number of the cell phone users is increased (such as downtown area), a simple-minded solution is to increase the number of the channels. A practical option is to reduce the cell size so that the number of channels per unit area could be kept comparable to the number of subscribers. It should be remembered that the cell area and the boundary length are important parameters that affect the handoff from a cell to an adjacent cell. Specific schemes to cope with increased traffic are considered later in more detail.

Table 5.1: ▶
Impact of Cell Shape and Radius on Service Characteristics

Shape of the Cell	Area	Boundary	Boundary Length/ Unit Area	Channels/Unit Area with N Channels/Cells	Channels/Unit Area when Number of Channels Increased by a Factor K	Channels/Unit Area when Size of Cell Reduced by a Factor M
Square cell (side = R)	R^2	$4R$	$\dfrac{4}{R}$	$\dfrac{N}{R^2}$	$\dfrac{KN}{R^2}$	$\dfrac{M^2N}{R^2}$
Hexagonal cell (side = R)	$\dfrac{3\sqrt{3}}{2}R^2$	$6R$	$\dfrac{4}{\sqrt{3}R}$	$\dfrac{N}{1.5\sqrt{3}R^2}$	$\dfrac{KN}{1.5\sqrt{3}R^2}$	$\dfrac{M^2N}{1.5\sqrt{3}R^2}$
Circular cell (radius = R)	πR^2	$2\pi R$	$\dfrac{2}{R}$	$\dfrac{N}{\pi R^2}$	$\dfrac{KN}{\pi R^2}$	$\dfrac{M^2N}{\pi R^2}$
Triangular cell (side = R)	$\dfrac{\sqrt{3}}{4}R^2$	$3R$	$\dfrac{4\sqrt{3}}{R}$	$\dfrac{4\sqrt{3}N}{3R^2}$	$\dfrac{4\sqrt{3}KN}{3R^2}$	$\dfrac{4\sqrt{3}M^2N}{3R^2}$

• • • • • • • • • • • • • ••••

5.3 Signal Strength and Cell Parameters

Cellular systems depend on the radio signals received by a MS throughout the cell and on the contours of signal strength emanating from the BSs of two adjacent cells i and j, as illustrated in Figure 5.2.

As discussed earlier, the contours may not be concentric circles and could be distorted by atmospheric conditions and topographical contours. One example of distorted tiles is shown in Figure 5.3.

It is clear that signal strength goes down as one moves away from the BS. The variation of received power as a function of distance is given in Figure 5.4. As the MS moves away from the BS of the cell, the signal strength weakens, and at some point a phenomenon known as *handoff* occurs (handoff is also written as hand-off or hand off, and known as handover outside North America). This implies a radio connection to another adjacent cell. This is illustrated in Figure 5.5, as the MS moves away from cell i and gets closer to cell j. Assuming that $P_i(x)$ and $P_j(x)$ represent the power received at the MS from BS_i and BS_j, the received signal strength at the MS can be approximated by curves shown in Figure 5.5 and the variations can be expressed by the empirical relations given in Chapter 3. At distance X_1, the received signal from BS_j is close to zero and the signal strength at the MS can be primarily attributed to BS_i. Similarly, at distance X_2, the signal from BS_i is negligible. To receive and interpret the signals correctly at the MS, the received signals must be at a given minimum power level P_{min}, and distances X_3 and X_4 represent two such points for BS_j and BS_i, respectively. This means that, between points X_3 and X_4, the MS can be served by either BS_i or BS_j, and the choice is left to the service provider and the underlying technology. If the MS has a radio link with BS_i and is continuously

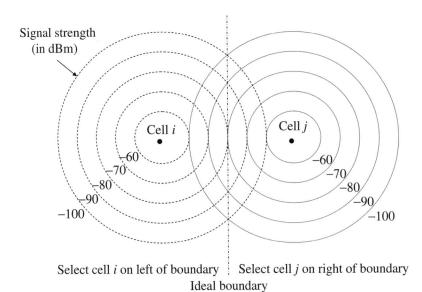

Figure 5.2
Signal strength contours around two adjacent cells i and j.

Select cell i on left of boundary ┊ Select cell j on right of boundary
Ideal boundary

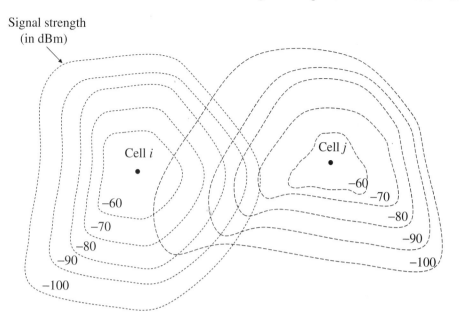

Figure 5.3
Received signal
strength indicating
actual cell tiling.

moving away toward BS_j, then at some point it has to be connected to the BS_j, and the change of such linkage from BS_i to BS_j is known as handoff. Therefore, region X_3 to X_4 indicates the handoff area. Where to perform handoff depends on many factors. One option is to do handoff at X_5 where two BSs have equal signal strength. A critical consideration is that the handoff should not take place too quickly to make the MS change the BS too frequently (e.g., ping-pong effect) if the MS moves back and forth between the overlapped area of two adjacent cells due to underlying terrain or intentional movements.

To avoid such a "ping-pong" effect, the MS is allowed to continue maintaining a radio link with the current BS_i, until the signal strength from BS_j exceeds that of BS_i by some prespecified threshold value E as is shown by point X_{th} in Figure 5.5. Thus, besides transmitting power, the handoff also depends on the mobility of the MS.

Figure 5.4
Variation of received
power from a base
station.

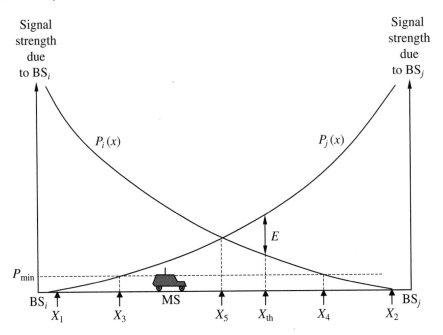

Figure 5.5
Handoff region.

Another factor that influences handoff is the area and the shape of the cell. An ideal situation is to have the cell configuration match the velocity of the MSs and to have a larger boundary where the handoff rate is minimal. The mobility of an individual MS is difficult to predict [5.1], with each MS having a different mobility pattern. Hence, it is impossible to have an exact match between the cell shape and subscriber mobility. Just to illustrate how handoff is related to the mobility and the cell area, consider a rectangular cell of area A and sides R_1 and R_2 shown in Figure 5.6. Assuming that N_1 is the number of MSs having handoff per unit length in the horizontal direction and N_2 is the similar quantity in the vertical direction, then the handoff could occur along the side R_1 of the cell or cross through the side R_2 of the cell. The number of MSs crossing along the R_1 side of the cell can be given by

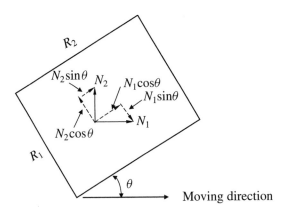

Figure 5.6
Handoff rate in a rectangular cell.

the component $R_1 (N_1 \cos\theta + N_2 \sin\theta)$, the number of MSs along the length R_2 can be expressed by $R_2 (N_1 \sin\theta + N_2 \cos\theta)$. Therefore, the total handoff rate λ_H can be given by equation (5.1):

$$\lambda_H = R_1(N_1 \cos\theta + N_2 \sin\theta) + R_2(N_1 \sin\theta + N_2 \cos\theta). \tag{5.1}$$

Assuming that the area $A = R_1 R_2$ is fixed, the question is how to minimize λ_H for a given θ. This is done by substituting the value of $R_2 = A/R_1$, differentiating with respect to R_1, and equating it to zero, which gives us

$$\frac{d\lambda_H}{dR_1} = \frac{d}{dR_1}\left[R_1 (N_1 \cos\theta + N_2 \sin\theta) + \frac{A}{R_1} (N_1 \sin\theta + N_2 \cos\theta) \right]$$

$$= N_1 \cos\theta + N_2 \sin\theta - \frac{A}{R_1^2} (N_1 \sin\theta + N_2 \cos\theta)$$

$$= 0. \tag{5.2}$$

Thus, we have

$$R_1^2 = A\frac{N_1 \sin\theta + N_2 \cos\theta}{N_1 \cos\theta + N_2 \sin\theta}. \tag{5.3}$$

Similarly, we can obtain

$$R_2^2 = A\frac{N_1 \cos\theta + N_2 \sin\theta}{N_1 \sin\theta + N_2 \cos\theta}. \tag{5.4}$$

Substituting these values in equation (5.1), we have

$$\lambda_H = \sqrt{A(\frac{N_1 \sin\theta + N_2 \cos\theta}{N_1 \cos\theta + N_2 \sin\theta})} \, (N_1 \cos\theta + N_2 \sin\theta)$$

$$+ \sqrt{A(\frac{N_1 \cos\theta + N_2 \sin\theta}{N_1 \sin\theta + N_2 \cos\theta})} \, (N_1 \sin\theta + N_2 \cos\theta)$$

$$= \sqrt{A(N_1 \sin\theta + N_2 \cos\theta)(N_1 \cos\theta + N_2 \sin\theta)}$$

$$+ \sqrt{A(N_1 \cos\theta + N_2 \sin\theta)(N_1 \sin\theta + N_2 \cos\theta)}$$

$$= 2\sqrt{A(N_1 \sin\theta + N_2 \cos\theta)(N_1 \cos\theta + N_2 \sin\theta)}. \tag{5.5}$$

The preceding equation can be simplified as

$$\lambda_H = 2\sqrt{A\left[N_1 N_2 + (N_1^2 + N_2^2)\cos\theta \sin\theta\right]}. \tag{5.6}$$

Equation (5.6) is minimized when $\theta = 0$. Hence, from equations (5.6), (5.3), and (5.4) we get

$$\lambda_H = 2\sqrt{A N_1 N_2} \tag{5.7}$$

and

$$\frac{R_2}{R_1} = \frac{N_1}{N_2}. \tag{5.8}$$

Intuitively, similar results can be expected for cells with other shapes. While it is relatively simple for rectangular cells, it is rather difficult to obtain similar analytical results for other types of cells. The only exception is the circular cell, where the rate of crossing the periphery is independent of direction because of its regular geometry. This means that the handoff is minimized if the rectangular cell is aligned with vertical and horizonntal axes and then the number of MSs crossing boundary is inversely proportional to the value of the other side of the cell. An attempt has been made for hexagonal cells in [5.2]—especially to quantify soft handoff wherein the handoff connection to the new BS is made first before breaking an existing connection in the handoff area.

When modeling handoff in cellular systems, it is sufficient to consider a single cell model for most analytical and planning purposes [5.3]. An empirical relation to compute the power received at the MS has been given in Chapter 3.

● ● ● ● ● ● ● ● ● ● ● ● ● ● ● ● ● ●

5.4 Capacity of a Cell

The offered traffic load of a cell is typically characterized by the following two important random parameters:

1. Average number of MSs requesting the service (average call arrival rate λ)

2. Average length of time the MSs requiring the service (average holding time T).

The offered traffic load is defined as

$$a = \lambda T. \tag{5.9}$$

For example, in a cell with 100 MSs, on an average, if 30 requests are generated during an hour, with average holding time $T = 360$ seconds, then the average request rate (or average call arrival rate) is

$$\lambda = \frac{30 \text{ requests}}{3600 \text{ seconds}}. \tag{5.10}$$

A servicing channel that is kept busy for an hour is quantitatively defined as one **Erlang**.

Hence, the offered traffic load for the preceding example by **Erlang** is

$$a = \frac{30 \text{ calls}}{3600 \text{ seconds}} \times 360 \text{ seconds}$$

$$= 3 \text{ Erlangs}. \tag{5.11}$$

The average arrival rate is λ and the average service (departure) rate is μ. When all channels are busy, an ariving call is turned away. Therefore, this system can be analyzed by a $M/M/S/S$ queing model. Since $M/M/S/S$ is a special case of $M/M/S/\infty$ introduced in Chapter 2, the steady-state probibilities $P(i)$s for this system have the same form as those for states $i = 0, \cdots, S$ in the $M/M/S/\infty$ model. Here, S is the number of channels in a cell. Thus, we have

$$P(i) = \frac{a^i}{i!} P(0), \tag{5.12}$$

where $a = \lambda/\mu$ is the offered load and

$$P(0) = \left[\sum_{i=0}^{S} \frac{a^i}{i!} \right]^{-1}. \tag{5.13}$$

Therefore, the probibility $P(S)$ of an arriving call being blocked is equal to the probability that all channels are busy, that is,

$$P(S) = \frac{\dfrac{a^S}{S!}}{\displaystyle\sum_{i=0}^{S} \frac{a^i}{i!}}. \tag{5.14}$$

Equation (5.14) is called the **Erlang B** formula and is denoted as $B(S, a)$. $B(S, a)$ is also called blocking probability, probability of loss, or probability of rejection. The **Erlang B** table is given in Appendix A

In the previous example, if S is given as 2 with $a = 3$, the blocking probability is

$$B\,(2, 3) = \frac{\dfrac{3^2}{2!}}{\displaystyle\sum_{k=0}^{2} \frac{3^k}{k!}}$$

$$= 0.529. \tag{5.15}$$

Therefore, a fraction of 0.529 calls is blocked, and we need to reinitiate the call. Thus the total number of blocked calls is about $30 \times 0.529 = 15.87$. The efficiency of the system can be given by

$$\text{Efficiency} = \frac{\text{Traffic nonblocked}}{\text{Capacity}}$$

$$= \frac{\text{Erlangs} \times \text{portion of used channel}}{\text{Number of channels}}$$

$$= \frac{3\,(1 - 0.529)}{2}$$

$$= 0.7065. \tag{5.16}$$

The probability of an arriving call being delayed is

$$C(S, a) = \frac{\frac{a^S}{(S-1)!(S-a)}}{\frac{a^S}{(S-1)!(S-a)} + \sum_{i=0}^{S-1} \frac{a^i}{i!}}$$

$$= \frac{SB(S, a)}{S - a[1 - B(S, a)]}, \qquad \text{for } a < S. \qquad (5.17)$$

This is called the **Erlang C** formula. In the previous example, if $S = 5$ and $a = 3$, we have $B(5, 3) = 0.11$. Therefore, the probability of an arriving call being delayed is

$$C(S, a) = \frac{SB(S, a)}{S - a[1 - B(S, a)]}$$

$$= \frac{5 \times B(5, 3)}{5 - 3 \times [1 - B(5, 3)]}$$

$$= \frac{5 \times 0.11}{5 - 3 \times [1 - 0.11]}$$

$$= 0.2360.$$

5.5 Frequency Reuse

Earlier cellular systems employed FDMA, and the range was limited to a radius from 2 to 20 km. The same frequency band or channel used in a cell can be "reused" in another cell as long as the cells are far apart and the signal strengths do not interfere with each other. This, in turn, enhances the available bandwidth of each cell. A typical cluster of seven such cells and four such clusters with no overlapping area is shown in Figure 5.7.

In Figure 5.7, the distance between the two cells using the same channel is known as the "reuse distance" and is represented by D. In fact, there is a close relationship between D, R (the radius of each cell), and N (the number of cells in a cluster), which is given by

$$D = \sqrt{3N} R. \qquad (5.18)$$

Therefore, the reuse factor q is

$$q = \frac{D}{R} = \sqrt{3N}. \qquad (5.19)$$

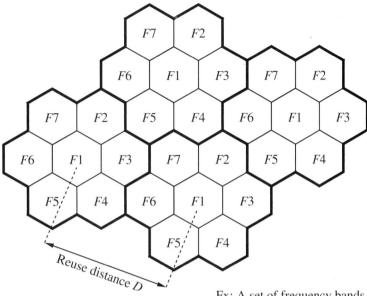

Figure 5.7
Illustration of frequency
reuse.

Fx: A set of frequency bands

Figure 5.8
Finding the center
of an adjacent
cluster using integers
i and *j* (directions
of *i* and *j* can be
interchanged).

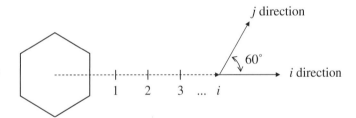

Another popular cluster size is with $N = 4$. In fact, the arguments made in selecting
a rectangular versus hexagonal shape of the cell are also applicable to the size of
the hex cell clusters such that multiple copies of such clusters should fit well with
each other, just like a puzzle. Additional areas can be covered by additional clusters
without having any overlapped area. In general, the number of cells N per cluster is
given by $N = i^2 + ij + j^2$. Here i represents the number of cells to be traversed along
direction i, starting from the center of a cell, and j represents the number of cells in
a direction $60°$ to the direction of i. Substituting different values of i and j leads to
$N = 1, 3, 4, 7, 9, 12, 13, 16, 19, 21, 28, \ldots$; the most popular values are 7 and 4.
Finding the center of all clusters around a reference cell for some selected values of
N, is illustrated in Figure 5.8. Repeating this for all six sides of the reference cell
leads to the center for all adjacent clusters. Unless specified, a cluster of size 7 is
assumed throughout this book.

● ● ● ● ● ● ● ● ● ● ● ● ● ● ● ●

5.6 How to Form a Cluster

In general, $N = i^2 + ij + j^2$, where i and j are integers. For computing convenience, we assume $i \geq j$. Based on the theory given in the article [5.4], we discuss a method to form a cluster of N cells as follows. (Note: this method is only for the case $j = 1$.)

First, select a cell, make the center of the cell as the origin, and form the coordinate plane as shown in Figure 5.9. The positive half of u-axis and the positive half of v-axis intersect at a 60-degree angle. Define the unit distance as the distance of centers of two adjacent cells. Then for each cell center, we can get an ordered pair (u, v) to mark the position.

Since this method is only for those cases $j = 1$ with a given N, integer i is also fixed by

$$N = i^2 + ij + j^2$$
$$= i^2 + i + 1. \tag{5.20}$$

Then using

$$L = [(i + 1)u + v] \bmod N, \tag{5.21}$$

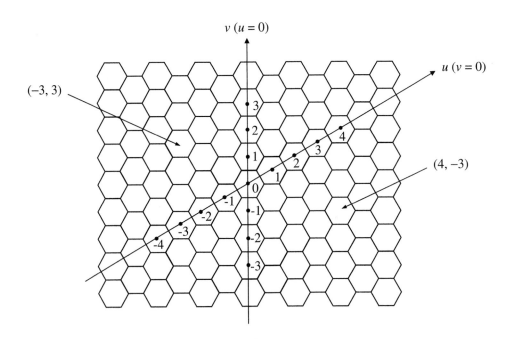

Figure 5.9
u and v coordinate plane.

Table 5.2: ▶
Some Cell Labels for N = 7

u	0	1	−1	0	0	1	−1
v	0	0	0	1	−1	1	1
L	0	3	4	1	6	2	5

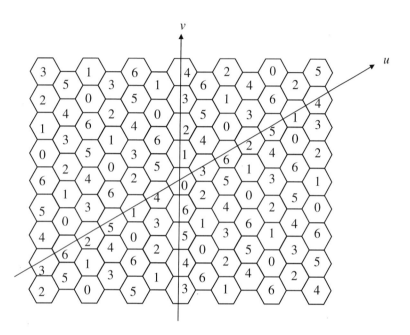

Figure 5.10
Cell label L for 7-cell cluster.

we can obtain the label L for the cell whose center is at (u, v). For the origin cell whose center is $(0, 0)$, $u = 0$, $v = 0$, using equation (5.21), we have $L = 0$ and label this cell as 0. Then we compute the labels of all adjacent cells. Finally, the cells with labels from 0 through $N - 1$ form a cluster of N cells. The cells with the same label can use the same frequency bands.

Now we give an example of $N = 7$ as follows. Using equation (5.20), we have $i = 2$. Then using equation (5.21), we have $L = (3u + v) \bmod 7$. We can compute label L for any cell using its center's positon (u, v). The results are shown in Table 5.2.

For each cell, we use its L values to label it. The results are shown in Figure 5.10. The cells with labels 0 through 6 form a cluster of 7 cells.

Using the same method, we also have the results for $N = 13$ is shown in Figure 5.11, with $i = 3$ and $j = 1$, giving $L = (4u + v) \bmod 13$. Some common reuse cluster patterns are given in Figure 5.12.

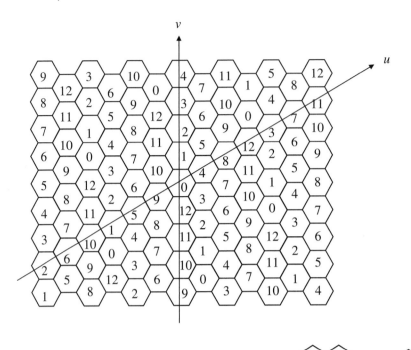

Figure 5.11
Cell label *L* for 13-cell cluster.

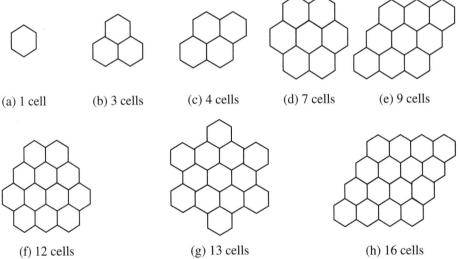

(a) 1 cell   (b) 3 cells   (c) 4 cells   (d) 7 cells   (e) 9 cells

Figure 5.12
Common reuse pattern of hex cell clusters.

(f) 12 cells   (g) 13 cells   (h) 16 cells

5.7 Cochannel Interference

As indicated earlier, there are many cells using the same frequency band. All the cells using the same channel are physically located apart by at least reuse distance. Even though the power level is controlled carefully so that such "cochannels" do not create a problem for each other, there is still some degree of interference due

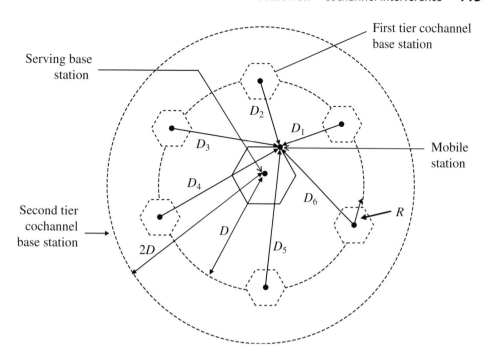

First tier cochannel base station

Serving base station

Mobile station

Second tier cochannel base station

D_2

D_1

D_3

D_4

D_6

D_5

D

$2D$

R

Figure 5.13
Cells with cochannels and their forward channel interference on transmitted signal.

to nonzero signal strength of such cells. In a cellular system, with a cluster of seven cells, there will be six cells using cochannels at the reuse distance; this is illustrated in Figure 5.13. The second tier cochannels, shown in the figure, are at two times the reuse distance apart, and their effect on the serving BS is negligible.

The cochannel interference ratio (CCIR) is given by

$$\frac{C}{I} = \frac{\text{Carrier}}{\text{Interference}} = \frac{C}{\sum\limits_{k=1}^{M} I_k}, \tag{5.22}$$

where I_k is cochannel interference from BS_k and M is the maximum number of cochannel interfering cells. For cluster size of 7, $M = 6$, CCIR is given by

$$\frac{C}{I} = \frac{1}{\sum\limits_{k=1}^{M} \left(\frac{D_k}{R}\right)^{-\gamma}}, \tag{5.23}$$

where γ is the propagation path loss slope and varies between 2 and 5.

When $D_1 = D_2 = D - R$, $D_3 = D_6 = D$, and $D_4 = D_5 = D + R$ (see Figure 5.14), the cochannel interference ratio in the worst case for the forward channel (downlink) is given as

$$\frac{C}{I} = \frac{1}{2(q-1)^{-\gamma} + 2q^{-\gamma} + 2(q+1)^{-\gamma}}, \tag{5.24}$$

where $q \ (= \frac{D}{R})$ is the frequency reuse factor.

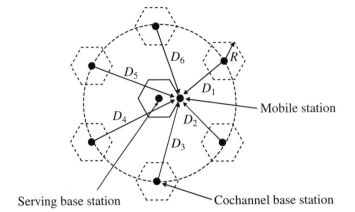

Figure 5.14
The worst case for forward channel interference (omnidirectional antenna).

There are many techniques that have been proposed to reduce interference. Here we consider only two specific ways: cell splitting and cell sectoring.

5.8 Cell Splitting

Until now, we have been considering the same size cell across the board. This implies that the BSs of all cells transmit information at the same power level so that the net coverage area for each cell is the same. At times, this may not be feasible, and, in general, this may not be desirable. Service providers would like to service users in a cost-effective way, and resource demand may depend on the concentration of users in a given area. Change in number of users could also occur over a period of time. One way to cope with increased traffic is to split a cell into several smaller cells; this is illustrated in Figure 5.15. This implies that additional BSs need to be established at the center of each new cell that has been added so that the higher

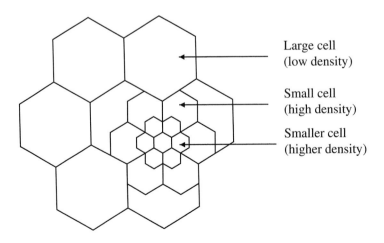

Figure 5.15
Illustration of cell splitting.

density of calls can be handled effectively. As the coverage area of new split cells is smaller, the transmitting power levels are lower, and this helps in reducing cochannel interference.

5.9 Cell Sectoring

We have been primarily concentrating on what is known as omnidirectional antennas, which allow transmission of radio signals with equal power strength in all directions. It is difficult to design such antennas, and most of the time, an antenna covers an area of 60 degrees or 120 degrees; these are called directional antennas, and cells served by them are called sectored cells. Different sizes of sectored cells are shown in Figure 5.16. From a practical point of view, many sectored antennas are mounted on a single microwave tower located at the center of the cell, and an adequate number of antennas is placed to cover the whole 360 degrees of the cell. For example, the 120 degree sectored cell shown in Figures 5.16(b) and 5.16(c) requires three directional antennas. In practice, the effect of an omnidirectional antenna can be achieved by employing several directional antennas to cover the whole 360 degrees.

The advantages of sectoring (besides easy borrowing of channels, which is discussed in Chapter 8) are that it requires coverage of a smaller area by each antenna and hence lower power is required in transmitting radio signals. It also helps in decreasing interference between cochannels, as discussed in Section 5.5. It is also observed that the spectrum efficiency of the overall system is enhanced. It is found that a quad-sector architecture of Figure 5.16(d) has a higher capacity for 90% area coverage than a tri-sector cell [5.5].

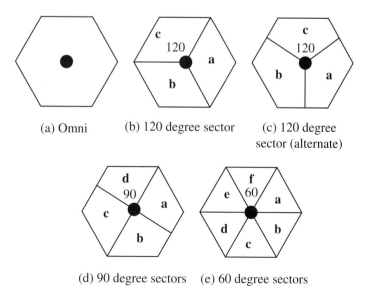

(a) Omni   (b) 120 degree sector   (c) 120 degree sector (alternate)

(d) 90 degree sectors   (e) 60 degree sectors

Figure 5.16
Sectoring of cells with directional antennas.

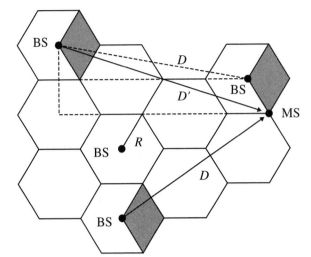

Figure 5.17
The worst case for
forward channel
interference in
three sectors
(directional antenna).

The cochannel interference for cells using directional antennas can also be computed. The worst case for the three-sector directional antenna is shown in Figure 5.17. From the figure, we have

$$D = \sqrt{\left(\frac{9}{2}R\right)^2 + \left(\frac{\sqrt{3}}{2}R\right)^2}$$

$$= \sqrt{21}R$$

$$\simeq 4.58R \tag{5.25}$$

and

$$D' = \sqrt{(5R)^2 + \left(\sqrt{3}R\right)^2}$$

$$= \sqrt{28}R$$

$$\simeq 5.29R$$

$$= D + 0.7R. \tag{5.26}$$

Therefore, CCIR can be obtained as

$$\frac{C}{I} = \frac{1}{q^{-\gamma} + (q+0.7)^{-\gamma}}. \tag{5.27}$$

The CCIR in the worst case for the six-sector directional antenna (see Figure 5.18) when $\gamma = 4$ can be given by

$$\frac{C}{I} = \frac{1}{(q+0.7)^{-\gamma}} = (q+0.7)^4. \tag{5.28}$$

Thus, we can see that the use of a directional antenna is helpful in reducing cochannel interference.

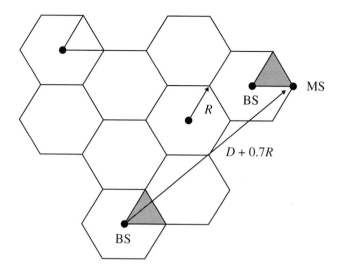

Figure 5.18
The worst case for forward channel interference in six sectors (directional antenna).

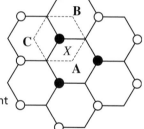

Figure 5.19
An alternative placement of directional antennas at three corners.

It is worth mentioning that there is an alternative way of providing sectored or omni-cell coverage, by placing directional transmitters at the corners where three adjacent cells meet (see Figure 5.19). It may appear that the arrangement of Figure 5.19 may require three times the transmitting towers as compared to a system with towers placed at the center of the cell. However, a careful consideration reveals that the number of transmitting towers remains the same, as the antennas for adjacent cells B and C could also be placed on the towers X, and for a coverage area with a larger number of cells, the average number of towers approximately remains the same.

● ● ● ● ● ● ● ● ● ● ● ● ● ● ● ● ●

5.10 Summary

This chapter provides an overview of various cell parameters, including area, load, frequency reuse, cell splitting, and cell sectoring. As limited bandwidth has been allocated for wireless communications, the reuse technique is shown to be useful for both FDMA and TDMA schemes. In the next chapter, we discuss how a control channel can be accessed by multiple MSs and how collision can be avoided.

• • • • • • • • • • • • • •

5.11 References

[5.1] A. Bhattacharya and S. Das, "LeZi-Update: An Information Theoretic Approach to Track Mobile Users in PCS Networks," *ACM/Kluwer Journal on Wireless Networks*, Vol. 8, No. 2-3, pp. 121–135 May 2002.

[5.2] J. Y. Kwan and D. K. Sung, "Soft Hand Off Modeling in CDMA Cellular Systems," *Proceedings of the IEEE Conference on Vehicular Tecnology (VTC'97)*, pp. 1548–1551, May 1997.

[5.3] P. V. Orlik and S. S. Rappaport, "On the Handoff Arrival Process in Cellular Communications," *Wireless Networks*, Vol. 7 No. 2, pp. 147–157, March/April 2001.

[5.4] V. H. MacDonald, "Advanced Mobile Phone Service—The Cellular Concept," *Bell System Technical Journal*, Vol. 58, No. 1, pp. 15–41, Jan. 1979.

[5.5] O. W. Ata, H. Seki, and A. Paulraj, "Capacity Enhancement in Quad-Sector Cell Architecture with Interleaved Channel and Polarization Assignments," *IEEE International Conference on Communications (ICC)*, Helsinki, pp. 2317–2321, June 2001.

• • • • • • • • • • • • • •

5.12 Problems

P5.1. An octagon-shaped cell is closer to a circle as compared to a hexagon. Explain why such a cell is not used as an ideal shape of the cell.

P5.2. A new wireless service provider decided to employ a cluster of 19 cells as the basic module for frequency reuse.
(a) Can you identify one such cluster structure?
(b) Repeat (a) for $N = 28$.
(c) Can you get an alternative cluster structure for part (a)?
(d) What is the reuse distance for the system of part (c)?
(e) Can you find the worst case cochannel interference in such a system?

P5.3. Two adjacent BSs i and j are 30 kms apart. The signal strength received by the MS is given by the following expressions:

$$P(x) = \frac{G_t G_r P_t}{L(x)},$$

where

$$L(x) = 69.55 + 26.16 \log_{10} f_c(\text{ MHz}) - 13.82 \log_{10} h_b(\text{m}) - a[h_m(\text{m})]$$
$$+ [44.9 - 6.55 \log_{10} h_b(\text{m})] \log_{10}(x),$$

and x is the distance of the MS from BS i. Assume unity gain for G_r and G_t, given that $P_i(t) = 10$ watts, $P_j(t) = 100$ watts, $f_c = 300$ MHz, $h_b = 40$ m, $h_m = 4$ m, $\alpha = 3.5$, $x = 1$ km, and $P_j(t)$ is the transmission power of BS j.

(a) What is the power transmitted by BS j, so that the MS receives signals of equal strength at x?

(b) If the threshold value $E = 1$ dB and the distance where handoff is likely to occur is 2 km from BS j, then what is the power transmitted by BS j?

P5.4. If each user keeps a traffic channel busy for an average of 5% time and an average of 60 requests per hour or generated, what is the Erlang value?

P5.5. Prove that $D = R\sqrt{3N}$.

P5.6. Prove that $N = i^2 + j^2 + ij$.

P5.7. The size and shape of each cluster in a cellular system must be designed carefully so as to cover adjacent spokes in a non-overlapped manner. Define such pattens for the following cluster sizes:

(a) 4-cell

(b) 9-cell

(c) 13-cell

(d) 37-cell

P5.8. A cellular scheme employed a cluster of size 16 cells. Later on, it was decided to use two different clusters of size 7 and 9 cells. Is it possible to replace each original cluster by two new clusters? Explain clearly.

P5.9. For the following cell pattern,

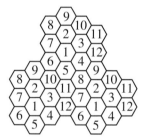

Figure 5.20
Figure for
Problem P5.9

(a) Find the reuse distance if the radius of each cell is 2 km.

(b) If each channel is multiplexed among 8 users, how many calls can be simultaneously processed by each cell if only 10 channels per cell are reserved for control, assuming a total bandwidth of 30 MHz available and each simplex channel of 25 kHz?

P5.10. A TDMA-based system, shown in the Figure 5.21, has a total bandwidth of 12.5 MHz and contains 20 control channels with equal channel spacing of 30 kHz. Here, the area of each cell is equal to 8 km^2 and cells required to cover a total area of 3600 km^2. Calculate the following:

(a) Number of traffic channels/cell

(b) Reuse distance

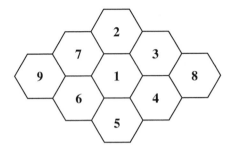

Figure 5.21
Figure for
Problem P5.10.

P5.11. During a busy hour, the number of calls per hour for each of the 12 cells of a cellular cluster is 2220, 1900, 4000, 1100, 1000, 1200, 1800, 2100, 2000, 1580, 1800 and 900. Assume that 75% of the car phones in this cluster are used during this period and that one call is made per phone.

(a) Find the number of customers in the system.

(b) Assuming the average hold time of 60 seconds, what is the total Erlang value of the system?

(c) Find the reuse distance D if $R = 5$ km.

P5.12. Given a bandwidth of 25 MHz and a frequency reuse factor of 1 and RF channel size of 1.25 MHz and 38 calls per RF channel, find:

(a) The number of RF channels for CDMA.

(b) The number of permissible calls per cell (CDMA).

P5.13. If a wireless service provider has 20 cells to cover the whole service area, with each cell having 40 channels, how many users can the provider support if a blocking probability p of 2% is required? Assume that each user makes an average of three calls/hour and each call duration is an average of three minutes. (**Erlang B** values are given in Appendix A.)

P5.14. The following figure shows a cellular architecture. Is there some specific reason why it could have been designed this way?

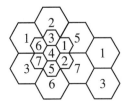

Figure 5.22
Figure for
Problem P5.14.

P5.15. The following figure shows the cell structure of a metro area. Can you explain why this might have been designed so?

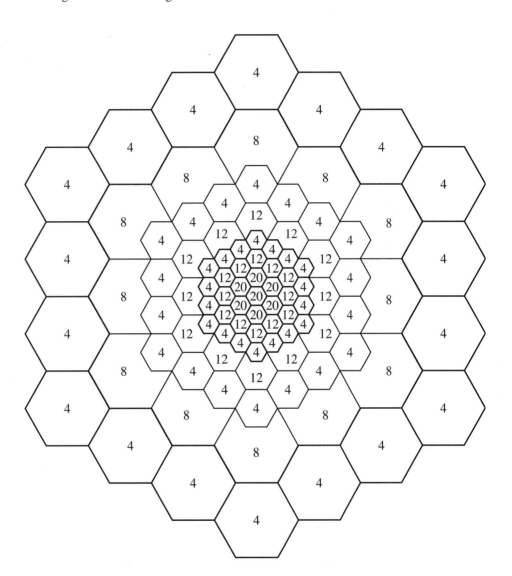

Figure 5.23
Figure for
Problem P5.15.

P5.16. Prove the following for a hexagonal cellular system with radius R, reuse distance D, and given the value of N:

(a) $N = 3$, prove $D = 3R$.

(b) $N = 4$, prove $D = \sqrt{12}R$.

(c) $N = 7$, prove $D = \sqrt{21}R$.

P5.17. In Figure 5.14, calculate the cochannel interference ratio in the worst case for the forward channel, given $N = 7$, $R = 3$ km, and $\gamma = 2$.

P5.18. What is meant by handoff interval and handoff region? Explain their usefulness with appropriate diagrams.

P5.19. What are the differences between adjacent channel interference and cochannel interference? Explain with suitable diagrams.

P5.20. What are the advantages of cell sectoring? Explain with suitable diagrams.

CHAPTER 6

Multiple Radio Access

● ● ● ● ● ● ● ● ● ● ● ● ⋯⋯

6.1 Introduction

Wireless devices have become both popular and cost-effective and have attracted considerable interest from the industry and the academia. Users of wireless networks, either walking on the street, driving a car, or operating a portable computer on an aircraft, enjoy the exchange of information without worrying about how technology makes such exchange possible. To make this communication feasible, a user needs access to a control channel, which can be exclusively assigned for this purpose or can be shared among numerous subscribers. As users need to access the channel at random times and for random periods, it is not desirable to allocate a control channel permanently. Such an expensive commodity is shared among users, as needed, using predefined rules or algorithms, if there is no central authority to handle such allocation. Even if there is a controller like a BS, MSs need to use a control channel to inform the BS before using the traffic channel or information channel. This makes it possible to assign a traffic channel to each MS by the BS for information transfer. Such exchange of facts necessitates shared access of a control channel, for which each MS has to compete. It may be noted that such contention is exclusively present in MANETs wherein the same frequency is used to enable all wireless devices to tune into the same band and listen to each other. It is helpful to study how shared channels can be accessed and what are the advantages and disadvantages of various rules and guidelines for their use, usually termed as protocols. In this chapter, we deal with some important multiple radio access protocols for wireless networks, describe their characteristics, and discuss their suitability.

A typical scenario in a wireless network is shown in Figure 6.1. MSs have to compete for a shared medium. Each MS has a transmitter/receiver that communicates with other MSs. In a general scheme, transmission from any MS can be received by all other MSs in the neighborhood. Therefore, if more than one MS attempts to transmit at one time, collision occurs, wherein signals in the medium (air in the case of wireless devices) are garbled, and MSs receiving the information cannot interpret or differentiate what is being transmitted. These situations are called collisions in the medium or multiple access issues. Collisions must be avoided, and we need to follow

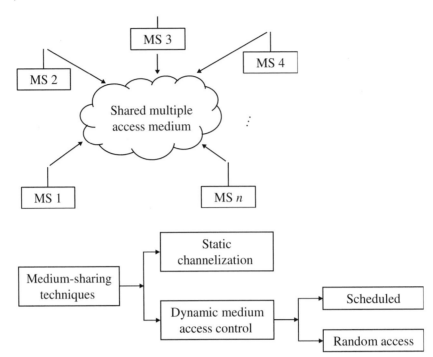

Figure 6.1
Multiple access of
shared medium in a
wireless network.

Figure 6.2
Medium-sharing
techniques.

some protocols to determine which MS has exclusive access to the shared medium
at a given time and for a given duration so that the MS can transmit and other MSs
can receive, understand, and interpret the received control information in a wireless
system. For handling multiple access issues, there are two different types of protocols: the contention-based protocol and the conflict-free (or collision-free) protocol.
Contention-based protocols resolve a collision after it occurs. These protocols execute a collision resolution protocol after detection of each collision. Collision-free
protocols (e.g., a bit-map protocol and binary countdown) ensure that a collision
never occurs.

Medium-sharing techniques can be classified into two methods: static channelization and dynamic medium access control. In static allocation, the channel assignment is done in a prespecified way and does not change with time. In dynamic technique, the channel is allocated as needed and changes with time. Dynamic medium
access control is classified into scheduled and random access protocols, as shown in
Figure 6.2.

● ● ● ● ● ● ● ● ● ● ● ● ● ● ● ● ● ●

6.2 Multiple Radio Access Protocols

For computer networks, a seven-layer ISO (International Standards Organization)
OSI (Open Systems Interconnection) reference model is widely used and is briefly
discussed in Chapter 12 [6.1]. The communication subnetwork can be described

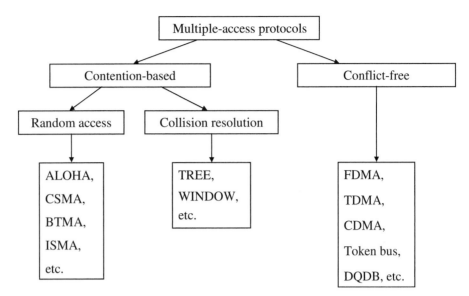

Figure 6.3
Classification of
multiple access
protocols.

by the lower three layers (i.e., physical, data link, and network layers). Existing LANs (local area networks), MANs (metropolitan area networks), PRNs (packet radio networks), PANs (personal area networks), and satellite networks do utilize broadcast channels rather than point-to-point channels for information transmission. Therefore, a simple modification of OSI model is done by adding the so-called MAC (medium access control) sublayer in data link layer. The MAC sublayer protocols, usually known as the multiple-access protocols, are primarily a set of rules that communicating MSs need to follow, and these are assumed to be agreed upon a priori.

Numerous multiple-access protocols have been proposed in the literature, and the list is fairly long. These can be categorized in many different ways. One of the most usual classifications (see Figure 6.3) is based on whether a protocol is contention-based or conflict-free [6.2] [6.3]. In this chapter, we concentrate on contention-based protocols. Conflict-free protocols for dedicated allocation of traffic channel to each MS will be discussed in Chapter 7.

● ● ● ● ● ● ● ● ● ● ● ● ● ● ● ● ●

6.3 **Contention-Based Protocols**

Since Abramson [6.4] proposed the well-known pure ALOHA scheme to enable exchange of messages between remote terminals and the central computer at the University of Hawaii, numerous alternative protocols have been proposed. Basically, a contention-based protocol differs from a conflict-free one on the guarantee aspect (i.e., it cannot assume responsibility for a successful transmission from a terminal at all the time). In a contention-based protocol, a terminal (MS) in the system may transmit its message at any time it wishes, hoping that no other terminals will transmit at the same time. Since collisions may exist in a contention-based protocol, the protocol has to have a provision to make collided messages retransmitted efficiently. Contention-based protocols can be classified into two groups according to the ways collisions are resolved: random access protocols and collision resolution protocols. In systems with one of the random access protocols, such as ALOHA-type protocols [6.4][6.5] [6.6], CSMA (carrier sense multiple access)-type protocols [6.7][6.8][6.9], BTMA (busy tone multiple access)-type protocols [6.8][6.10][6.11][6.12], ISMA (idle signal multiple access) -type protocols [6.13][6.14], and so on, a terminal is allowed to transmit the collided message only after a random delay. On the other hand, rather than using a random delay, collision resolution protocols, such as TREE [6.15] and WINDOW [6.16], employ a more sophisticated way to control the retransmission process.

6.3.1   Pure ALOHA

Pure ALOHA was developed in the 1970s for a packet radio network at the University of Hawaii. It is a single-hop system which consists of infinite users. Each user generates packets according to a Poisson process with arrival rate λ (packets/sec), and all packets have the same fixed length T. In this scheme, when a MS has a packet to transmit, it transmits the packet right away. The sender side also waits to see whether transmission is acknowledged by the receiver; no response within a specified period of time indicates a collision with another transmission. If the presence of a collision is determined by the sender, it retransmits after some random wait time, as illustrated in Figure 6.4, where the arrows indicate the arrival instants. Successful transmissions are indicated by blank rectangles, and collided packets are hatched. Because there are an infinite number of users in the system, we can assume that each packet is generated from different users, which means each new arrival packet can be considered as generated from an idle user that has no packet to retransmit. Using this method, we can consider that the packets and the users are identical and we only need to consider the time point at which the packet transmission attempts are made. Now considering the channel over the time, the scheduling time includes both the generation times of new packets and the retransmission times of previously collided packets. Let the rate of the scheduling be g (packets/sec). The parameter g is referred to as the offered load to the channel. Clearly, since some packets are transmitted more than once before they are successfully transmitted, $g > \lambda$. The exact characterization of the scheduling process is extremely complicated. To overcome this complexity and make the analysis of ALOHA-type systems tractable, it is assumed that this scheduling process is also a Poisson process with arrival rate g. This assumption is a good approximation which has been verified by simulation.

Consider a new or retransmitted packet scheduled for transmission at some instant t (see Figure 6.4). This packet can be successfully transmitted if there are no other packets scheduled for transmission between the instants $t - T$ and $t + T$ (this period of duration $2T$ is called the vulnerable period). Therefore, the probability P_s of successful transmission is the probability that no packet is scheduled in an interval

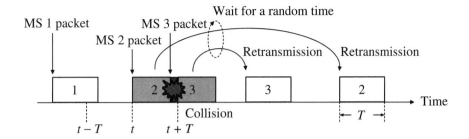

Figure 6.4
Collision mechanism
in pure ALOHA.

of length $2T$. Since the distribution of scheduling time is assumed to be a Poisson process, we have

$$P_s = P \text{ (no collision)}$$

$$= P \text{ (no transmission in two packets time)}$$

$$= e^{-2gT}. \tag{6.1}$$

Since packets are scheduled at a rate of g packets per second with only a fraction P_s successful, the rate of successful transmission is gP_s. If we define the throughput as the fraction of time during which the useful information is carried on the channel, we get the throughput of pure ALOHA as

$$S_{\text{th}} = gTe^{-2gT}, \tag{6.2}$$

which gives the channel throughput as a fraction of the offered load. Defining $G = gT$ to be the normalized offered load to the channel, we have

$$S_{\text{th}} = Ge^{-2G}. \tag{6.3}$$

Using equation (6.3), we can find the maximum throughput $S_{\text{th max}}$ by differentiating equation (6.3) with respect to and equaling to zero as

$$\frac{dS_{\text{th}}}{dG} = -2Ge^{-2G} + e^{-2G} = 0. \tag{6.4}$$

The equation (6.4) indicates that the maximum throughput $S_{\text{th max}}$ occurs at the offered load $G = 1/2$. Therefore, substituting $G = 1/2$ in equation (6.3), we have

$$S_{\text{th max}} = \frac{1}{2e} \approx 0.184. \tag{6.5}$$

This value can be improved if we impose some restrictions on how to select scheduling time, which is discussed next.

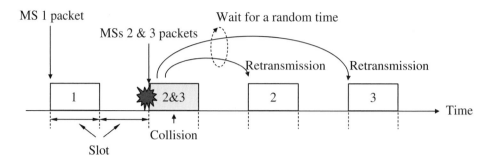

Figure 6.5
Collision mechanism
in slotted ALOHA.

6.3.2 Slotted ALOHA

Slotted ALOHA is a modification of pure ALOHA having slotted time with the slot size equal to the duration of packet transmission T. If a MS has a packet to transmit, before sending it waits until the beginning of the next slot. Thus, the slotted ALOHA is an improvement over pure ALOHA by reducing the vulnerable period for packet collision to a single slot. It means that a transmission will be successful if and only if exactly one packet is scheduled for transmission for the current slot. Figure 6.5 shows a collision mechanism in slotted ALOHA where a collision is observed to be a full collision; thus, no partial collision is possible.

Since the process composed of newly generated and retransmitted packets is Poisson, the probability of successful transmission is given by

$$P_s = e^{-gT} \qquad (6.6)$$

and the throughput S_{th} becomes

$$S_{\text{th}} = gTe^{-gT}. \qquad (6.7)$$

Using the definition of the normalized offered load $G = gT$, equation (6.7) can be rewritten as

$$S_{\text{th}} = Ge^{-G}. \qquad (6.8)$$

The maximum throughput maximum throughput $S_{th\,\text{max}}$ is obtained by

$$\frac{dS_{\text{th}}}{dG} = e^{-G} - Ge^{-G} = 0. \qquad (6.9)$$

The equation (6.9) indicates that the maximum throughput $S_{\text{th}\,max}$ occurs at the offered load $G = 1$. Therefore, substituting $G = 1$ in equation (6.8), we have

$$S_{\text{th}\,max} = \frac{1}{e} \approx 0.368. \qquad (6.10)$$

Figure 6.6 shows the throughputs of pure ALOHA and slotted ALOHA.

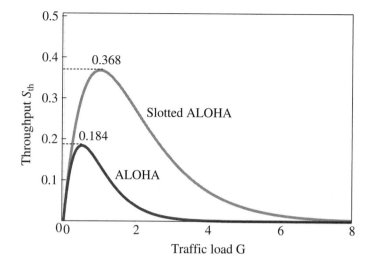

Figure 6.6
Throughputs of
pure ALOHA and
slotted ALOHA.

6.3.3 CSMA

Looking at the performance curves of pure and slotted ALOHA protocols, we see that the maximum throughputs are equal to 0.184 and 0.368, respectively. We need to find another way of improving throughputs and supporting high-speed communication networks. We could achieve better throughput if we can prevent potential collision by simply listening to the channel before transmitting a packet. In this way, collisions could be avoided; this is known as a carrier sense multiple access (CSMA) protocol. Each MS can sense the transmission of all other terminals, and the propagation delay is small as compared with the transmission time. Figure 6.7 shows the collision process in the CSMA protocol. There are several variants of the basic CSMA protocols, which are summarized in Figure 6.8.

Nonpersistent CSMA Protocol

In this protocol, the MS senses the medium first whenever the MS has a packet to send. If the medium is busy, the MS waits for a random amount of time and senses

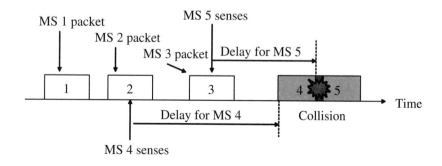

Figure 6.7
Collision mechanism
in CSMA.

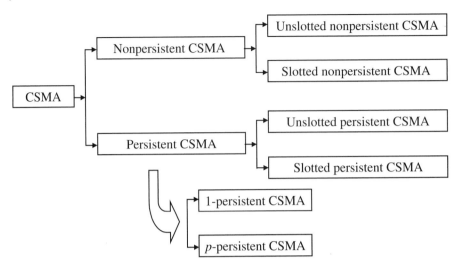

Figure 6.8
Types of CSMA
protocols.

the medium again. If the medium is idle, the MS transmits the packet immediately. If a collision occurs, the MS waits for a random amount of time and starts all over again. The packets can be sent during a slotted period or can be transmitted at any arbitrary time. This leads to two different subcategories: slotted nonpersistent CSMA and unslotted nonpersistent CSMA.

To understand and quantify the throughputs for all kinds of CSMA protocols, we define the following system parameters: S_{th} (throughput), G (offered traffic rate), T (packet transmission time), τ (propagation delay through the air), and p (p-persistent parameter). Without loss of generality, we choose $T = 1$. This is equivalent to expressing time in units of T. We express τ in the normalized time unit as $\alpha = \tau/T$.

For unslotted nonpersistent CSMA, the throughput is given by [6.7]

$$S_{th} = \frac{Ge^{-\alpha G}}{G(1+2\alpha) + e^{-\alpha G}}.$$ (6.11)

For slotted nonpersistent CSMA, the throughput is given by [6.7]

$$S_{th} = \frac{\alpha Ge^{-\alpha G}}{\left(1 - e^{-\alpha G}\right) + \alpha}.$$ (6.12)

1-Persistent CSMA Protocol

In this protocol, the MS senses the medium when the MS has a packet ready to send. If the medium is busy, the MS keeps listening to the medium and transmits the packet immediately after the medium becomes idle. This protocol is called 1-persistent because the MS transmits with a probability of 1 whenever it finds the medium to be idle. However, in this protocol, there will always be a collision if two or more MSs have ready packets, are waiting for the medium to become free, and start transmitting at the same time.

Given the system parameters G and α, the throughput for unslotted 1-persistent CSMA is given by [6.7]

$$S_{\text{th}} = \frac{G\left[1 + G + \alpha G(1 + G + \frac{\alpha G}{2})\right]e^{-G(1+2\alpha)}}{G(1+2\alpha) - (1 - e^{-\alpha G}) + (1 + \alpha G)e^{-G(1+\alpha)}}. \tag{6.13}$$

For slotted 1-persistent CSMA, the throughput is given by [6.7]

$$S_{\text{th}} = \frac{G\left(1 + \alpha - e^{-\alpha G}\right)e^{-G(1+\alpha)}}{(1+\alpha)(1 - e^{-\alpha G}) + \alpha e^{-G(1+\alpha)}}. \tag{6.14}$$

p-Persistent CSMA Protocol

In this protocol, the time is slotted. Let the size of slot be the contention period (i.e., the round trip propagation time). In this protocol, the MS senses the medium when it has a packet to send. If the medium is busy, the MS waits until the next slot and checks the medium again. If the medium is idle, the MS transmits with probability p or defers transmission with probability $(1-p)$ until the next slot. If a collision occurs, the MS waits for a random amount of time and starts all over again. Intuitively, this protocol is considered as an optimal access strategy.

There is a tradeoff between 1-persistent and nonpersistent CSMA protocols. Assuming the presence of three terminals A, B, and C in the system, let us consider the situation when terminals B and C become ready in the middle of MS A's transmission. For the 1-persistent CSMA protocol, terminals B and C will collide. For the nonpersistent CSMA protocol, terminals B and C may not collide. If only MS B becomes ready in the middle of MS A's transmission, for the 1-persistent CSMA protocol, MS B succeeds as soon as MS A ends. But for the nonpersistent CSMA protocol, MS B may have to wait.

For the p-persistent CSMA protocol, we must consider how to select the probability p. If N terminals have a packet to send, Np, the expected number of terminals will attempt to transmit once the medium becomes idle. If $Np > 1$, then a collision is expected. Therefore, the network must make sure that $Np \leq 1$.

Given the system parameters G, α, and $g = \alpha G$, the throughput for p-persistent CSMA is given by [6.7]

$$S_{\text{th}}(G, p, \alpha) = \frac{(1 - e^{-\alpha G})\left[P'_s \pi_0 + P_s(1 - \pi_0)\right]}{(1 - e^{-\alpha G})\left[\alpha \bar{t'} \pi_0 + \alpha \bar{t}(1 - \pi_0) + 1 + \alpha\right] + \alpha \pi_0}, \tag{6.15}$$

where P'_s, P_s, $\bar{t'}$, $\bar{t}$, and π_0 are given by the following equations, respectively:

$$P'_s = \sum_{n=1}^{\infty} P_s(n)\pi'_n, \tag{6.16}$$

$$P_s = \sum_{n=1}^{\infty} P_s(n) \frac{\pi_n}{1 - \pi_0}, \tag{6.17}$$

$$\bar{t}' = \sum_{n=1}^{\infty} \bar{t}_n \pi'_n, \tag{6.18}$$

$$\bar{t} = \sum_{n=1}^{\infty} \bar{t}_n \frac{\pi_n}{1 - \pi_0}, \tag{6.19}$$

and

$$\pi_n = \frac{[(1+\alpha)G]^n}{n!} e^{-(1+\alpha)G}, \quad n \geq 0, \tag{6.20}$$

where

$$P_s(n) = \sum_{l=n}^{\infty} \frac{lp(1-p)^{l-1}}{1 - (1-p)^l} \Pr\{L_n = l\}, \tag{6.21}$$

$$\pi'_n = \frac{g^n e^{-g}}{n! (1 - e^{-g})}, \quad n \geq 1, \tag{6.22}$$

and

$$\bar{t}_n = \sum_{k=0}^{\infty} \Pr\{\bar{t}_n > k\}$$

$$= \sum_{k=0}^{\infty} (1-p)^{(k+1)n} e^g \left\{ \frac{(1-p)\left[1-(1-p)^k\right]}{p} - k \right\}, \tag{6.23}$$

where

$$\Pr\{L_n = l\} = \sum_{k=1}^{\infty} \frac{(kg)^{l-n}}{(l-n)!} e^{-kg} \Pr\{t_n = k\} + \left[1 - (1-p)^n\right] \delta_{l,n}, \quad l \geq n, \tag{6.24}$$

$$\Pr\{\bar{t}_n = k\} = (1-p)^{kn} \left[1 - (1-p)^n e^{-g\left[1-(1-p)^k\right]}\right] e^g \left\{ \frac{(1-p)\left[1-(1-p)^{k-1}\right]}{p} - (k-1) \right\}, k > 0, \tag{6.25}$$

and $\delta_{i,j}$ is the **Kronecker** delta.

The throughputs of different ALOHA and CSMA protocols depend on the scheme and are illustrated in Figure 6.9.

6.3.4 CSMA/CD

In a typical CSMA protocol, if two terminals begin transmitting at the same time, each will transmit its complete packet, even though they collide. This wastes the medium for an entire packet time and can be addressed by a new protocol called CSMA

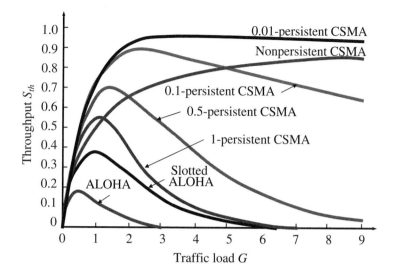

Figure 6.9
Throughput for different ALOHA and CSMA protocols with $\alpha = 0.01$.

with collision detection (CSMA/CD). The main idea is to terminate transmission immediately after detection of a collision.

In this protocol, the terminal senses the medium when the terminal has a packet to send. If the medium is busy, the terminal transmits its packet immediately. If the medium is busy, the terminal waits until the medium becomes idle. If a collision is detected during the transmission, the terminal aborts its transmission immediately and it attempts to transmit later after waiting for a random amount of time. Figure 6.10

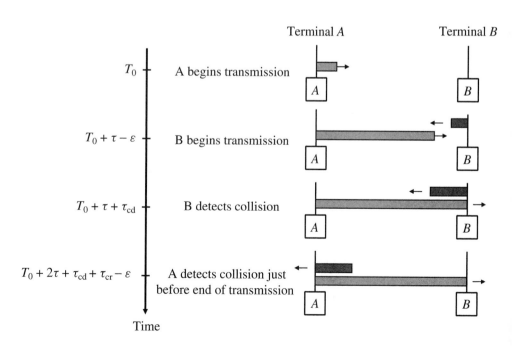

Figure 6.10
Collision mechanism in CSMA/CD.

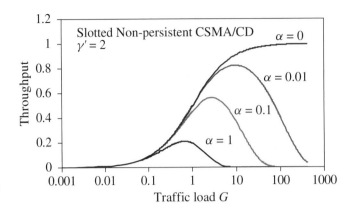

Figure 6.11
Throughput of
slotted non-persistent
CSMA/CD.

shows a collision mechanism in the CSMA/CD. In this figure, we consider two terminals A and B, the propagation delay between them is τ. Suppose that terminal A starts transmission at time T_0 when the channel is idle, then its transmission reaches terminal B at time $T_0 + \tau$. Suppose that terminal B initiates a transmission at time $T_0 + \tau - \varepsilon$ (here ε is small time period and $0 < \varepsilon \leq \tau$). It takes τ_{cd} for a terminal to detect the collision so that at time $T_0 + \tau + \tau_{cd}$ terminal B detects the collision. In LANs such Ethernet, whenever a collision is detected by a terminal, a consensus reinforcement procedure is initiated. Subsequently, the channel is jammed with a collision signal for a period of τ_{cr} longer enough for all network terminals to detect the collision. Thus, at time $T_0 + \tau + \tau_{cd} + \tau_{cr}$ terminal B completed the consensus reinforcement procedure which reaches terminal A at time $T_0 + 2\tau + \tau_{cd} + \tau_{cr}$. From terminal A's standpoint this transmission period lasted $\gamma = 2\tau + \tau_{cd} + \tau_{cr}$.

If given the system parameters G, τ, $\alpha = \tau/T$, and $G = gT$, the throughput for slotted non-persistent CSMA/CD protocol is given by [6.9]

$$S_{th} = \frac{\alpha G e^{-\alpha G}}{\alpha G e^{-\alpha G} + \left(1 - e^{-\alpha G} - \alpha G e^{-\alpha G}\right)\gamma' + \alpha}, \qquad (6.26)$$

where γ' is the ratio between γ and the transmission time of a packet ($\gamma' = \gamma/T$). Notice that when $\gamma' = 1$ the result in equation (6.26) is identical to slotted non-persistent CSMA.

We can see that the CSMA protocol minimizes the number of collisions while the CSMA/CD can further reduce the effect of a collision as it renders the medium ready to be used as soon as possible. The collision detection time is two times end-to-end propagation delay. Figure 6.11 depicts the throughput of slotted non-persistent CSMA/CD. The improvement in performance is readily apparent.

6.3.5   CSMA/CA

A modified version of CSMA/CD has been adopted by the IEEE 802.11 MAC and is called the distributed foundation wireless MAC (DFWMAC). The access mechanism is based on the CSMA/CD access protocol and is called CSMA with collision avoidance (CSMA/CA). The IEEE 802.11 wireless LAN standard supports operation in

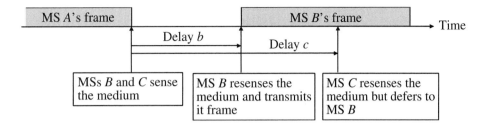

Figure 6.12
A basic collision avoidance scheme.

two separate modes: a distributed coordination and a centralized point-coordination mode. Figure 6.12 shows a general mechanism of collision avoidance protocol.

Basic CSMA/CA

Under basic CSMA/CA technique, all MSs watch at the medium in the same way as CSMA/CD. A MS that is ready to transmit data senses the medium and the collision instead of starting traffic immediately after the medium becomes idle; transmission is deferred to minimize. The MS will transmit its data if the medium is idle for a time interval that exceeds the distributed interframe space (DIFS). Otherwise, it waits for an additional predetermined time period, denoted as DIFS, and then picks a random backoff period within its contention window to wait before transmitting its data. The backoff period is used to initialize the backoff counter. The backoff counter can count down only when the medium is idle. Otherwise, it is frozen as soon as the medium gets busy. After the busy period, counting down of the backoff counter resumes only after the medium has been free longer than DIFS. The MS can start transmitting its data when the backoff counter becomes zero. Collisions can occur only when two or more terminals select the same time slot in which to transmit their frames. Figure 6.13 illustrates this basic mechanism of CSMA/CA.

CSMA/CA with ACK

In this scheme, an immediate positive acknowledgment (ACK) is employed to indicate a successful reception of each data frame (note that explicit ACKs are required since a transmitter cannot listen while transmitting and hence cannot determine if the data frame was successfully received as in the case of wired LANs). This is

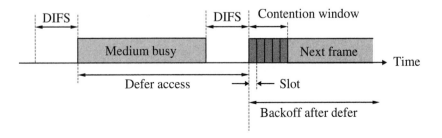

Figure 6.13
Basic CSMA/CA.

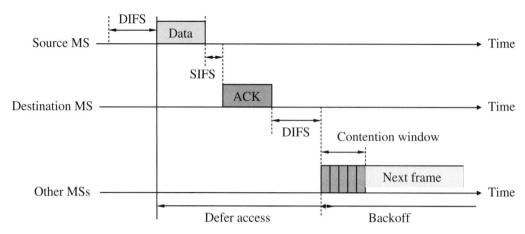

Figure 6.14 CSMA/CA with ACK.

accomplished by making the receiver send an acknowledgment frame immediately after a time interval of short interframe space (SIFS). SIFS is smaller than DIFS, and following the reception of the data frame, the receiver transmits acknowledgment without sensing the state of the medium, as no other MS or device is expected to use the shared medium at that time. In case an ACK is not received, the data frame is presumed to be lost and a retransmission is automatically scheduled by the transmitter. This access method is summarized in Figure 6.14.

CSMA/CA with RTS and CTS

The distributed coordination function (DCF) also provides an alternative way of transmitting data frames by using a special hand-shaking mechanism. It sends request to send (RTS) and clear to send (CTS) frames prior to the transmission of the actual data frame. A successful exchange of RTS and CTS frames attempts to reserve the medium for the entire time duration required to transfer the data frame under consideration within the transmission ranges of sender and receiver. The rules for the transmission of an RTS frame are the same as those for a data frame under basic CSMA/CA (i.e., the transmitter sends an RTS frame after the medium has been idle for a time interval exceeding DIFS). On receiving an RTS frame, the receiver responds with a CTS frame (the CTS frame acknowledges the successful reception of an RTS frame), which can be transmitted after the medium has been idle for a time interval exceeding SIFS. After the successful exchange of RTS and CTS frames, the data frame can be sent by the transmitter after waiting for a time interval SIFS. RTS is retransmitted following the backoff rule as specified in the CSMA/CA with ACK procedures outlined previously. The medium access method using RTS and CTS frames is shown in Figure 6.15.

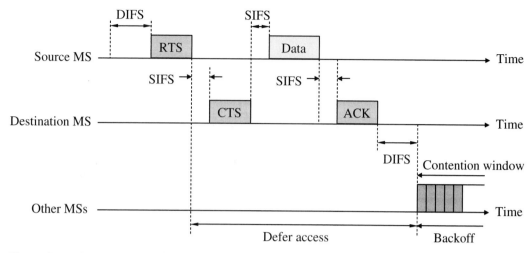

Figure 6.15   CSMA/CA with RTS and CTS.

• • • • • • • • • • • • • • •

6.4 Summary

Controlling access to a shared medium is important from the point of view that, at any given time, only one MS is allowed to talk while the rest of the MSs listen. This kind of scheme is important to avoid the presence of garbled information; such kind of transmission simply wastes bandwidth. This chapter considers numerous ways of minimizing collisions among more than one MS using the same channel in a wireless environment. Such an efficient use of resources is especially important in requesting access to the BS so that the BS can assign exclusive access to individual traffic channels to each requesting MS using one of the multiplex techniques in a wireless cellular system, and this is discussed in the next chapter.

• • • • • • • • • • • • • • •

6.5 References

[6.1] A. S. Tanenbaum, *Computer Networks*, Prentice Hall, Upper Saddle River, NJ, 1988.

[6.2] R. Rom and M. Sidi, *Multiple Access Protocols Performance and Analysis*, Springer-Verlag, New York, 1990.

[6.3] V. O. K. Li, "Multiple Access Communication Networks," *IEEE Communications Magazine*, Vol. 25, No. 6, pp. 41–48, June 1987.

[6.4] N. Abramson, "The ALOHA System—Another Alternative for Computer Communications," *Proc. 1970 Fall Joint Comput. Conf.*, AFIPS Press, Vol. 37, pp. 281–285, 1970.

[6.5] L. G. Roberts, "ALOHA Packet Systems with and Without Slots and Capture," *Computer Communications Review*, Vol. 5, No. 2, pp. 28–42, April 1975.

[6.6] S. S. Lam, "Packet Broadcast Networks—a Performance Analysis of the R-ALOHA Protocol," *IEEE Transactions on Computers*, Vol. 29, No. 7, pp. 596–603, July 1980.

[6.7] L. Kleinrock and F. A. Tobagi, "Packet Switching in Radio Channels: Part I—Carrier Sense Multiple Access Modes and Their Throughput Delay Characteristics," *IEEE Transactions on Communications*, Vol. 23, No. 12, pp. 1400–1416, December 1975.

[6.8] F. A. Tobagi and L. Kleinrock, "Packet Switching in Radio Channels: Part II—The Hidden Terminal Problem in Carrier Sense Multiple Access and the Busy Tone Solution," *IEEE Transactions on Communications*, Vol. 23, No. 12, pp. 1417–1433, December 1975.

[6.9] F. A. Tobagi and V. B. Hunt, "Performance Analysis of Carrier Sense Multiple Access with Collision Detection," *Computer Networks*, No. 4, pp. 245–259, 1980.

[6.10] A. Murase and K. Imamura, "Idle-Signal Casting Multiple Access with Collision Detection (ICMA-CD) for Land Mobile Radio," *IEEE Transactions on Vehicular Technology*, Vol. 36, No. 1, pp. 45–50, February 1987.

[6.11] Z. C. Fluhr and P. T. Poter, "Advance Mobile Phone Service: Control Architecture," *BSTJ*, Vol. 58, No. 1, pp. 43–69, January 1979.

[6.12] S. Okasaka, "Control Channel Traffic Design in a High-Capacity Land Mobile Telephone System," *IEEE Transactions on Vehicular Technology*, Vol. 27, No. 4, pp. 224–231, November 1978.

[6.13] K. Mukumoto and A. Fukuda, "Idle Signal Multiple Access (ISMA) Scheme for Terrestrial Packet Radio Networks," *IEICE Transactions on Communications*, Vol. J64-B, No. 10, October 1981 (in Japanese).

[6.14] G. Wu, K. Mukumoto, and A. Fukuda, "An Integrated Voice and Data Transmission System with Idle Signal Multiple Access—Static Analysis," *IEICE Transactions on Communications*, Vol. E76-B, No. 9, pp. 1186–1192, September 1993.

[6.15] J. I. Capetanakis, "Tree Algorithms for Packet Broadcast Channels," *IEEE Transactions on Information Theory*, Vol. 25, No. 9, pp. 505–515, September 1979.

[6.16] M. Paterakis and P. Papantoni-Kazakos, "A Simple Window Random Access Algorithm with Advantageous Properties," *Proceedings of INFOCOM'88*, pp. 907–915, 1988.

● ● ● ● ● ● ● ● ● ● ● ● ● ● ● ●

6.6 Problems

P6.1. What is the key issue for contention-based access protocols? How is it solved? Give an example to explain your answer.

P6.2. How does slotted ALOHA improve the throughput as compared to pure ALOHA?

P6.3. Is it impractical to use ALOHA or slotted ALOHA for MSs to access a control channel associated with the BS? Explain clearly.

P6.4. What is meant by a collision in data transfer, and why is it not possible to decipher information from collided data? Explain clearly.

P6.5. In a given system with shared access, the probability of n terminals communicating at the same time is given by

$$p(n) = \frac{(1.5G)^n e^{-1.5G}}{(n-1)!},$$

where G is the traffic load in the system. What is the optimal condition for p?

P6.6. What are relative advantages and disadvantages of persistent and non-persistent CSMA protocols? What makes you select one over the other? Explain.

P6.7. Describe the advantages and disadvantages of 1-persistent CSMA and p-persistent CSMA.

P6.8. Can we use CSMA/CD in cellular wireless networks? Explain your answer with solid reasoning.

P6.9. What are the major factors affecting the throughput of CSMA/CA?

P6.10. What is the difference between collision detection and collision avoidance?

P6.11. What are the purposes of using RTS/CTS in CSMA/CA?

P6.12. What are the relative advantages and disadvantages of basic CSMA/CA and CSMA/CA with RTS/CTS protocols? What makes you select one over the other?

P6.13. What in your opinion should be the criteria to select the value of the contention window? Also explain how you will decide the value of the time slot for CSMA/CA.

P6.14. In a CSMA/CA scheme, a random delay is allowed whenever a collision occurs. What is the guarantee that future collision between previously collided terminals will not occur? Explain the rationale behind your answer.

P6.15. Why does the contention window need to be changed sometimes? Explain clearly.

P6.16. In CSMA/CA, why do you need a contention window even after DIFS? What is the typical size of the contention window?

P6.17. Suppose propagation delay is α, SIFS is α, DIFS is 3α, and RTS and CTS are 5α respectively for CSMA/CA with RTS/CTS.

(a) What is the earliest time for the receiver to send the CTS message?

(b) If data packet is 100α long, what is the shortest time for the receiver to send the ACK signal?

(c) Explain why SIFS is kept smaller than DIFS period?

(d) Can you make SIFS $= 0$?

P6.18. In an experiment, the persistent value p is varied as a function of load G, from 1 to 0.5 to 0.1 to 0.01. For what value of G would you have such a transmission? Are there any specific advantages in having such changes? Be specific in your answer.

P6.19. Under the CSMA/CA protocol, suppose there are n users and the contention window for each user is W, then what is the collision probability?

P6.20. The IEEE 802.11x is the popular CSMA/CA protocol employed for wireless LANs and ad hoc networks. Briefly describe all the current 802.11 standards and explain clearly how each is distinct from the other.

P6.21. Look at your favorite Web site and find out what is meant by the hidden terminal problem and the exposed terminal problems. Explain clearly how can you address them.

Multiple Division Techniques

7.1 Introduction

Multiple radio access schemes for wireless networks, discussed in Chapter 6, are primarily used for exchanging control information between a BS and a MS. One of the important control messages sent by a MS is its readiness to send information to the BS and the BS, in turn, advises the MS which particular traffic channel is to be used exclusively by that MS for actual information. Such channel allocation is done for the duration of a call from the MS, and such an assignment is done dynamically as needed so that wireless resources can be used effectively and efficiently. In a wireless environment, a BS needs a radio connection between a BS and all the MSs in their transmission range. Since wireless communication is characterized by wide propagation, there is a need to address the issue of simultaneous multiple access by numerous users in the transmission range. Users can also receive signals transmitted by other users in the system. In fact, many users access the traffic channels when the reverse (uplink) path from MS to BS is to be established. Therefore, it is important for users to distinguish among the different signals. To accommodate a number of users, many traffic channels need to be made available. In principle, there are three basic ways to have many channels within an allocated bandwidth: frequency, time, or code. They are addressed by three multiple division techniques—that is, frequency division multiple access (FDMA), time division multiple access (TDMA), and code division multiple access (CDMA). Two other variants known as orthogonal frequency division multiplexing (OFDM) and space division multiple access (SDMA) have recently being introduced. In this chapter, we introduce these techniques and discuss their relative advantages and disadvantages.

7.2 Concepts and Models for Multiple Divisions

There may be many MSs located in the radio range serviced by a BS. A MS must distinguish which signal is meant for itself among many signals being transmitted by other users or BSs, and the BS should be able to recognize the signal sent by a particular user. In other words, in a wireless cellular system, each MS not only can distinguish a signal from the serving BS but also can discriminate the signals from an adjacent BS. Therefore, a multiple access technique is important in mobile cellular systems. Multiple access techniques are based on the orthogonalization of signals. A radio signal can be presented as a function of frequency, time, or code as

$$s\left(f, t, c\right) = s\left(f, t\right) c\left(t\right), \tag{7.1}$$

where $s\left(f, t\right)$ is a function of frequency and time and $c\left(t\right)$ is a function of code.
When $c\left(t\right) = 1$, equation (7.1) can be replaced by

$$s\left(f, t, c\right) = s\left(f, t\right). \tag{7.2}$$

This constitutes a well-known general expression for the signal as a function of frequency and time.

If a system employs different carrier frequencies to transmit the signal for each user, it is called a FDMA system. If a system uses distinct time slots to transmit the signal for different users, it is a TDMA system. If a system uses different code to transmit the signal for each user, it is a CDMA system. Let $s_i(f, t)$ and $s_j(f, t)$ be two signals being transmitted in the cell space. The orthogonality conditions can be given by using a general mathematical model, and we formally consider them as follows.

In wireless communications, it is necessary to utilize limited frequency bands at the same time, allowing multiple users (MSs) to share radio channels simultaneously. The scheme that is used for this purpose is called multiple access. To provide simultaneous two-way communications (duplex communications), a forward channel (downlink) from the BS to the MS and a reverse channel (uplink) from the MS to the BS are necessary. Two types of duplex systems are utilized: frequency division duplexing (FDD) divides the frequency used, and time division duplexing (TDD) divides the same frequency by time. FDMA mainly uses FDD, while TDMA and CDMA systems use either FDD or TDD. A number of channels can be simultaneously used to transfer data at a much higher rate and such an effective technique is known as OFDM. We consider how these concepts are employed in a mobile communication system.

7.2.1   FDMA

The orthogonality condition of the two signals in FDMA is given by

$$\int_F s_i\left(f, t\right) s_j\left(f, t\right) df = \begin{cases} 1, & i = j \\ 0, & i \neq j \end{cases}, \quad i, j = 1, 2, \dots, k. \tag{7.3}$$

Equation (7.3) indicates that there is no overlapping frequency in frequency domain F for the signals $s_i\left(f, t\right)$ and $s_j(f, t)$ and the two signals do not interfere with each other.

Frequency

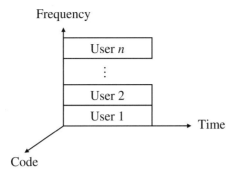

Figure 7.1
The concept of
FDMA.

FDMA is a multiple access system that has been widely adopted in existing analog systems for portable and automobile telephones. The BS dynamically assigns a different carrier frequency to each active user (MS). A frequency synthesizer is used to adjust and maintain the transmission and reception frequencies. The concept of FDMA is shown in Figure 7.1.

Figure 7.2 shows the basic structure of a FDMA system, consisting of a BS and many MSs. There is a pair of channels for the communication between the BS and the MS. The paired channels are called forward channel (downlink) and reverse channel (uplink). Different frequency bandwidths are assigned to different users. This implies that there is no frequency overlapping between the forward and reverse channels. For example, the forward and reverse channels for MS #1 are f_1 and f_1', respectively. The radio antenna is at a much higher elevation and the MSs are shown at the same level in Figure 7.2 while these are not necessarily at the same relative height of each other. Also, if the physical separation between the BS and MSs is drawn to the scale, the MSs will become too small to be represented by a point and all other details will be lost.

The structure of forward and reverse channels in FDMA is shown in Figure 7.3. A protecting bandwidth is used between the forward and reverse channels and a guard band W_g between two adjacent channels (Figure 7.4) is used to minimize adjacent channel interference between them. The frequency bandwidth for each user is called subband W_c. If there are N channels in a FDMA system, the total bandwidth is equal to $N \cdot W_c$.

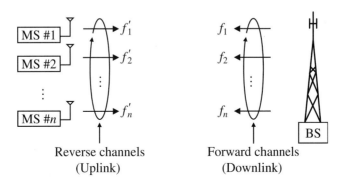

Figure 7.2
The basic structure
of a FDMA system.

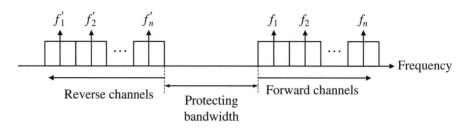

Figure 7.3
Structure of forward
and reverse channels
in FDMA.

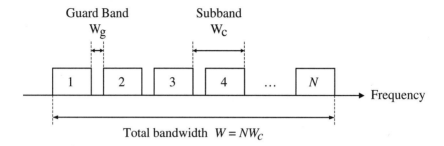

Figure 7.4
Guard band in
FDMA.

7.2.2 TDMA

The orthogonality condition for the signals in TDMA is

$$\int_T s_i\,(f,t)\,s_j\,(f,t)\,dt = \begin{cases} 1, & i=j \\ 0, & i \neq j \end{cases}, \quad i,j = 1,2,\dots,k. \tag{7.4}$$

Equation (7.4) indicates that there is no overlapping time in time axis T for signals $s_i\,(f,t)$ and $s_j(f,\,t)$.

TDMA splits a single carrier wave into several time slots and distributes the slots among multiple users, as shown in Figure 7.5. The communication channels essentially consist of many units, i.e., time slots, over a time cycle, which makes it possible for one frequency to be efficiently utilized by multiple users as each utilizes a different time slot (Figure 7.6). This system is widely used in the field of digital portable and automobile telephones and mobile satellite communication systems.

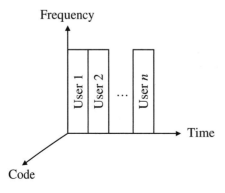

Figure 7.5
The concept of
TDMA.

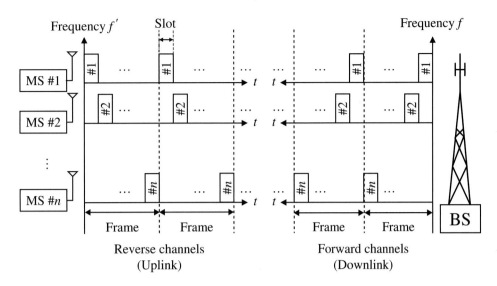

Figure 7.6
The basic structure
of a TDMA system.

A TDMA system may be in either of two modes: FDD (in which the forward/reverse or uplink/downlink communication frequencies differ) and TDD (in which the forward/reverse communication frequencies are the same). That is, TDMA/FDD and TDMA/TDD systems may be as shown in Figures 7.7 and 7.8. Figure 7.9 shows a frame structure of TDMA. For a TDMA system, there is guard time between the slots so that interference due to propagation delays along different paths can be minimized.

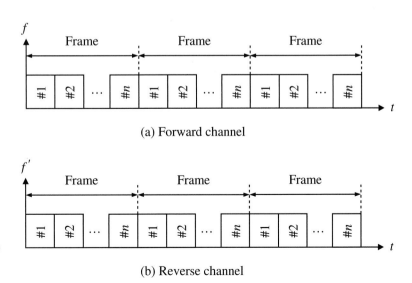

Figure 7.7
Structure of forward
and reverse channels
in a TDMA/FDD
system.

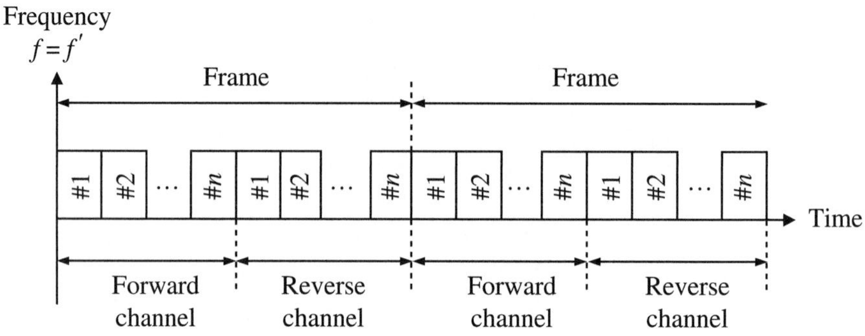

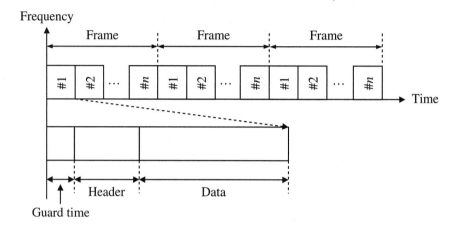

Figure 7.8
Structure of forward and reverse channels in a TDMA/TDD system.

Figure 7.9
Frame structure of TDMA.

A wideband TDMA enables high-speed digital transmissions, in which selective frequency fading due to the use of multiple paths can become a problem. This requires that bandwidth be limited to an extent such that selective fading can be overcome, or appropriate measures such as adaptive equalization techniques could be adopted for improvement. A high-precision synchronization circuit also becomes necessary on the MS side to carry out intermittent burst signal transmission.

7.2.3 CDMA

The orthogonality condition for the signals in CDMA is

$$\int_C s_i(t)\, s_j(t)\, dt = \begin{cases} 1, & i = j \\ 0, & i \neq j \end{cases}, \quad i, j = 1, 2, \ldots, k. \tag{7.5}$$

Equation (7.5) indicates that there is no overlapping of signals in code axis C for signals $s_i(t)$ and $s_j(t)$ and implies that the signals do not have any common codes in the code space.

Thank you for your interest the title overleaf. Please complete the following questi(

teaching needs. Please return your completed form to :-

Thomson Learning, Cheriton House, North Way, Andover, Hants, SP10 5BE, U.K.

☐   I have decided to use the above book for teaching purposes, and it will b

☐   Essential

☐   Recommended

☐   Supplementary

For use in the following course : -

Course Name _____

Course Start Date _____

No of Students per Year _____

Which other titles do you use on this course ? _____

_____

_____

Name(s) and Address(es) of your student's usual Bookseller(s) _____

_____

What do you consider the main strengths and weaknesses of this book?

_____

I can be contacted at work on telephone _____

Email (PLEASE PRINT) _____

☐   Please tick here if you DO NOT wish to receive information from Thoms(

Thank you for completing this form.

Inspectn Advice

Inspection Copy

Page Number	1
AREA	1187

IN CASE OF QUERY PLEASE QUOTE.

Account Number	M010025397 R
Document Number	4334589Y ST
Document Date	13DEC05
Document Type	Inspectn Advice
Payment to be Received By	Payment due 11-FEB-06
VAT No. GB198923209	

W/H

	Price	Disc	Net Value	Despatch Value	Tax	Total Value
	39.99					

Gratis - Payment not required

Total Net Value	Total Desp Val	Tax Total	TOTAL

Terms and Conditions of Supply are available on request

NAIRE ON THE REVERSE

mpshire, SP10 5BE, UK

umber - M010025397   Invoice Number - 4334589Y ST   **Amount GBP**

ns as fully and clearly as possible so we can more adequately cater for your

e :-

_Course Level (eg. UG, PG, MBA) _____Course Year_____

_Course End Date _____

_I estimate _____% of my students are likely to buy this book.

_____

_____

_____

_____

_____

(I may / may not be quoted) _____

_____

_____   extn   _____

_____

n Learning in the future.

THOMSON

Thomson Learning (EMEA) L
Distributed by - **Thomson Publishing Services**, Cheriton House, North Way,
Customer Service Contacts.
Telephone - 01264 342932 , Fax - 01264 3

INVOICE TO *Rep ½ 0 a note of it in my book mala*

DELIVER TO

LEEDS METROPOLITAN UNIVERSITY
SCHOOL OF COMPUTING
BECKETT PARK CAMPUS
LEEDS
WEST YORKSHIRE LS6 3QS

Fash Safdar
Leeds Metro
School of C
The Grange
Beckett Parl
Leeds
LS6 3QS

12GU5A

Shipper -

Line	Order Reference	ISBN	QTY	Title / Author Details
1		1401886590	1	Customer Tel No: +4401132832600 WIRELESS COMMUNICATION/MULLETT 1-64A6TC

Order Lines - 1	**Net Weight -** 1.141	**Total Quantity -** 1
OP CCC Batch 88943 /001	**Document Type - ICS2**	**Entered Date 13DEC05**
Despatch Method - SECURICOR B		**Site - Andover - T.L.**

PLEASE COMPLETE THE QUESTION

All Orders, Claims, Returns and Enquiries should be addressed to:

Thomson Publishing Services, Dept ICS, Cheriton House, North Way, Andover, H

PAYMENT METHODS (Please ensure you quote your account number.) **Account**

CHEQUE
Please make payable to Thomson Publishing Services and attach to this slip.
GIROBANK
Sterling Account Number - 1028466
CREDIT TRANSFER
Please Quote Thomson Publishing Services and Invoice Number.
Bankers - The Royal Bank of Scotland PLC Corporate Banking Office, PO Box 450,
5 - 10 Great Tower Street, London, EC3P 3HX.
  STERLING Sort Code: 16-04-00 , Account Number: 20121434
  STERLING IBAN number : GB03RBOS16040020121434
  EURO Sort Code: 16-04-00 , Account Number: ITHPUSE EURC
  EURO IBAN number : GB37RBOS16106510002043

SWIFT
Please Quote SWIFT Number: RB
Bankers - The Royal Bank of Sco
P O Box 348, 42 Is
Acco
Bra
Co bra

CREDIT or CHARGECARD -

☐☐☐☐ - ☐☐

Expiry Date _____ Am

Please send your payment to - Thomson Publishing Services, P O Box 1633, Cheriton House, North Way, Andover, Ham

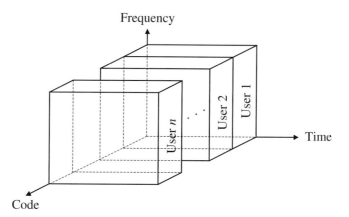

Figure 7.10
The concept of CDMA.

In a CDMA system, different spread-spectrum codes are selected and assigned to each user, and multiple users share the same frequency, as shown in Figures 7.10 and 7.11. A CDMA system is based on spectrum-spread technology, which makes it less susceptible to the noise and interference by substantially spreading over the bandwidth range of modulated signal. In addition, because of its broadband characteristics, fading resistance can be achieved by the RAKE multipath synthesis. Reserving a wider bandwidth for a single communication channel was once regarded as disadvantageous in terms of effective frequency utilization. However, high efficiency of frequency usage has been demonstrated by using CDMA since the introduction of power control enable us to adjust the antenna emitting power so that the near-far problem could be solved. In a general CDMA system, received signals at the BS from a far away MS could be masked by signals from a close-by MS in the reverse channel. As a consequence, CDMA is the multiple access system that is now attracting the most attention as a core technology for the next generation mobile communications

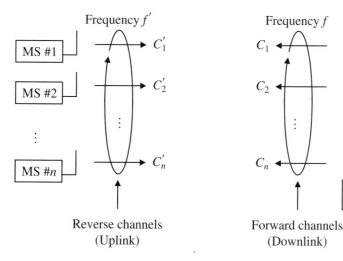

Figure 7.11
Structure of a CDMA system.

system. A CDMA system is usually quantified by the chip rate which is defined as the number of bits changed per second. Chip rate is usually applied to CDMA systems.

There are two basic types of CDMA implementation methodologies: direct sequence (DS) and frequency hopping (FH). Since it is difficult to use the FH on a practical basis unless a super-fast synthesizer is employed, the DS is considered the most feasible generic method when the code is selected and assigned dynamically to each MS.

Spread Spectrum

Spread spectrum is a transmission technique wherein data occupy a larger bandwidth than necessary. Bandwidth spreading is accomplished before the transmission through the use of a code that is independent of the transmitted data. The same code is used to demodulate the data at the receiving end. Figure 7.12 illustrates the spreading done on the data signal $s(t)$ by the code signal $c(t)$ resulting in the message signal to be transmitted, $m(t)$. That is,

$$m(t) = s(t) \otimes c(t). \tag{7.6}$$

Originally designed for military use to avoid jamming (interference created intentionally to make a communication channel unusable), spread spectrum modulation is now used in personal communication systems also due to its superior performance in an interference dominated environment.

Direct Sequence Spread Spectrum (DSSS)

In a DSSS method, the radio signal is multiplied by a pseudorandom sequence whose bandwidth is much greater than that of the signal itself, thereby spreading its bandwidth (Figure 7.13). This is a modulation technique wherein a pseudorandom sequence directly phase modulates a (data-modulated) carrier, thereby increasing the bandwidth of the transmission and lowering the spectral power density (i.e., the power level at any given frequency). The resulting RF signal has a noiselike spectrum, and in fact can be intentionally made to look like noise to all but the intended radio receiver. The received signal is despread by correlating it with a local pseudorandom sequence identical to and in synchronization with the sequence used to spread the carrier at the radio transmitting end.

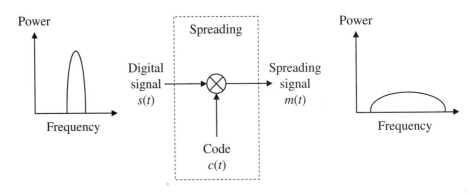

Figure 7.12
Spread spectrum.

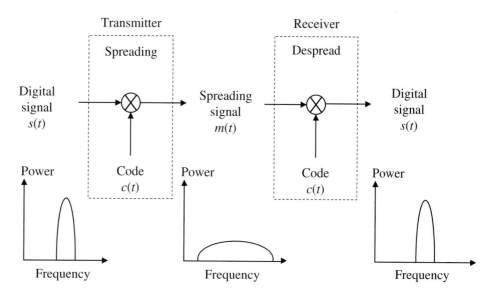

Figure 7.13
Concept of direct sequence spread spectrum.

Frequency Hopping Spread Spectrum (FHSS)

In a FH method, a pseudorandom sequence is used to change the radio signal frequency across a broad frequency band (Figure 7.14) in a random fashion. A spread spectrum modulation technique implies that the radio transmitter frequency hops from channel to channel in a predetermined but pseudorandom manner. The RF signal is dehopped at the receiver end using a frequency synthesizer controlled by

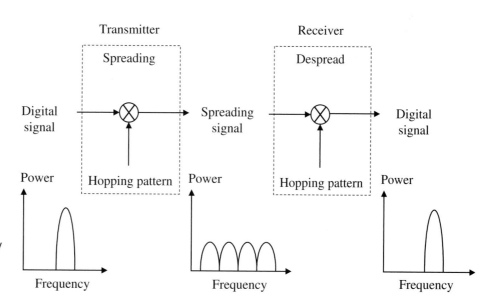

Figure 7.14
Concept of frequency hopping spread spectrum system.

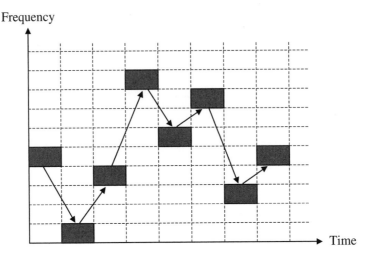

Frequency

Time

Figure 7.15
An example of frequency hopping pattern.

a pseudorandom sequence generator synchronized to the transmitter's pseudorandom sequence generator. A frequency hopper may be fast hopped, where there are multiple hops per data bit, or slow hopped, where there are multiple data bits per hop. Figure 7.15 shows an example of a frequency hopping pattern. Multiple simultaneous transmission from several users is possible using FH, as long as each uses different frequency hopping sequences and none of them "collide" (no more than one unit using the some band) at any given instant of time.

Walsh Codes

In CDMA, each user is assigned one or many orthogonal waveforms derived from one orthogonal code. Since the waveforms are orthogonal, users with different codes do not interfere with each other. CDMA requires synchronization among the users, since the waveforms are orthogonal only if they are aligned in time. An important set of orthogonal codes is the Walsh set (see Figure 7.16).

Walsh functions are generated using an iterative process of constructing a Hadamard matrix starting with $H_0 = [0]$. The Hadamard matrix is built by using the function

$$H_n = \begin{pmatrix} H_{n-1} & H_{n-1} \\ H_{n-1} & H_{n-1} \end{pmatrix}. \tag{7.7}$$

Near-Far Problem

The near-far problem stems from a wide range of signal levels received in wireless and mobile communication systems. We consider a system in which two MSs are communicating with a BS, as illustrated in Figure 7.17. If we assume the transmission power of each MS to be the same, received signal levels at the BS from the MS_1 and MS_2 are quite different due to the difference in the path lengths. Let us assume that the MSs are using adjacent channels, as shown in Figure 7.18. Out-of-band radiation of the signal from the MS_1 interferes with the signal from the MS_2 in the

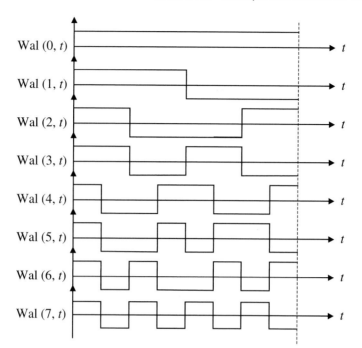

Figure 7.16
Walsh codes.

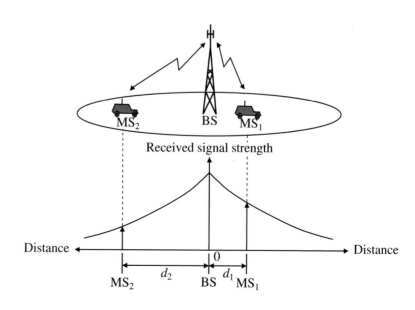

Received signal strength

Figure 7.17
Near-far problem.

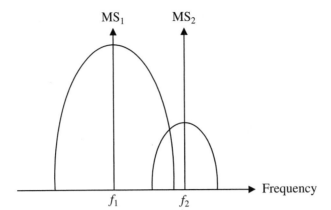

Figure 7.18
Adjacent channel
interference.

adjacent channel. This effect, called adjacent channel interference, becomes serious when the difference in the received signal strength is high. For this reason, the out-of-band radiation must be kept small. The tolerable relative adjacent channel interference level can be different depending on the system characteristics. If power control technique is used, the system can tolerate higher relative adjacent channel interference levels. The near-far problem becomes more important for CDMA systems where spread spectrum signals are multiplexed on the same frequency using low crosscorrelation codes, as shown in Figure 7.19. In CDMA, a real question is how to address the near-far problem. One simple solution is power control and is considered next.

Power Control

Power control is simply the technique of controlling the transmit power so as to affect the received power, and hence the CIR. For example, in free space, the propagation

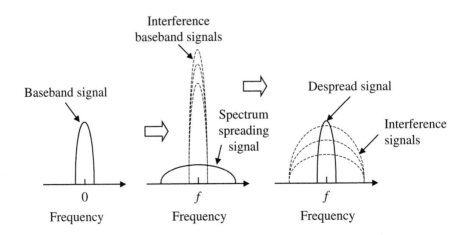

Figure 7.19
Interference in
spread spectrum
system.

path loss depends on the frequency of transmission, f, and the distance between transmitter and receiver, d, as follows:

$$\frac{P_r}{P_t} = \frac{1}{\left(\frac{4\pi df}{c}\right)^\alpha}, \tag{7.8}$$

where P_t is the transmitted power, P_r is the received power in free space, c is the speed of light, and α is an attenuation constant.

Assuming that the interference remains constant, a desired P_r (and thus a desired CIR) can be attained by adjusting the transmit power P_t appropriately. Note that this can be done by observing currently transmitted and received power, if we assume that the distance d does not change significantly between the time of observation and the adjustment of P_t.

While power control can often be effective, there are some disadvantages. First, since battery power at a MS is a limited resource that needs to be conserved, it may not be possible or desirable to set transmission powers to higher values. Second, increasing the transmitted power on one channel, irrespective of the power levels used on other channels, can cause inequality of transmission over other channels. As a result, there is also the possibility that a set of connections using a pure power control scheme can suffer from unstable behavior, requiring increasingly higher transmission powers. Finally, power control techniques are restricted by the physical limitations on the transmitter power levels.

7.2.4 OFDM

The basic strategy in OFDM is to split high rate radio signals into multiple lower rate subsignals that are then simultaneously transmitted over multiple orthogonal carrier frequencies. The orthogonality condition of the two signals in OFDM can be given by [?prasad-98?]

$$\int_F s_i\,(f,t)\,s_j^*\,(f,t)\,dt = \begin{cases} 1, & i = j \\ 0, & i \neq j \end{cases}, \quad i, j = 1, 2, \ldots, k, \tag{7.9}$$

where $*$ means a complex conjugate relation.

It has been proved mathematically that sinusoidal waves are orthogonal over an interval of integer number of periods T. Figure 7.20 illustrates the spectrum of an OFDM signal; if there is no crossing of other channels at the center frequency of each subcarrier in the frequency domain, the ISIs (intersymbol interferences) would be zero.

The transmitter of OFDM converts high-speed data streams into n parallel low-speed bit streams and then modulated, mixed with inverse discrete Fourier transform (IDFT) and then guard time inserted to reduce ISI. The reverse actions are taken at the receiver side. Figure 7.21 illustrates the modulation operation of the OFDM transmitter, while Figure 7.22 shows the demodulation steps of the OFDM receiver, with explicit use of the discrete Fourier transform (DFT).

In all these systems, the information is first modulated before being transmitted over a channel. In the next section, we consider several useful modulation techniques.

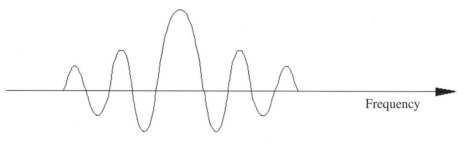

(a) Single OFDM subchannel

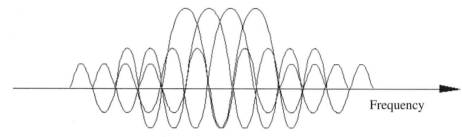

Figure 7.20
The frequency spectrum
of an OFDM signal.

(b) An OFDM signal with multiple subchannels

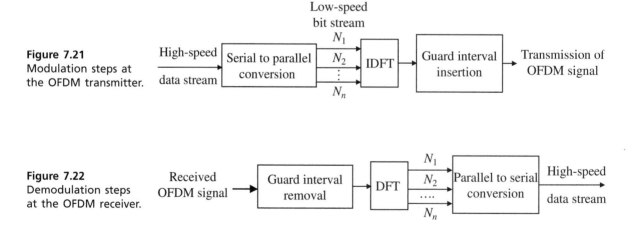

Figure 7.21
Modulation steps at
the OFDM transmitter.

Figure 7.22
Demodulation steps
at the OFDM receiver.

7.2.5 SDMA

In SDMA, the omni-directional communication space is divided into spatially separable sectors. This is possible by having a BS use smart antennas, allowing multiple MSs to use the same channel simultaneously. The communication characterized by

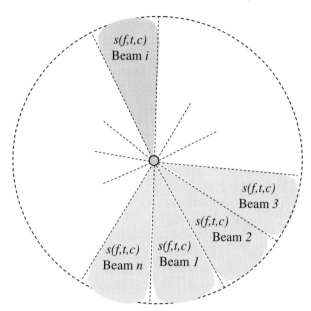

Figure 7.23
The concept of SDMA.

either time slot, carrier frequency or spreading code can be used as shown in Figure 7.23. Use of a smart antenna maximizes the antenna gain in the desired direction and directing antenna gain in a particular direction leads to range extension, which reduces the number of cells required to cover a given area. Moreover, such focused transmission reduces the interference from undesired directions by placing minimum radiation patterns in the direction of interferers.

A simplified version of transmission using SDMA is illustrated in Figure 7.24. As the BS forms different beams for each spatially separable MS on the forward and reverse channels, noise and interference for each MS and BS is minimized. This enhances the quality of the communication link significantly and increases overall

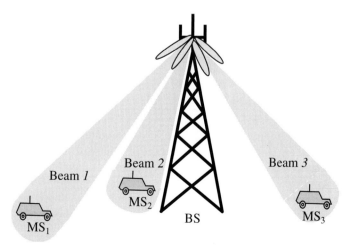

Figure 7.24
The basic structure of a SDMA system.

system capacity. Also, by creating separate spatial channels in each cell intra-cell reuse of conventional channels can be easily exploited. Currently, this technology is still being explored and its future looks quite promising.

7.2.6   Comparison of Multiple Division Techniques

SDMA is generally used in conjunction with other multiple access schemes as there can be more than one MS in one beam. With TDMA and CDMA, different areas can be covered by the antenna beam, providing frequency reuse. Also, when used with TDMA and FDMA, the higher CIR ratio due to smart antennas can be exploited for better frequency channel reuse. With CDMA the user can transmit less power for each link, thereby reducing MAC interference and hence supporting more users in the cell. However, there will be more intra-cell handoffs in SDMA as compared to TDMA or CDMA systems, requiring a closer watch at the network resource management. Table 7.1 shows a comparison of various multiple access schemes.

Table 7.1: ▶
Comparison of Various Multiple Division Techniques

Technique	FDMA	TDMA	CDMA	SDMA
Concept	Divide the frequency band into disjoint subbands	Divide the time into non-overlapping time slots	Spread the signal with orthogonal codes	Divide the space into sectors
Active terminals	All terminals active on their specified frequencies	Terminals are active in their specified slot on same frequency	All terminals active on same frequency	Number of terminals per beam depends on FDMA/TDMA/CDMA
Signal separation	Filtering in frequency	Synchronization in time	Code separation	Spatial separation using smart antennas
Handoff	Hard handoff	Hard handoff	Soft handoff	Hard and soft handoffs
Advantages	Simple and robust	Flexible	Flexible	Very simple, increases system capacity
Disadvantages	Inflexible, available frequencies are fixed, requires guard bands	Requires guard space, synchronization problem	Complex receivers, requires power control to avoid near-far problem	Inflexible, requires network monitoring to avoid intracell handoffs
Current applications	Radio, TV and analog cellular	GSM and PDC	2.5G and 3G	Satellite systems, other being explored

7.3 **Modulation Techniques**

7.3.1 AM

Amplitude modulation (AM) is the first method ever used to transfer voice information from one place to another. The amplitude of a carrier signal with a constant frequency is varied as the information signal required to transmit. The total power of the transmitted wave varies in amplitude in accordance with the power of the modulating signal. Mathematically, the modulated carrier signal $s(t)$ is

$$s(t) = [A + x(t)] \cos{(2\pi f_c t)}, \tag{7.10}$$

where $A \cos{(2\pi f_c t)}$ is the carrier signal with amplitude A and carrier frequency f_c, and $x(t)$ is the modulating signal. A is the direct current (dc) portion of the signal. We know that $x(t) \cos{(2\pi f_c t)}$ represents a double sideband (DSB) signal. Figure 7.25 shows the AM waveforms.

The bandwidth of an AM scheme—that is, the amount of space that it occupies in the Fourier domain—is twice that of the modulating signal. This double sideband nature of AM halves the number of independent signals that can be sent using a given range of transmission frequencies. By suppressing one sideband before transmission, single sideband (SSB) modulation doubles the number of transmissions that can fit into a given transmission band.

At the receiver end, the carrier signal is filtered out, rebuilding the information signal (speech, data, etc.). When a carrier is amplitude modulated with a pure sine

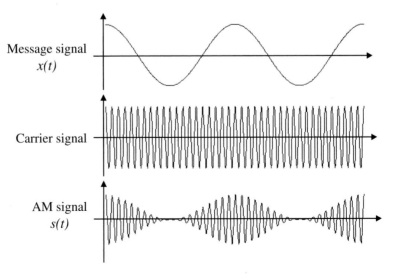

Figure 7.25
Amplitude modulation.

wave, up to one-third (33.3%) of the overall signal power is contained in the side-bands. The other two-thirds of the signal power are contained in the carrier, which does not contribute to the transfer of data. This makes AM an inefficient mode of communication.

7.3.2   FM

Frequency modulation (FM) is a method of integrating the information signal with an alternating current (ac) wave by varying the instantaneous frequency of the wave. The carrier is stretched or squeezed by the information signal, and the frequency of the carrier is changed according to the value of the modulating voltage. Thus, the signal that is transmitted is of the form

$$s(t) = A \cos \left(2\pi f_c t + 2\pi f_\Delta \int_{t_0}^{t} x(\tau)\, d\tau + \theta_0 \right), \tag{7.11}$$

where f_Δ is the peak frequency deviation that is the farthest away from the original frequency that the FM signal can be with the condition $f_\Delta \ll f_c$. Figure 7.26 shows the FM waveforms.

The carrier frequency varies between the extremes of $f_c + f_\Delta$ and $f_c - f_\Delta$. The index of modulation of FM is defined as $\beta = f_\Delta / f_m$, where f_m is the maximum modulating frequency used. In FM, the total wave power does not change when the frequency alters. To recover the signal, the receiver rebuilds the information wave by checking how the known carrier signal has modified the information. A FM system provides a better SNR than an AM system, which implies that it has less noise content. Another advantage is that it needs less radiated power. However, it does require a larger bandwidth than AM. The bandwidth (BW) of a FM signal may be determined using

$$\text{BW} = 2(\beta + 1) f_m. \tag{7.12}$$

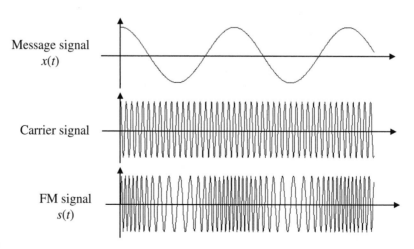

Figure 7.26
Frequency modulation.

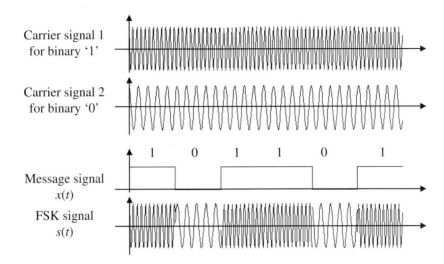

Figure 7.27
Frequency shift
keying.

7.3.3   FSK

Frequency shift keying (FSK) is used for modulating a digital signal over two carriers by using a different frequency for a "1" or a "0". The difference between the carriers is known as the frequency shift. The waveforms of FSK are shown in Figure 7.27.

One obvious way to generate a FSK signal is to switch between two independent oscillators according to whether the data bit is a "1" or a "0". This type of FSK is called discontinuous FSK since the waveform generated is discontinuous at the switching time. The phase discontinuity poses several problems, such as spectral spreading and spurious transmissions. A common method of generating an FSK signal is to frequency modulate a single carrier oscillator using the message waveform. This type of modulation is similar to FM generation, except that the modulating signal is in binary [7.1]. FSK has high signal-to-noise ratio (SNR) but low spectral efficiency. It was used in all early low bit-rate modems.

7.3.4   PSK

In digital transmission, the phase of the carrier is discretely varied with respect to a reference phase and according to the data being transmitted. Phase shift keying (PSK) is a method of transmitting and receiving digital signals in which the phase of a transmitted signal is varied to convey information. For example, when encoding, the phase shift could be $0°$ for encoding a "0" and $180°$ for encoding a "1" thus making the representations for "0" and "1" apart by a total of $180°$. This kind of PSK is also called binary phase shift keying (BPSK) since 1 bit is transmitted in a single modulation symbol. Figure 7.28 shows the waveforms of BPSK.

PSK has a perfect SNR but must be demodulated synchronously, which means a reference carrier signal is required to be received at the receiver to compare with the phase of the received signal, which makes the demodulation circuit complex.

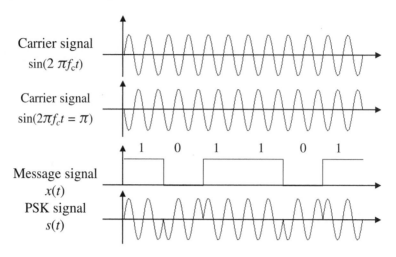

Carrier signal $\sin(2\pi f_c t)$

Carrier signal $\sin(2\pi f_c t = \pi)$

Message signal $x(t)$

PSK signal $s(t)$

Figure 7.28
Phase shift keying.

7.3.5 QPSK

Quadrature phase shift keying (QPSK) moves the concept of PSK a step further as it assumes that the number of phase shifts is not limited to only two states. The transmitted carrier can undergo any number of phase changes. This is indeed the case in quadrature phase shift keying. With QPSK, the carrier undergoes four changes in phase and can thus represent four binary bit patterns of data, effectively doubling the bandwidth of the carrier. The following are the phase shifts with the four different combinations of input bits [7.2].

$$\begin{cases} \phi_{0,0} = 0 \\ \phi_{0,1} = \frac{\pi}{2} \\ \phi_{1,0} = \pi \\ \phi_{1,1} = \frac{3\pi}{2} \end{cases} \quad \text{or} \quad \begin{cases} \phi_{0,0} = \frac{\pi}{4} \\ \phi_{0,1} = \frac{3\pi}{4} \\ \phi_{1,0} = -\frac{3\pi}{4} \\ \phi_{1,1} = -\frac{\pi}{4} \end{cases}$$

Normally, QPSK is implemented using I/Q modulation with I (in-phase) and Q (quadrature) signals summarized with respect to the same reference carrier signal (in other words, from the same local oscillator). A 90° phase offset is placed in one of the carriers. Suppose input sequence d_k ($k = 0, 1, 2, \ldots$) arrives at the modulator at a rate of R_b and is separated into two data streams $d_I(t)$ and $d_Q(t)$ containing odd and even bits, respectively. Then, $d_I(t)$ and $d_Q(t)$ have a bit rate of $R_s = R_b/2$. For example, if $d_k = [1, 0, 1, 1]$, then $d_I(t) = [d_0, d_2] = [1, 1]$ and $d_Q(t) = [d_1, d_3] = [0, 1]$.

We can consider each of the two binary sequences to be a BPSK signal. The two binary sequences are separately modulated by the two quadrate signals. The summation of the two modulated waveforms is the QPSK waveform, and the phase shift also has four states corresponding to every two adjacent input bits. Figure 7.29 shows the constellations of BPSK and QPSK.

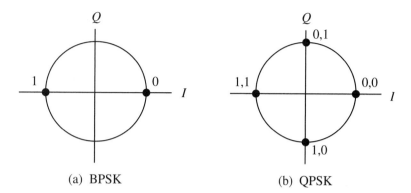

Figure 7.29
Signal constellations
of BPSK and QPSK.

(a) BPSK   (b) QPSK

7.3.6 $\pi/4$QPSK

In QPSK and BPSK, the input sequence is encoded in the absolute position in the constellation. In $\pi/4$QPSK, the input sequence is encoded by the changes in the amplitude and direction of the phase shift and not in the absolute position in the constellation. $\pi/4$QPSK uses two QPSK constellations offset by $\pm\pi/4$. Signaling elements are selected in turn from the two QPSK constellations. Transitions must occur from one constellation to the other one. This ensures that there will always be a phase change for each symbol. Therefore, $\pi/4$QPSK can be noncoherently demodulated, which simplifies the design of the demodulator.

In $\pi/4$QPSK, the phase of the carrier is

$$\theta_k = \theta_{k-1} + \phi_k, \tag{7.13}$$

where ϕ_k is the carrier phase shifts corresponding to the input bit pairs [7.1].

For example, if $\theta_0 = 0$, input bit steam is $[1011]$, then

$$\theta_1 = \theta_0 + \phi_1 = -\frac{\pi}{4},$$

$$\theta_2 = \theta_1 + \phi_2 = -\frac{\pi}{4} + \frac{\pi}{4} = 0.$$

From the preceding example, we can see that the information in the input sequence is completely contained in the phase difference of the modulated waveform corresponding to two adjacent symbols. (In the preceding example, the two adjacent symbols are $[1, 0]$ and $[1, 1]$.)

Figure 7.30 shows all possible state transitions in $\pi/4$QPSK.

$\pi/4$QPSK is popular in most second-generation systems, such as North American Digital Cellular (IS-54) and Japanese Digital Cellular (JDC).

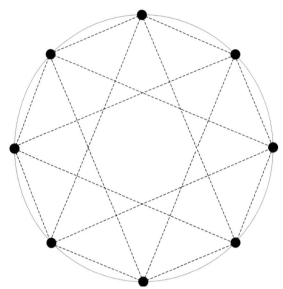

Figure 7.30
All possible state
transitions in $\pi/4$QPSK.

7.3.7 QAM

Quadrature amplitude modulation (QAM) is simply a combination of AM and PSK, in which two carriers out of phase by 90° are amplitude modulated. If the baud rate is 1200 Hz, 3 bits per baud, a signal can be transmitted at 3600 bps. We modulate the signal by using two measures of amplitude and four possible phase shifts. Combining the two, we have eight possible waves (Table 7.2).

Mathematically, there is no limit to the data rate that may be supported by a given baud rate in a perfectly stable, noiseless transmission environment. In practice,

Table 7.2: ▶
A Representative QAM Table

Bit sequence represented	Amplitude	Phase shift
000	1	0
001	2	0
010	1	$\pi/2$
011	2	$\pi/2$
100	1	π
101	2	π
110	1	$3\pi/2$
111	2	$3\pi/2$

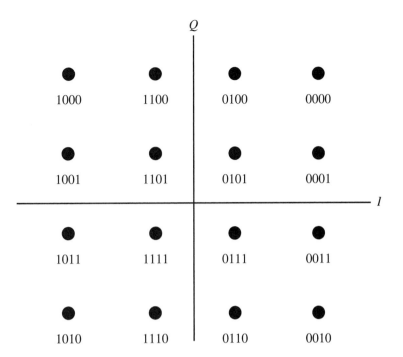

Figure 7.31
Rectangular
constellation of
16QAM.

the governing factors are the amplitude (and, consequently, phase) stability, and the amount of noise present, in both the terminal equipment and the transmission medium (carrier frequency, or communication channel) involved.

7.3.8   16QAM

16QAM involves splitting the signal into twelve different phases and three different amplitudes for a total of 16 different possible values, each encoding 4 bits. Figure 7.31 shows the rectangular constellation of 16QAM.

16QAM is used in applications including microwave digital radio, DVB-C (Digital Video Broadcasting—Cable), and modems. 16QAM or other higher-order QAMs (64QAM, 256QAM) are more bandwidth efficient than BPSK, QPSK, or 8PSK and are used to gain high-speed transmission. However, there is a tradeoff, and the radio becomes more complex and is more susceptible to errors caused by noise and distortion. Error rates of higher-order QAM systems degrade more rapidly than QPSK as noise or interference is introduced. A measure of this degradation would be a higher BER.

7.4 Summary

Communication channels are used by system subscribers, and there are many ways they can be used effectively using different multiplexing techniques. Problems and limitations using such resources have been discussed and their relative advantages and disadvantages have been outlined in this chapter. Various modulation techniques have also been described. It is important to understand how the overall system works and how traffic from multiple MSs is supported by the limited number of channels available in a wireless system. These topics are considered in the next chapter.

● ● ● ● ● ● ● ● ● ● ● ● ● ● ●

7.5 **References**

[7.1] T. S. Rappaport, *Wireless Communications: Principles & Practice*, Prentice Hall, NJ, 1996.

[7.2] J. G. Proakis and M. Salehi, *Communication System Engineering*, Prentice Hall, NJ, 1994.

● ● ● ● ● ● ● ● ● ● ● ● ● ● ●

7.6 **Problems**

P7.1. What is the difference between the guard band and the guard time and why are they important in a cellular system? Explain clearly.

P7.2. A TDMA system uses a 270.833 kbps data rate to support eight users per frame.
(a) What is the raw data rate provided for each user?
(b) If guard time and synchronization occupy 10.1 kbps, determine the traffic efficiency.
(c) If (7, 4) code is used for error handling, what is the overall efficiency?

P7.3. Radio signal travels from the BS to an MS along different paths, some direct, some reflected, and some deflected. If the worst-case difference in the path length traversed by a signal is 2 km, what is the minimum value of guard time that must be used? Assume a signal propagation rate of 512 kbps.

P7.4. Repeat Problem P7.2 if only four users per frame can be supported.

P7.5. Repeat Problem P7.3 if the difference in path length is 4 km.

P7.6. Find the Walsh functions for 16-bit code.

P7.7. What are the orthogonal Walsh codes? Why is synchronization among the users required for CDMA?

P7.8. Is it possible to jam CDMA? Explain clearly.

P7.9. To address the service to be increased in the number of MSs in a CDMA system, it was decided to use TDMA as well. Is it possible to do so? If yes, how; and if no, why not?

P7.10. The number of Walsh codes determines the maximum number of MSs that can be serviced simultaneously. Then, why not use a large Walsh code? What are the limitations or disadvantages? Explain clearly (range of Walsh code is 28–128 bits).

P7.11. What are the nonmilitary applications of frequency hopping? Why is Bluetooth used in home devices and for a wireless computer mouse?

P7.12. What frequency band is used in biomedical devices for surgical applications? How does that limit the use of wireless devices?

P7.13. What is FSK/QPSK?

P7.14. Why does power control become one of the main issues for the efficient operation of CDMA?

P7.15. How do you decide the range of a guard channel? Is it a function of the carrier frequency? Explain clearly.

P7.16. The message signal $x(t) = \sin(100t)$ modulates the carrier signal $c(t) = A\cos(2\pi f_c t)$. Using amplitude modulation, find the frequency content of the modulated signal.

P7.17. A signal shown below amplitude modulates a carrier $c(t) = \cos(50t)$. Precisely plot the resulting modulated signal as a function of time.

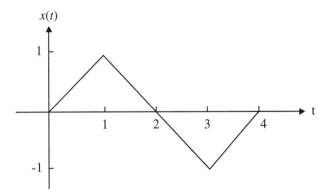

Figure 7.32
Figure for
Problem P7.17.

P7.18. The message signal is given by $x(t) = \cos(20\pi t)$ and the carrier is given by $c(t) = \cos(2\pi f_c t)$. Use frequency modulation. The modulation index is 5.

(a) Write an expression for the modulated signal.

(b) What is the maximum frequency deviation of the modulated signal?

(c) Find the bandwidth of the modulated signal.

P7.19. Besides BPSK and QPSK, 8PSK is another kind of phase shift keying. Try to give the constellation for 8PSK.

P7.20. Use 16QAM to transmit a binary sequence, if the baud rate is 1200 Hz, how many bits can be transmitted in one second?

P7.21. Increasing the amount of amplitude level and phase shift, we can gain higher level xQAM, such as 64QAM and 256QAM. It seems the transmission rate can be as high as we want by this kind of modulation. Is that true? Explain briefly.

Channel Allocation

● ● ● ● ● ● ● ● ● ● ● ● ● ● ● ● ●

8.1 Introduction

Channel allocation in a cellular system is important from the performance point of view. Channel allocation usually covers how a BS should assign traffic channels to the MSs. Here we used the term channel instead of traffic channel. As the channels are only managed by the BS of a cell, A MS attempting to make a new call needs to submit a request for a channel. The BS can grant such an access to the MS provided that a channel is readily available for use by the BS. If this is possible, most of the time the probability that a new call will be blocked or the blocking probability for call originated in a cell can be minimized. One way to ascertain such a radio resource to be free is to increase the number of channels per cell. If this is done, then every cell would expect to have a larger number of channels. However, because a limited frequency band is allocated for wireless communication, there is a limit to the maximum number of channels, thereby restricting the number of available channels that can be assigned to each cell, especially for FDMA/TDMA–based systems. Channel allocation implies that a given radio spectrum is to be divided into a set of disjoint channels, which can be used simultaneously by different MSs, while interference in adjacent channels could be minimized by having good channel separation. One simplistic approach is to divide the channels equally among the cells in FDMA/TDMA–based systems and use appropriate reuse distance to minimize interference. Such an allocation could easily handle a user's calls if the system load is uniformly distributed. Consider a case where channels are equally partitioned among cells of a cluster. If S_{total} is the total number of channels and N is the size of the reuse cluster, then the number of channels per cell is

$$S = \frac{S_{total}}{N}. \tag{8.1}$$

For example, if $S_{total} = 413$ and reuse cluster size $N = 7$ (*i.e.*, seven cells make up a cluster), then $S = 59$, the number of channels per cell. Looking at such relation, we may think that reducing the value of N (which goes against the philosophy of reuse) might increase the number of channels per cell. This, in turn, reduces the reuse distance, which can also increase interference. Therefore, another option is to allocate channels to different cells according to their traffic load. However, it is hard to predict instantaneous traffic, even if we have past statistical information about the calls made in each cell. Therefore, it is reasonable to assign an equal number of channels to each cell. In the ideal situation, all parameters would be assumed to be the same, appropriate action could be taken later on. This means that the location of MSs over an area is considered uniformly distributed and the probability of each MS making a call is also assumed to be the same; external conditions such as terrain and presence of hills, tall buildings, and valleys are also assumed to be of the same type. Such assumptions are unrealistic, and alternative solutions must be explored to address the irregular traffic load present in any real wireless system. An excellent survey dealing with channel assignment schemes has been published [8.1]. It may be noted that the CDMA–based system could be equated to FDMA/TDMA–based systems if the number of possible codes, reflecting the number of possible simultaneous calls per cell, can be said to the number of channels in FDMA/TDMA. Therefore, many of the conclusions are equally applicable to CDMA as well.

● ● ● ● ● ● ● ● ● ● ● ● ● ● ⋯

8.2 Static Allocation versus Dynamic Allocation

There are two ways by which traffic channels can be allocated to different cells in a FDMA/TDMA cellular system: static and dynamic. In static allocation, a fixed number of channels is allocated to each cell, while dynamic allocation implies that allocation of channels to different cells is done dynamically, as needed, possibly from a central pool. There are many possible variations of channel allocation, each having specific characteristics and offering different advantages. Even within a static scheme, an equal number of channels can be allocated to each cell, or nonuniform fixed channel allocation (FCA) could be done based on the amount of traffic in different cells (which is based on past statistical information). Another alternative is to combine some aspects of both FCA and dynamic channel allocation (DCA) schemes.

In brief, channel allocation schemes can be classified as follows:

1. Fixed channel allocation (FCA)

2. Dynamic channel allocation (DCA)

3. Hybrid channel allocation (HCA) [8.2]

There are many alternatives within each scheme, and some of the important ones are considered in this chapter.

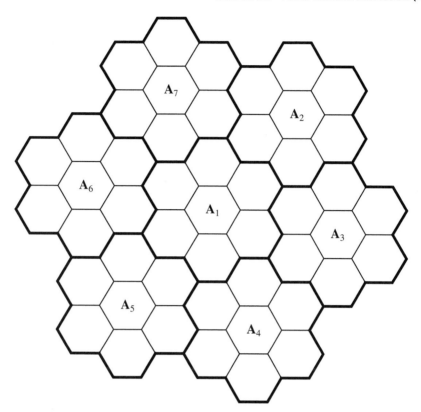

Figure 8.1
Impact of channel
borrowing by cluster
A_1 on adjacent
clusters within reuse
distance.

• • • • • • • • • • • • • • • •

8.3 Fixed Channel Allocation (FCA)

In FCA schemes, a set of channels is permanently allocated to each cell of the system. If the total number of available channels in the system is divided into sets, the minimum number of channel sets required to serve the entire coverage area is related to the frequency reuse distance D and radius R of each cell as follows:

$$\sqrt{N} = \frac{D}{\sqrt{3}R}. \tag{8.2}$$

One approach to address increased traffic of originating and handoff calls in a cell is to borrow free channels from neighboring cells. For example, in the seven-cell-based cluster scheme shown in Figure 8.1, if a cell of a cluster A_1 borrows channels from cells of adjacent clusters, we need to make sure that there is no interference with cells associated with clusters A_2, A_3, A_4, A_5, A_6, and A_7, which are within reuse distance of cluster A_1. There are many possible channel-borrowing schemes, from simple to complex, and they can be selected based on employed controller software and the feasibility of borrowing under given conditions.

8.3.1 Simple Borrowing Schemes

A simple borrowing scheme implies that if all channels allocated to a cell have already been used, then additional channels can be borrowed from any cell that has some free unused channels. Such a cell is called a donor cell. An obvious choice is to select a donor from among adjacent cells that has the largest number of free channels. This is known as borrowing from the richest. A further consequence is to return the borrowed channel to the donor if a channel becomes available in the cell that initially borrowed a channel. Such an algorithm is defined as basic algorithm with reassignment. Another alternative is to select the first free channel found for borrowing when the search follows a predefined sequence; this is known as the borrow-first-available scheme.

8.3.2 Complex Borrowing Schemes

The basic strategy for complex schemes is to divide the channels into two groups, one group assigned to each cell permanently and the second group kept reserved as donors to be borrowed by neighboring cells. The ratio between the two groups of channels is determined a priori and can be based on estimated traffic in the system. An alternative, known as borrowing with channel ordering, is to assign priorities to all channels of each cell, with highest priority channels being used in a sequential order for local calls in the cell while channel borrowing is done starting from lowest priority channels.

As mentioned earlier, every attempt must be made to minimize interference. Therefore, if channel borrowing is done such that a particular channel is available in nearby cochannel cells, then that channel can be borrowed. Such a scheme is known as borrowing with directional channel locking. Since this scheme imposes additional constraints, the number of channels available is reduced.

The basic sectoring technique discussed in Chapter 5 can be used to allocate channels temporarily. In the following section, we look into channel borrowing in such a scenario and discuss why cell sectoring is useful, how it influences the selection of donor cells, and what kind of impact it has on channel interference. One way of using the sector cell method is to share with bias, which implies borrowing of channels from one of the two adjacent sectors of neighboring cells. This can be further enhanced by a scheme known as channel assignment with borrowing and reassignment, by ensuring that borrowing causes minimum impact on future call blocking probability in neighboring cells and reassignment of borrowed channels is done to provide maximum help to the neighborhood. The channels can also be ordered based on which channels provide better performance; this can be useful in selecting lower-order channels for borrowing. In addition, borrowed channels can be returned to the donor cells if the channels become available in the borrowing cell. This scheme is known as an ordered channel assignment scheme with rearrangement.

There are relative advantages and disadvantages of different complex schemes in terms of total channel utilization, total carried traffic, and allocation complexity, and decisions are made based on the traffic behavior and system specifications.

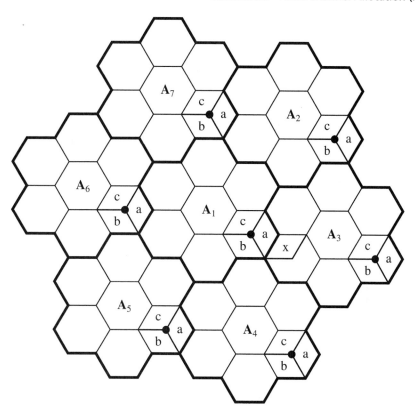

Figure 8.2
Impact of channel borrowing in sectored cell-based wireless system.

Seven adjacent clusters that could have cochannel interference are shown in Figure 8.2. Let us assume that each sector of all clusters uses the same frequency bands or channels to maintain reuse distance. With such an arrangement and fixed distribution of channels to different sectors, interference could be kept to a minimum desired level. Let us assume that sector "x" of cluster A_3 needs to borrow channels from an adjacent cell, let us say from sector "a" of cluster A_1. But, when some channels are borrowed from sector "a" of A_1 to sector "x" of A_3, there could be potential violation of reuse distance, and there could be interference between the borrowed channel in sector "x" with the same channels of all "a" channels of clusters A_2, A_3, A_4, A_5, A_6, and A_7. Looking at the distance between "x" and sector "a" of other clusters, only clusters A_5, A_6, and A_7 satisfy the reuse distance requirements, while clusters A_2, A_3, and A_4 violate the reuse distance from "x." Therefore, we need to look at the directions of sector "a" for clusters A_2, A_3, and A_4 with respect to "x." Clearly, the "a" sectors for both clusters A_2 and A_4 are in different directions from "x," and simultaneous use of the same channels in these areas will not cause any additional interference, as normally expected. The only question that needs to be addressed is the interference between sector "x" and sector "a" of cluster A_3, and even though the reuse distance is violated (they belong to the same cluster A_3), their directions are such that they would most likely not interfere with each other. If the cells are not sectored, then in Figure 8.2, borrowed channels will be used in the cell marked "x" and would cause interference with the cell "abc" of clusters A_2, A_3, and A_4. These borrowed channels cannot be used in these clusters as well. Therefore, we can see the obvious advantage of sectored cells.

Similar analysis needs to be performed if channels are borrowed from adjacent cells belonging to the same cluster. Therefore, the two steps of verifying potential interference and possible prohibition of those borrowed channels from other cells are as follows: first checking the reuse distance with other nearby clusters using those borrowed channels, and second looking at the sector directions of all cells not satisfying the reuse distance. Such checking would determine any potential interference with other cells and ensure smooth operation of the overall system.

8.4 Dynamic Channel Allocation (DCA)

DCA implies that channels are allocated dynamically as new calls arrive in the system and is achieved by keeping all free channels in a central pool. This also means that when a call is completed, the channel currently being used is returned to the central pool. In this way, it is fairly straightforward to select the most appropriate channel for any new call with the aim of minimizing the interference, as allocation of different traffic channels for current traffic is known. In this way, a DCA scheme overcomes the problem of a FCA scheme. In fact, a free channel can be allocated to any cell, as long as interference constraints in that cell, can be satisfied. The selection of a channel could be very simple or could involve one or more considerations, including future blocking probability in the vicinity of the cell, reuse distance, usage frequency of the candidate channel, average blocking probability of the overall system, and instantaneous channel occupancy distribution. The control could be centralized or distributed, and accordingly, DCA schemes are classified into two types—centralized and distributed schemes—with many important alternatives in each type.

8.4.1 Centralized Dynamic Channel Allocation Schemes

In these schemes, a channel is selected for a new call from a central pool of free channels, and a specific characterizing function is used to select one among candidate free channels. The simplest scheme is to select the first available free channel that can satisfy the reuse distance. An alternative is to pick a free channel that can minimize the future blocking probability in the neighborhood of the cell that needs an additional channel; this is defined as locally optimized dynamic assignment. Another scheme of channel reuse optimization maximizes the use of every channel in the system by appropriate allocation of channels, thereby maximizing system efficiency.

For a given reuse distance, cells can be identified that satisfy minimum reuse distance, all these cells could be allocated the same channel and are defined as cochannel cells. These cochannel cells can form a set, and each group is looked at carefully while allocating channels. If a cell needs to support a new call, then a free channel from the central pool is selected that would maximize the number of members in its cochannel set. A further modification is to select a channel that would minimize the mean square of the distance between cells using the same channel. Global optimization can be achieved if channel allocation can be evaluated using a graph theoretic model by representing each cell by a vertex and by placing an edge between two vertices as an indication of no cochannel interference. Maximization of the number of edges indicates availability of many vertices after current selection and, in turn, reflects a low blocking probability.

DCA schemes handle randomly generated new calls and hence cannot maximize overall channel reuse. Therefore, these schemes are observed to carry less traffic as compared to FCA, especially for higher traffic rates. Therefore, suggestions have been made to reassign channels and change channels for existing calls if that minimizes the distance between cells using the same channel and hence influencing the reuse distance.

8.4.2 Distributed Dynamic Channel Allocation Schemes

Centralized schemes can theoretically provide near optimal performance, but the amount of computation and communication among the BSs leads to excessive system latencies and makes them impractical. Therefore, schemes have been proposed that involve scattering channels across a network. However, centralized schemes are still used as a benchmark to compare various decentralized schemes.

Distributed DCA schemes are primarily based on one of the three parameters: cochannel distance, signal strength measurement, and SNR (signal-to-noise ratio). In a cell-based distributed scheme, a table indicates if other cochannel cells in the neighborhood are not using one or more channels and are selecting, one of the free channels for the requesting cell. In an adjacent channel interference constraint scheme, in addition to cochannel interference, adjacent channel interference is taken into account while choosing a new channel. The main limitation of this scheme is that a maximum packing of channels may not be possible as the MS's location is not taken into account.

In a signal strength measurement–based distributed scheme, channels are allocated to a new call if the anticipated CCIR (cochannel interference ratio) is above a threshold. This could cause the CCIR for some existing calls to deteriorate and hence those would require finding new channels that could satisfy a desired CCIR. Otherwise, those interrupted calls could be dropped prematurely or may also have a further ripple effect, possibly leading to system instability.

A comparison of fixed versus DCA schemes, taken from [8.1], is shown in Table 8.1.

Table 8.1: ▶
Comparison of Fixed and Dynamic Channel Allocation Schemes
Credit: From "Channel Assignment Schemes for Cellular Mobile Telecommunications Systems: A Comparative Study," by I. Katzela and M. Naglshinen, 1996, *IEEE Personal Communications* (now *IEEE Wireless Communications*), pp. 10-30. Copyright 1996 IEEE.

FCA	DCA
Performs better under heavy traffic	Performs better under light to moderate traffic
Low flexibility in channel assignment	Flexible channel allocation
Maximum channel reusability	Not always maximum channel reusability
Sensitive to time and spatial changes	Insensitive to time and time spatial changes
Unstable grade of service per cell in an interference cell group	Stable grade of service per call in an interference cell group
High forced call termination probability	Low to moderate forced call termination probability
Suitable for large cell environment	Suitable in microcellular environment
Low flexibility	High flexibility
Radio equipment covers all channels assigned to the cell	Radio equipment covers the temporary channel assigned to the cell
Independent channel control	Fully centralized to fully distributed control dependent on the scheme
Low computational effort	High computational effort
Low call setup delay	Moderate to high call setup delay
Low implementation complexity	Moderate to high implementation complexity
Complex, labor-intensive frequency planning	No frequency planning
Low signaling load	Moderate to high signaling load
Centralizing control	Centralized, distributed control depending on the scheme

● ● ● ● ● ● ● ● ● ● ● ● ● ● ● ● ●

8.5 Hybrid Channel Allocation (HCA)

Many other channel allocation schemes have been suggested, and each is based on different criteria employed as a way to optimize performance. Some of the important considerations include HCA, flexible channel allocation, and handoff allocation schemes, and they are discussed here.

8.5.1 Hybrid Channel Allocation Schemes

HCA schemes are a combination of fixed and DCA schemes, with the channels divided into fixed and dynamic sets. This means that each cell is given a fixed number of channels that is exclusively used by the cell. A request for a channel from the dynamic set is initiated only when a cell has exhausted using all channels in the fixed set. A channel from the dynamic set can be selected by employing any of the DCA schemes. The real question is what should be the ratio between the number of fixed and dynamic channels. The value of the optimal ratio depends on traffic characteristics, and it may be desirable to vary this value as per estimates of instantaneous load distributions. It has been observed [8.1] that for a fixed to dynamic channel ratio of 3:1, the hybrid allocation leads to better service than the fixed scheme for traffic up to 50%; beyond that load, fixed schemes perform better. Doing a similar comparison with dynamic schemes, when the load varies from 15% to 40%, the corresponding best values vary from most to medium to no dynamic channels. A lot of computation time is required if simulation is to determine the behavior of a large system, and an analytical approach is desirable. However, exact analytical models are much more difficult to define for hybrid schemes, and if data traffic also needs to be incorporated, it is almost impossible to have even an approximate model. This is an interesting area that needs further investigation.

8.5.2 Flexible Channel Allocation Schemes

The idea behind a flexible channel allocation scheme is similar to a hybrid scheme having available channels divided into fixed and flexible (emergency) sets, with fixed sets assigned to each cell to handle lighter loads effectively. The flexible channels are used by the cells only when additional channels are needed after exhausting the fixed set. Flexible schemes require centralized control, with up-to-date traffic pattern information, to assign flexible channels effectively. There are two different strategies used in allocating channels: scheduled and predictive. In scheduled assignment, a priori estimates about variation in traffic (i.e., peaks in time and space) are needed to schedule emergency channels at predetermined peaks of traffic change. In a predictive strategy, the traffic intensity and blocking probability is monitored in each cell at all the time so that flexible channels can be assigned to each cell according to its needs. This is similar to allocating additional channels as needed, rather than assigning extra channels during the office hours of 8 A.M. to 5 P.M. when there is peak load.

● ● ● ● ● ● ● ● ● ● ● ● ● ● ● ● ● ●

8.6 Allocation in Specialized System Structure

Allocation of channels also depends on some inherent characteristics of the system structure. For example, if a cellular system is specifically designed for a freeway, then allocation of channels to several mobile units moving in one direction can be assigned effectively and correlated to one-dimensional motion.

8.6.1 Channel Allocation in One-Dimensional Systems

Consider a one-dimensional microcellular system for a highway, shown in Figure 8.3, wherein handoff and forced termination of call do occur frequently due to the small size of cells and the speed of MSs located inside fast moving cars.

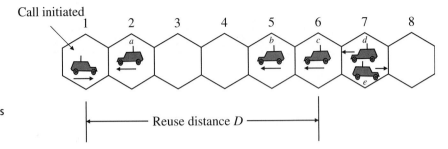

Figure 8.3
Allocation of channels in one-dimensional moving direction.

To understand this type of channel assignment in such an environment, consider the example shown in Figure 8.3. A new call is initiated in cell 1, with current allocation of channels "a," "b," "c," "d," and "e" as shown in the diagram. Looking at the reuse distance and direction of other moving vehicles, it is better to select a channel in a cell at least $(D+1)$ distance apart. This rule allows us to assign channel "e" to the MS in cell 1. This is based on an assumption that by the time the MS of cell 1 moves to cell 2, the MS in cell 7 would also have moved to cell 8, and both these MSs can continue to use the same channel "e," even after moving to the next cell. This would minimize forced termination when handoff occurs in terms of access to a new BS (but not changing the channel) of the next cell. It should be obvious why a cell at D distance is not used as MSs are moving at different speeds and are located at different parts of the cell. Therefore, by adopting $(D+1)$ cells apart, even if the MS in cell 1 moves to cell 2 while the MS in cell 7 is still in that cell, the distance D is maintained. In this way, it is unlikely that two MSs using the same channel could violate the reuse distance requirements, as long as the speeds of the two MSs are similar.

8.6.2   Reuse Partitioning–Based Channel Allocation

In a reuse partitioning-based allocation strategy, each cell is divided into multiple concentric, equal-size zones, as illustrated in Figure 8.4.

The basic idea is that the inner zone, being closer to the BS, would require lesser power to attain a desired CIR or signal-to-interference ratio (SIR). This is equally

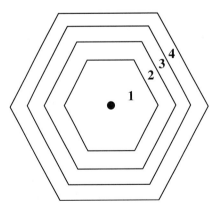

Figure 8.4
Concentric zone of a cell

true for the CDMA–based scheme. When applied to the FDMA/TDMA–based scheme, due to lower SIR, it is possible to use a lower value of reuse distance for inner zones as compared to outer zones, thereby enhancing spectrum efficiency. Such reuse partitioning schemes can be based on either fixed or adaptive allocation. In simple reuse partitioning, mobile subscribers with the best SIR are assigned a group of channels that have the smallest reuse distance. A similar strategy is used to allocate channels with the largest reuse distance and worst SIR.

Appropriate adjustment in reuse group channels needs to be performed whenever the SIR for a MS changes. An alternative is to measure the SIR of all the MSs in the cell, sort them, and assign channels starting from the inner zone to the outer zone in descending SIR values of the MS.

The concentric zones are formed to help enhance channel utilization and the number of zones and the size of each zone are not fixed. Moreover, in actual practice, the zone shape and size may not exactly correspond to a given SIR value. Therefore, many dynamic reuse partitioning schemes have been proposed, and details can be found in [8.1].

8.6.3   Overlapped Cells–Based Channel Allocation

One such example is shown in Figure 8.5, wherein a cell is split into seven microcells, with separate BS and microwave tower placed at the center of each microcell. There are many different alternatives for allocating channels. One way to assign channels for the cell and the microcells is to characterize the mobility of each MS into fast-moving and slow-moving groups. For slow-moving MSs, channels are assigned from one of the microcells, based on the current location. Fast-moving MSs would have more frequent handoffs if channels associated with the microcells are assigned for the same. For this reason, fast-moving MSs are given channels from the cell. Therefore, channel allocation from the cells/microcells is matched with the speed of the MSs. In

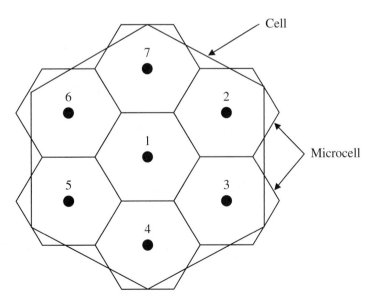

Figure 8.5
Illustration of cell splitting.

such a multitier cellular system, the number of channels allocated to each tier depends on the total number of channels, the area to be covered, the average moving speed of the MSs in each tier, the call arrival rate and duration of information in each tier, desirable blocking and dropping probabilities, and the number of channels set aside for handoffs. Optimization of such a system is fairly complex and beyond the scope of this chapter. One approach to handle increased traffic in a cell is to split it into a number of smaller cells inside a cell, and such partitioned smaller cells are called microcells and picocells.

An alternative to using cells and microcells as shown in Figure 8.5 is to change the logical structure dynamically, starting with only the main cell being used and other microcells being switched off under the control of the cell for low traffic. As traffic increases in one or more parts of the cell, the corresponding microcells are turned on if an unacceptable level of cochannel interference or unavailability of resources leads to forced call blocking. Switching on the microcell nearest to the MSs requesting channels makes the microcell BS physically closer to them, thereby enhancing the CIR values. If traffic decreases, then the cell switches off selected BSs located at the microcells, thereby automatically adapting to instantaneous call traffic density and lowering the probability of calls being terminated. Simulation results from such a multitier network approach [8.2][8.3][8.4] [8.5] indicate a drastic reduction in the number of handoffs, and optimal partitioning of channels among the cell and microcells is a complex function of numerous parameters, including the rate at which switching on and off can be done and threshold parameters. Another possibility is to have an overlap of cell areas between two adjacent cells as shown in Figure 8.6 [8.6]. In such an overlapped-cells scheme, either directed retry or directed handoff can be used. In directed retry, if a MS located in the shaded area cannot find any free channels from cell A, then it can use a free channel from cell B, if the signal quality is acceptable. In directed handoff, another extreme step is taken to free up a channel by forcing some of the existing connections in the shaded area of cell A to do forced handoff to cell B, if new calls in cell A do not find a free channel. A similar measure can be taken for other parts of cell A as well. Both of these approaches are observed to improve system performance, and many factors, including the ratio of overlapped area to total cell area, influence the blocking probabilities of originating calls. A detailed investigation is needed to determine an appropriate overlap so that all calls can be served and unavailability of free channels can be minimized. Given the number of channels, we next consider how the rates for new originating and handoff calls can influence blocking probability and hence system performance.

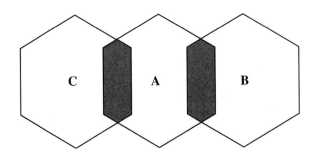

Figure 8.6
Use of overlapped
cell areas.

● ● ● ● ● ● ● ● ● ● ● ● ● ● ● ● ● ●

8.7 System Modeling

As described above, in order to fulfill specific needs, different channel allocation schemes are used. To evaluate the channel allocation schemes, mathematical models are developed in this section. Among many QoS parameters considered important in wireless networks, blocking probability of originating call and forced termination probability are the two most critical ones. This is different from a wired network, in which the delay and jitter are given higher priority. Appropriate models for evaluating these parameters, are considered next.

8.7.1 Basic Modeling

If S channels are allocated to a cell, then they have to be used both for the originating calls in the cell and the handoff calls from adjacent cells. These call rates influence the probability of call acceptance. Since it is relatively difficult to model an exact scenario, some simplistic assumptions are made to obtain an approximate model of the system:

1. All MSs are assumed to be uniformly distributed through the cell.

2. Each MS moves at a random speed and in a random direction.

3. The average arrival rate of originating calls is given by λ_O.

4. The average arrival rate of handoff calls is given by λ_H.

5. The average service rate for calls is given by μ.

6. Originating and handoff calls are given equal priority.

7. All assumptions are equally applicable to all cells in the system.

8. The arrival processes of both originating and handoff calls are assumed to be Poisson processes while an exponential service time is assumed.

9. $P(i)$: the probability of i channels to be busy

10. B_O: the blocking probability of originating calls

11. B_H: the blocking probability of handoff calls

12. S: the total number of channels allocated to a cell

As both originating and handoff calls are treated equally by S channels in a cell, the calls are served as they arrive if there are channels available, and both kinds of requests are blocked if all S channels are busy. The system of a cell can be modeled as shown in Figure 8.7. The cell state can be represented by the $(S+1)$ states Markov model, with each state indicating the number of busy channels within the cell. The total request rate becomes $\lambda_O + \lambda_H$. This leads to a state transition diagram of the $M/M/S/S$ model, as shown in Figure 8.8.

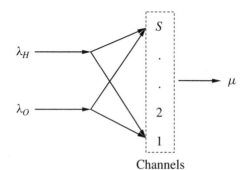

Figure 8.7
A generic system
model for a cell.

Channels

Figure 8.8
State transition
diagram for Figure 8.7.

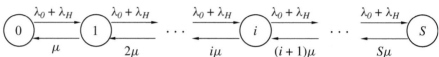

From Figure 8.8, the state equilibrium equation for state i can be given as

$$P(i) = \frac{\lambda_O + \lambda_H}{i\mu} P(i-1), \qquad 0 \le i \le S. \tag{8.3}$$

Using the preceding equation recursively, along with the assumption that the system will be in one of the $(S+1)$ states, the sum of all states must be equal to one:

$$\sum_{i=0}^{S} P(i) = 1. \tag{8.4}$$

The steady-state probability $P(i)$ is easily found as follows:

$$P(i) = \frac{(\lambda_O + \lambda_H)^i}{i!\mu^i} P(0), \qquad 0 \le i \le S, \tag{8.5}$$

where

$$P(0) = \left[\sum \frac{(\lambda_O + \lambda_H)^i}{i!\mu^i} \right]^{-1}. \tag{8.6}$$

The blocking probability for an originating call can be expressed by

$$B_O = P(S) = \frac{\dfrac{(\lambda_O + \lambda_H)^S}{S!\mu^S}}{\displaystyle\sum_{i=0}^{S} \dfrac{(\lambda_O + \lambda_H)^i}{i!\mu^i}}. \tag{8.7}$$

The blocking probability of a handoff request or the forced termination probability of a handoff call is

$$B_H = B_O. \tag{8.8}$$

Equation (8.7) is known as the **Erlang B** formula, as covered in Chapter 5.

8.7.2 Modeling for Channel Reservation

It is well known that if an originating call is unsuccessful due to blocking, that is not as disastrous as a handoff call being dropped. Therefore, it is important to provide a higher priority to an existing call that goes through the handoff process so that ongoing calls can be continued [8.7][8.8][8.9][8.10]. One way of assigning priority to handoff requests is by reserving S_R channels exclusively for handoff calls among the S channels in a cell. The remaining Sc $(= S - S_R)$ channels are shared by both originating and handoff calls. An originating call is blocked if channels have been allocated. A handoff request is blocked if no channel is available in the cell. The system model must be modified to reflect priorities, as shown in Figure 8.9.

The probability $P(i)$ can be determined in a similar way, with the state transition diagram shown in Figure 8.10. The state balance equations can be obtained as

$$\begin{cases} i\mu P(i) = (\lambda_O + \lambda_H)P(i-1), & 0 \le i \le S_C \\ i\mu P(i) = \lambda_H P(i-1), & S_C < i \le S. \end{cases} \tag{8.9}$$

Using these equations recursively and with the addition of all $(S+1)$ states as

$$\sum_{i=0}^{S} P(i) = 1, \tag{8.10}$$

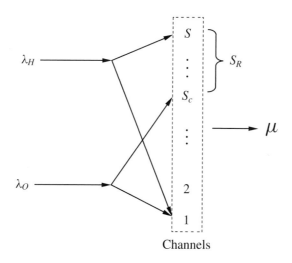

Figure 8.9
System model with reserved channels for handoff calls.

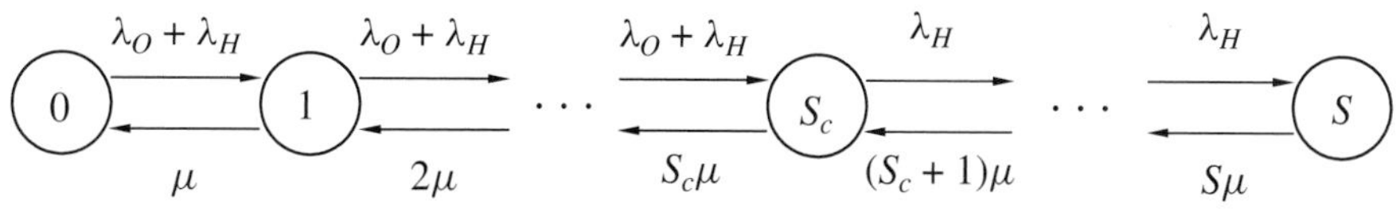

Figure 8.10
State transition diagram for Figure 8.9.

the steady-state probability $P(i)$ can be obtained:

$$
P(i) = \begin{cases} \dfrac{(\lambda_O + \lambda_H)^i}{i!\mu^i} P(0), & 0 \le i \le S_c \\[2ex] \dfrac{(\lambda_O + \lambda_H)^{S_c} \lambda_H^{i-S_c}}{i!\mu^i} P(0), & S_c < i \le S, \end{cases} \tag{8.11}
$$

where

$$
P(0) = \left[\sum_{i=0}^{S_c} \frac{(\lambda_O + \lambda_H)^i}{i!\mu^i} + \sum_{i=S_c+1}^{S} \frac{(\lambda_O + \lambda_H)^{S_c} \lambda_H^{i-S_c}}{i!\mu^i} \right]^{-1}. \tag{8.12}
$$

The blocking probability B_O for an originating call is given by

$$
B_O = \sum_{i=S_c}^{S} P(i). \tag{8.13}
$$

The blocking probability of a handoff request or the forced termination probability of a handoff call is

$$
B_H = P(S) = \frac{(\lambda_O + \lambda_H)^{S_c} \lambda_H^{S-S_c}}{S!\mu^S} P(0). \tag{8.14}
$$

The relations of equations (8.13) and (8.14) clearly show that the two probabilities are not equal as priority is given to handoff calls. In fact, another possible improvement in servicing handoff calls is to provide buffers for such calls so that B_H can be minimized and serviced later even if no channels are available instantaneously. This is discussed in Chapter 15, along with the possibility of adding buffers for originating calls as well. There are some limitations of the simplified model, such as even distribution of MSs, their random speed and moving direction, and exponential call rates. These need careful attention.

$\bullet \; \bullet \; \bullet \; \bullet \; \bullet \; \bullet \; \bullet \; \bullet \; \bullet \; \cdot \; \cdot \; \cdot$

8.8 Summary

Resource allocation is important for the system performance of wireless networks, and assigning priority for handoff calls provides substantial enhancements. Any wireless system consists of both wireless components as well as underlying wired networks as a backbone, and any changes in overall performance require enhancing both of these components. In this chapter, we have considered how traffic channels can be allocated in FDMA/TDMA–based cellular systems while many issues are equally applicable to CDMA–based schemes as well. The information may have to go through the backbone wireline network, and such routing should be changed when handoff occurs. This opens up the issue of authentication, which is covered in Chapter 9.

● ● ● ● ● ● ● ● ● ● ● ● ● ●●●●

8.9 **References**

[8.1] I. Katzela and M. Naghshineh, "Channel Assignment Schemes for Cellular Mobile Telecommunication Systems: A Comparative Study," *IEEE Personal Communications*, pp. 10–31, June 1996.

[8.2] H. Jiang and S. S. Rappaport, "Hybrid Channel Borrowing and Directed Retry in Highway Cellular Communications," *Proceedings of 1996 IEEE 46th VTC*, pp. 716–720, Altanta, GA, USA, April 28–May 1, 1996.

[8.3] S. S. Rappaport and L-R. Hu, "Microcelluar Communication Systems with Hierarchical Macrocell Overlays: Traffic Performance Models and Analysis," *Proceedings of the IEEE Vol.* 82, No. 9, September 1994, pp. 1383–1397.

[8.4] H. Furakawa and Y. Akaiwa, "A Microcell Overlaid with Umbrella Cell System," *Proceedings of 1994 IEEE 44th VTC*, pp. 1455–1459, June 1994.

[8.5] A. Ganz, Z. J. Haas, and C. M. Krishna, "Multi-Tier Wireless Networks for PCS," *Proceedings of IEEE VTC'96*, pp. 436–440, 1996.

[8.6] S. A. El-Dolil, W-C. Wong, and R. Steele, "Teletraffic Performance of Highly Microcells with Overlay Macrocell," *IEEE Journal on Selected Areas in Communications*, Vol. 7, No. 1, pp. 71–78, January 1989.

[8.7] K. Pahlavan, P. Krishnmurthy, A. Hatami, M. Ylianttila, J-P. Makela, R. Pichna, and J. Vallstrom, "Handoff in Hybrid Mobile Data Networks," *IEEE Personal Communications*, pp. 34–47, April 2000.

[8.8] G. Cao and M. Singhal, "An Adaptive Distributed Channel Allocation Strategy for Mobile Cellular Networks," *Journal of Parallel and Distributed Computing*, Vol. 60, No. 4, pp. 451–473, April 2000.

[8.9] Q-A. Zeng and D. P. Agrawal, "Modeling of Handoffs and Performance Analysis of Wireless Data Networks," *IEEE Transactions on Vehicular Technology*, Vol. 51, No. 6, pp. 1469–1478, November 2002.

[8.10] B. Jabbari, "Teletraffic Aspects of Evolving and Next-Generation Wireless Communication Networks," *IEEE Personal Communications*, Vol. 3, No. 6, pp. 4–9, December 1996.

● ● ● ● ● ● ● ● ● ● ● ● ● ●●●●

8.10 **Problems**

P8.1. What are the specific advantages of static channel allocation over dynamic channel allocation strategies?

P8.2. Are there collisions present in traffic or information channels in a cellular system? Explain clearly.

P8.3. What are the differences in channel allocation problems in FDMA/TDMA–based systems versus CDMA–based systems? Explain clearly.

P8.4. If you do not sector the cells, can you still borrow channels from adjacent cells? Explain clearly.

P8.5. In a cellular system with omnidirectional antennas, a 7-cell cluster is employed. The cell at the center of the cluster has a lot more traffic than others and needs to borrow some channels from adjacent cells. Explain the strategy you would employ to determine a donor cell

(a) Within the cluster.

(b) Outside the cluster.

P8.6. Which cell(s) may borrow channels and which could be an appropriate donor(s) in Problem P5.11?

P8.7. What are the advantages of cell sectoring? How do you compare this with SDMA?

P8.8. In a cellular system with 7-cell clusters, the average number of calls at a given time is given as follows:

Cell number	Average number of calls/unit time
1	900
2	2000
3	2500
4	1100
5	1200
6	1800
7	1000

If the system is assigned 49 traffic channels, how would you distribute the channels if

(a) Static allocation is used.

(b) Simple borrowing scheme is used.

(c) Dynamic channel allocation scheme is used.

P8.9. Each cell is divided in a slightly different way as 3-sectors as follows:

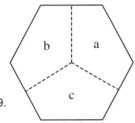

Figure 8.11
Figure for Problem P8.9.

What will be the impact of such sectoring on channel borrowing and its effect on cochannel interference? Explain carefully.

P8.10. Each cell of a wireless system is partitioned into six-sector format as shown below:

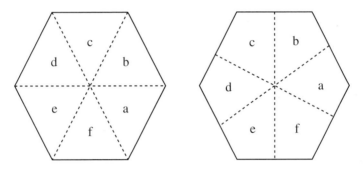

Figure 8.12
Figure for
Problem P8.10.

(i) 6-sectors of a cell   (ii) Alternative sectoring scheme

(a) What will be the impact of channel-borrowing and cochannel interference if sectoring scheme (i) is used?

(b) Repeat (a) if scheme (ii) is used.

(c) How do you compare (a) with (b)?

(d) Is it possible or desirable to use a combination of the sectoring schemes of (i) and (ii)? Explain carefully.

P8.11. In a cellular system with a 7-cell cluster, 48 traffic channels are assigned. Show the assignment of channels to each cell if:

(a) Omnidirectional antennas are used.

(b) 3-sector directional antennas are used.

(c) 6-sector directional antennas are used.

P8.12. A service provider decided to restructure allocation of channels by selecting a cluster with 4-cell as its basic building block. What would be the impact of channel borrowing if each cell employs (a) 3-way sectoring or (b) 6-way sectoring.

P8.13. How do you compare hybrid versus flexible channel allocation? Which one would you prefer and why?

P8.14. For a wireless network with integrated services, e.g., including both voice and data applications, there are two basic channel allocation schemes: complete sharing (CS) and complete partitioning (CP). The CS policy allows all users to equally access the channels available at all times. The CP policy, on the other hand, divides up the available bandwidth into separate sub-pools according to user type. Compare both advantages and disadvantages of these two schemes.

P8.15. What kind of technique(s) could you possibly use to serve a new call if all the channels in the current cell have been occupied and no channel can be borrowed from neighboring cells.

P8.16. A service provider decided to split each hexagonal cell of 20 km radius to seven microcells of appropriate size.
 (a) What is the size of each microcell?
 (b) How is the signal strength influenced by such a redesign?
 (c) What is CCIR compared to the original design, assuming the propagation path loss slope $\zeta = 4.5$?

P8.17. Providing cellular service along a freeway is a tough job, and such a scenario is illustrated in the following figure. A typical road-width varies from 200 m to 400 m. If you select 1000 m as the radius of each cell, then one cell is required for each km, while the radius of a conventional normal cell is about 20 km. From the freeway usage point of view, only a very small segment of each cell is useful. Do you have any suggestions for alternative designs? What are the tradeoffs? Do you suggest the use of SDMA technique?

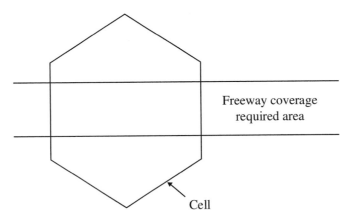

Figure 8.13
Figure for
Problem P8.17.

P8.18. In a cellular system with four channels, one channel is reserved for handoff calls.
 (a) What is the value of B_O and B_H, given $\lambda_O = \lambda_H = 0.001$ and $\mu = 0.0003$?
 (b) What are the values of probabilities $P(0)$, $P(1)$, $P(2)$, $P(3)$, and $P(4)$?
 (c) What is the average number of occupied channels in this problem?

P8.19. Repeat Problem P8.18 for the case that if the number of channels is increased to ten.

P8.20. In a cellular system, the total number of channels per cell, is given as six, and two channels are reserved exclusively for handoff calls. What are the blocking probabilities for originating if the handoff request rate is 0.0001, the originating call rate is 0.001, and the service rate $\mu = 0.0003$?

P8.21. What is the impact on the answer for Problem P8.20 if the number of reserved channels is changed to

(a) 1?

(b) 3?

Mobile Communication Systems

• • • • • • • • • • • • • • • •

9.1 Introduction

A wireless system implies support for subscriber mobility by the communication infrastructure, and such movement includes not only from one cell to another but also from cell's mobile switching center (MSC) and areas controlled by other service providers. In an ideal situation, any MS should be able to communicate with the rest of the world by accessing local wireless infrastructure facilities. Therefore, handoff and roaming among cells and MSCs that are serviced by the same service provider or different service providers needs to be supported. In this chapter, we consider handoff schemes, allocation of resources, and routing in the backbone network as well as security considerations in wireless networks.

• • • • • • • • • • • • • • • •

9.2 Cellular System Infrastructure

A cellular system requires a fairly complex infrastructure. A generic block diagram built on our earlier discussions in previous chapters is shown in Figure 9.1. Each BS consists of a base transceiver system (BTS) and a BS controller (BSC). Both tower and antenna are part of the BTS and all associated electronics are contained in the BTS. The authentication center (AUC) unit provides authentication and encryption parameters that verify the user's identity and ensure the confidentiality of each call. The AUC protects network operators from different types of frauds and spoofing found in today's cellular world. The equipment identity register (EIR) is a database that contains information about the identity of mobile equipment. Both AUC and EIR can be implemented as individual stand-alone units or as a combined AUC/EIR unit. The home location register (HLR) and visitor location register (VLR) are two sets of pointers that support mobility and enable the use of the same cell phone number (or mobile phone) over a wide range. The HLR is located at the MSC where the MS is initially registered and is the initial home location for billing and access information. In simple words, the mobility of MS support can be explained by a simple and well-known example of post offices forwarding the mail. If someone moves, he or she informs the post office serving the old

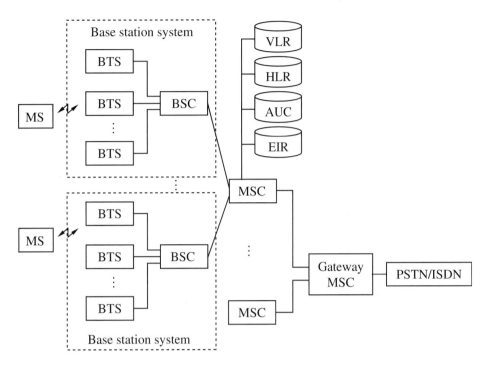

Figure 9.1
A detailed block
diagram of a cellular
system.

location about the new address (and hence the new serving post office). This way,
all mail coming to the old post office serving the prior location, is forwarded to the
new post office taking care of mail for the current address so that it can be delivered
to the current address of the person. This is equivalent to one-way pointer and
redirection. Such a scenario is illustrated in Figure 9.2. In post office system, there
is no use for having a backward pointer from the new post office to the old one. In
a similar way, in a cellular system, two-way pointers are established using HLRs and

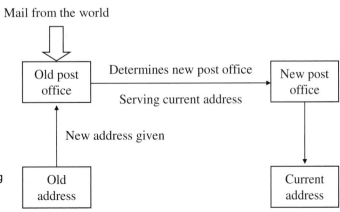

Figure 9.2
Classical mail forwarding
done by mail service.

VLRs. Any incoming call, based on the calling number, is directed to the HLR of the home MS where the MS is registered (similar to the old post office). The HLR then points to the VLR of the MSC where the MS is currently located (similar to the new serving post office). The VLR contains information about all MSs visiting that particular MSC and hence points to the HLR of the visiting MSs for exchanging related information about the MS. Such a pointer allows calls to be routed or rerouted to the MS, wherever it is located. In cellular systems, a reverse direction pointer is needed that allows traversal of many control signals back and forth between the HLR and VLR (including billing and access permissions maintained at the home MSC); such bidirectional HLR–VLR pointers help in carrying out various functionalities, as illustrated in Figure 9.3 and are more versatile than unidirectional pointers used in a post office setup. In the next section, we discuss how these pointers between HLR–VLR pairs are automatically set during the initial phase of roaming.

This works very well if the destination MS has moved from one cell to another. If a call is initiated from a residential telephone, the call is forwarded through the backbone network to the gateway closest to the home MSC where the MS being called is registered. Thereafter, a similar routing enables connection to the MS. In the same way, a reverse path connection can be established (i.e., from a MS to a home telephone subscriber).

As indicated earlier, the home MSC also maintains access information about all MSs registered, including state of the MS (active/nonactive), type of allowed service (local and/or long distance calls), and billing information (past credit, current charges, chronological order of calls made and timings of each call, etc.). For simplicity of understanding, we have described a simple control mechanism of forwarding calls to a MS in a visiting area.

Such call forwarding works very well if the MS has moved from the registered sector of a cell to another sector within the cell, or within the area controlled by either the same BSC or by the same MSC. For these cases, the redirection mechanism shown in Figure 9.3 is adequate. This would work even if the home MSC is different from the visiting MSC, as long as the two MSCs have information about how to forward messages to each other. Mobility can also be supported in an unknown territory as long as there is a mechanism in place to reach the intended destination. A complex forwarding scheme for this is discussed in a later section.

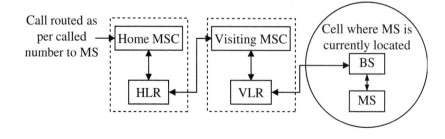

Figure 9.3
Redirection of a call to MS at a visiting location.

• • • • • • • • • • • • • •
9.3 **Registration**

The MSs must be registered at one of the MSCs for successful operation of numerous system functionalities. This is maintained not only for billing, but also for authentication and verification, as well as for access privileges. In addition to this permanent information, the wireless system needs to know whether the MS is currently located in its own home area or is visiting some other area. This enables incoming calls to be routed to an appropriate location and assures desirable support for outgoing calls.

This is done by exchanging signals known as "beacon signals" between the BS and the MS [9.1]. BSs periodically broadcast beacon signals to determine and test nearby MSs (see Figure 9.4). Each MS listens for beacon signals, and if it hears from a new BS, it adds it to the active beacon kernel table. This information is then used by a MS to locate the nearest BS and establish an appropriate rapport to initiate dialogue with the outside world through the BS as a gateway. Some of the information carried by the beacon signals includes cellular network identifier, timestamp, gateway address, ID (identification) of the paging area (PA), and other parameters of the BS.

The following steps are used by MSs outside their own subscription areas:

1. A MS listens for new beacons, and if it detects one, it adds it to the active beacon kernel table. If the device determines that it needs to communicate via a new BS, kernel modulation initiates the handoff process.

2. The MS locates the nearest BS via user-level processing.

3. The visiting BS performs user-level processing and determines who the user (MS) is, the user's registered home site (MSC) for billing purposes, and what kind of access permission the user has.

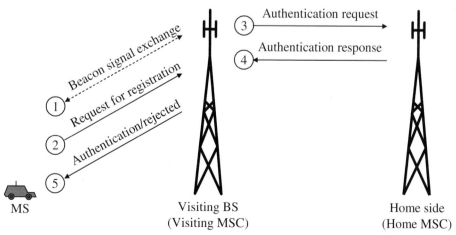

Figure 9.4
Using a mobile phone outside the subscription area. *Courtesy*: From "Beacon Signals: What, Why, How, and Where," by S.Gerasenko, A.A. Joshi, S. Rayaprolu, K. Ponnavaikko, and D.P. Agrawal, 2001. *IEEE Computer*, 34, pp. 108-110. Copyright 2001 IEEE.

4. The home site sends an appropriate authentication response to the BS currently serving the user and is stored in the corresponding VLR of serving MSC (two way pointers between HLR–VLR pairs).

5. The BS at the visited location approves or disapproves user access.

In the United States, these signals are transmitted in the Advanced Mobile Phone System (AMPS) and the Cellular Digital Packet Data (CDPD) system. A similar technique is used in second generation GSM, the cellular standard used throughout Europe and Asia.

Figure 9.4 illustrates how a cellular network uses beacon signals when a cell (or mobile) phone user is in a location outside his or her subscription area (for example, just after getting off an airplane). When the user switches on the handheld device, the beacon signal activates a roaming service, and the user registers and communicates through the closest BS. Implementation of the system typically occurs at three levels: user-level processing at the BS, user-level processing at the MS, and kernel modulation at the MS.

Although transparent to the user community, beacon signals have made wireless systems more intelligent and humanlike. As Table 9.1 shows, they are an integral part of numerous scientific and commercial applications ranging from mobile networks to search and rescue operations and location tracking systems.

Table 9.1: ▶
Applications and Characteristics of Beacon Signals
Courtesy: From "Beacon Signals: What, Why, How, and Where," by S.Gerasenko, A.A. Joshi, S. Rayaprolu, K. Ponnavaikko, and D.P. Agrawal, 2001. *IEEE Computer*, 34, pp. 108-110. Copyright 2001 IEEE.

Application	Frequency band	Information carried
Cellular networks	824–849 MHz (AMPS/CDPD), 1,850–1,910 MHz (GSM)	Cellular IP network identifier, gateway IP address, paging area ID, timestamp
Wireless LANs (discussed in Chapter 14)	902–928 MHz (industrial, scientific, and medical band for analog and mixed signals) 2.4–2.5 GHz (ISM band for digital signals)	Traffic indication map
MANETs (discussed in Chapter 13)	902–928 MHz (ISM band for analog and mixed signals) 2.4–2.5 GHz (ISM band for digital signals)	Network node identity
GPS	1575.42 MHz	Timestamped orbital map and astronomical information
Search and rescue	406 and 121.5 MHz	Registration country and ID of vessel or aircraft in distress
Mobile robotics	100 kHz–1 MHz	Position of pallet or payload
Location tracking	300 GHz–810 THz (infrared)	Digitally encoded signal to identify user's location
Aid to the impaired	176 MHz	Digitally coded signal uniquely identifying physical locations

Beacon signals help synchronize, coordinate, and manage electronic resources using minuscule bandwidth for a very short duration. Researchers continue to improve their functionality by increasing signal coverage while optimizing energy consumption. Beacon signals' perceptibility and usefulness in minimizing communication delays and interference are spurring exploratory efforts in many domains, ranging from home to outer space.

● ● ● ● ● ● ● ● ● ● ● ● ● ● ● ● ●

9.4   Handoff Parameters and Underlying Support

Handoff basically involves change of radio resources from one cell to another adjacent cell. From a handoff perspective, it is important that a free channel is available in a new cell whenever handoff occurs, so that undisrupted service is available.

9.4.1   Parameters Influencing Handoff

As discussed in Chapter 5, handoff depends on cell size, boundary length, signal strength, fading, reflection and refraction of signals, and man-made noise. If we make a simplistic assumption that the MSs are uniformly distributed in each cell, we can also say that the probability of a channel being available in a new cell depends on the number of channels per unit area. From Table 5.1, it can be easily observed that the number of channels per area increases if the number of channels allocated per cell is increased or if the area of each cell is decreased. The radio resources and hence the number of assigned channels are limited and may not be changed to a great extent. However, the cell coverage area could be decreased for a given number of channels per cell. This leads to a smaller cell size and may be good for the availability of free channel perspectives. However, this would cause more frequent handoffs, especially for MSs with high mobility and speed.

Handoff can be initiated either by the BS or the MS, and it could be due to

1. The radio link
2. Network management
3. Service issues

Radio link–type handoff is primarily due to the mobility of the MS and depends on the relative value of the radio link parameters. Radio link–type handoff depends on

- Number of MSs that are in the cell
- Number of MSs that have left the cell
- Number of calls generated in the cell
- Number of calls transferred to the cell from neighboring cells by the handoff
- Number and duration of calls terminated in the cell
- Number of calls handed off to neighboring cells
- Cell dwell time

Network management may cause handoff if there is a drastic imbalance of traffic over adjacent cells, and optimal balance of channels and other resources are required. Service-related handoff is due to degradation of quality of service (QoS), and handoff could be invoked when such a situation is detected.

The factors that define the right time for handoff are

- Signal strength
- Signal phase
- Combination of the above two
- Bit error rate (BER)
- Distance

The need for handoff is determined in two different ways:

1. Signal strength
2. Carrier-to-interference ratio (CIR)

An example of handoff based on received power has been covered in Figure 5.5. In addition to the power level of the received signal, another important aspect is the value of CIR in a cell at a given location. A low value of CIR may force the BS to change the channel currently being used between the BS and the MS. Handoff could also occur if directional antennas are employed in a cell and a MS moves from one sector to another sector of the cell (or one beam area to another in a SDMA system). The handoff procedure and associated steps depend on the cellular systems, and the specific units involved in setting up a call are as follows:

1. Base station controller (BSC)
2. Mobile station (MS)
3. Mobile switching center (MSC)

9.4.2   Handoff Underlying Support

Handoff can be classified into two different types: hard and soft handoffs. Hard handoff, also known as "break before make," is characterized by releasing current radio resources from the prior BS before acquiring resources from the next BS. Both FDMA and TDMA employ hard handoff. In CDMA, as the same channel is used in all the cells (as you recall, the reuse distance is 1), if the code is not orthogonal to other codes being used in the next BS, the code could be changed. Therefore, it is possible for a MS to communicate simultaneously with the prior BS as well as the new BS, just for some short duration of time. Such a scheme is called soft handoff (or "make before break"). These handoffs are illustrated in Figures 9.5 and 9.6, respectively.

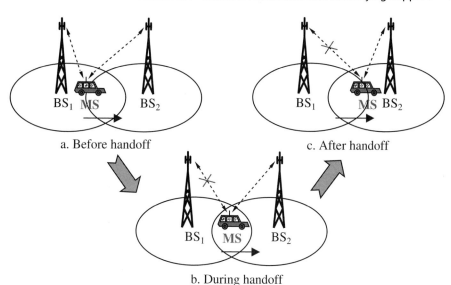

Figure 9.5
Hard handoff.

a. Before handoff

b. During handoff

c. After handoff

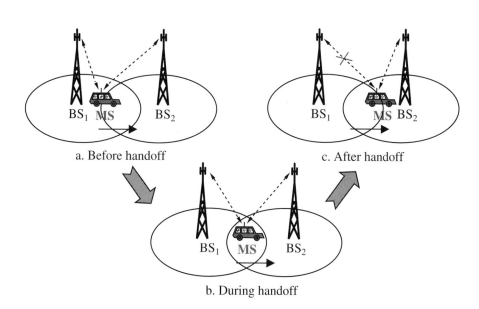

Figure 9.6
Soft handoff.

a. Before handoff

b. During handoff

c. After handoff

It is also possible to move from a cell controlled by one MSC area to a cell connected to another MSC. In fact, beacon signals and the use of the HLR–VLR pair allow MSs to roam anywhere as long as the same service provider, using the particular frequency band present in that area. This is illustrated in Figure 9.7.

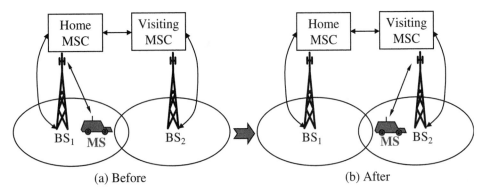

Figure 9.7
Handoff between
MSCs.

(a) Before   (b) After

9.5 Roaming Support

In earlier sections, emphasis has been on allocating channels to different calls so that handoff can be efficiently supported as much as possible and blocking probability of both originating and handoff calls can be minimized. We do need to worry about what happens when channel and hence radio contact is changed from one cell to another for successful handoff. As discussed in Chapter 1, a number of cells are controlled by a MSC, and depending on the destination, the signals go through the backbone network, interconnecting MSCs with the PSTN, which serves as a basic infrastructure between MSs and existing home or commercial telecommunication systems. The hardwired network is primarily supported by ultra-high-speed fiber optic cables, and information transfer is in terms of packet scheduling, reflecting the bandwidth allocation to different users.

The MSCs are connected to the backbone network via different gateways. Therefore, with mobility support, the real problem in routing becomes that of moving packets to appropriate endpoints of the backbone network. Various possible handoff scenarios are illustrated in Figure 9.8.

Assuming MSC_1 to be the home of the MS for registration, billing, authentication, and all access information, when the handoff is from location "a" to location "b," the routing of messages meant for the MS can be performed by MSC_1 itself. However, when the handoff occurs from location "b" to location "c," then bidirectional pointers are set up to link the HLR of MSC_1 to the VLR of MSC_2 so that information can be routed to the cell where the MS is currently located (Figure 9.9). The call in progress can be routed by HLR of MSC_1 to VLR of MSC_2 and to the corresponding BS to eventually reach the MS at location "c."

The situation is different and slightly more complicated when handoff occurs at locations "d" and "e" in Figure 9.8, and routing of information using simply the HLR–VLR pair of pointers may not be adequate. The paging area (PA) is the area covered by one or several MSCs in order to find the current location of the MS [9.2]. This concept is similar to the Internet network routing area [9.3][9.4], and to understand how the connection is established and maintained, let us concentrate on

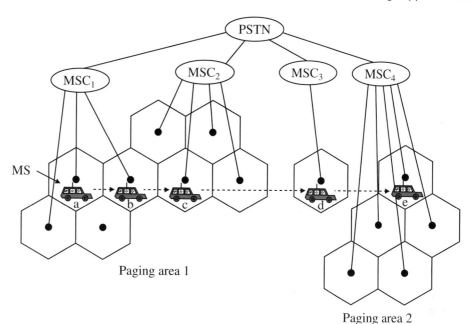

Figure 9.8
Handoff scenarios
with different degrees
of mobility.

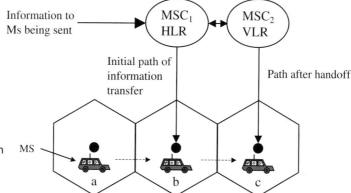

Figure 9.9
Information transmission
path when MS
hands off from "b"
to "c."

an example backbone network that interconnects various MSCs to the Internet and the rest of the world. For illustration, only a small portion of the backbone is shown in Figure 9.10.

Basically, there are two issues involved. One determines the path along the shortest path, and the second ascertains the path according to the current location of the MS. Selecting a new path and making changes to an existing path of the MS would largely depend on the topology of the backbone network. A part of connections between two MSCs is shown in Figure 9.10. Assume that an incoming call is being routed to the backbone along a link as shown in Figure 9.10. Paths needed to reach different backbone networks, and MSCs to be used are shown by the dotted lines for different MS locations and the controlling MSCs. The movement

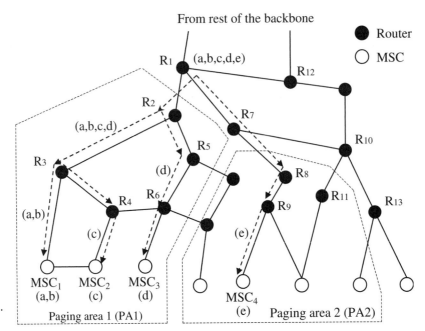

Figure 9.10
Illustration of MSC connections to backbone network and routing/rerouting.

from "a" to "c" can be supported effectively by HLR–VLR, wherein MSC_1 knows how to route the data to MSC_2. One option is to let all the messages reach MSC_1 and forward the messages from there to the MS, wherever it happens to be. But on a long-term basis, this is not the best way to deliver messages. Another option is to find a router along the original path, from where a new path needs to be used to reach the destination MSC along the shortest path. If this is done, then part of the message in the pruned tree could be lost if a hard handoff is performed, which breaks the connection before it makes. Therefore, after handoff, it may be desirable to forward messages from an old location to a new one, for a short duration of time. For MSC_3 and MSC_4 (corresponding to MS locations "d" and "e"), the "break-off" router points are different, and partial pruning of the existing path may be useful in minimizing the delay, avoiding unnecessary forwarding of messages and enhancing utilization of network resources. Similar observation is applicable to the system if the MS is the source of the initiating message. A more complex situation is when both the source and the destination are mobile nodes and a communication path needs to be set up between two such MSs.

9.5.1 Home Agents, Foreign Agents, and Mobile IP

As discussed earlier, depending on the current location and mobility, a MS may have to change its current point of attachment while maintaining its connection to other hosts and the rest of the world. In mobile Internet protocol (Mobile IP), two important agents are associated with the routers: home agent (HA) and foreign agent (FA) [9.5][9.6]. A MS is also registered with a router and for simplicity, a router closest to the home MSC can be selected to serve as its HA. Routers serving

Table 9.2: ▶
Home MSC and Home Agent for Figure 9.9

Home MSC	MSC_1	MSC_2	MSC_3	MSC_4
Selected router for maintaining its home agent	R_3	R_4	R_6	R_9

as HAs for all MSs registered in different MSCs of Figure 9.9 are shown in Table 9.2. It should be noted that routers may have different capabilities, and a router other than the closest one could also serve as the HA router.

Once a MS moves from the home network (where it is registered) to a foreign network, a software agent in the new network known as the FA assists the MS by forwarding packets for the MS. The functionality of HA–FA is somewhat analogous to the HLR–VLR pair, except that it supports mobility in a much broader sense and even in an unknown territory as long as there is an agreement and understanding about "roaming" charges between different service providers of the home network and the foreign network. This way of forwarding packets between HA and FA is also known as "tunneling" between the two involved networks. The way it works is as follows: Whenever a MS moves into a new network, its HA remains unchanged. A MS can detect the FA of the current network domain by the periodic beacon signals that the FA transmits. On the other hand, the MS can itself send agent solicitation messages, to which the FA responds. When the FA detects that a new MS has moved into its domain, it allocates a care-of-address (CoA) to the MS. The CoA can either be the address of the FA itself, or it may be a new address called colocated CoA (C-CoA) that the FA allocates to the MS using the dynamic host configuration protocol (DHCP) [9.7].

Once the MS receives the CoA, it registers this CoA with its HA and the time limit for its binding validity. Such a registration is initiated either directly by the MS to its HA or indirectly through the FA at the current location (Figure 9.11). The HA then confirms this binding through a reply to the MS. A message sent from an arbitrary source to the MS at the home address is received by the HA, binding for the MS is checked, without which the message will be lost, as it will remain unknown where to send or forward the packets. The HA encapsulates the packet with the CoA of the MS and forwards it to the FA area. If the C-CoA address is used, the MS receives the packet directly and is decapsulated to interpret the information. If CoA for the FA is used, then the packet reaches the FA, which decapsulates the packet and passes it on to the MS at the link layer. This registration and message forwarding process is illustrated in Figures 9.11 and 9.12. In an Internet environment, this is known as Mobile IP.

If after expiry of the binding the MS still wants to have packets forwarded through HA, it needs to renew its registration request. When the MS returns to its home network, it sends a registration request to its HA so that the HA need not forward to the FA anymore. If the MS moves to another foreign network, it has to go through another registration process so that the HA can update the location of the currently serving FA.

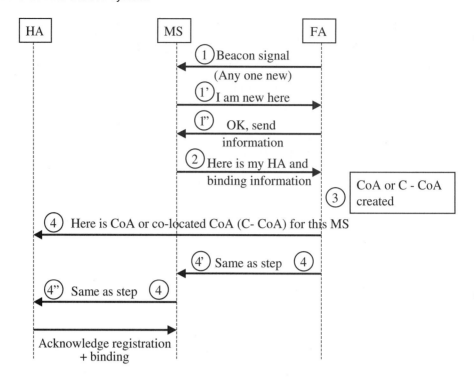

Figure 9.11
Registration process between FA, MS, and HA when the MS moves to a new paging area.

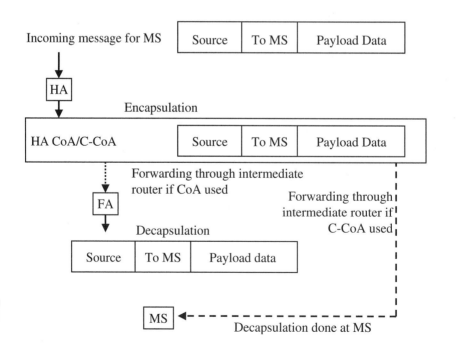

Figure 9.12
Message forwarding to the MS using the HA–FA pair.

9.5.2   Rerouting in Backbone Routers

As discussed in an earlier section, rerouting is needed whenever a MS moves to a new connecting point of the backbone network or moves to a new PA so that the FA–HA pair can exchange control information. The MS still has the same HA, even if it travels to a new network, so that the FA can get information about the closest router attachment point to its HA. However, the question is how a FA in another area can locate the HA. There are many ways to achieve this in the backbone router network. A simplistic approach is to have a global table at all routers of the network so that the route from FA to HA (associated with the MS) can be found. But this kind of one-step global table may become excessively large, and one network provider may not like to furnish information about all its routers to another network enterprise, but may provide information about how to access that network at some selected router (commonly known as a gateway router). This practical limitation necessitates the use of a distributed routing scheme, and one such approach is shown in Figure 9.13. Only gateway routers that support routing within the backbone are shown, and other intermediate routers have been eliminated as they do not help in routing within the backbone. The distributed routing table given in Table 9.3 is made available at different gateway routers so that different PAs and hence the HA can be located in a distributed manner from one router to another until the FA is reached. The process of creating indirect links and having virtual bidirectional paths between HA and FA is known as "tunneling" and is very useful in supporting indirection in such a mobile environment.

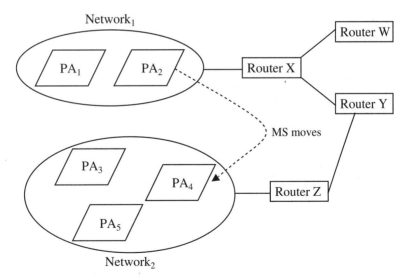

Figure 9.13
Illustration of paging areas (PAs) and backbone router interconnect.

Table 9.3: ▶
Distributed Routing Table and Location of PAs

Table at Router W		Table at Router X		Table at Router Y		Table at Router Z	
Route to PA	Next Hop	Route to PA	Next Hop	Route to PA	Next Hop	Route to PA	Next Hop
1	X	1	–	1	X	1	Y
2	X	2	–	2	X	2	Y
3	X	3	Y	3	Z	3	–
4	X	4	Y	4	Z	4	–
5	X	5	Y	5	Z	5	–

● ● ● ● ● ● ● ● ● ● ● ● ● ● ● ● ●

9.6 Multicasting

Multicasting [9.8] is the process of transmitting messages from a source to multiple recipients by using a single address known as a group address. It greatly reduces the number of messages that need to be transmitted as compared to multiple unicasting for each member, thereby optimizing the bandwidth utilization. Multicasting is found to be an extremely valuable technology for video/audio conferencing, distance learning, and multiparty games that are anticipated to be available with wireless capabilities in the near future.

Generally, multicasting is performed either by building a source-based tree or by using a core-based tree. In a source-based tree approach, for each source of the group, a shortest path tree is created, encompassing all the members of the group, with the source being at the root of the tree; whereas in a core-based tree approach a particular router is chosen as a core. Every source forwards the packet to the core router, which takes care of forwarding the packet to all members of the multicast group.

Multicasting requires grafting and pruning of the tree due to members continuously joining and leaving the group. Users can dynamically join a multicast group to receive multicast packets. However no subscription is needed to send multicast packets to a given group.

In the Internet, multicast has been supported by adding multicast-capable routers (MROUTERs) which are connected through dedicated paths, called tunnels. Tunnels connect one MROUTER to another, and carry multicast packets via other regular routers. MROUTERs encapsulate the multicast packet as a regular IP packet and send it through the tunnel to other MROUTERs as a unicast packet, which is decapsulated at the other end. This MROUTER arrangement in the Internet is generally referred to as multicast backbone (MBONE).

In a wireless network, because of the movement of group members, packet forwarding is much more complex. There is a need to design an efficient scheme to

Figure 9.14
Packet duplication
in BT approach
[9.10].*Courtesy*:
Siddesh Kamat,
"Handling Source
Movement over
Mobile-IP and Reducing
the control overhead
for a secure, scalable
Multicast Framework"
M.S. Thesis, University
of Cincinnati, October
2002.

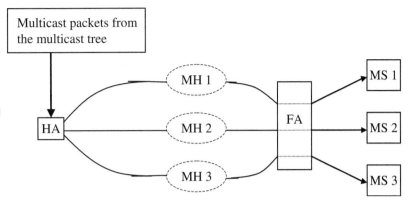

address problems like nonoptimal path length, avoid packet duplication, and prevent disruption of packet delivery during multicast tree generation.

The Internet Engineering Task Force (IETF) has proposed two methods for providing multicast over Mobile IP [9.9]: bidirectional tunneling (BT) and remote subscription. In the BT approach, whenever a MS moves into a foreign network, the HA creates a bidirectional tunnel to the FA that is currently serving the MS and encapsulates the packets for the MS. The FA then forwards the packets to the MS through the reverse tunnel as shown in Figure 9.14. On the other hand, in the remote subscription approach, whenever a MS moves into a foreign network, the FA (if not a member of the multicast tree) sends a tree join request. The MS then directly receives the multicast packets through the FA. Although this approach is simple and prevents packet duplication and nonoptimal path delivery, it needs the FA to join the multicast tree and hence can cause data disruption until the FA is connected to the tree. It also results in frequent tree updates when the MSs move frequently.

The BT approach prevents data disruption due to movement of the MS, but it causes packet duplication if several MSs of the same HA, which are subscribed to the same multicast group, move to the same FA. For each MS that has moved into the FA, each of their respective HAs forwards a copy of the multicast packet to the subscribed group. It may happen that MSs under different HAs move into the same foreign domain. Hence, the FA would receive duplicate packets from the HAs for their MSs located in the foreign domain. This is generally referred to as the tunnel convergence problem (Figure 9.15).

The mobile multicast (MoM) protocol [9.11] tries to address the issue of the tunnel convergence problem by forcing a HA to forward only one multicast packet for a particular group to the FA irrespective of the number of its MSs being present in the FA network for that group. Here the FA selects a designated multicast service provider (DMSP) for each group among the given set of HAs. Here, only the DMSP is responsible for forwarding a multicast packet to the FA for that group. This scheme is illustrated in Figure 9.16.

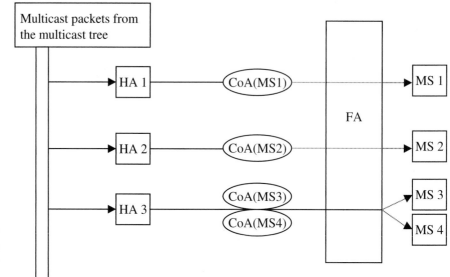

Figure 9.15
Tunnel convergence problem [9.10].
Credit: Siddesh Kamat, "Handling Source Movement over Mobile-IP and Reducing the control overhead for a secure, scalable Multicast Framework" M.S. Thesis, University of Cincinnati, October 2002.

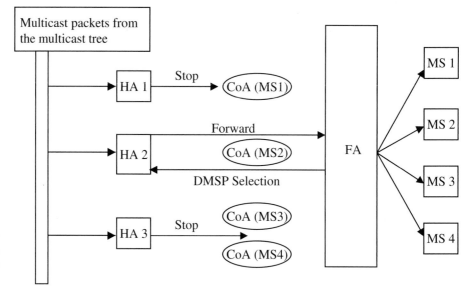

Figure 9.16
Illustration of MoM protocol [9.10].
Credit: Siddesh Kamat, "Handling Source Movement over Mobile-IP and Reducing the control overhead for a secure, scalable Multicast Framework" M.S. Thesis, University of Cincinnati, October 2002.

However, if the MS of the serving DMSP moves out, then the DMSP may stop forwarding packets to the FA. It will result in data disruption until the FA reselects a new DMSP. To handle this issue, the scheme employs more than one DMSP for a particular group (which may result in data duplication). In the MoM protocol, packet duplication can also occur if the FA itself is a tree node (Figure 9.16). A comprehensive review of the multicast routing protocols that have been proposed in the literature has been given in [9.12].

9.7 **Security and Privacy**

In all the network communication, whether implementing unicast or multicast, it is extremely important to ensure authenticity of all the messages. In a wireless system, transfer through an open-air medium makes messages vulnerable to many additional types of attack. If the problem is that of "jamming" by a very powerful transmitting station at one frequency band, then that could be easily overcome by using the frequency-hopping (FH) technique. We can ask why the jamming transmitter does not also use the same hopping sequence. First, it is relatively difficult to do such hopping for a powerful station whose primary objective is to overcome jamming by its own powerful signal. Second, the FH sequence is known only to the authorized wireless transmitters and the corresponding receivers, and if the sequence is known to an intruder, then many other things can be done. Therefore, the real challenge is how to ensure that unauthorized users cannot easily interpret the signals going through the air. In this section, we discuss many possible encryption techniques. The other issue is how to check the authenticity of all users, and we also explain this in detail.

9.7.1 Encryption Techniques

Encryption of a message can be provided by simply permuting the bits in a pre-specified manner before being transmitted, and one such example of perfect shuffle is shown in Figure 9.17. Transformation from input to output is fixed, and input WIRELESS at input terminal pins 1, 2, 3, 4, 5, 6, 7, and 8 is changed to WLIERSES at output side. Any other fixed permutation can be used for encryption as long as the transformations are also known at the receiver for decryption. In other words, such permuted information, received by a legitimate receiver, can easily be reconstructed by performing a backward operation as long as the process is reversible. One such data encryption standard (DES) on input bits is shown in Figure 9.18. Given a block

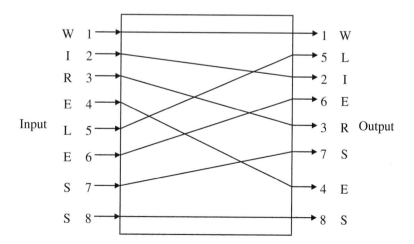

Figure 9.17
Simple permutation function.

1	2	3	4	5	6	7	8	57	49	41	33	25	17	9	1	8	24	40	56	16	32	48	64
9	10	11	12	13	14	15	16	61	53	45	37	29	21	13	5	7	23	39	55	15	31	47	63
17	18	19	20	21	22	23	24	58	50	42	34	26	18	10	2	6	22	38	54	14	30	46	62
25	26	27	28	29	30	31	32	62	54	46	38	30	22	14	6	5	21	37	53	13	29	45	61
33	34	35	36	37	38	39	40	59	51	43	35	27	19	11	3	4	20	36	52	12	28	44	60
41	42	43	44	45	46	47	48	63	55	47	39	31	23	15	7	3	19	35	51	11	27	43	59
49	50	51	52	53	54	55	56	60	52	44	36	28	20	12	4	2	18	34	50	10	26	42	58
57	58	59	60	61	62	63	64	64	56	48	40	32	24	16	8	1	17	33	49	9	25	41	57

(a) Information sequence to be transmitted

(b) Permutation of information sequence before transmission

(c) Permutation to be performed on received information sequence

Figure 9.18
Initial bit pattern and effect of permutation before transmission and after reception using DES.

of 64 input bits of information shown in Figure 9.18(a), the bits are permuted as shown in Figure 9.17(b) before they are sent out. This means that the 57th bit is transmitted first, then the 49th bit, and so on. At the receiving end, a reverse operation on received bits is performed as shown in Figure 9.18(c). This implies that the 8th received bit is moved as the first information bit, then the 24th received bit is considered as the second bit, and so on, till the 8th bit is sent at the end. It may be noted that the first bit of information is transmitted as the 8th bit and the second information bit is sent as the 24th bit. In this way, the original information before permutation can be reconstructed at the receiving side by applying the same permutation to each group of 64 bits before transmission and by doing reverse operation at the receiving side. As it is important to know the permutation in order to get the original information bits in the right order, permuted patterns going through the air received by other MSs cannot easily decrypt the message. Of course, trying different possible combinations of permutations could break the encrypted information. It may be noted that the permutation can be done at the level of group of bits and not just necessarily for each bit.

A complex encryption scheme can involve transforming input blocks to some encoded form, which can be difficult for others to understand and interpret. However, it should be done in such a way that the encoded information could be uniquely mapped back to the initial information. Such a generic process is shown in Figure 9.19. The simplest transformation can involve logical, arithmetic, or both operations, and the selection of such functions depends on whether there is one-to-one correspondence and how difficult it is to encode and decode. Both EX–OR and its complementary Boolean functions do translate uniquely, and decoding also leads to a unique solution. Among arithmetic functions, either addition or subtraction can achieve similar results, and there is no need to look at more complex arithmetic operations of

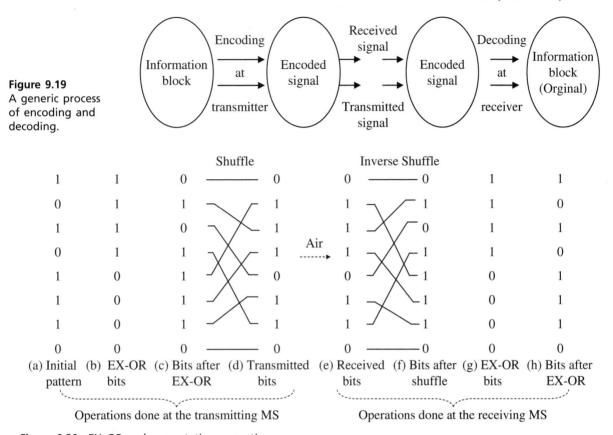

Figure 9.19
A generic process of encoding and decoding.

Figure 9.20   EX–OR and permutation operations.

multiplication and division. In fact, a combination of logical, arithmetic, or permutation operations could be employed to make the encryption process robust and secure.

For example, consider a combination of arithmetic and permutation operations illustrated in Figure 9.20. The initial information pattern 10101110 is first EX–OR with 1111000; then a perfect shuffle permutation is performed, and the resultant bits are ready for transmission through the air, which can be received by all MSs in the receiving range. To get the original information back, a bitwise reverse permutation must be performed and then EX–OR with 1111000 leads to the original message. It may be noted that the sequence of operation at the receiving end ought to be done in exactly the reverse sequence, which should be known to the receiver. Otherwise, it is not easy to decrypt the transmitted message. These steps are shown in Figure 9.20. A generic procedure and more complex steps are illustrated in Figure 9.21.

9.7.2 Authentication

Authentication of a subscriber basically implies making sure that the user is genuine. There are many ways to ascertain this, and one simple technique is to use a hash function (just like a password) from an associated user's unique identification. But

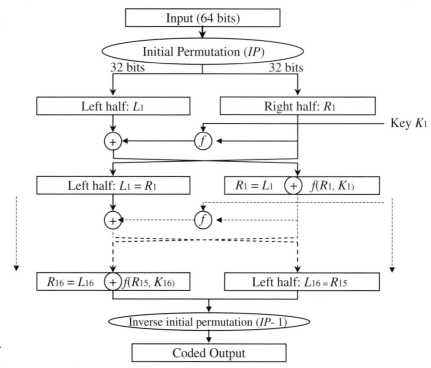

Figure 9.21
Permutation and
coding of information.

this is not foolproof, as many key words can be mapped to the same hashing function and there is no unique correspondence when decoded. Another approach is to use two different interrelated keys, the first key known only to the system generating it and the second used for sending to the outside world. Such private and public key pairs are extensively used in numerous authentication applications. The popularity of such a scheme hails from the fact that it is relatively difficult and computationally complex to determine the private key, even if the public key is known to everyone.

The steps for public–private key authentication are shown in Figure 9.22. The system selects a private key for an arbitrary user, i, such that it is difficult to guess

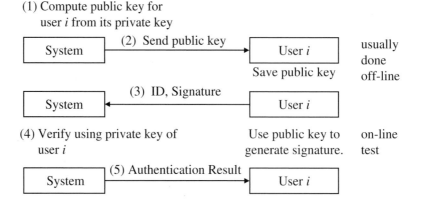

Figure 9.22
Public–private key
authentication
steps.

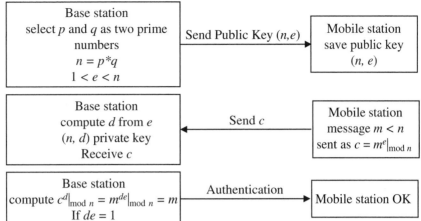

Figure 9.23
Message authentication using public–private key.

for other users. One approach is to utilize a large prime number as the private key at the time of initial setup of the user. This prevents other users from knowing even the public key. The RSA algorithm (named after its inventors, Ron Rivest, Adi Shamir, and Len Adleman of the Massachusetts Institute of Technology) is the best known public–private key pairing system.

In the RSA method (see Figure 9.23), two large prime numbers p and q are picked and n is obtained from the multiplication of the two $(n = p * q)$. Then a number e is selected appropriately to use (n, e) as the public key and is transmitted by the system to the user. The user stores that, and whenever a message $m < n$ needs to be transmitted, the user computes $c = m^e|_{\mod n}$ and sends that to the system. After receiving c, the system computes $c^d|_{\mod n}$, where d is computed using the public key (n, e). To reconstruct m at the system, some specific condition needs to be satisfied. As

$$c = m^e|_{\mod n}, \text{give}$$

$$c^d|_{\mod n} = \left(m^e|_{\mod n}\right)^d|_{\mod n} = \left(m^e\right)^d|_{\mod n}$$

$$= m^{ed}|_{\mod n}.$$

To make this equal to m, $e*d$ needs to be equal to 1. That means e and d need to multiplicative inverse using mod n; or mod $(p*q)$. This can be satisfied provided e is prime with respect to $(p-1)(q-1)$. Therefore, imposing this restriction, the original message can be reconstructed. For example, let us have $p = 3, q = 11, n = pq = 33$. The number e is selected so that e is relatively prime to $(p-1)(q-1) = 20$. Therefore, $e = 7$. The number d satisfies $de = 1 \ |_{\mod (p-1)(q-1)}$. Therefore, we have $d = 3$. For the message $m = 4$, the user i computes $c = m^e \ |_{\mod n} = 4^7 \ |_{\mod 33} = 16$. After receiving the number $c = 16$, the system reconstructs the message by calculating $m' = c^d \ |_{\mod n} = 16^3 \ |_{\mod 33} = 4$. A similar scheme can be used between independent users i and j by using each other's public keys.

(a) Authentication based on ID

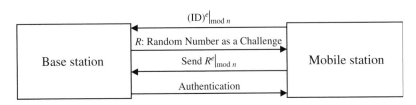

Figure 9.24
Authentication of
a mobile station by
the base station.

(b) Authentication using a challenge

In a wireless environment, such a scheme can be used to check each MS's ID, and a simple scheme using public–private keys is given in Figure 9.24(a). Whenever this scheme is used, the MS having a fixed ID will always send the same bit pattern for authentication by the BS. This signal goes through the air, and other MSs observe this specific response bit pattern and could try to pose as someone else by using $(ID)^e|_{mod\,n}$ of the MS. This problem could be easily solved by having an additional level of security, as shown in Figure 9.24(b). The modification is in step 2; when the BS verifies the ID of a MS, it sends a random number as a challenge message R to the MS. The MS, using its public key, computes $R^e|_{mod\,n}$ and returns that value to the BS. The BS checks that using the private key and can finally send the authentication message to the MS. The public key of the MS given by the BS is assumed to be retained only by the MS for future use, and the BS serves as the central authority for authentication.

The public–private keys can also be used to encrypt/decrypt messages between BS and the MS, or vice versa. This works very well unless someone tries to break in and do some nasty things. Security in wireless systems is extremely important and is discussed in the next section.

9.7.3 Wireless System Security

A secure wireless system needs to be capable of protecting its associated constituents with respect to confidentiality, integrity, and nonrepudiation [9.13][9.14]. Security maintenance services enhance the security of all information transfers. These are basically implemented to counter different possible intruder attacks. The services of security can be classified in the following categories:

1. **Confidentiality**: Only the authorized party can access the information in the system and transmit data.

2. **Nonrepudiation**: The sender and receiver cannot deny the transmission.

3. **Authentication**: Ensures that the sender of the information is correctly identified. It enables partial nonrepudiation, but use of additional features (e.g., timestamping services) is also required to protect the routing traffic from tamper attacks, such as the replaying or delaying of routing messages.

4. **Integrity**: The content of the message or information can only be modified by the authorized users.

5. **Availability**: Computer system resources should be available only to the authorized users.

Similarly, security mechanisms can also be divided into three categories:

1. **Security prevention**: Enforces security during the operation of a system by preventing security violations. It is implemented to counter security attacks.

2. **Security detection**: Detects attempts both to violate security and to address successful security violations. An intrusion detection system (IDS) comes under this category.

3. **Recovery**: Used to restore the system to a presecurity violation state after a security violation has been detected.

Security requirements of wireless systems depend on the amount of investment and the characteristics of applications running on the system. For example, electronic funds transfer, reservation systems, and typical control systems all have different levels of demands and expectations. Absolute security and reliability are relatively abstract terms, and a "secure system" can be defined as one where an intruder has to spend an unacceptable amount of time and effort in interpreting the system. For most systems, the cost for security increases exponentially with an increase in the security level toward the 100% level (Figure 9.25). Hence, there is a tradeoff between the level of increasing system security and the potential cost incurred.

Threats to security can be viewed as potential violations of security, and they exist mainly due to weak points in a system. Threats can be broadly classified in two types: accidental threats and intentional threats. Accidental threats result due to operational mistakes of system and hardware/software failure. Intentional threats or attacks can be seen as any action performed by an entity with an intention to violate security. These attacks can be categorized as shown in Figure 9.26.

1. **Interruption**: An intruder attacks availability by blocking or interrupting system resources.

2. **Interception**: System resources are rightfully accessed by the illegal party. This attacks confidentiality.

3. **Modification**: To create an anomaly in the network, an illegal party transmits spurious messages. This affects authenticity.

4. **Fabrication**: An unauthorized party transmits counterfeit objects into the system and causes an attack on authenticity.

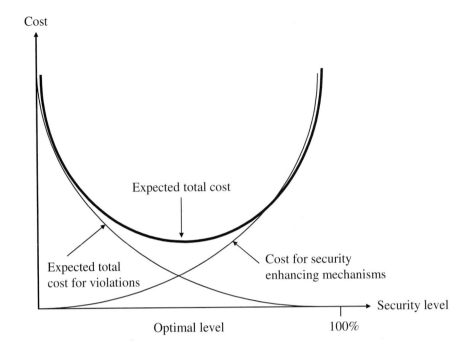

Figure 9.25
Cost function of
a secured wireless
system.

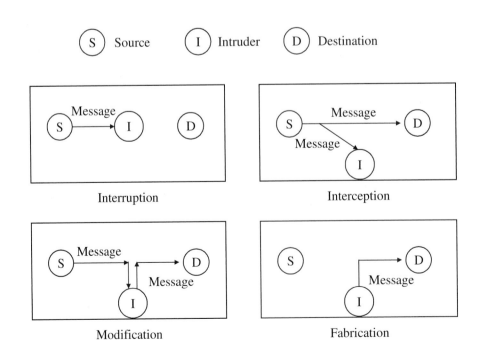

Figure 9.26
Security threat
categories.

Furthermore, attacks can be classified into active and passive attacks. Active attacks include transmission of data to the parties, or unauthorized user blocking the data stream. The following are different types of active attacks:

1. **Masquerade**: Attacker poses as an authorized party to make privileged changes to the messages.

2. **Replay**: Playing back of previously sent data to create undesirable effects.

3. **Modification of data**: Original message is tampered with to create inconsistency in the network.

4. **Denial of service (DoS)**: Involves hijacking of network resources, thereby preventing other authorized users from utilizing network resources.

Passive attacks are done when an unauthorized attacker monitors or listens to the communication between two parties. In general, it is very hard to detect passive attacks since they do not disturb the system. Examples of passive attacks are monitoring network traffic, CPU, and disk usage. Encrypting messages can partly solve the problem since even the traffic flow on a network may reveal some information. Traffic analysis, such as measuring the length, time, and frequency of transmissions, can help in predicting or guessing network activities.

● ● ● ● ● ● ● ● ● ● ● ● ● ● ⋯

9.8 Firewalls and System Security

As more and more wireless systems get deployed, security emerges as the single most important feature that needs to be taken care of. As opposed to traditional wired systems, wireless systems are more vulnerable because the signal goes through the open air. Hence, it is essential to implement control access to the network through design of robust firewall mechanisms. A network firewall can be defined as a black box that resides between the World Wide Web and the network. It keeps out malicious and unwanted traffic while also preventing inside users from accessing prohibited locations on the Web. There are mainly two types of firewalls: network firewalls and host-based firewalls. Network firewalls protect the network by monitoring and controlling incoming and outgoing traffic. Host-based firewalls, on the other hand, protect individual hosts irrespective of the network to which they are connected.

A firewall mainly carries out traffic filtering, Web authentication, and other security mechanisms. Traffic filtering is the process of monitoring traffic based on certain parameters. The way traffic filtering works is that the firewall blocks everything that has not been explicitly allowed by the administrator. The way a filtering mechanism can be configured is by fixing the values for one or more of the following:

- Source IP
- Destination IP
- Source TCP/UDP port
- Destination TCP/UDP port
- Arrival interface
- Destination interface
- IP protocol

In a typical WLAN environment, a firewall resides at a wireless access point. A wireless AP is the single point of connectivity to the Internet for all wireless users within the domain of the access point. The AP carries out authentication by the help of an authentication server (also called an authentication authorization and accounting (AAA) server), which resides somewhere in the same administrative domain. The most popular protocol used by the AAA server is known as the remote authentication dial-in user service (RADIUS) [9.15] protocol, although a new protocol called DIAMETER [9.16] has been proposed, which seeks to solve many of the problems that the RADIUS protocol had. In the case of 3G networks, a cell phone is connected to a base station subsystem (BSS), which connects to a MSC (mobile switching center). The MSC connects to a gateway-MSC (G-MSC) which connects to a wireless application protocol (WAP) gateway. It is between the WAP Gateway and the Internet that a firewall typically resides. Again, it is up to the system administrator to place a firewall further inside the network, depending upon the level of security and control required by the system.

● ● ● ● ● ● ● ● ● ● ● ● ● ● ● ●

9.9 Summary

Resource allocation is important in influencing the wireless network performance. As any wireless system consists of wireless components as well as an underlying wired network as a backbone, overall performance requires enhancing both of these infrastructures. In this chapter, we have considered how channels can be allocated in wireless systems and how routing/rerouting/security is provided effectively in the wired backbone so that there is minimal impact on the performance of a wireless system.

In the following chapters, we consider details of existing wireless systems in the United States, Europe, and Japan as well as other relevant areas, such as satellite communications, wireless ad hoc networks, wireless LANs, and wireless PANs, and illustrate their usefulness.

● ● ● ● ● ● ● ● ● ● ● ● ● ● ● ●

9.10 References

[9.1] S. Gerasenko, A. Joshi, S. Rayaprolu, K. Ponnavaikko, and D. P. Agrawal, "Beacon Signals: What, Why, How, and Where," *IEEE Computer*, Vol. 34, No. 10, pp. 108–110, October 2001.

[9.2] D. Chung, H. Choo, and H. Y. Youn, "Reduction of Location Update Traffic Using Virtual Layer in PCS," *Proceedings of the 2001 International Conference on Parallel Processing*, pp. 331–338, September 2001.

[9.3] P. P. Mishra and M. Srivastava, "Effect of Virtual Circuit Rerouting on Application Performance," *Proceedings of the 17th International Conference on Distributed Computing Systems*, pp. 374–383, May 27–30 1997.

[9.4] J. C. Chen, K. M. Sivalingam, and R. Acharya, "Comparative Analysis of Wireless ATM Channel Access Protocols Supporting Multimedia Traffic," *Mobile Networks and Applications*, Vol. 3, pp. 293–306, 1998.

[9.5] A. Acharya, J. Li, F. Ansari, and D. Raychaudhari, "Mobility Support for IP Over Wireless ATM," *IEEE Communications Magazine*, pp. 84–88, April 1998.

[9.6] R. H. Glitho, E. Olougouna, and S. Pierre, "Mobile Agents and Their Use for Information Retrieval: A Brief Overview and an Elaborate Case Study," *IEEE Networks*, pp. 34–41, January/February 2002.

[9.7] R. Droms, "Dynamic Host Configuration Protocol (DHCP)," *IETF RFC 2131*, March 1997.

[9.8] S. Deering, "Multicast Routing in a Datagram Internetwork," Ph.D. Thesis, Stanford University, Palo Alto, California, December 1991.

[9.9] C. Perkins, "IP Mobility Support," *IETF RFC 2002*, IBM, October 1996.

[9.10] S. Kamat, "Handling Source Movement over Mobile-IP and Reducing the Control Overhead for a Secure, Scalable Multicast Framework," M.S. Thesis, University of Cincinnati, 2003.

[9.11] V. Chikarmane, C. Williamson, R. Bunt, and W. Mackrell, "Multicast Support for Mobile Hosts Using Mobile IP: Design Issues and Proposed Architecture," *ACM/Baltzer Mobile Networks and Applications*, Vol. 3, no. 4, pp. 365–379, 1998.

[9.12] H. Gossain, C. M. Cordeiro, and D. P. Agrawal, "Multicast: Wired to Wireless," *IEEE Communications Magazine*, pp. 116–123, June 2002.

[9.13] L. Venkataraman and D. P. Agrawal, "Strategies for Enhancing Routing Security in Protocols for Mobile Ad Hoc Networks," *JPDC Special Issue on Mobile Ad Hoc Networking and Computing*, Vol. 63, No. 2, pp. 214–227, February 2003.

[9.14] S. Bhargava and D. P. Agrawal, "Security Enhancements in AODV Protocol for Wireless Ad Hoc Networks," *IEEE Vehicular Technology Conference (VTC)*, Fall 2001, pp. 2143–2147, October 2001.

[9.15] C. Rigney, S. Willens, A. Rubens, and W. Simpson, "Remote Authentication Dial In User Service (RADIUS)," *IETF RFC 2865*, June 2000.

[9.16] P. Calhoun, J. Loughney, E. Guttman, G. Zorn, and J. Arkko, "Diameter Base Protocol," *IETF RFC 3588*, September 2003.

• • • • • • • • • • • • • • •

9.11 Problems

P9.1. From a local wireless service provider, find out what kind of EIR information is retained for each subscriber.

P9.2. You have temporarily moved to a new area and you would like to use your cell phone. What alternatives do you have if:

(a) There is no service provider in that area?

(b) There is no agreement between your wireless phone service provider and the service provider in the new area?

(c) The area is covered only by a satellite phone service?

P9.3. What is the bandwidth and the power level used by the "beacon signals" in your area?

P9.4. Like the cellular system, the IEEE 802.11 wireless LANs also have the "beacon signals." Search for an IEEE 802.11 specification online and find out what information is included in a beacon signal.

P9.5. From your favorite Web site, find out the acceptable bit error rate for the following applications:

(a) Voice communication

(b) Video communication

(c) Defense applications

(d) Sensor data communication in a nuclear plant

(e) Sensor measuring paper thickness in a plant

(f) Sensor measuring temperature for different steps of a chemical process

(g) Sensor measuring accuracy of a lathe machine

P9.6. Assuming that you just got out of an airplane and you switched on your cell phone. If the closest BS is located at a distance of 5 km, what are the minimum and the maximum delay before a contact is established between your cell phone and the nearest BS, given that the BS transmits beacon signals every one second?

P9.7. In the backbone network, it is desirable to find out the shortest path from the source to the destination. How do you do this in a wireless network environment, where the subscribers have finite mobility? Explain clearly.

P9.8. What is the use of "attachment points" from one network to another network? Explain their significance in wireless network routing?

P9.9. In a wireless network, the radio signal is broadcast through the air. Therefore, what is the significance of multicasting in this context? Explain in detail.

P9.10. What is meant by bidirectional tunneling? Why do you need HA–FA in addition to the HLR–VLR pair? Explain clearly.

P9.11. The function of a 10×10 permutator is given by the following table:

Input	1	2	3	4	5	6	7	8	9	10
Output	1	6	2	7	3	8	4	9	5	10

(a) Find out the output message going through the air if the input message sequence is given by:

> I WANT TO LEARN ABOUT PERMUTATION FUNCTION
> IN WIRELESS DEVICES AND APPLICATIONS. . . .

(b) Assume that the message is transmitted as a group of ten characters. What are the advantages and disadvantages if two such permuters are used?

P9.12. Consider the word "wireless," composed of eight symbols, each symbol being a letter of the English alphabet. This word is encrypted by first applying a permutation function and then a substitution function. The permutation function is applied on a 4-symbol half word as follows: (1234) => (4132), i.e., every half word with input symbols 1234 is transformed to an output half word, which is 4132. The word is interpreted as a sequence of two half words. The substitution function is as follows:

Input Symbol:	w	i	r	e	l	s
Output Symbol:	i	r	e	l	s	w

(a) What is the final output?

(b) What would be the output if the substitution function were applied before the permutation function instead of after it?

P9.13. The function of an 8x8 permutator is given by the following table:

Input	1	2	3	4	5	6	7	8
Output	8	4	2	1	7	5	3	6

The following message is to be sent through the air:

> I AM DONE WITH MY FINALS AND NOW CAN TAKE A
> BREAK OR A VACATION.

Find the message going through the air if

(a) An 8-way interleaving is done before using the permutator.

(b) An 8-way interleaving is done after using a permutator and before it is transmitted.

P9.14. In the RSA algorithm, the public key is transmitted to all MSs through the air by the BS. How is the security ascertained?

P9.15. Given two prime numbers, $p = 37$ and $q = 23$, define the private and public keys by selecting appropriate values of the number "e."

P9.16. Answer the following:

    (a) Using the public key of Problem P9.15, find the sequence of values transmitted through the air if the ASCII values corresponding to the following message are sent by the BS:

I LIKE THIS CLASS.

    (b) Verify how this message is recovered back at the MS by using the public key.

P9.17. Answer the following:

    (a) How do you differentiate between privacy and security?

    (b) What are the differences between authentication and encryption?

P9.18. Why do you need to send a random number to test a MS? Explain clearly.

P9.19. Answer the following:

    (a) Do you recall any recent event that is an example of denial of service?

    (b) Can denial of service (DoS) be more effective if you have information about the traffic? Explain clearly.

P9.20. A linear permutation $i \rightarrow (i + m) \bmod n$ is used as follows: Does the encryption depend on the value of m? What is the impact of increasing the value of m? Explain clearly.

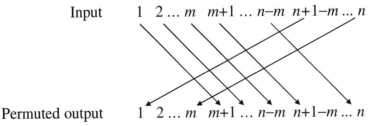

Figure 9.27
Figure for
Problem P9.20.

P9.21. In organizing a conference, a single key has to be used by all program committee members to encrypt and decrypt the message. Assuming this key has to be changed every year, how can you set up such a "session key" using public–private key pairs? Explain clearly.

CHAPTER 10

Existing Wireless Systems

10.1 Introduction

A wireless system needs to take many factors into account such as call rate, call duration, distribution of MSs, traffic in an adjacent cell, the terrain, and atmospheric conditions. To get an idea of how a wireless system could behave in the real world, it is important to study various characteristics of existing cellular systems and how they support seamless mobile communication. In this chapter, we study the details of some of these existing systems.

It is important to emphasize that communication between any two devices is successful only when the receiver gets the intended information from the sender, and this is possible if both the sender and the receiver follow a set of rules called the communications protocol. To facilitate easy transfer of information, the protocol employs many steps of seven-layers as described in International Organization for Standardization (ISO)–Open Systems Interconnection (OSI) model that is widely employed for wired communication (Chapter 12). In a wireless environment, similar steps are followed, except that a few steps or layers are not used for the sake of efficiency. On the other hand, some layers may be subdivided into a number of successive operations and are given in conjunction with the specific cellular systems. From a historical point of view, we consider AMPS (Advanced Mobile Phone System) as the first representative of wireless systems.

10.2 AMPS

AMPS is the first-generation cellular system used in the United States. It transmits speech signals employing FM, and important control information is transmitted in digital form using FSK. AMPS is the first cellular phone technology created by AT&T Bell Labs with the idea of dividing the entire service area into logical divisions called cells. Each cell is allocated one specific band in the frequency spectrum.

To explore a reuse pattern, the frequency spectrum is divided among seven cells, improving the voice quality as each user is given a larger bandwidth. Typically, AMPS uses a cell radius of 1 to 16 miles, depending on various factors such as density of users and traffic intensity. However, there is a tradeoff between the cell area and the quality of service. Larger cells tend to have more thermal noise and less interference, while smaller cells have more interference and less thermal noise. One important aspect of AMPS is that it allows both cell sectoring and splitting. It is also sufficient to have a low-power MS (about 4 watts or less) and a medium-power BS (about 100 watts). AMPS is capable of supporting about 100,000 customers per city, and the system is aimed to reduce blocking probability to about 2% during busy hours.

10.2.1   Characteristics of AMPS

AMPS uses the frequency band from 824 MHz to 849 MHz for transmissions from MSs to the BS (reverse link or uplink) and the frequency band between 869 MHz to 894 MHz from the BS to MS (forward link or downlink). The 3 kHz analog voice signal is modulated onto 30 kHz channels. In transmitting data, the system uses **Manchester** frequency modulation at the rate of 10 kbps, while the control parameters remain the same as in voice transfer. Separate channels are used for transmitting control information and data. Since fewer control messages are exchanged between the MS and the BS as compared with voice or data messages, a smaller number of control channels are employed than voice antennas. In AMPS, there is one control transreceiver for every eight voice transreceivers.

Frequency allocation in AMPS is done by dividing the entire frequency spectrum into two bands—Band A and Band B. Frequencies are allocated to these bands, as shown in the Table 10.1 [10.1].

The non-wireline providers are given Band A and Bell wireline providers are given Band B. A total of 666 channels (which was later increased to 832 channels) is divided among these two bands, and a cluster of seven cells allows many users to employ the same frequency spectrum simultaneously. AMPS's use of directional radio

Table 10.1: ▶
Band Allocation in AMPS

Band	MS Transmitter (MHz)	BS Transmitter (MHz)	Channel Numbers	Total Number of Channels
A	825.03–834.99	870.03–879.99	1–333	333
A'	845.01–846.48	890.01–891.48	667–716	50
A"	824.04–825.00	869.04–870.00	991–1023	33
B	835.02–844.98	880.02–889.98	334–666	333
B'	846.51–848.97	889.51–893.97	717–799	83
Not used	824.01	869.01	990	1

propagation enables different frequencies to be transmitted in different directions, thereby reducing radio interference considerably.

10.2.2   Operation of AMPS

A general state diagram of how an AMPS system handles calls and various other responsibilities is shown in Figure 10.1. At the powerup, all MSs in the range of a BS have to go through a registration with AMPS before actual service begins. Thereafter, any incoming or outgoing call is handled according to the state of the system. Each MS also goes through the registration process when handoff to an adjacent cell occurs.

Three identification numbers are included in the AMPS system to perform various functions [10.1]:

1. **Electronic serial number (ESN)**: A 32-bit binary number uniquely identifies a cellular unit or a MS and is established by the manufacturer at the factory. Since it is unique, any MS can be precisely identified by this number. For security reasons, this number should not be alterable and should be present in all MSs.

2. **System identification number (SID)**: A unique 15-bit binary number assigned to a cellular system. The Federal Communications Commission (FCC) assigns one SID to every cellular system, which is used by all MSs registered in the service region. A MS should first transmit this number before any call can be handled. The SID serves as a check and can be used in determining if a particular MS is registered in the same system or if it is just roaming.

3. **Mobile identification number (MIN)**: A digital representation of the MS's 10-digit directory telephone number.

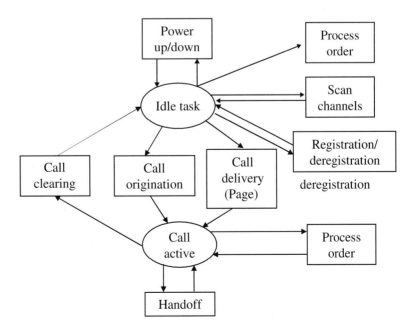

Figure 10.1
General operation
of MS in AMPS.

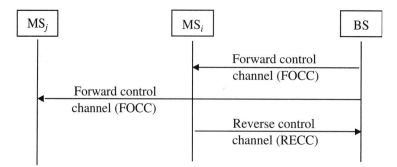

Figure 10.2
Forward and reverse channels.

The location of a particular MS is not predictable. Then the question is, how does a MS know when it receives a call? The answer lies in the messages passed on the control channels. Whenever the MS is not in service, it tunes to the strongest channel to find out useful control information. The same happens at the BS as well. There are two important control channels: forward control channel (FOCC) from BS to the MS, and reverse control channel (RECC) from MS to BS, both operating at 10 kbps, as shown in Figure 10.2. Various channels used by the AMPS are as follows:

- **Forward control channel (FOCC)**: FOCC is primarily used by the BS to page and locate the MSs using the control information in three-way time division multiplexing mode (Figure 10.3). The busy/idle status shows if the RECC is busy, and stream A and stream B allow all the MSs to listen to the BS. Stream A is for MSs having least significant bit (LSB) of MIN as zero, while stream B is for those MSs with LSB of MIN as one. As a part of control information, BS also allocates voice channels to MSs. Each data frame consists of several components, starting with a dotting sequence (alternating 1s and 0s), continues with a word-sync pattern, and is followed by five repeats of word-A and word-B data. The BS forms each word by encoding 28 content bits into a (40, 28) BCH code. Figure 10.3 shows the detailed FOCC frame format. The first busy/idle bits are inserted at the beginning of the dotting sequence. The second is inserted at the beginning of the word sync, and the third is inserted at the end of the word sync. After the third busy/idle bit, a busy/idle bit is inserted every 10 bits through the

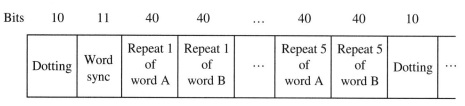

Figure 10.3
Format of FOCC.

Dotting = 1010…101
Word sync = 11100010010

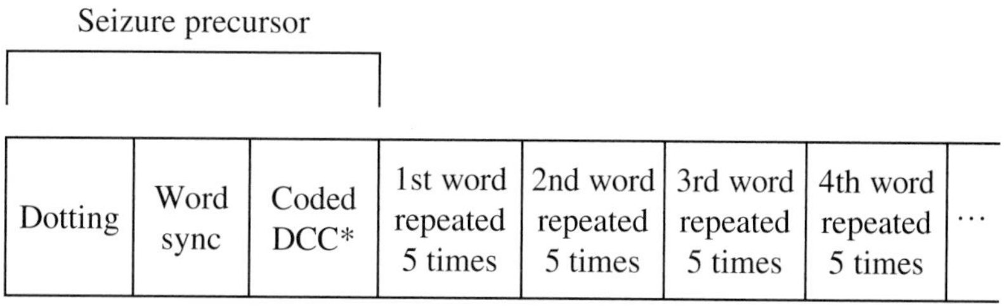

Seizure precursor

Dotting	Word sync	Coded DCC*	1st word repeated 5 times	2nd word repeated 5 times	3rd word repeated 5 times	4th word repeated 5 times	⋯

Dotting = 1010…101

Word sync = 11100010010

* DCC = Digital color code

Figure 10.4
Format of RECC.

five repeats of word-A and word-B data. The busy/idle bits indicate the control channel availability with the BS. An idle-to-busy transition coordinates messages sent on the control channel.

- **Reverse control channel (RECC)**: Control for the reverse direction is little involved as this information comes from one or more MSs using the RECC channel. This could also be in response to the page sent by the BS. There could be several MSs responding to queries. A simple mechanism to indicate whether RECC is busy or idle, is to model it after the slotted ALOHA packet radio channel. Figure 10.4 shows a typical format of the RECC message, which begins with the RECC seizure precursor of 30 bits of dotting, 11 bits of word sync, and the 7-bit coded digital color code (DCC). DCC is primarily used to detect if any cochannel interference is occurring in the specified region. For a single-word transmission following the seizure precursor, a single RECC message word repeats itself five times. The seizure precursor fields are used for synchronization and identification. For a multiple-word transmission following the seizure precursor, the first RECC message word repeats itself five times; then the second RECC message word is repeated five times.

- **Forward voice channel (FVC)**: FVC is used for one-to-one communication from the BS to each individual MS. A limited number of messages can be sent on this channel. A 101-bit dotting pattern represents the beginning of the frame. The forward channel supports two different tones—continuous supervisory audio, in which the BS transmits beacon signals to check for the live MSs in the service area, and discontinuous data stream, which is used by the BS to send orders or new voice channel assignments to the MS.

- **Reverse voice channel (RVC)**: Reverse voice channel is used for one-to-one communication from MS to the BS during calls in progress and is assigned by the BS to a MS for its exclusive use.

10.2.3   General Working of AMPS Phone System

When a BS powers up, it has to know its surroundings before providing any service to the MSs. Thus, it scans all the control channels and tunes itself to the strongest channel. Then it sends its system parameters to all the MSs present in its service area. Each MS updates its SID and establishes its paging channels only if its SID matches the one transmitted by the BS. Then the MS goes into the idle state, responding only to the beacon and page signals.

If a call is placed to a MS, the BS locates the MS through the IS-41 message exchanges (discussed in the next section). Then the BS pages the MS with an order. If the MS is active, it responds to the page with its MIN, ESN, and so on. The BS then sends the control information necessary for the call, for which the MS has to confirm with a supervisory audio tone (SAT), indicating completion of a call. If a call is to be placed from a MS, the MS first sends the origination message to the BS on the control channel. The BS passes this to the IS-41 and sends the necessary control signals and orders to the MS. Thereafter, both MS and BS shift to the voice channels. A FVC and RVC control message exchange follows to confirm the channel allocation. Then the actual conversation starts.

10.3 IS-41

10.3.1 Introduction

IS-41 is an interim standard that allows handoffs between BSs under control of different MSCs and allows roaming of a MS outside its home system. In order to facilitate this, the following services need to be provided:

■ Registering of the MS with a visiting MSC

■ Allowing for call origination in a foreign MSC

■ Allowing the MS to roam from one foreign system to another

The basic elements involved in the IS-41 architecture are shown in Figure 10.5. The key terms and concepts are shown in Table 10.2.

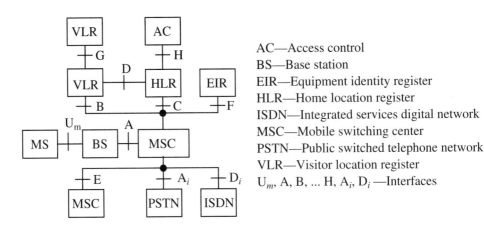

AC—Access control
BS—Base station
EIR—Equipment identity register
HLR—Home location register
ISDN—Integrated services digital network
MSC—Mobile switching center
PSTN—Public switched telephone network
VLR—Visitor location register
U_m, A, B, ... H, A_i, D_i —Interfaces

Figure 10.5
IS-41 architecture.

Table 10.2: ►
Key Terms and Concepts

Term	Definition
Anchor MSC	MSC serving as an initial contact point for MS when an originated call is initiated by the MS or when it is received from the PSTN to MS.
Originating MSC	MSC contact for destination MS when a call is originated from PSTN.
Candidate MSC	MSC that provides service during handoff.
Homing MSC	MSC that owns MS when initially put in service.
Serving MSC	MSC currently serving the MS at a cell under control of the MSC.
Target MSC	Selected MSC that can service MS with the best signal quality.

In addition to the three identification numbers described for AMPS, a switch number (SWNO) is used to identify a particular switch within a group of switches with which it is associated. It is the parameter derived from the concatenation of the SID and switch identification (SWID).

The relationship between the IS-41 and the OSI protocol stack is depicted in Figure 10.6.

It is worth noting that most of the IS-41 functionality is in the application layer, to support the mobile application part, the application service element, and the transaction capabilities application part. The association control service element (ACSE) is used to correlate two applications together (i.e., setting up an association between the two entities A and B). ROSE is invoked during the exchange of IS-41

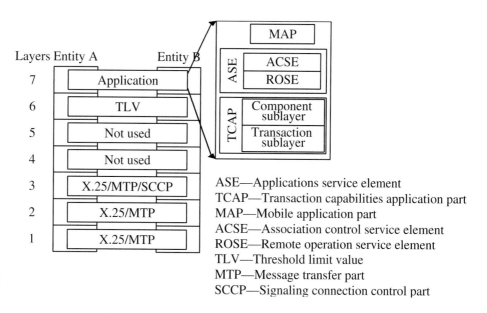

Figure 10.6
Relationship between IS-41 and OSI protocol stack.

ASE—Applications service element
TCAP—Transaction capabilities application part
MAP—Mobile application part
ACSE—Association control service element
ROSE—Remote operation service element
TLV—Threshold limit value
MTP—Message transfer part
SCCP—Signaling connection control part

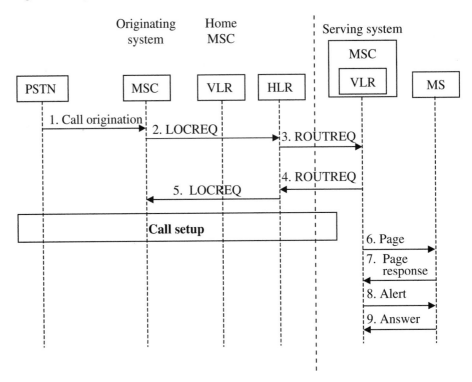

Figure 10.7
Internetworking of
IS-41 and AMPS.

messages using an asymmetric client/server–based model in which a client requests a service and the server responds with an appropriate reply. Various combinations of success and failure are possible, and the server provides a corresponding response for either synchronous or asynchronous communication. Internet working of IS-41 and AMPS can be defined easily, as shown in Figure 10.7.

10.3.2 Support Operations

The various operations supported by IS-41 are as follows:

■ **Registration in a new MSC** When a mobile terminal moves into a new area (served by a different MSC), it has to register with the new serving MSC. IS-41 messages (registration notification) are used to inform the home MSC of the current location of the MS, so that future calls can be routed to the MS via the serving MSC.

■ **Calling an idle MS in a new system** When a call is to be routed to a MS in a new system, the HLR of the home MSC contacts the VLR of the latest visiting system and, after appropriate authentication and exchange of IS-41 messages, allows the call to be routed to the MS in the visiting system.

- **Call with unconditional call forwarding** In case the visiting MS has unconditional call forwarding in effect, the visiting MSC sends a location request response to the home MSC, which contains the identifier of the telephone to which this call is to be forwarded. It is the responsibility of the home MSC to forward this call to the specified number using appropriate IS-41 procedures.

- **Call with no answer** In case the visiting MS does not answer the call, the calling terminal is issued an appropriate response and the call is disconnected.

- **Calling a busy MS** This follows the same pattern as for call with no answer, except that a busy tone is conveyed to the calling terminal in case the MS does not have call waiting. If the MS has call waiting, the MS is informed of the second incoming call.

- **Handoff measurement request** A serving MSC can sometimes request an adjacent MSC for a handoff measurement. In case the response requires a handoff to be performed, the MSC informs the HLR of this handoff and the HLR updates its database to indicate this change.

- **Recovery from failure at the HLR** This IS-41 procedure is used in the event of an HLR failure. In case of failure the HLR sends an UNRELDIR (Unreliable Roamer Data Directive INVOKE) to all the VLRs in its database. On receiving this message, all the VLRs remove all the associated data regarding this HLR and go through the registration process again.

10.4 GSM

GSM (Global System for Mobile communications or Groupe Speciale Mobile) communications, initiated by the European Commission, is the second generation mobile cellular system aimed at developing a Europe-wide digital cellular system. GSM was created in 1982 to have a common European mobile telephone standard that would formulate specifications for a pan-European mobile cellular radio system operating at 900 MHz. The main objective of GSM is to remove any incompatibility among the systems by allowing the roaming phenomenon for any cell phone. It also supports speech transmissions between MSs, emergency calls, and digital data transmissions.

A block diagram representation of the GSM infrastructure is given in Figure 10.8, with various interfaces clearly marked [10.2]. The radio link interface through the air is between the MS and the base transceiver station (BTS). A MS interfaces only with the BTS. Many BTSs are controlled by a BS controller (BSC), which in turn has an interface to a MSC. Specific functions of different constituents are as follows:

- **Base station controller (BSC):** The main function of the BSC is to look over a certain number of BTSs to ensure proper operation. It takes care of handoff from one BTS to the other, maintains appropriate power levels of the signal, and administers frequency among BTSs.

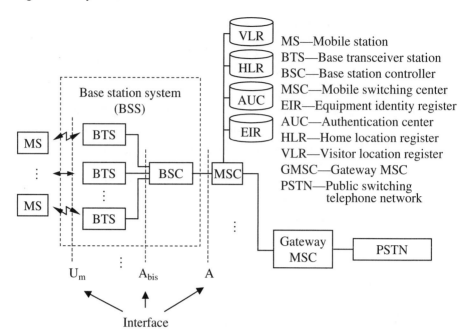

Figure 10.8
GSM infrastructure.

- **Mobile switching center (MSC)**: The MSC basically performs the switching functions of the system by controlling calls to and from other telephone and data systems. It also does functions such as network interfacing and common channel signaling. If the MSC has an interface to the PSTN, then it is called a gateway MSC. GSM uses two important databases called HLR and VLR, to keep track of current location of a MS.

- **Authentication center (AUC)**: AUC unit provides authentication and encryption parameters that verify the user's identity and ensure the confidentiality of each call. The AUC protects network operators from different types of frauds and spoofing found in today's cellular world.

- **Equipment identity register (EIR)**: EIR is a database that contains information about the identity of mobile equipment that prevents calls from being stolen and prevents unauthorized or defective MSs. Both AUC and EIR can be implemented as individual stand-alone nodes or as a combined AUC/EIR node.

10.4.1 Frequency Bands and Channels

GSM has been allocated an operational frequency from 890 MHz to 960 MHz. To reduce possible interference, the MS and the BS use different frequency ranges (i.e., MSs employ 890 MHz to 915 MHz and BS operates in 935 MHz to 960 MHz). GSM follows FDMA and allows up to 124 MSs to be serviced at the same time (i.e., the frequency band of 25 MHz is divided into 124 frequency division multiplexing (FDM) channels, each of 200 kHz as shown in Figure 10.9). A guard frame of 8.25 bits is used in between any two frames transmitted either by the BS or the MS.

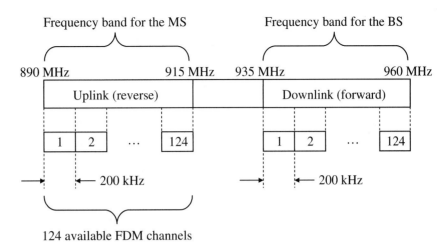

Figure 10.9
Frequency band used by GSM.

GSM uses a variety of multiplexing techniques to create a collection of logical channels. The channels used by a GSM system are shown in Table 10.3.

GSM system uses a variety of control channels to ensure uninterrupted communication between MSs and the BS. Three control channels are used for broadcasting some information to all MSs:

■ **Broadcast control channel (BCCH)**: Used for transmitting system parameters, (e.g., the frequency of operation in the cell, operator identifiers) to all the MSs.

Table 10.3: ▶
Channels in GSM

Channel	Group	Channel	Direction
Control channel	BCCH (Broadcast control channel)	BCCH (Broadcast control channel)	BS → MS
		FCCH (Frequency correction channel)	BS → MS
		SCH (Synchronization channel)	BS → MS
	CCCH (Common control channel)	PCH (Paging channel)	BS → MS
		RACH (Random access channel)	BS ← MS
		AGCH (Access grant channel)	BS → MS
	DCCH (Dedicated control channel)	SDCCH (Stand-alone dedicated control channel)	BS ↔ MS
		SACCH (Slow associated control channel)	BS ↔ MS
		FACCH (Fast associated control channel)	BS ↔ MS
Traffic channel	TCH (Traffic channel)	TCH/f (Full-rate traffic channel)	BS ↔ MS
		TCH/s (Half-rate traffic channel)	BS ↔ MS

- **Frequency correction channel (FCCH)**: Used for transmission of frequency references and frequency correction burst of 148 bits length.

- **Synchronization channel (SCH)**: Used to provide the synchronization training sequences burst of 64 bits length to the MSs.

Three common control channels are used for establishing links between the MS and the BS, as well as for any ongoing call management:

- **Random-access channel (RACH)**: Used by the MS to transmit information regarding the requested dedicated channel from GSM.

- **Paging channel**: Used by the BS to communicate with individual MS in the cell.

- **Access-grant channel**: Used by the BS to send information about timing and synchronization.

Two dedicated control channels are used along with traffic channels to serve for any control information transmission during actual communication:

- **Slow associated control channel (SACCH)**: Allocated along with a user channel, for transmission of control information during the actual transmission.

- **Stand-alone dedicated control channel (SDCCH)**: Allocated with SACCH; used for transfer of signaling information between the BS and the MS.

- **Fast associated control channel (FACCH)**: FACCH is not a dedicated channel but carries the same information as SDCCH. However, FACCH is a part of the traffic channel, while SDCCH is a part of the control channel. To facilitate FACCH to steal certain bursts from the traffic channel, there are 2 bits, called the flag bits in the message.

10.4.2   Frames in GSM

The GSM system uses the TDMA scheme shown in Figure 10.10 with a 4.615 ms–long frame, divided into eight time slots each of 0.557 ms. Each frame measured in terms of time is 156.25 bits long, of which 8.25 period bits are guard bits for protection. The 148 bits are used to transmit the information. Delimited by tail bits (consisting of 0s), the frame contains 26 training bits sandwiched between two bursts of data bits. These training bits allow the receiver to synchronize itself. Many such frames are combined to constitute multiframe, superframe, and hyperframes.

10.4.3   Identity Numbers Used by a GSM System

Several identity numbers are associated with a GSM system, as follows:

- **International mobile subscriber identity (IMSI)**: When a cell phone attempts a call, it needs to contact a BS. The BS can offer its service only if it identifies the cell phone (MS) as a valid subscriber. For this, the MS needs to store certain values uniquely defined for the MS, like the country of subscription, network type, and subscriber ID, and so on. These values are called the international mobile subscriber identity (IMSI). This number is usually 15 digits or less. The structure of an IMSI is shown in Figure 10.11. The first three digits specify the country code, the next two specify the network provider code, and the rest are

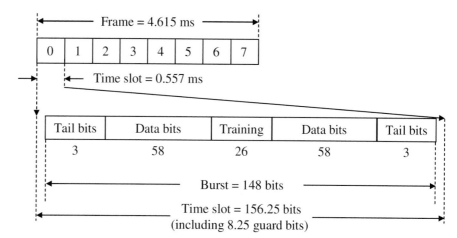

Figure 10.10
Frame structure in TDMA.

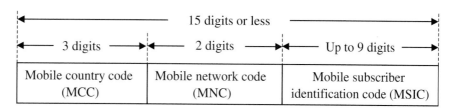

Figure 10.11
Format of IMSI.

the mobile subscriber identification code (the customer ID number). Another use of IMSI (similar to the MSC/VLR pair) is to find the information about the subscriber's home public land mobile network (PLMN). All such information is placed on a subscriber identity module (SIM), also known as a SIM card.

■ **Subscriber identity module (SIM)**: Every time the MS has to communicate with a BS, it must correctly identify itself. A MS does this by storing the phone number (or the number used to contact the MS), personal identification number for the station, authentication parameters, and so on in the SIM card. Smart SIM cards also have a flash memory that can be used to store small messages sent to the unit. The main advantage of SIM is that it supports roaming with or without a cell phone, also called SIM roaming. All a person needs to do is carry the card, and he or she can insert it into any phone to make it work as his or her customized MS. In other words, the SIM card is the heart of a GSM phone, and the MS is unusable without it.

■ **Mobile system ISDN (MSISDN)**: MSISDN is the number that identifies a particular MS's subscriber, with the format shown in Figure 10.12. Unlike other standards, GSM actually does not identify a particular cell phone, but a particular HLR. It is the responsibility of the HLR to contact the cell phone.

■ **Location area identity (LAI)**: As shown in Figure 10.13, the GSM service area is usually divided into a hierarchical structure that facilitates the system to access any MS quickly, irrespective of whether it is in home agent territory or roaming. Each PLMN is divided into many MSCs. Each MSC typically contains a VLR to tell the system if a particular cell phone is roaming, and if it is roaming, the VLR

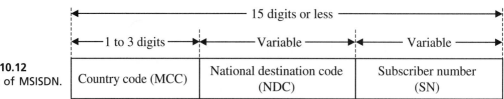

Figure 10.12
Format of MSISDN.

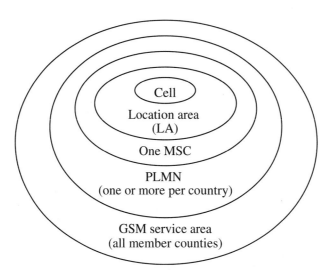

Figure 10.13
GSM layout.

of the MSC, in which the cell phone is, reflects the fact. Each MSC is divided into many location areas (LAs). A LA is a cell or a group of cells and is useful when the MS is roaming in a different cell but the same LA. Since any LA has to be identified as the part of the hierarchical structure, the identifier should contain the country code, the mobile network code, and the LA code.

■ **International MS equipment identity (IMSEI)**: Each manufactured GSM unit is assigned a 15-bit long identification number to contain manufacturing information, as shown in Figure 10.14. Conceptually, when the unit passes the interoperability tests, it is assigned a type approval code (TAC). Since a single unit may not be manufactured at the same place, a field in IMSEI, called the final assembly code (FAC), identifies the final assembly place of the unit. To identify uniquely a unit manufactured, a serial number (SNR) is assigned. A spare digit is available to allow further assignment depending on requirements.

■ **MS roaming number (MSRN)**: When a MS roams into another MSC, that unit has to be identified based on the numbering scheme format used in that MSC. Hence, the MS is given a temporary roaming number called the MS roaming number (MSRN), with the format shown in Figure 10.15. This MSRN is stored by the HLR, and any calls coming to that MS are rerouted to the cell where the MS is currently located.

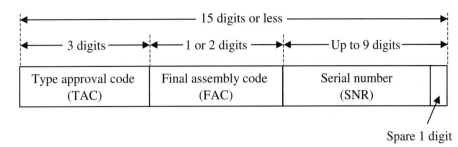

Figure 10.14
Format of IMSEI.

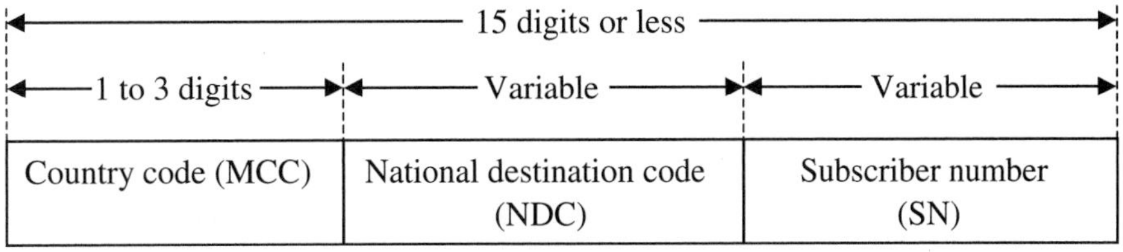

Spare 1 digit

Figure 10.15
Format of MSRN.

- **Temporary mobile subscriber identity (TMSI)**: As all transmission is sent through the air interface, there is a constant threat to the security of information sent. A temporary identity is usually sent in place of IMSEI.

10.4.4   Interfaces, Planes, and Layers of GSM

In a cellular network, possible interfaces are air interface U_m between MS and BTS; interface A_{bis} between BSC and BTS; interface A between BSC and MSC; and MAP (mobile application part), which defines operation between the MSC and the telephone network (Table 10.4).

Table 10.4: ▶
Interfaces of GSM

Interface Designation		Between
U_m		MS–BTS
A_{bis}		BTS–BSC
A		BSC–MSC
MAPn	B	MSC–VLR
	C	MSC–HLR
	D	HLR–VLR
	E	MSC–MSC
	F	MSC–EIR
	G	VLR–VLR

Sending
entity

Receiving
entity

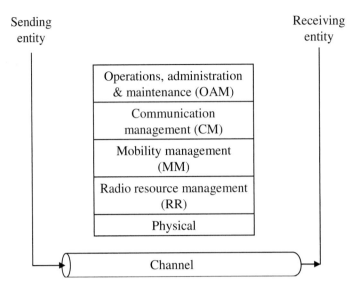

| Operations, administration & maintenance (OAM) |
| Communication management (CM) |
| Mobility management (MM) |
| Radio resource management (RR) |
| Physical |

Channel

Figure 10.16
Functional planes in GSM.

Functionally, the GSM system can be divided into five planes, as shown in Figure 10.16.

The physical plane provides the means to carry user information (speech or data) on all segments along the communication path and to carry signaling messages between entities [10.3]. Radio resource management (RR) establishes and releases stable connections between MSs and a MSC and maintains them despite user movements. The RR functions are mainly performed by the MS and the BSC. Mobility management (MM) functions are handled by the MS (or SIM), the HLR/AUC, and the MSC/VLR. These also include management of security functions. Communication management (CM) is used to set up calls between users and maintain and release resources. In addition to call management, it includes supplementary services management and short message management. Operation, administration and maintenance (OAM) enables the operator to monitor and control the system at any time.

For a MS to operate in a MSC, it must be registered by accessing the BSS, which allocates the channels, after authenticating the MS by accessing the VLR through the MS's HLR. The MSC then assigns a TMSI to the MS and updates the VLR and HLR.

To make a call from a telephone in the PSTN, the packets travel through the gateway MSC to the terminating MSC (the place where the MS is located) after getting the information from the home HLR of the MS. Then the MS is contacted through the BSS, where the MS is roaming. If it is the same MSC, there is no problem. But if it is not, then the VLR of the current MSC contacts the HLR of the MS's home MSC, which notifies the prior MSC about relocation of the MS. Hence these three registers are updated with the new information.

Authentication in GSM is done with the help of a fixed network that is used to compare the IMSI of the MS reliably (Figure 10.17). When the MS asks for any request, the fixed network sends it a random number and it also uses an authentication

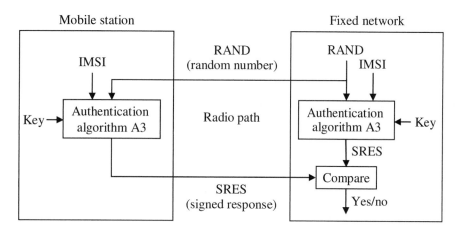

Figure 10.17
Authentication
process in GSM.

algorithm to encrypt with the IMSI and the key stored in its memory. In the MS, the received random number is encrypted using IMSI and the same key is transmitted to the fixed network, which compares it with the original value sent by the fixed network. If they match, then the MS is authentic.

10.4.5   Handoff

Handoff in GSM is divided into four major categories:

- **Intracell/intra-BTS handoff**: The channel for the connection is changed within the cell (usually when higher interference occurs). The change can apply to another frequency of the same cell or to another time slot of the same frequency. The change is initiated by sending out a page to the MS.

- **Intercell/intra-BSC handoff**: In this case, the change is in the radio channel between two cells that are served by the same BSC. This handoff from one BSS to another has to go through a series of steps. Initially, the handoff request is initiated by one BSS to the serving MSC. The MSC transmits the request to the destination BSS. When it is acknowledged back to MSC, the MSC gives the handoff command, which is transmitted to the MS. To inform the MSC that the handoff has been successful, the MS transmits a handoff complete message to the second BSS, which relays it to the MSC. The MSC then issues the command to the first BSS to clear the channel allocated to the MS.

- **Inter-BSC/intra-MSC handoff**: A connection is changed between two cells that are served by different BSCs but operate in the same MSC. A handoff is required when the measured value of the signal strength at the MS is lower than the threshold. This value is sent to the first BSC which actually initiates the handoff command. When the handoff from one BSC to another is instigated,

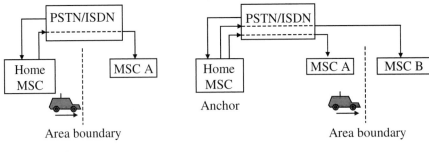

Figure 10.18
Inter-MSC handoff.

(a) Basic handoff   (b) Subsequent handoff

the command from the BSC relays to the MSC of that area. The MSC relays the request to the BSC, which sends a channel activation request to its BTS. It is then possible for the handoff call to be handled in the new area. This information is relayed to the MS, which then receives the information about the allocated channels from the new BTS. Then the MSC is notified of the change, and it sends out a clear command to the old BSC to clear the channel previously occupied by the MS.

- **Inter-MSC handoff**: A connection is changed between two cells that are in different MSCs. When the cell phone enters the state of roaming, this handoff occurs. If we take a closer look at this, we find that there are two possible handoffs (see Figure 10.18):

- **Basic handoff**: When the MS travels from its home MSC to a foreign MSC

- **Subsequent handoff**: When the MS travels from one foreign MSC to another foreign MSC

Either of these handoffs occurs through the PSTN or the ISDN, wherein the home MSC is notified of the handoff condition through the PSTN, and the home MSC sends the necessary data to the new MSC through the PSTN again. If a subsequent handoff situation occurs, the home MSC also sends the clear commands to the previous MSC through the PSTN.

10.4.6 Short Message Service (SMS)

The short message service (SMS) is the ability to send or receive a text message to or from mobile phones. It is widely used in GSM system outside North America (e.g., Europe, Asia, Australia, the Middle East, and Africa) and some parts of North America. GSM system supports SMS messages using unused bandwidth and has several unique features. SMS features confirmation of message delivery. This means the sender of the short message can receive a return message back, notifying them whether the SMS has been delivered or not. SMS can be sent and received simultaneously with GSM voice, data, and Fax calls. This is possible because voice, data, and Fax calls utilize dedicated channels for the duration of the call, while short messages travel over the control channels. As such, the message can be stored if the recipient is not available. SMS is basically a store and forward service. In other words, SMS text is not sent directly from a sender to the receiver, but always processed via an SMS center instead. Each mobile phone network that supports SMS has one or more messaging centers to handle and manage the short messages.

A single SMS can be up to 160 characters of text in length, and these 160 characters comprise a combination of words, numbers, or alphanumeric characters. Non-text–based SMSs (for example, in binary format) are also supported. There are ways of sending multiple SMS. For example, SMSs concatenation (stringing several short messages together) and SMS compression (getting more than 160 characters of information within a single short message) have been defined and incorporated in the GSM SMS standards

10.5 PCS

PCS (personal communications services) employs an inexpensive, lightweight, and portable handset to communicate with a PCS BS. PCS encompasses the whole spectrum of communication services ranging from an ordinary cellular telephone to cable television. The FCC view PCS is shown in Figure 10.19.

The PCS can be classified into high-tier and low-tier standards. High-tier systems include high-mobility units with large batteries, such as a MS in a car. The PCS high-tier standards are given in Table 10.5. Low-tier systems include systems with low mobility, capable of providing high-quality portable communication service over a wide area. The PCS low-tier standards based on personal access communications systems (PACS) and digital European cordless telecommunications (DECT) are given in Table 10.6.

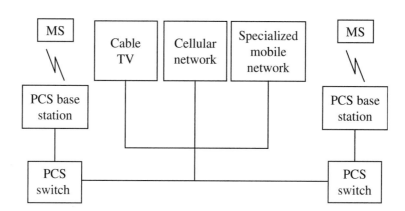

Figure 10.19
FCC view of PCS.

Table 10.5: ▶
PCS High-Tier Standards

	IS-54 based	IS-95 based	DCS (Digital Cellular System) based
MAC	TDMA	CDMA	TDMA
Duplexing	FDD	FDD	FDD
Carrier BW	30 kHZ	1.25 MHz	200 kHz
Cannels/carrier	3	20	8
× AMPS times	3	10	8
Modulation	$\pi/4$ DQPSK	QPSK	GMSK
Frequency reuse	7	1	4
Power	100 mW	200 mW	125 mW
Frame length	40 ms	20 ms	4.165 ms
Equalizer	Yes	Rake filters	Yes
Vocoder	8/4 kbps	8/4/2/1 kbps	13/6.5 kbps

Table 10.6: ▶
PCS Low-Tier Standards

	PACS	W-CDMA	DECT based
MAC	TDMA	W-CDMA	TDMA
Duplexing	FDD	FDD	TDD
Carrier BW	300 kHZ	>5 MHz	1728 kHz
Cannels/carrier	8	128	12
× AMPS times	0.8	16	0.2
Modulation	$\pi/4$ DQPSK	QPSK	GMSK
Frequency reuse	7	1	9
Power	100 mW	500 mW	20.8 mW
Frame length	40 ms	10 ms	10 ms
Equalizer	Yes	No	No
Vocoder	32 kbps	>32 kbps	32 kbps

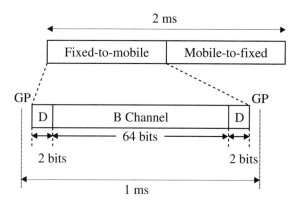

Figure 10.20
CT2 TDD slot (first
generation).

10.5.1   Chronology of PCS Development

CT2

CT2 (Cordless Telephone) operates using FDMA with a speech rate of 32 kbps using adaptive differential pulse code modulation (ADPCM). The transmitter data rate is 72 kbps. CT2 uses TDD, which allows BS and MS to share one channel. A CT2 TDD slot is shown in Figure 10.20. Here, D is called D-channel which includes 4 bits of control information.

DECT

The DECT (Digital European Cordless Telecommunications) standard is a second-generation cordless telephone system. DECT operates on frequencies ranging from 1880 MHz to 1900 MHz and uses ADPCM with 32 kbps speech rate. DECT uses TDD with two frames (BS to MS and MS to BS) with 10 ms periods. The control channel operates at a rate of 4 kbps. A typical DECT TDD slot is shown in Figure 10.21. DECT supports both voice and data communications.

10.5.2   Bellcore View of PCS

The Bellcore view of PCS is based on five different access services provided between the Bellcore client company (BCC), BCC network, and the PCS wireless provider network as follows:

- PCS access service for networks (PASN) is a connection service to and from the PCS service provider (PSP).
- PCS access service for controllers (PASC) is a service for use with PCS wireless provider (PWP) across radio channels and some type of automatic link transfer capability.
- PCS access service for ports (PASP) is an interface into PWP.
- PCS service for data (PASD) is a database information transport service.
- PCS access service for external service providers (PASE) is used to support specialized PCS services like voice mail, paging, and so on.

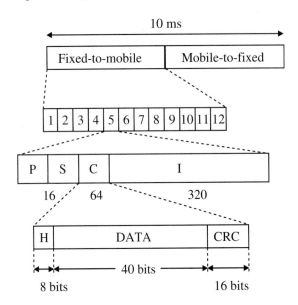

Figure 10.21
DECT TDD slot
(second-generation).

Bellcore PCS Reference Architecture

Figure 10.22 depicts the Bellcore PCS architecture. The air interface A connects the MS with the radio port (RP) which is used among other things to convert the air interface to or from a wire or fiber signal. The RPs are connected through the port (P) interface to the radio port control unit (RPCU). The other connections and interfaces are self explanatory. The advanced intelligent network view shows the collection of SS7 (signaling system 7), AM (access manager), VLR, and HLR as tailored for PCS architecture as illustrated in Figure 10.22.

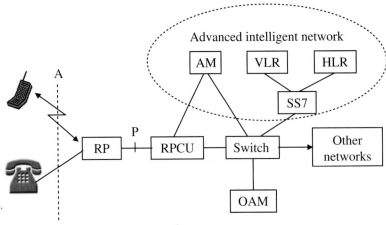

Figure 10.22
Bellcore PCS architecture.

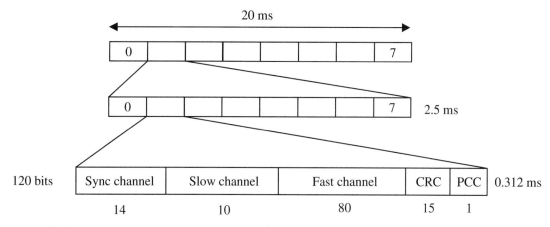

Figure 10.23   Forward TDMA frame for PCS.

Description of the PCS Air Interface

PCS uses TDMA for channel access. The reverse frame format for PCS, with a duration of 2.5 ms, is shown in Figure 10.23. Eight frames are multiplexed together to create a superframe 20 ms in duration. The downlink slot duration is 312.5 μs, and eight such slots are present in a frame to give a frame of 2.5 ms. The superframe consists of eight such frames for a total duration of 20 ms, which is similar to the uplink superframe. 15 bits CRC (cycle redundancy check) is calculated from slow and fast channels for each burst. Also, a 1 bit PCC (power control channel) is set according to individual systems.

Various messages are exchanged in a PCS call session between the MS and the BS. This is almost analogous to AMPS and GSM. A number of PCSs can be connected together by a backbone using a technique called distributed queue dual bus (DQDB). The network primarily employs two unidirectional buses each transmitting in opposite directions with data transfer rates between 34 and 150 Mbps.

● ● ● ● ● ● ● ● ● ● ● ● ● ● ● ●

10.6 **IS-95**

IS-95 uses the existing 12.5 MHz cellular bands to derive 10 different CDMA bands (1.25 MHz per band). Because the same frequency can be used even in adjacent cells, the frequency reuse factor is 1. The channel rate is 1.228 Mbps (in chips per second). CDMA takes advantage of multipath fading by providing for space diversity. RAKE receivers are used to combine the output of several received signals. Sixty-four-bit orthogonal Walsh codes (W_0 to W_{63}) are used to provide 64 channels in each frequency band. In addition to Walsh codes, long pseudonoise (PN) codes and short PN codes are also used.

The logical channels of CDMA are the control and traffic channels, as illustrated in Figure 10.24. The control channels are the pilot channel (forward), the paging

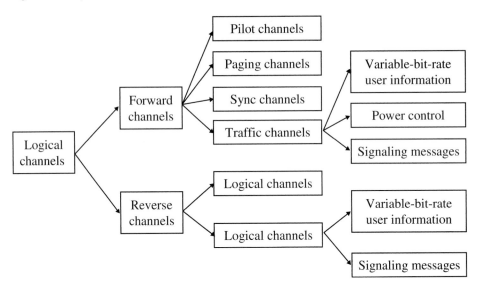

Figure 10.24
Logical channels in
IS-95.

channels (forward), the sync channels (forward), and the access channels (reverse). The traffic channels are used to carry user information between the BS and the MS, along with signaling traffic. Four different rates are used. When the user speech is replaced by the associated signal, it is called blank and burst. When part of the speech is replaced by signaling information, it is called dim and burst. The downlink or forward link has a power control subchannel that allows the mobile to adjust its transmitted power by ± 1 dB every 1.25 ms. The pilot channel W_0 is always required. There can be one sync channel and seven paging channels; the remaining fifty-six channels are called traffic channels [10.4].

■ **Pilot channel**: The pilot channel is used by the base station as a reference for all MSs. It does not carry any information and is used for strength comparisons and to lock onto other channels on the same RF carrier. Pilot channel processing is shown in Figure 10.25.

The signals (pilot, sync, paging, and traffic) are spread using high frequency spread signals I and Q using modulo 2 addition. This spread signal is then modulated over a high frequency carrier and sent to the receiver, where the entire process is inverted to get back the original signal.

■ **Sync channel**: The sync channel is an encoded, interleaved, and modulated spread-spectrum signal that is used with the pilot channel to acquire initial time synchronization. It is assigned the Walsh code W_{32}.

■ **Paging channel**: As the name suggests, the paging channel is used to transmit control information to the MS. When the MS is to receive a call, it will receive a page from the BS on an assigned paging channel. There is no power control for the paging channel on a per-frame basis. The paging channel provides the MSs system information and instructions. The paging channel processing is shown in Figure 10.26.

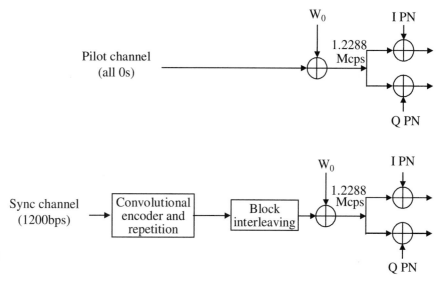

Figure 10.25
Pilot and sync channels
in IS-95.

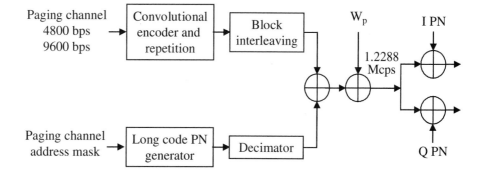

Figure 10.26
Paging channel
generation in IS-95.

- **Access channel**: Figure 10.27 shows the processing of the access channel. The access channel is used by the MS to transmit control information to the BS. The access rate is fixed at 4800 bps. All MSs accessing a system share the same frequency. When any MS places a call, it uses the access channel to inform the BS. This channel is also used to respond to a page.

- **Forward traffic channels**: Forward traffic channels are grouped into rate sets. Rate set 1 has four elements; 9600, 4800, 2400, and 1200 bps. Rate set 2 has four elements: 14400, 7200, 3600, and 1800 bps. Walsh codes that can be assigned to forward traffic channels are available at a cell or sector (W_2 through W_{31}, and W_{33} through W_{63}). Only 55 Walsh codes are available for forward traffic channels. The speech is encoded using a variable rate encoder to generate forward traffic data depending on voice activity. The power control subchannel is continuously transmitted on the forward traffic channel. The forward channel processing is as shown in Figures 10.28 and 10.29.

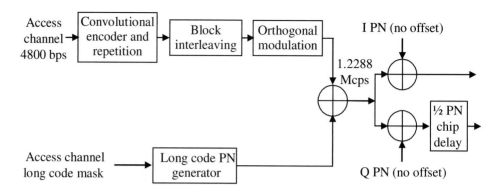

Figure 10.27
Access channel
generation
in IS-95.

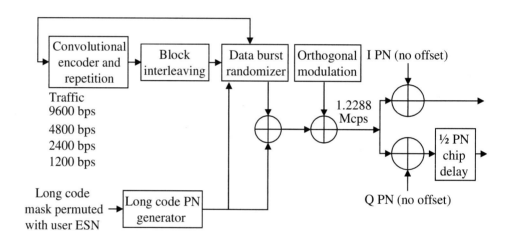

Figure 10.28
Rate set 1 forward
traffic channel
generation in IS-95.

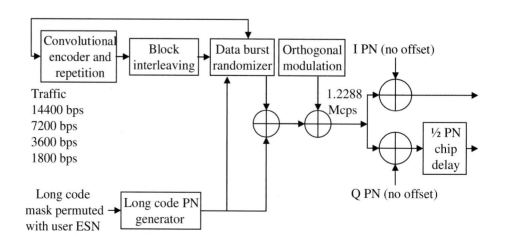

Figure 10.29
Rate set 2 forward
traffic channel
generation in IS-95.

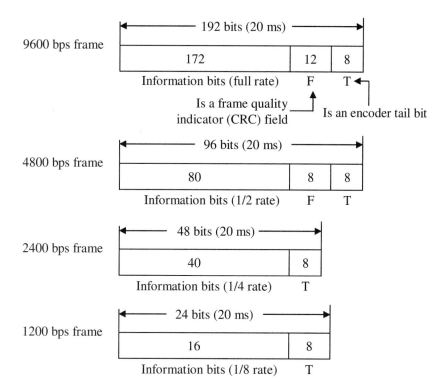

Figure 10.30
Forward/reverse
traffic channel
frame structure for
rate set 1.

The forward and reverse channel frame structure is as shown in Figures 10.30 and 10.31:

- **Reverse traffic channels**: For rate set 1, the reverse traffic channel uses 9600, 4800, 2400, or 1200 data rates for transmission. The duty cycle for transmission varies proportionally with the data rate being 100% at 9600 bps to 12.5% at 1200 bps. The reverse traffic channel processing is similar to the access channel except for the fact that the reverse channel uses a data burst randomizer. Reverse channel processing is shown in Figures 10.32 and 10.33.

10.6.1 Power Control

Power control plays an important role in view of the fact that every receiver gets the signals transmitted by all the transmitters. To ensure maximum efficiency, the power received at the BS from all the MSs must be nearly equal. If the received power is too low, there is a high probability of bit errors, and if the received power is too high, interference increases. Power control is applied at both the MSs as well as the BS. There are several different mechanisms that are used for power control initiated either by the MS or the BS, and the control can be based on the signal strength perceived by the BS or can depend on other parameters.

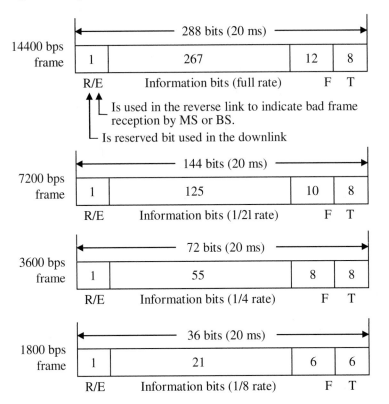

Figure 10.31
Forward/reverse
traffic channel
frame structure for
rate set 2.

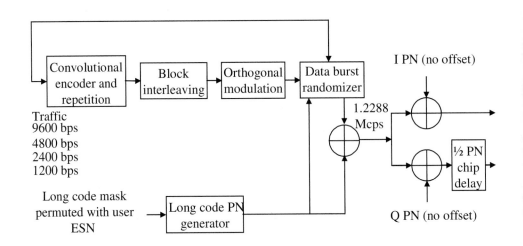

Figure 10.32
Rate set 1 reverse
traffic generation.

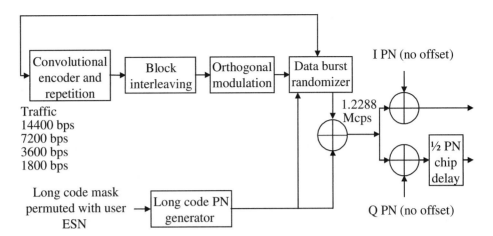

Figure 10.33
Rate set 2 reverse traffic generation.

In open-loop power control at the MS, the MS senses the strength of the pilot signal and can adjust its power based on that. If the signal is very strong, it can be assumed that the MS is too close to the BS and the power level should be dropped. In closed-loop power control at the MS, power control information is sent to the MSs from the BS. This message indicates either a transition up or transistion down in power. In open-loop power control at the BS, the BS decreases its power level gradually and waits to hear the frame error rate (FER) from the MS. If the FER is high, it increases its power level.

● ● ● ● ● ● ● ● ● ● ● ● ● ● ● ● · · · ·

10.7 IMT-2000

The International Telecommunications Union-Radio communications (ITU-R) developed the 3G specifications to facilitate a global wireless infrastructure, encompassing terrestrial and satellite systems providing fixed and mobile access for public and private networks. IMT-2000 is a general name used for all 3G systems. It includes new capabilities and provides a seamless evolution from existing 2G wireless systems. The key features of the IMT-2000 system are as follows:

- High degree of commonality of design worldwide
- Compatibility of services within IMT-2000 and with fixed networks
- High quality
- Small terminal for worldwide use, including pico, micro, macro, and global satellite cells
- Worldwide roaming capability
- Capability for multimedia applications and a wide range of services and terminals

10.7.1   International Spectrum Allocation

In 1992 the World Administration Radio Conference (WARC) specified the spectrum for the 3G mobile radio system, as illustrated in Figure 10.34.

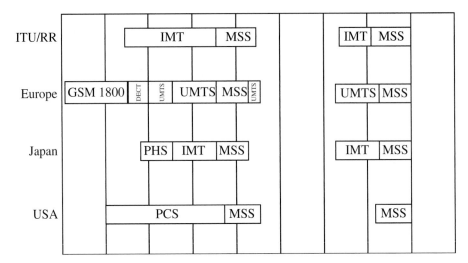

Figure 10.34
Spectrum allocation.

Europe and Japan followed the FDD specification. The lower-band parts of the spectrum are currently used for DECT and PHS (Personal Handyphone System), respectively. The FCC in the United States has allocated a significant part of the spectrum in the lower band to 2G PCS systems. Most of the North American countries are following the FCC frequency allocation. Currently no common spectrum is available for 3G systems worldwide.

10.7.2   Services Provided by Third-Generation Cellular Systems

The following services are provided by third-generation cellular systems:

- High bearer rate capabilities, including

  - 2 Mbps for fixed environment

  - 384 kbps for indoor/outdoor and pedestrian environment

  - 144 kbps for vehicular environment

- Standardization work

  - Europe (ETSI: European Telecommunications Standardization Institute) $\Rightarrow$ UMTS (W-CDMA)

  - Japan (ARIB: Association of Radio Industries and Businesses) $\Rightarrow$ W-CDMA

  - USA (TIA: Telecommunications Industry Association) $\Rightarrow$ cdma2000 [10.6]

- Scheduled service

  - Service started in October 2001 (Japan's W-CDMA)

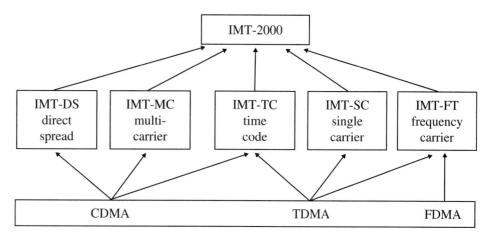

Figure 10.35
Approved radio
interfaces.

The radio interfaces for IMT-2000 as approved by the ITU meeting in Helsinki, Finland are shown in Figure 10.35.

10.7.3 Harmonized 3G Systems

A harmonized 3G system based on the Operators Harmonization Group (OHG) [10.5] recommendation is required to support the following:

- High-speed data services, including Internet and intranet applications

- Voice and nonvoice applications

- Global roaming

- Evolution from the embedded base of 2G systems

- ANSI-41 (American National Standards Institute-41) and GSM-MAP core networks

- Regional spectrum needs

- Minimization of mobile equipment and infrastructure cost

- Minimization of the impact of intellectual property rights (IPRs)

- The free flow of IPRs

- Customer requirements on time

A diagram representing the terrestrial component of the harmonization efforts for IMT-2000 is shown in Figure 10.36.

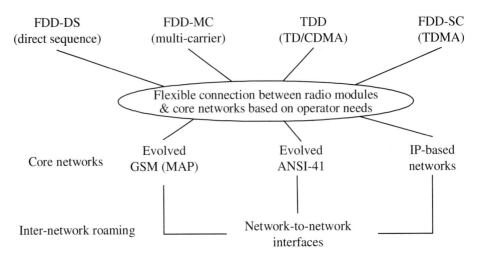

Figure 10.36
Modular IMT-2000
harmonization.

10.7.4 Multimedia Messaging Service (MMS)

The multimedia messaging service (MMS) [10.7] is an open industry specification developed by the WAP forum for the 3rd Generation Partnership Program (3GPP). The service is a significant enhancement to the current SMS service which allows only text. MMS has been designed to allow rich text, color, icons and logos, sound clips, photographs, animated graphics, and video clips and works over the broadband wireless channels in 2.5G and 3G networks. MMS and SMS are similar in the sense that both are store-and-forward services where the message is first sent to the network which then delivers it to the final destination. But unlike SMS, which can only be sent to another phone, the MMS service can be used to send messages to a phone or may be delivered as an email.

The main components of MMS architecture are:

■ MMS Relay—It transcodes and delivers messages to mobile subscribers.

■ MMS Server—It provides the "store" in the store-and-forward MMS architecture.

■ MMS User Agent—An application server gives users the ability to view, create, send, edit, delete, and manage their multimedia messages.

■ MMS User Databases—Containing records of user profiles, subscription data, etc.

The content of MMS messages is defined by the MMS conformance specification version 2.0.0, which specifies SMIL 2.0 (synchronization multimedia integration language) basic profile for the format and the layout of the presentation.

Although MMS is targeted towards 3G networks, carriers all over the world have been deploying MMS on networks like 2.5G using WAP, and it helps in generating revenue from existing older networks.

Some of the possible application scenarios are as follows:

■ Next generation voicemail—By which it is now possible to leave text, pictures, and even video mail.

- Immediate Messaging—MMS features "push" capability that enables the message to be delivered instantly if the receiving terminal is on and avoids the need for "collection" from the server. This "always-on" characteristic of the terminals opens up the exciting possibility of multimedia "chat" in real time.

- Choosing how, when, and where to view the messages—Not everything has to be instant. With MMS, users have an unprecedented range of choices about how their mail is to be managed. They can predetermine what categories of messages are to be delivered instantly, stored for later collection, redirected to their PCs, or deleted. In other words, they posses dynamic ability to make ad hoc decisions about whether to open, delete, file, or transfer messages as they arrive.

- Mobile fax—Using any fax machine to print out any MMS message.

- Sending multimedia postcards—A clip of holiday video could be captured through the integral video cam of a user's handset or uploaded via Bluetooth from a standard camcorder, then combined with voice or text messages and mailed instantly to family members and friends.

10.7.5 Universal Mobile Telecommunications System (UMTS)

Network Reference Architecture

The latest UMTS architecture is shown in Figure 10.37. It is partly based on the 3G specification, while some 2G elements have been kept [10.8]. UMTS Release'99 architecture inherits a lot from the global system for mobile (GSM) model on the core network (CN) side. The MSC basically has very similar functions both in GSM and

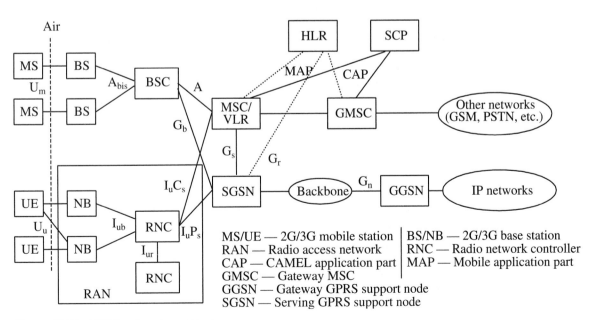

Figure 10.37 UMTS network architecture.

UMTS. Instead of circuit-switched services for packet data, a new packet node, packet data access node (PDAN), or 3G serving general packet radio services (GPRS) support node (SGSN) is introduced. This new element is capable of supporting data rates up to 2 Mbit/s. CN elements are connected to the radio network via the I_u interface, which is very similar to the A-interface used in GSM. The major changes in the new architecture are in the radio access network (RAN), which is also called UMTS terrestrial RAN (UTRAN). There is a totally new interface called I_{ur}, which connects two neighboring radio network controllers (RNCs). This interface is used for combining macrodiversity, which is a new WCDMA-based function implemented in the RNC. BSs are connected to the RNC via the I_{ub} interface [10.9]. Throughout the standardization process, extra effort has been made so that most of the 2G core elements can smoothly support both generations, and any potential changes are kept minimum. In 2G, the RAN is separated from the CN by an open interfaces, called "A" in circuit-switched (CS) and G_b in packet-switched (PS) networks. The former uses time division multiplex (TDM) transport, while packet data are carried over frame relay. In 3G, the corresponding interfaces are called $I_u C_s$ and $I_u P_s$. The circuit-switched interface will utilize ATM, while the packet switched interface will be based on IP.

UTRAN Architecture

UTRAN consists of a set of radio network subsystems (RNSs) [10.5]. as shown in Figure 10.38 The RNS has two main elements: Node B and a RNC. The RNS is responsible for the radio resources and transmission/reception in a set of cells.

A RNC is responsible for the use of and allocation of all radio resources of the RNS to which it belongs. The responsibilities of the RNC include

- Intra-UTRAN handoff
- Macrodiversity combining and splitting of the I_{ub} datastreams
- Frame synchronization
- Radio resource management
- Outer loop power control

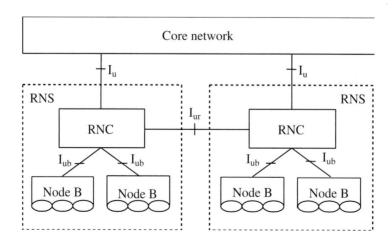

Figure 10.38
UTRAN architecture.

- Serving RNS relocation
- UMTS radio link control (RLC) sublayers function execution

UTRAN Logical Interfaces

In UTRAN, the protocol structure is designed so that the layers and planes are logically independent of each other and, if required, parts of protocol structure can be changed in the future without affecting other parts. The protocol structure contains two layers: the radio network layer (RNL) and the transport network layer (TNL). In the RNL, UTRAN-related functions are visible, whereas the TNL deals with transport technology selected to be used for UTRAN but without any UTRAN-specific changes. A general protocol model for UTRAN interfaces is shown in Figure 10.39. Here RANAP is radio access network application protocol.

Channels

Three types of channels are defined in UMTS: transport, logical, and physical channels. Transport channels are described by how the information is transmitted on the radio interface. Logical channels are described by the type of information they carry. On the other hand, physical channels are defined differently for FDD and TDD. For

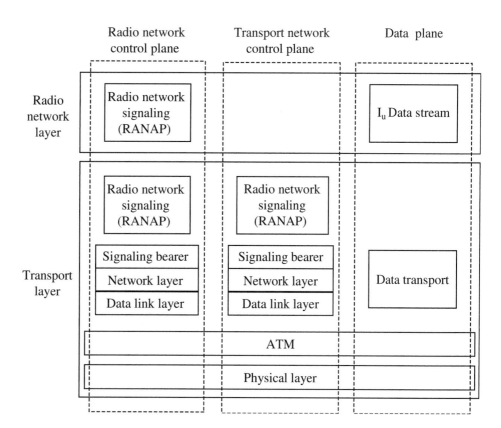

Figure 10.39
General protocol model for UTRAN interfaces.

FDD, a physical channel is identified by its carrier frequency, its access code, and the relative phase of the signal in the uplink (either the In-phase or Quadrature component). Similarly, TDD identifies a physical channel by its carrier frequency, access code, relative phase for the uplink, and the time slot in which it is transmitted.

Transport Channels

Transport channels are the services offered by the physical layer to the higher layers. A general classification of transport channels is into two groups:

1. Common transport channels (where there is a need for in-band identification of the UEs when particular UEs are addressed)

2. Dedicated transport channels (where the UEs are identified by the physical channel, i.e., code, time slot, and frequency)

In the following, we describe the transport channels in detail:

■ **Common transport channel types**:

– **Random access channel (RACH)**: A contention-based uplink channel used for transmission of relatively small amounts of data (e.g., for initial access or non–real-time dedicated control or traffic data).

– **ODMA (Opportunity driven multiple access) random access channel (ORACH)**: A contention-based channel used in relay link.

– **Common packet channel (CPCH)**: A contention-based channel used for transmission of bursty data traffic. This channel only exists in FDD mode and only in the uplink direction. The common packet channel is shared by the user equipment (UE or MS) in a cell, and therefore is a common resource. The CPCH is fast power controlled.

– **Forward access channel (FACH)**: Common downlink channel without closed-loop power control used for transmission of relatively small amount of data.

– **Downlink shared channel (DSCH)**: A downlink channel shared by several UEs carrying dedicated control or traffic data.

– **Uplink shared channel (USCH)**: An uplink channel shared by several UEs carrying dedicated control or traffic data, used in TDD mode only.

– **Broadcast channel (BCH)**: A downlink channel used for broadcast of system information into an entire cell.

– **Paging channel (PCH)**: A downlink channel used for broadcast of control information into an entire cell allowing efficient UE sleep mode procedures. Currently identified information types are paging and notification. Another use could be UTRAN notification of change in BCCH information.

- **Dedicated transport channel types**:
    - **Dedicated channel (DCH)**: A channel dedicated to one UE used in uplink or downlink.
    - **Fast uplink signaling channel (FAUSCH)**: An uplink channel used to allocate dedicated channels in conjunction with FACH.
    - **ODMA dedicated channel (ODCH)**: A channel dedicated to one UE used in relay link.

Logical Channels

Two types of logical channels are defined: traffic and control channels. Traffic channels (TCH) are used to transfer user and/or signaling data. Signaling data consists of control information related to the process of a call. Control channels carry synchronization and information related to the radio transmission. UTRAN logical channels are described in Figure 10.40.

- **Control channels**:
    - **Broadcast control channel (BCCH)**: A downlink channel for broadcasting system control information.
    - **Paging control channel (PCCH)**: A downlink channel that transfers paging information. This channel is used when the network does not know the location cell of the UE, or the UE is in the cell-connected state (utilizing UE sleep mode procedures).

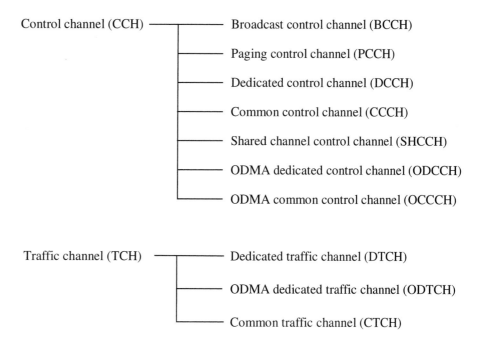

Figure 10.40
Logical channels in UTRAN.

- **Common control channel (CCCH)**: Bidirectional channel for transmitting control information between network and UEs. This channel is commonly used by the UEs having no RRC connection with the network and by the UEs using common transport channels when accessing a new cell after cell reselection.

- **Dedicated control channel (DCCH)**: A point-to-point bidirectional channel that transmits dedicated control information between a UE and the network. This channel is established through the RRC connection setup procedure.

- **Shared channel control channel (SHCCH)**: Bidirectional channel that transmits control information for uplink and downlink shared channels between the network and UEs. This channel is for TDD only.

- **ODMA common control channel (OCCCH)**: Bidirectional channel for transmitting control information between UEs.

- **ODMA dedicated control channel (ODCCH)**: A point-to-point bidirectional channel that transmits dedicated control information between UEs. This channel is established through the RRC connection setup procedure.

- **Traffic channels**:
  - **Dedicated traffic channel (DTCH)**: A DTCH is a point-to-point channel, dedicated to one UE, for the transfer of user information. A DTCH can exist in both uplink and downlink.

  - **ODMA dedicated traffic channel (ODTCH)**: An ODTCH is a point-to-point channel, dedicated to one UE, for the transfer of user information between UEs. An ODTCH exists in relay link.

  - **Common traffic channel (CTCH)**: A point-to-multipoint unidirectional channel for transfer of dedicated user information for all or a group of specified UEs.

Physical Channels

All physical channels follow four-layer structure of superframes, radio frames, subframes, and time slots/codes. Depending on the resource allocation scheme, the configurations of subframes or time slots are different. All physical channels need guard symbols in every time slot. The time slots or codes are used as a TDMA component so as to separate different user signals in the time and the code domain.

10.8 Summary

This chapter presents an overview of the first-, second-, and third-generation wireless systems used in various parts of the world. Different countries use different standards for cellular communication. A global standard for wireless communication has not yet been conceived because of differences in infrastructure and facilities in different countries. The recent development of 3G mobile cellular systems (IMT-2000) represents an attempt to create a global cellular standard. Although some countries have already introduced the standard (Japan W-CDMA), it is still in the test phase and may take time before it is extended throughout the world. Countries use different networks as their backbones and different technologies for wireless communications. Satellite communication is the simplest way to cover the entire world, and various issues associated with it are discussed in Chapter 11.

● ● ● ● ● ● ● ● ● ● ● ● ● ● ● ●

10.9 References

[10.1] U. Black, *Mobile and Wireless Networks*, Prentice Hall, Upper Saddle River, NJ, 1996.

[10.2] T. S. Rappaport, *Wireless Communications—Principles & Practice,* Prentice Hall, Upper Saddle River, NJ, 1996.

[10.3] *hhp://www.iec.org/online/tutorials.*

[10.4] V. K. Garg, *Wireless Network Evolution 2G to 3G,* Prentice Hall, Upper Saddle River, NJ, 2002.

[10.5] *http://www.3gpp.org.*

[10.6] *http://www.3gpp2.org.*

[10.7] *http://www.symbian.com/technology/mms.html.*

[10.8] *http://www.wiley.co.uk/wileychi/commstech/472_ftp.pdf.*

[10.9] "UMTS Protocols and Protocol Testing," Tektronix Company, *http://www.tek.com/Measurement/App_Notes/2F_14251/eng/2FW_14251_1.pdf.*

● ● ● ● ● ● ● ● ● ● ● ● ● ● ●

10.10 Problems

P10.1. What is meant by logical channel, and how is the concept useful? Explain.

P10.2. How do you differentiate between different types of handoff? Explain.

P10.3. Where does the MAC sublayer lie on the ISO-OSI layer hierarchy? What issues are handled in this sublayer?

P10.4. What is the role of different functional planes in GSM? Explain each one clearly.

P10.5. How are PCS systems different from conventional cellular systems like AMPS?

P10.6. What are the important functionalities of SS7? Explain their use.

P10.7. What are the similarities and the differences between AMPS and GSM? Explain clearly.

P10.8. How do you compare AMPS and GSM systems in terms of coverage area, transmitting power, and error control? Explain.

P10.9. Why is a smart card needed in GSM, while it is not required in AMPS? Explain the logic behind this.

P10.10. What is the function of ACSE and ROSE service elements? Explain clearly.

P10.11. A cellular system employs the CDMA scheme. Is it possible to use a composite TDMA/CDMA scheme? If not, why not; and if yes, what may be the potential advantages? Explain clearly.

P10.12. One approach to using Walsh code in a CDMA system is to assign a code permanently to each subscriber. What are advantages, disadvantages, or limitations of such an approach?

P10.13. In IMSI, why is a temporary ID used? Explain clearly.

P10.14. What is the rationale behind the traffic channel indicating the reverse control channel to be busy in AMPS?

P10.15. Why is the near-far problem present in CDMA and not in FDMA?

P10.16. A large company consists of 10,000 employees, and an infrastructure needs to be created to broadcast messages to all the employees. If an AMPS system is to be used for such a broadcast, what may be the possible alternate scheme if:
   (a) All employees are located in the same city?
   (b) 50% of employees are in one location, while the remaining 50% are in another place?
   (c) Twenty-five percent of employees are located in four different locations?
   (d) People are spread all over the world?

P10.17. How would you address Problem P10.16 if a GSM scheme is to be employed?

P10.18. Repeat Problem P10.17 for IMT2000 system.

P10.19. Search the various Web sites and find why IS-41 message transfer employs X.25.

P10.20. What is the fundamental principle and use of spread spectrum?

P10.21. Find out the SMS service providers in your area? How can you compare their performance parameters?

P10.22. What is the future of SMS services, and how do you compare them with paging? Explain clearly.

Satellite Systems

● ● ● ● ● ● ● ● ● ● ● ● ● ●
11.1 Introduction

Satellite systems have been in use for several decades. There is a long history of the development of satellite systems from a communications point of view. Important events related to satellite systems are shown in Table 1.10. Possible application areas are outlined in Table 1.11. Satellites, which are far away from the surface of the earth, can cover a wider area on the surface of the earth, and several satellite beams are controlled and operated by each satellite. The information to be transmitted from a mobile user (MS) must be correctly received by a satellite and forwarded to one of the earth stations (ESs). Thus, only LOS communication between the mobile user and the satellite should be possible.

● ● ● ● ● ● ● ● ● ● ● ● ● ●
11.2 Types of Satellite Systems

Satellites have been put in space for various purposes [11.1], and their placement in space and orbiting shapes have been determined as per their specific requirements. Four different types of satellite orbits have been identified:

1. GEO (geostationary earth orbit) at about 36,000 km above the earth's surface

2. LEO (low earth orbit) at about 500–1500 km above the earth's surface

3. MEO (medium earth orbit) or ICO (intermediate circular orbit) at about 6000–20,000 km above the earth's surface

4. HEO (highly elliptical orbit)

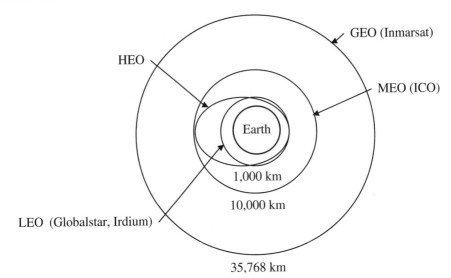

Figure 11.1
Orbits of different
satellites.

Satellite orbiting paths and distances from the surface of the earth are illustrated in Figure 11.1. The orbits can be elliptical or circular, and the complete rotation time (and hence frequency) is related to the distance between the satellite and the earth and the mass of the satellite and the gravitational acceleration. For satellites following circular orbits (Figure 11.2), **Newton**'s gravitational law can be applied to compute attractive force F_g and centrifugal force F_c as follows:

$$F_g = mg \left(\frac{R}{r} \right)^2, \tag{11.1}$$

$$F_c = mr\varpi^2 \tag{11.2}$$

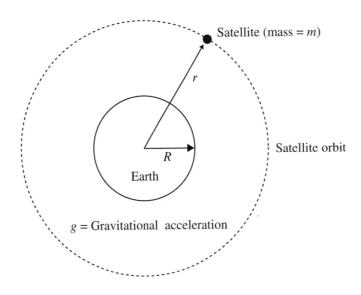

Figure 11.2
Earth-satellite
parameters for a
stable orbiting path.

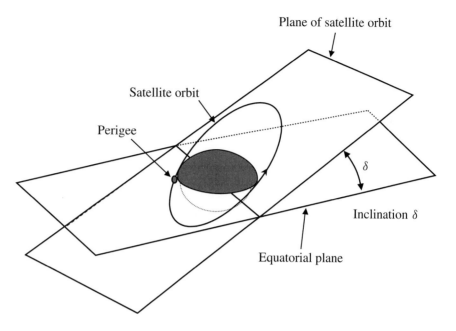

Plane of satellite orbit

Satellite orbit

Perigee

δ

Inclination δ

Equatorial plane

Figure 11.3
Inclination δ of a
satellite orbit.

with

$$\varpi = 2\pi f_r, \tag{11.3}$$

where m is the mass of the satellite, g is the gravitational acceleration of the earth ($g = 9.81$ m/s^2), R is the radius of the earth ($R = 6370$ km), r is the distance of the satellite from the center of the earth, ϖ is the angular velocity of the satellite, and f_r is the rotational frequency.

For the orbit of the satellite to be stable, we need to equate the two forces, giving

$$r = \sqrt[3]{\frac{gR^2}{(2\pi f_r)^2}}. \tag{11.4}$$

The plane of the satellite orbit with respect to the earth is shown in Figure 11.3. The plane of the satellite orbit will primarily dictate part of the earth that is covered by the satellite beam in each rotation. The elevation angle between the satellite beam and the surface of the earth has an impact on the illuminated area (known as the footprint) and is shown in Figure 11.4. The elevation angle θ of the satellite beam governs the distance of the satellite with respect to the MS. The intensity level of a footprint is given in Figure 11.5, with a circle corresponding to 0 dB intensity clearly marked. The area inside this circle is considered to be an isoflux region and this constant intensity area is usually taken as the footprint of a beam. A satellite consists

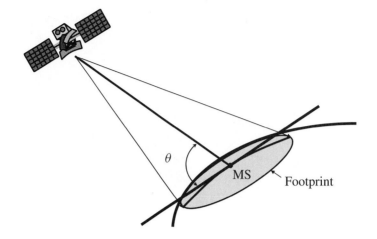

Figure 11.4
Elevation angle θ
and footprint.

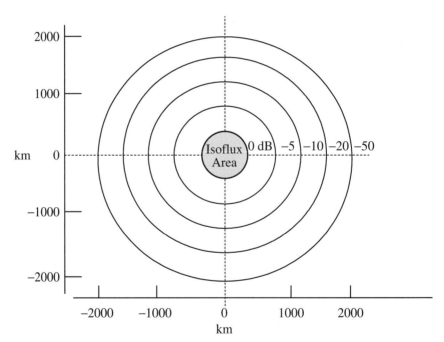

Figure 11.5
GEO satellite beam
footprint.

of several illuminated beams, and one such example of beam geometry is illustrated in Figure 11.6. These beams could be considered as cells of the conventional wireless system.

Figure 11.7 shows the path d taken for communication from a MS to the satellite. The time delay for the signal to travel from the satellite to a MS is a function of various parameters and can be obtained using the geometry of Figure 11.7 as:

$$\text{Delay} = \frac{d}{c} = \frac{1}{c}\left[\sqrt{(R+h)^2 - R^2\cos^2\theta} - R\sin\theta\right], \qquad (11.5)$$

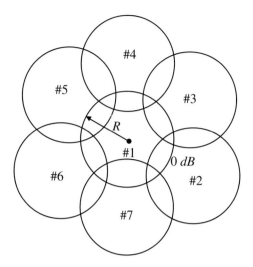

Figure 11.6
Satellite beam
geometry.

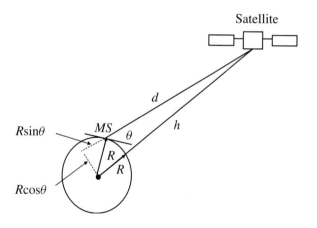

Figure 11.7
Satellite
communication
display.

where R is the radius of the earth (6370 km), h is the orbital altitude, θ is the satellite elevation angle, and c is the speed of light.

Figure 11.8 shows the variation of delay as a function of the elevation angle θ of a MS when a satellite is at an elevation of 10,355 km. The satellites operate at different frequencies for the uplink (MS to satellite) and downlink (satellite to MS). The frequency bands used for most satellite systems are shown in Table 11.1.

C band frequencies have been used in first-generation satellites. This band has become overcrowded because of terrestrial microwave networks that employ these frequencies. The Ku and Ka bands are becoming more popular even though rain causes a high level of attenuation. Satellites receive signals at very low power levels, typically less than 100 picowatts, which is one to two orders of magnitude lower than terrestrial receivers (typical range 1 to 100 microwatts). Signals from the satellite

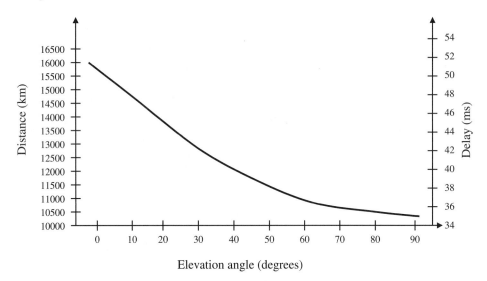

Figure 11.8
Variation of delay
in MS as a function
of elevation angle.

Table 11.1: ▶
Frequency Range for Different Bands

Band	Uplink (GHz)	Downlink (GHz)
C	3.7–4.2	5.925–6.425
Ku	11.7–12.2	14.0–14.5
Ka	17.7–21.7	27.5–30.5
LIS	1.610–1.625	2.483–2.50

travel to MSs through the open space and are affected by the atmospheric conditions. The received power is determined by the following four parameters:

■ Transmitting power

■ Gain of the transmitting antenna

■ Distance between the satellite transmitter and the receiver

■ Gain of the receiving antenna

Atmospheric conditions cause attenuation of the transmitted signal, and the loss L at the MS is given by a generic relationship

$$L = \left(\frac{4\pi r f_c}{c} \right)^2,$$
(11.6)

where f_c is the carrier frequency and r is the distance between the transmitter and the receiver. The impact of rain on the signal attenuation is illustrated in Figure 11.9.

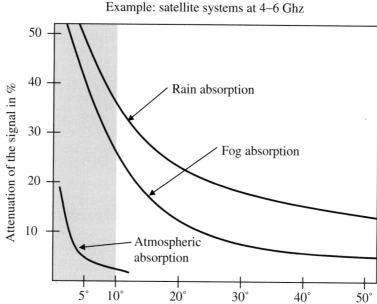

Example: satellite systems at 4–6 Ghz

Figure 11.9
Atmospheric attenuation.

● ● ● ● ● ● ● ● ● ● ● ● ● ● ● ●

11.3 Characteristics of Satellite Systems

As discussed earlier, satellites have been launched for various applications and are placed at different altitudes. Moreover, their weights are also dissimilar. The GEO satellites, which are at an altitude of 35,768 km, orbit in the equatorial plane with $0°$ inclination and complete exactly one rotation in a day. The antennas are at fixed positions, and an uplink band (reverse band) of 1634.5 to 1660.5 MHz and a downlink band (forward band) in the range of 1530 to 1559 MHz, are employed. Ku band frequencies (11 and 13 GHz) are employed for connection between the base station (earthbase) and the satellites. A satellite typically has a large footprint, which can be up to 34% of the earth's surface covered, and therefore it is difficult to reuse frequencies. The elevation areas with latitude above $60°$ have become undesirable due to their relative position above the equator. The global coverage of small mobile phones and data transmission typically cause high latency in the range of about 275 ms.

LEO satellites are divided into little and big satellites. Little LEOs are smaller in size and are in the frequency range of 148 to 150.05 MHz (uplink represented by ↑) and 137 to 138 MHz (downlink shown by ↓). They use alphanumeric displays at low bit rates (of the order of 1 kb/s) for two-way message and positioning information. Big LEO satellites have adequate power and bandwidth to provide various global mobile services (i.e., data transmission, paging, facsimile, and position location) along with good quality voice services for mobile systems such as handheld devices and vehicular transceivers. Big LEOs transmit in the frequency range of 1610 to 1626.5 MHz (uplink) and 2483.5 to 2500 MHz (downlink) and orbit at about 500 to 1,500 km above the earth's surface. The latency is around 5 to 10 ms, and the satellite is visible for about 10 to 40 ms. The smaller the footprint, the better it is from a frequency reuse point of view. Several satellites are needed to ensure global coverage. The same frequency spectrum is also used by MEO and GEO. In MEO systems, the slow-moving satellites orbit at a height of about 5,000 to 12,000 km above the earth and have a latency of about 70 to 80 ms. Specialized antennas are used to provide smaller footprints and higher transmitting power. A detailed comparison of LEO/MEO satellites is given in Tables 11.2(a) and (b).

11.4 Satellite System Infrastructure

There are many ensembles that enable a satellite infrastructure to work. A detailed examination is needed to understand the operation of the overall system. An example diagram representation of a satellite system is shown in Figure 11.10, with numerous components shown explicitly. Once a contact has been established between a mobile system and a satellite using a LOS beam, almost everyone in the

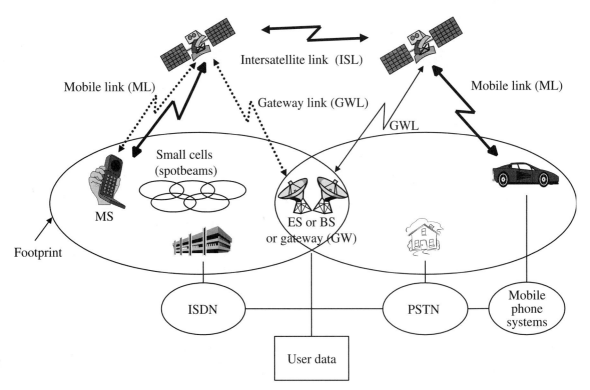

Figure 11.10 A typical satellite system.

Table 11.2: ▶
Comparison of LEO/MEO Satellites

Characteristics	Little-LEO			
	LEO SAT	ORBCOM	STARNET	VITASAT
Number of satellites	18	26	24	2
Altitude (km)	1000	970	1300	800
Coverage	Global	United States	Global	Global
Minimum elevation	42°	2 polar, 3 inclined	60°	99°
Frequencies (GHz)	148–149↑ 137–138↓	148–149↑ 137–138↓	148–149↑ 137–138↓	148–149↑ 137–138↓
Services	Nonvoice 2-way message, positioning	Nonvoice 2-way message, positioning	Nonvoice 2-way message, positioning	Nonvoice 2-way message, positioning
Mass (kg)	50	40	150	700
Orbitalvelocity (km/s)	7.35	7.365	7.205	7.45
Orbital period	1h45 m7.58s	1h44 m29.16s	1h51 m36.16s	1h40 m52.87s

(a)

Characteristics	Big-LEO			MEO
	Iridium (Motorola)	Globalstar (Qualcomm)	Teledesic	ICO (Global Communications)
Number of satellites	66+6*	48+4*	288	10 active and 2 in-orbit spares
Altitude (km)	780	1414	Ca. 700	10355 (changed to 10390 in 1998)
Coverage	Global	+70° latitude	Global	Global
Minimum elevation	8°	20°	40°	10°
Frequencies (GHz)	1.6 MS↓ 29.2↑ 19.5↓ 23.3 ISL	1.6 MS↑ 2.5 MS↓ 5.1↑ 6.9↓	19 28.8↑ 62 ISL	2 MS↑ 2.2 MS↓ 5.2 MS↑ 7↓
Access method	FDMA/TDMA	CDMA	FDMA/TDMA	FDMA/TDMA
ISL (Inter-satellite link)	Yes	No	Yes	No
Bit rate	2.4 kbit/s	9.6 kbit/s	64 Mbit/s↓ 2/64 Mbit/s↑	4.8 kbit/s
No. of channels	4000	2700	2500	4500
Lifetime (years)	5–8	7.5	10	12
Cost etimation	$4.4B	$2.9B	$9B	$4.5B
Services	Voice, data, fax, paging, messaging, position location, RDSS	Voice, data, fax, paging, position location, RDSS	Voice, data, fax, paging, video—as network-borne, RDSS	Voice, data, fax, short message RDSS
Mass (kg)	700	450	771	2600 (was listed in 1925)
Orbital velocity (km/s)	7.46	7.15	7.5	4.88 (changed to 4.846)
Orbital period	1h40 m27.59s	1h54 m5.83s	1h38 m46.83s	5h59 m2.25s (changed to 6h0 m9.88s)

* "+" indicates reserve.

(b)

world can be accessed, using the underlying hardware backbone network on the surface of the earth. To keep the weight of each satellite to a reasonable level, a minimum amount of electronic circuitry is kept in the satellite so that received incoming messages can be relayed to other satellites and mobile users. The satellites are controlled by the BS located at the surface of the earth, which serves as a gateway. Intersatellite links can be used to relay information from one satellite to another, but they are still controlled by the ground BS (also known as earth station or ES). The illuminated area of a satellite beam, called a footprint, is the area within which a mobile user can communicate with the satellite; many beams are used to cover a wide area.

There are losses in free space due to atmospheric absorption of transmitting satellite beams. Rain also causes substantial attenuation of signal strength, especially when frequency bands in the range of 12 to 14 GHz and 20 to 30 GHz are used in satellite communication to minimize orbital congestion. Therefore, it is important to consider availability of links. In addition, satellites are constantly rotating around the earth, and a beam may be temporarily blocked either due to other flying objects or the terrain of the earth's surface. Therefore, a redundancy concept, known as diversity, is used to transmit the same message through more than one satellite, as shown in Figure 11.11.

The basic idea behind path diversity is to provide a mechanism that can combine two or more correlated information signals (primarily the same copy) traveling along different paths and hence having uncorrected noise and/or fading characteristics. Such a combination of two signals improves signal quality, which enables the receiver to have flexibility in selecting a better quality signal. This can easily take care of problems due to temporary LOS problem or excessive noise and/or attenuation. The primary interest is with path diversity, though other forms of diversity such as antenna, time, frequency, field, or code, are possible. Path diversity will depend on the technology that is used to transmit and receive messages. The net effect of diversity is utilization of at least twice the bandwidth, and therefore it is desirable to employ diversity in as a small fraction of time as possible. On the other hand, diversity must be used as frequently as needed to ensure that the effect of link disconnection

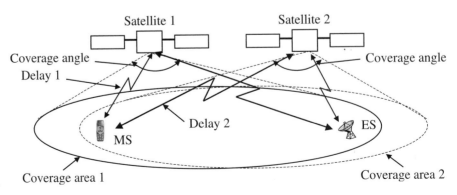

Figure 11.11
Satellite path diversity.

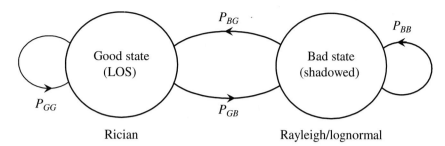

Figure 11.12
Channel model for
the MS.

is minimized. The channel of a satellite system is usually represented by a two-state Markov model, with the MS in the good state having Rician fading while a bad or shadowed state indicates Rayleigh/lognormal fading. The channel model is illustrated in Figure 11.12. Here, P_{ij} (i = G, B; j = B, G) is the transition probability.

The use of diversity can be initiated by either the MS or the BS located on the earth. The diversity request from the BS (ES) enables the MS to locate and scan unshadowed satellite paging channels for unobstructed communication. This kind of situation cannot be detected or determined by the BS, even though the MS's location is known to the BS. The use of satellite path diversity may be primarily due to the following conditions:

1. **Elevation angle**: Higher elevation angle decreases shadowing problems. One approach is to initiate path diversity when the elevation angle becomes less than some predefined threshold.

2. **Signal quality**: If the average signal level (in dB), quality (in BER), or fade duration goes beyond some threshold, then path diversity can be used. Signal quality is a function of parameters such as elevation angle, available capacity, current mobility pattern of the MS, and anticipated future demand.

3. **Stand-by option**: A channel can be selected and reserved as a stand-by for diversity whenever a threshold crossing is detected by the MS. Such a stand-by channel is used only when the primary channel is obstructed. Since the use of diversity is considered a rare event, several MSs can share the same stand-by channel.

4. **Emergency handoff**: Whenever a connection of a MS with a satellite is lost, the MS tries to have an emergency handoff.

Once the allocated channel(s) is(are) no longer used by the MS, the BS can release the channel and make it available for other MSs.

● ● ● ● ● ● ● ● ● ● ● ● ● ● ●
11.5 Call Setup

A generic satellite system architecture is shown in Figure 11.13, with the ES (BS) constituting the heart of the overall system control. The ES performs functions similar to the BSS of a cellular wireless system. The ES keeps track of all MSs located in the area and controls the allocation and deallocation of radio resources.

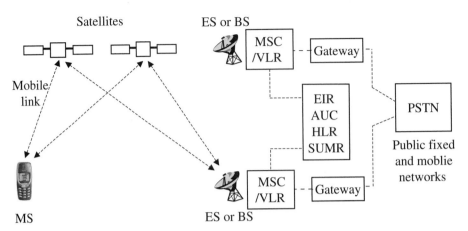

Figure 11.13
Satellite system
architecture.

This includes the use of frequency band or channel in FDMA, time slot for TDMA, and code assignment for CDMA. Both MSC and VLR are important parts of the BS and provide functions similar to those for the cellular network. The databases EIR, AUC, and HLR also perform the same operations as in conventional wireless systems and are an integral part of the overall satellite system. The HLR–VLR pair supports the basic process of mobility management. A satellite user mapping register (SUMR) is also maintained at the BS to note the locations of all satellites and to indicate the satellite assigned to each MS. All these systems are associated with the BS to minimize the weight of satellites. In fact, satellites can be considered to function as relay stations with a worldwide coverage, given that most of the intelligence and decision-making process is performed by the BS. These BSs are also connected to the PSTN and ATM backbone through the appropriate gateway so that calls to regular household phones as well as to cellular devices can be routed and established.

For an incoming call from the PSTN, the gateway helps to reach the closest BS, which, in turn, using the HLR–VLR pair, indicates the satellite serving the most recently known location of the MS. The satellite employs a paging channel to inform the MS about an incoming call and the radio resource to be used for the uplink channel.

For a call originating from a MS, it accesses the shared control channel of an overhead satellite and the satellite, in turn, informs the BS for authentication of the user/MS. The BS then allocates a traffic channel to the MS via the satellite and informs the gateway about additional control information, if it is necessary to route the call through the backbone. Thus, there may be an exchange of control signaling between the MS, the satellite beam, the ES, and the PSTN gateway. Call setup may involve satellite communication before the actual traffic can be exchanged and can vary in the range of a few hundred nanoseconds (~ 300 ns).

Similar to cellular systems, whenever a MS moves to a new area served by another satellite, then the MS has to go through the registration process; the only difference here is the use of ES in all intermediate steps. A typical system timing

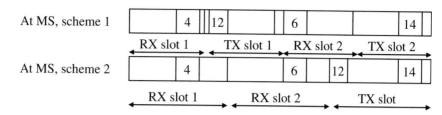

Figure 11.14
System timings for
the satellite.

for a TDMA-based satellite system with different possible schemes is shown in Figure 11.14. Scheme 1 employs half of the 16-burst half-rate while the second half is for the TDMA frame of satellite 2. Diversity is employed in scheme 2, and the TDMA frame is split into three parts, the first two for reception from satellites 1 and 2 and the third for communication with the satellite that has the best signal after employing the required timing adjustment.

Several additional situations are present for handoff in satellite systems as compared with cellular wireless networks, primarily due to the movement of satellites and the wider coverage area. Various types of handoffs can be summarized as follows:

1. **Intrasatellite handoff**: There could be handoff from one spot beam to another due to relative movement of the MS with respect to the satellites because the MS needs to be in the footprint area to communicate with a satellite. Therefore, if the MS moves to the footprint path of another beam, there would be an intrasatellite handoff.

2. **Intersatellite handoff**: Since the MS is mobile and most satellites are not geosynchronous, the beam path may change periodically. Therefore, there could be a handoff from one satellite to another satellite under control of the BS.

3. **BS handoff**: A rearrangement in frequency may be desirable to balance the traffic in neighboring beams or the interference with other systems. There could be situations in which satellite control may change from one BS to another because of their relative locations. This may cause a handoff at the BS level, even though the MS may still be in the footprint of the current satellite.

4. **Intersystem handoff**: There could be a handoff from a satellite network to a terrestrial cellular network, which would be cheaper and would have a lower latency.

The handoff is termed seamless if the communication path between MS and ES is not broken during the handoff process. TDMA schemes with and without diversity support seamless handoff. In case of diversity, one of the channels is released for handoff, and attempts are made to find a new channel for maintaining the diversity.

● ● ● ● ● ● ● ● ● ● ● ● ● ● ● ●

11.6 Global Positioning System

Global positioning systems, widely known as GPSs, have been of great importance since the days of World War II. Although the initial focus was mainly on military targeting, fleet management, and navigation, commercial usage began finding relevance as the advantages of radiolocation were extended to (but not limited to) tracking down stolen vehicles and guiding civilians to the nearest hospital, gas station, hotel, and so on. Present-day wireless service providers are expected to indicate an exact location of callers for 911 emergency assistance.

A GPS system consists of a network of 24 orbiting satellites [11.2], called NAVSTAR (Navigation System with Time and Ranging), placed in space in six different orbital paths with four satellites in each orbital plane and covering the entire earth under their signal beams (Figure 11.15). The orbital period of these satellites is 12 hours. The satellite signals can be received anywhere in the world and at any time. The spacing of the satellites is arranged such that a minimum of five satellites are in view from every point on the globe. The first GPS satellite was launched in February 1978, and the twenty-fourth block II satellite, deployed in March 1994, completed the GPS constellation. Each satellite is expected to last approximately 7.5 years, and replacements are constantly being built and launched into orbit. The GPS system is currently funded with replacements through 2006. Each satellite is placed at an altitude of about 10,900 nautical miles and weighs about 862 kg (1900 lb). The satellites extend to about 5.2 m (17 ft) in space including the solar panels. Each satellite transmits on three frequencies. Civilian GPS uses the L1 frequency of 1575.42 MHz.

The GPS control, or the ground segment, consists of unmanned monitor base stations located around the world (Hawaii and Kwajalein in the Pacific Ocean; Diego Garcia in the Indian Ocean; Ascension Island in the Atlantic Ocean; and a master

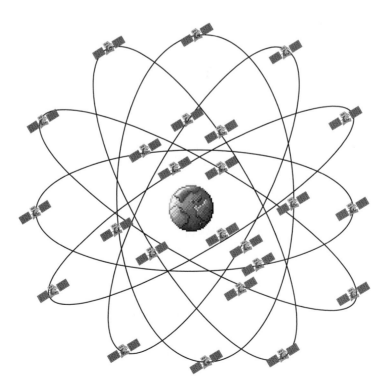

Figure 11.15
GPS nominal constellation of 24 satellites in six orbital planes [11.2].

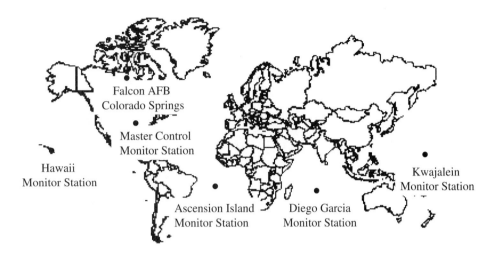

Figure 11.16
GPS master control
and monitor station
network [11.2].

base station at Schriever [Falcon] Air Force Base in Colorado Springs, Colorado [Figure 11.16]); along with four large ground antenna stations that broadcast signals to the satellites. The stations also track and monitor the GPS satellites.

These monitor stations measure signals from the space vehicles (SVs) that are incorporated into orbital models for each satellite. The models compute precise orbital data (ephemeris) and SV clock corrections for each satellite. The master control station uploads ephemeris data to GPS receivers.

GPS is based on a well-known concept called the triangulation technique [11.3]. The concept is illustrated in Figure 11.17. Consider the GPS receiver MS to be placed on one point on an imaginary sphere of radius equal to the distance between satellite "A" and the receiver on the ground (with the satellite "A" as the center of the sphere). Now the GPS receiver MS is also a point on another imaginary sphere with a second satellite "B" at its center. We can say that the GPS receiver is somewhere on the circle formed by the intersection of these two spheres. Then, with a measurement of distance from a third satellite "C," the position of the receiver is narrowed down to just two points on the circle, one of which is imaginary and is eliminated from the calculations. As a result, the distance measured from three satellites suffices to determine the position of the GPS receiver on earth. Therefore, the measured parameters are the distances between the satellites in space and the receiver on earth. The distance is calculated from the speed of these radio signals and the time taken for these signals to reach earth. With a distance so large, an error of even a few milliseconds can cause an error of about 200 miles from the actual position of the GPS receiver on earth.

Let us look at how the travel time is measured. Two signals, say signal $X(T)$ and signal $Y(T)$, are synchronously transmitted: Signal $X(T)$ is generated in the satellite

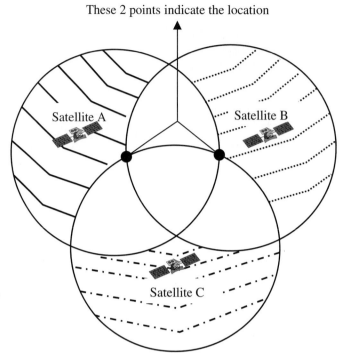

These 2 points indicate the location

Satellite A

Satellite B

Satellite C

Figure 11.17
The triangulation
technique.
Credit: From "GPS:
Location-tracking
Technology," by
R. Bajaj, S. Rannweera,
and D.P. Agrawal,
2002, *IEEE Computer*,
35, 92-94. Copyright
2002 IEEE.

while signal $Y(T)$ is generated in the receiver on earth. The time taken by signal $X(T)$ to reach earth is what needs to be found. This signal is basically a function of $T + t$, where t is the travel time of signal $X(T)$ from the satellite to earth. This time can also be calculated from the difference between signals $Y(T)$ (both signals are synchronous in time) and $X(T + t)$. The time t multiplied by the speed of the radio signal (the speed of light) gives the distance of the satellite from the receiver on earth. The clocks used by the satellites are atomic to provide a very high degree of accuracy. Receiver MS clocks, on the other hand, do not have to be very accurate because an extra satellite range measurement can eliminate errors.

The GPS signal is composed of a pseudorandom code, ephemeris and navigation data. Ephemeris data (this is part of the data message used to predict the current satellite position transmitted to the user) correct errors (called ephemeris errors) caused by gravitational pulls from the moon and sun and by the pressure of solar radiation on the satellites. Navigation data constitute the information about the located position of the GPS receiver, which is relayed back to the satellite itself. The pseudorandom number (PRN) code (an ID code) identifies which satellite is transmitting. Satellites are referred to by their PRN, from 1 through 32, and this is the number displayed on the GPS receiver (MS) to indicate the satellites with which there is an interaction going on, depending on the MS's position. The use of more than 24 PRNs simplifies the maintenance of the GPS network. A replacement satellite can be launched, turned on, and used before the satellite it was intended to replace is actually taken out of service. Ephemeris data are constantly transmitted by each satellite and contain important information, such as status of the satellite (healthy or unhealthy), current date, and time. This part of the message indicates to the GPS receiver the satellites nearest to it. The GPS receiver reads the message and saves the ephemeris and almanac data for continual use. This information can also be used to set (or correct) the clock within the GPS receiver.

11.6.1   Limitations of GPS

There are several factors [11.3] that introduce error to GPS position calculations and prevent us from achieving the best possible accuracy. A major source of error arises from the fact that the speed of the radio signals is constant only in a vacuum, which means that distance measurements may vary as the values of the signal speed vary in the atmosphere. The atmosphere, as we know, is composed of the ionosphere and the troposphere. The presence of the troposphere (essentially composed of water vapor) is known to cause errors due to variation of temperature and pressure, and the particles in the ionosphere are known to cause significant measurement errors (as would be the case with bad clocks!). Factors affecting accuracy are shown in Table 11.3.

Another source of error is the multiple paths that signals take between the satellite and the MS ground receivers. The effects of multipath fading and shadowing are significant due to the absence of a direct LOS path. In other words, multipath is the result of a radio signal being reflected off an object. Multipath is what causes "ghost" images on a television set. These effects are not seen on television sets much nowadays since they are most likely to occur with those old-style "rabbit ear" antennas, not on cable. With GPS, multipath fading occurs when the signal bounces off a building or terrain before reaching the GPS receiver's antenna. The signal takes longer to reach the receiver than if it travels along a direct path. This added time makes the GPS receiver think that the satellite is farther away, which adds to the

Table 11.3: ▶
Factors Affecting Accuracy of GPS Position Calculations

Error Factor	Accuracy level (in meters)	
	Standard GPS	Differential GPS (DGPS)
Atmospheric conditions (troposphere)	0.5–0.7	0.2
Atmospheric conditions (ionosphere)	5–7	0.4
Multipath fading and shadowing effects	0.6–1.2	0.6
Receiver noise	0.3–1.5	0.3
Selective availability	24–30	0
Atomic clock errors	1.5	0
Ephemeris errors	2.5	0

error in the overall position determination. When they occur, multipath errors may typically add 0.6 to 1.2 meters of error to the overall position.

Another factor affecting the precision is satellite geometry (i.e., locations of the satellites relative to each other). If a GPS receiver is locked with four satellites and all four of these satellites are in the sky to the north and west of the receiver, satellite geometry is relatively poor. This is because all the distance measurements are from the same general direction. This implies that triangulation is poor and the common area where these distance measurements intersect is fairly wide (i.e., the area where the GPS receiver determines its position covers a large space, so pinpoint positioning is not possible). In this scenario, even if the GPS receiver does report a position, accuracy will not be very good (maybe as much as 0.9 to 1.5 m). If the same four satellites are spread out in all directions, the position accuracy is known to improve dramatically. When these four satellites are separated equally at approximately 90° intervals (north, east, south, west), the satellite geometry is very good, since distance measurements are from all directions. The common area where all four distance measurements intersect is much smaller. Satellite geometry also becomes an issue when using a GPS receiver (MS) in a vehicle, near tall buildings, or in mountainous or canyon areas because propagation delay due to atmospheric effects can affect accuracy. The internal clock can also cause small errors.

Propagation delay is the slowing down of the GPS signal as it passes through the earth's ionosphere and troposphere. In space, radio signals travel at the speed of light, but they are significantly slower once they enter our atmosphere.

The largest source of position error is selective availability (SA), which is an intentional degradation of civilian GPS by the U.S. Department of Defense. The idea behind intentionally induced errors due to SA is to make sure that no hostile force or terrorist group can use GPS to make accurate weapons. As mentioned, GPS was originally designed and built for military applications, and as the system has evolved, it is being used for numerous civilian applications as well. All current GPS satellites are capable of and subject to SA degradation.

In addition to the aforementioned errors, there are ephemeris errors, already mentioned, and unaccounted atomic clock errors, which may lack precision of the desired level due to the absence of continuous monitoring. Another limitation is that a GPS receiver's needs can prove to be a limitation for existing mobile devices, including cell phones, because they are normally not equipped with GPS capability.

There are a number of free subscription services available to provide DGPS corrections. The U.S. Coast Guard and U.S. Army Corps of Engineers (and many foreign governmental departments as well) transmit DGPS corrections through marine beacon stations. These beacons operate in the 283.5 to 325.0 kHz Industrial, Scientific, and Medical (ISM) frequency range, and no licensing is required. The cost to use the service is the purchase of a DGPS beacon receiver. This receiver is then coupled to the user's GPS receiver via a three-wire connection, which relays the corrections in a standard serial data format called RTCM SC-104. Some GPS receivers provide a timing pulse accurate to within a microsecond, while more expensive models can offer accuracies within a nanosecond. Subscription DGPS services are available on FM radio station frequencies or via a satellite. In fact, the requirements vary with the type of DGPS applications, and hence different solutions may be applicable. Some may not need the radio link because an instantaneous precise positioning may not be needed. For example, trying to position a drill bit over a particular spot on the ocean floor from a pitching boat is different from trying to record the track of a new road for inclusion on a map. For applications like the latter, the mobile GPS receiver needs to record all of its measured positions and the exact time it made each measurement. These data are then combined with the corrections recorded at a reference receiver. The radio link that is present in real-time systems is not needed. In the absence of a reference receiver, there may be an alternative source (such as the Internet) for distributing corrections to the recorded data.

11.6.2   Beneficiaries of GPS

First and foremost, GPS has proved to be a most valuable aid to U.S. military forces. Picture the desert, with its wide, featureless expanses of sand, with the terrain looking much the same for miles. Without a reliable navigation system like GPS, the US forces could not have performed the maneuvers of Operation Desert Storm. With GPS, soldiers were able to maneuver in sandstorms at night. At the start of Desert Storm, more than 1000 portable commercial GPS receivers were purchased for military use. The demand was so great that, before the end of the conflict, more than 9000 commercial GPS receivers were in use in the Gulf region. They were carried by soldiers and attached to vehicles, helicopters, and aircraft instrument panels. GPS receivers were used in several aircrafts, including F-16 fighters, KC-135 aerial refuelers, and B-2 bombers; Navy ships used them for rendezvous, minesweeping, and aircraft operations. GPS has become important for nearly all military operations and weapons systems.

In addition, GPS benefits nonmilitary operations. It is used on satellites to obtain highly accurate orbit data and to control spacecraft orientation. During construction of the English channel tunnel (the "Chunnel"), British and French crews started digging from opposite ends: one from Dover, England, and another one from Calais, France. They relied on GPS receivers outside the tunnel to check their positions along the way and to make sure they met exactly in the middle. GPS has a variety of applications on land, at sea, and in the air. GPS can be used everywhere except indoors and places where a GPS signal cannot be received because of natural or man-made obstructions. Both military and commercial aircraft use GPS for navigation purposes. It is also used by commercial fishermen and boaters to aid in navigation. The precision timing capability provided by GPS is used by the scientific community for research purposes. The GPS enables survey units to help surveyors to set up their survey sites fairly quickly. GPS is also used for noncommercial purposes by car racers, hikers, hunters, mountain bikers, and cross-country skiers. GPS also helps in providing emergency roadside assistance, by allowing an accident victim to transmit his or her position to the nearest response center at the push of a button. Vehicle tracking has become one of the major GPS applications. GPS-equipped fleet vehicles, public transportation systems, delivery trucks, and courier services use receivers to monitor their locations at all times. GPS is also helping to save lives. Many police, fire, and emergency medical service units are using GPS receivers to determine the location of a police car, a fire truck, or an ambulance nearest to an emergency, enabling the

Table 11.4: ▶
Applications of GPS

User Group	Application Area
U.S. military	Maneuvering in extreme conditions and navigating planes, ships, etc.
Building the English channel tunnel	Checking positions along the way and making sure that they meet in the middle
General aviation and commercial aircraft	Navigation
Recreational boaters and commercial fishermen	Navigation
Surveyors	Reducing setup time at survey sites and offering precise measurements
Recreational users (e.g., hikers, hunters, snowmobilers, mountain bikers)	Keeping track of where they are and finding a specified location
Automobile services	Emergency roadside assistance
Fleet vehicles, public transportation systems, delivery trucks, and courier services	Monitoring locations at all times
Emergency vehicles	Determining location of car, truck, or ambulance closest to the emergency, allowing quick response time
Automobile manufacturers	Display of maps in moving cars that can be used to plan a trip
Carrier companies	Positioning and navigation

quickest possible response in life-or-death situations. Automobile manufacturers are offering moving-map displays guided by GPS receivers as an option on new vehicles. The displays can be removed and taken into a home to plan a trip. Among the latest important developments, it is observed that several carrier companies have already informed the FCC that they have opted for a handset-based 911 system, which means using a satellite-based global positioning system. It is surveyed that more than 118,000 calls a day are made in the United States to 911 and other emergency numbers from wireless phones. GPS offers other features and applications for handset subscribers. Applications of GPS are summarized in Table 11.4.

● ● ● ● ● ● ● ● ● ● ● ● ● ● ● ●
11.7 A-GPS and E 911

Tracking the location of mobile users using GPS is one of the fastest growing application areas. In this receiver-based approach, an MS directly contacts the constellation of GPS satellites and downloads information necessary to determine its position. Therefore there is a lot of delay in obtaining all the information from the satellites (data is transferred at 50 bps). Another problem is that when the MS is indoors, it may not be possible to contact the GPS satellites. An alternative is to use a network-based approach wherein the MS triangulates its position using information from three or more BSs. This approach has the disadvantage that the location information obtained may not possess the desired accuracy. The A-GPS is a hybrid solution to this problem whereby information from both the satellites and the network is used to accurately determine the location of an MS. Information about the satellite positions may be downloaded and precalculated by powerful A-GPS servers located at the BSs, and this is fed to the MSs, which use this information along with the encoded signals obtained from the satellites to accurately and quickly obtain its location. A-GPS also addresses the problem of weak GPS signals indoors. Different companies providing A-GPS solutions address this problem differently [11.4][11.5]. The basic idea is to increase the sensitivity of the GPS receiver, using massively parallel correlation techniques [11.5].

Enhanced 911 (E 911) is a location technology mandated by the FCC that will enable mobile, or cellular, phones to process 911 emergency calls and enable emergency services to locate the geographic position of the caller. In a traditional wired phone, the 911 call is routed to the nearest public safety answering point (PSAP) that then distributes the emergency call to the proper services and the exact location of the phone is determined. E 911 is a specific application built over the A-GPS technology described above.

● ● ● ● ● ● ● ● ● ● ● ● ● ● ● ● ●

11.8 Summary

Satellite systems possess global connectivity and provide a lot more flexibility than a conventional land-based wireless system. However, the delay involved in traversing back and forth from the earth to the satellite and the complexity of the handheld transmitter/receiver makes its satellite systems acceptable for personal and commercial use. Satellites, because of so many practical considerations, are still controlled by the earth station. In the near future, it is unlikely that the delay will be reduced by any noticeable level, but advances in signal processing and VLSI (very large scale integration) design may minimize the complexity of handheld devices. However, the usefulness of GPS is yet to be explored fully, and the future of the satellite systems seems promising. All wireless devices follow a set of predefined rules and guidelines so that two entities can successfully communicate with each other. These are discussed in Chapter 12.

● ● ● ● ● ● ● ● ● ● ● ● ● ● ● ● ●

11.9 References

[11.1] W. W. Wu, E. F. Miller, W. L. Pritchard, and R. L. Pickholtz, "Mobile Satellite Communications," *Proceedings of the IEEE*, Vol. 82, No. 9, pp. 1431–1448, September 1994.

[11.2] *http://www.colorado.edu/geography/gcraft/notes/gps*.

[11.3] R. Bajaj, S. Ranaweera, and D. P. Agrawal, "GPS: Location-Tracking Technology," *IEEE Computer*, Vol. 35, No. 4, pp. 92–94, April 2002.

[11.4] *www.snaptrack.com*.

[11.5] *www.globallocate.com*.

● ● ● ● ● ● ● ● ● ● ● ● ● ● ● ●

11.10 Problems

P11.1. What is the rationale behind using highly elliptical orbits? Explain.

P11.2. What will be the propagation delay between a satellite and an earth-based mobile station if the satellite is located at a distance of 850 km and if its inclination angle is 35°?

P11.3. The beam footprint depends on the inclination angle. What will be the impact on the coverage if the angle is changed from 35° to 30°? Explain clearly.

P11.4. What should be the velocity of the satellite if it orbits around earth at a distance of 1000 km and weighs 2000 kg?

P11.5. If the isoflux area boundary is fuzzy, what should you do and what will be the overall impact on the system performance? Explain clearly.

P11.6. Setting up a path for a satellite phone subscriber requires a comprehensive hand-shaking mechanism between the MS, the satellite, and the BS. Prepare the steps that are desirable in setting up such a path and comment on how you could minimize traversal of signals between the satellite and the MS/BS.

P11.7. What is the information content if an average of two-way diversity is used in a satellite system 10% of time by 50% of the traffic and 5% of time by the rest of the traffic?

P11.8. In Problem P11.7, if $(128, 32)$ code is used for error correction, what is the fraction of information contents?

P11.9. A code (n, k, t) is defined by k information bits and $(n - k)$ redundant bits so as to correct t errors in the resulting word of n bits. Given a channel bit error rate of p, what is the word error rate (WER)?

P11.10. What are the differences between orbital and elevation angles of a satellite?

P11.11. What are the advantages and disadvantages of LEO and GEO?

P11.12. How do you compare delays in a satellite system versus a cellular system, versus an inter-terrestial satellite system? How about the power level, coverage area, and transmission rates?

P11.13. How is the call setup in a satellite system different from a cellular system?

P11.14. In the satellite system, there is some degree of free space loss. Besides this loss, does it have any other source of loss? Explain.

P11.15. Why can there be more than one satellite orbiting in a single orbiting path of GPS?

P11.16. Why are errors inherent in the triangulation technique? Explain clearly.

P11.17. Is it possible to find a precise location inside a building or a room where GPS will not work? Explain.

P11.18. From your local wireless service provider, find out if emergency 911 service is provided in your area and what kind of technique is used in location determination.

P11.19. What are the different alternative techniques for determining location using cell phones? Explain the role of beacon signals.

P11.20. How is the location of packets and parcels updated by United Parcel Service (UPS) or Federal Express (Fedex)? Explain clearly.

P11.21. What are some unconventional uses of GPS?

P11.22. How do you compare the functionality of an earth station with the corresponding unit in the cellular system?

CHAPTER 12

Network Protocols

12.1 Introduction

A common language is needed between two people so that each can understand what the other person means, and their actions can confirm a desired response. By the same token, two devices exchanging information need to follow some simple rules so that the information can be interpreted correctly. Therefore, there is a need to define a set of rules or guidelines so that all digital communicating entities can follow them for their successful operation. In a wireless network, handshaking and routing are required as the signal travels through the backbone landline as well as through wireless infrastructures, as discussed in Chapter 9. This chapter primarily deals with rules applied to wireless and mobile networks.

Communications between entities over a network can take place only if entities have a common understanding between them. In technical terms, this understanding is called a network protocol. A network protocol gives a set of rules that are to be followed by entities situated on different parts of a network. In this chapter a brief overview of the OSI (Open Systems Interconnection) reference model is given. The OSI reference model was created out of a need for a common reference for protocol development. A practical implementation of the Transmission Control Protocol/Internet Protocol (TCP/IP) protocol stack is briefly described. TCP/IP is by far the most popular network protocol stack, and its study provides an understanding of the needs of a practical network. The Internet, which has become extremely popular, uses the TCP/IP stack as its backbone. It uses several algorithms to route data from one system to another across the network. A brief look at these protocols is a good starting point in the study of computer networks. The TCP protocol used over wireline networks has several features that make it inefficient when used in exactly the same form over wireless networks. A study of the various mechanisms used to fine-tune the TCP for wireless networks is useful in understanding the difficulty in internetworking wired and wireless networks. The existing version of Internet Protocol (IP), known as Internet Protocol version 4(IPv4), uses only 32 bits for representing a host on a network. This space is very limited, and an enhanced version Internet Protocol version 6 (IPv6), using 128 bits, has been created to overcome this problem it also incorporates several additional features required for future protocol development.

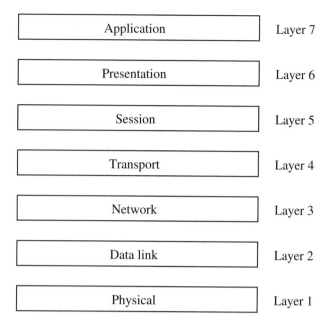

Figure 12.1
OSI model.

• • • • • • • • • • • • • • •

12.2 OSI Model

The standard model for networking protocols and distributed applications is the International Standards Organization's (ISO's) OSI model. The work on OSI was initiated in the late 1970s and came to maturity in the late 1980s and early 1990s. The OSI represents the totality of protocol definitions and along with associated additional documents that provide international standardization of many aspects of data communication and networking. In principle, it extends from the lowest level of signaling techniques between two entities to high-level interactions in support of specific applications.

The OSI model is a layered framework for the design of network systems that allows communication between all types of data systems (see Figure 12.1). The OSI model is composed of seven ordered layers: physical layer (layer 1), data link layer (layer 2), network layer (layer 3), transport layer (layer 4), session layer (layer 5), presentation layer (layer 6), and application layer (later 7). These layers are defined to be modular in nature so that compatibility could be easily maintained. We briefly describe the functions of each layer in the OSI model before considering how it has been modified and adopted for the wireless world.

12.2.1 Layer 1: Physical Layer

The physical layer supports the electrical or mechanical interface to the physical medium and performs services requested by the data link layer. The major functions and services performed by the physical layer are as follows:

1. Establishment and termination of a connection to a communications medium
2. Participation in the process whereby the communication resources are effectively shared among multiple users (e.g., contention resolution and flow control)
3. Conversion between the representation of digital data in the end user's equipment and the corresponding signals transmitted over a communications channel

The physical layer is concerned with the following:

1. Physical characteristics of interfaces and media
2. Representation of bits, transmission rate, synchronization of bits
3. Link configuration
4. Physical topology, and transmission mode

12.2.2 Layer 2: Data Link Layer

The data link layer provides the functional and procedural means to transfer data between network entities and to detect and possibly correct errors that may occur in the physical layer. This layer responds to service requests from the network layer and issues service requests to the physical layer. Specific responsibilities of the data link layer include the following:

1. Framing
2. Physical addressing
3. Flow control
4. Error control
5. Access control

12.2.3 Layer 3: Network Layer

The network layer provides the functional and procedural means of transferring variable-length data sequences from a source to a destination via one or more networks while maintaining the QoS requested by the transport layer. The network layer performs network routing, flow control, segmentation and reassembly, and error control functions. This layer responds to service requests from the transport layer and issues service requests to the data link layer. Specific responsibilities of the network layer include the following:

1. Logical addressing
2. Routing

12.2.4 Layer 4: Transport Layer

The purpose of the transport layer is to provide transparent transfer of data between end users, thus relieving the upper layers from any concern with providing reliable and cost-effective data transfer. This layer responds to service requests from the session layer and issues service requests to the network layer. Specific responsibilities of the transport layer include the following:

1. Service-point addressing
2. Segmentation and reassembly
3. Connection control and flow control
4. Error control

12.2.5   Layer 5: Session Layer

The session layer provides the mechanism for managing a dialog between end-user application processes. It supports either duplex or half-duplex operations and establishes checkpointing, adjournment, termination, and restart procedures. This layer responds to service requests from the presentation layer and issues service requests to the transport layer. Specific responsibilities of the session layer include the following:

1. Dialog control
2. Synchronization

12.2.6   Layer 6: Presentation Layer

The presentation layer relieves the application layer of concern regarding syntactical differences in data representation within the end-user systems. This layer responds to service requests from the application layer and issues service requests to the session layer. Specific responsibilities of the presentation layer include the following:

1. Translation
2. Encryption
3. Compression

12.2.7   Layer 7: Application Layer

The application layer is the highest layer. This layer interfaces directly to and performs common application services for the application processes and also issues requests to the presentation layer. The common application services provide semantic conversion between associated application processes. Specific services provided by the application layer include the following:

1. Network virtual terminal
2. File transfer, access, and management
3. Mail services
4. Directory services

● ● ● ● ● ● ● ● ● ● ● ● ● ● ● ● ● ●
12.3 TCP/IP Protocol

Transfer of information between two entities (e.g., e-mail) involves transfer of data over the Internet, and with this in mind TCP/IP has been defined. The TCP/IP protocol suite provides service to transfer data from one network device to another using the Internet. The TCP/IP protocol suite is composed of five layers: physical, data link, network, transport, and application. The lower four layers of the TCP/IP correspond to the lower four layers of the OSI model, while the application layer in TCP/IP represents the three topmost layers of the OSI model of Figure 12.1. The TCP/IP protocol stack is shown in Figure 12.2. Unlike the OSI model, which specifies different functions belonging to various layers, TCP/IP consists of independent protocols that can be mixed and matched depending on requirements.

12.3.1 Physical and Data Link Layers

The physical and data link layers are responsible for communicating with the actual network hardware (e.g., the Ethernet card). Data received from the physical medium are handed over to the network layer, and data received from the network layer are sent to the physical medium. The TCP/IP does not specify any specific protocol at this layer and supports all standard and proprietary protocols.

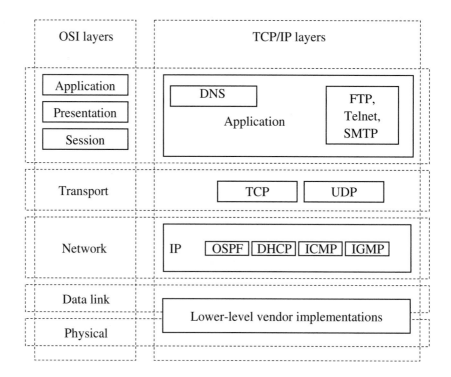

Figure 12.2
TCP/IP protocol stack.

12.3.2 Network Layer

The network layer is responsible for delivering data to the destination. It does not guarantee the delivery of data and assumes that the upper layer will handle this issue. This layer consists of several supporting protocols.

Internet Protocol (IP)

The Internet protocol (IP) [12.1] is a network layer protocol that provides a connectionless, "best effort" delivery of packets through an internetwork. The term *best effort* means that there is no error checking or tracking done for the sequence of packets being transmitted. It assumes that the higher-layer protocol takes care of the reliability of packet delivery. The packets being transmitted are called datagrams. Each of these datagrams is transmitted independently and may take different routes to reach the same destination. IP supports a mechanism of fragmentation and reassembly of datagrams to handle data links with different maximum-transmission unit (MTU) sizes.

Internet Control Message Protocol (ICMP)

The Internet control message protocol (ICMP) [12.2] is a companion protocol to IP that provides a mechanism for error reporting and query to a host or a router. The query message is used to probe the status of host or a router by the network manager whereas the error-reporting message is used by the host and routers to report errors.

Internet Group Management Protocol (IGMP)

The Internet group management protocol (IGMP) [12.3] is used to maintain multicast group membership within a domain. Similar to ICMP, it uses query and reply messages to maintain multicast group membership in its domain. A multicast router sends a periodic IGMP query message to find out the multicast session members in its domain. If a new host wants to join a multicast group, it sends an IGMP join message to its neighboring multicast router, which takes care of adding the host to the multicast delivery tree.

Dynamic Host Configuration Protocol (DHCP)

The dynamic host configuration protocol (DHCP) [12.4] is designed to handle dynamic assignments of IP addresses in a domain. This protocol is an extension of the bootstrap protocol (BOOTP) and provides a way for the mobile nodes to request an IP address from a DHCP server in case nodes move to a different network. This dynamic assignment of IP address is also applicable to the hosts that attach to the network occasionally. It saves precious IP address space by utilizing the same IP address for needed hosts. DHCP is fully compatible with BOOTP, which supports only static binding of physical address to IP address.

Internet Routing Protocols

Some of the widely used routing protocols at the network layer are routing information protocol (RIP) [12.5], open shortest path first (OSPF) [12.6], and border gateway protocol (BGP) [12.7].

■ **Routing information protocol (RIP)**: RIP is a distance vector–based interior routing protocol. It uses the Bellman-Ford algorithm (discussed in the following subsection) to calculate routing tables. In distance vector routing, each router periodically shares its knowledge about other routers in the network with its neighbors. Each router also maintains a routing table consisting of each destination IP address, the shortest distance to reach the destination in terms of hop count, and the next hop to which the packet must be forwarded. The current RIP message contains the minimal amount of information necessary for routers to route messages through a network and is meant for small networks. RIP version 2 [12.8] enabled RIP messages to carry more information, which permits the use of a simple authentication mechanism to update routing tables securely. More important, RIP version 2 supports subnet masks, a critical feature that was not available in RIP.

■ **Open shortest path first (OSPF)**: OSPF is an interior routing protocol developed for IP networks. This protocol is based on the shortest path first (SPF) algorithm, which sometimes is referred to as the Dijkstra algorithm. OSPF supports hierarchical routing, in which hosts are partitioned into autonomous systems (AS). Based on the address range, an AS is further split into OSPF areas that help border routers to identify every single node in the area. The concept of OSPF area is similar to subnetting in IP networks. Routing can be limited to a single OSPF or can cover multiple OSPFs. OSPF is a link-state routing protocol that requires sending link-state advertisements (LSAs) to all other routers within the same hierarchical area. As OSPF routers accumulate link-state information, they use the SPF algorithm to calculate the shortest path to each node. As a link-state routing protocol, OSPF contrasts with RIP, which is a distance vector routing protocol. Routers running the distance vector algorithm send all or a portion of their routing tables in routing-update messages to their neighbors.

■ **Border gateway protocol (BGP)**: BGP is an interdomain or interautonomous system routing protocol. Using BGP, interautonomous systems communicate with each other to exchange reachability information. BGP is based on the Path Vector Routing Protocol, wherein each entry in the routing table contains the destination network, the next router, and the path to reach the destination. The path is an ordered list of autonomous systems that a packet should travel to reach the destination.

12.3.3 TCP

TCP [12.9][12.10] is a connection-oriented reliable transport protocol that sends data as a stream of bytes. At the sending end, TCP divides the stream of data into smaller units called segments. TCP marks each segment with a sequence number. The sequence number helps the receiver to reorder the packets and detect any lost packets. If a segment has been lost in transit from source to destination, TCP retransmits the data until it receives a positive acknowledgment from the receiver. TCP can also recognize duplicate messages and can provide flow control mechanisms in case the sender is transmitting at a faster speed than the receiver can handle.

12.3.4   Application Layer

In TCP/IP the top three layers of OSI—session, presentation, and application layers—are merged into a single layer called the application layer. Some of the applications running at this layer are, domain name server (DNS), simple mail transfer protocol (SMTP), Telnet, file transfer protocol (FTP), remote login (Rlogin), and network file system (NFS).

12.3.5   Routing Using Bellman-Ford Algorithm

One step that could take substantial amount of time is the selection of a route between the source and destination. This is important as appropriate path selection is critical for minimizing communication delays. The Bellman-Ford algorithm [12.11] is one of the routing algorithms designed to find shortest paths between two nodes of a given graph (Figure 12.3), representing an abstract model of a communication network, with communicating entities indicated by nodes and links represented by graph edges. In such a graph, a routing table is maintained at each node, indicating the best known distance to each destination and the next hop to get there. Such tables are updated by exchanging information with the neighbors. Let n be the number of nodes in the network. $w(u, v)$ is the cost (weight) associated with each edge uv between nodes u and v. $d(u)$ is the distance between node u and the root node under consideration and is initialized to ∞. For each edge uv in the network, set $d(v) = \min[d(v), d(u) + w(u, v)]$. Edges can be taken in any order. This algorithm is repeated $n - 1$ times, constituting the Bellman-Ford algorithm. After each step, tables are exchanged and updated between adjacent nodes. Figure 12.4 shows the results of each pass for the sample network in Figure 12.3.

The complexity of the Bellman-Ford algorithm is $O(VE)$, where V and E are the number of nodes and edges in the graph, respectively.

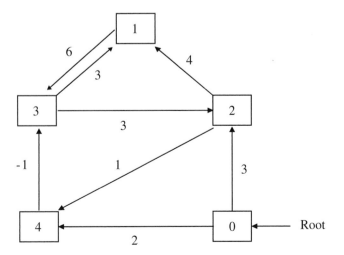

Figure 12.3
Abstract model of a wireless network in the form of a graph.

To Node	0	1	2	3	4
Pass 0	0	8	8	8	8
Pass 1	0	8	3	8	2
Pass 2	0	7	3	1	2
Pass 3	0	4	3	1	2
Pass 4	0	4	3	1	2

To Node	0	1	2	3	4
Pass 0	*	8	8	8	8
Pass 1	*	8	0	8	0
Pass 2	*	2	0	4	0
Pass 3	*	3	0	4	0
Pass 4	*	3	0	4	0

Figure 12.4
Steps in the Bellman-Ford algorithm for the sample network.

(a) Successive calculation of distance $d(u)$ from node 0

(b) Predecessor from node 0 to other network nodes

● ● ● ● ● ● ● ● ● ● ● ● ● ● ● ● ●

12.4 TCP over Wireless

12.4.1 Need for TCP over Wireless

The existing Internet employs TCP/IP as its protocol stack. Many of the existing applications require TCP as the transport layer for reliable transfer of data packets. Accessing the Internet is essential for commercial applications, while voice and other data communications utilize the underlying Internet backbone. For wireless networks to become popular, support for the existing applications and compatibility with the wired Internet must be provided. Therefore, it is imperative that wireless networks also adopt and support TCP for reliable transfer of data.

12.4.2 Limitations of Wired Version of TCP

The primary concern in the use of conventional TCP over wireline networks is packet loss, because congestion can be present at various nodes in the network. In such systems where congestion is the only source for errors, TCP congestion-avoidance mechanisms are extremely useful. However, the same cannot be said about wireless networks, as errors can be introduced due to inherent use of air as a medium of packet transport. Errors can also be attributed to the mobility of users in the network. In such cases, TCP's congestion-avoidance and error-recovery mechanisms lead to unnecessary retransmissions, thereby leading to inefficient use of available wireless bandwidth. In the following subsection, a summary of the various approaches used to improve the efficiency of TCP over wireless networks is given. These strategies range from modifying link layer modules to using split TCP.

12.4.3 Solutions for Wireless Environment

The scarce spectrum imposes a fundamental limit on the performance of the wireless channel, and MSs have limited computing resources and severe energy constraints. Due to these characteristics, a lot of work has been done to optimize the performance of protocol stack. The network layering principle provides good abstraction in the network design and its effectiveness has been most demonstrated by the Internet. However, this leads to a noticeable loss in overall efficiency. Wireless networks are interference limited, and the information delivery capability of a transmission link is closely dependent on the current channel quality. As a result, the notion of congestion is quite different from that in a wired network. Physical and link layer characteristics have important impact on network congestion. Wireless networks operate in an inherent broadcast medium. Hence, adoption of physical and link layer broadcast can very often lead to transmission schemes that are efficient in resource usage, (e.g., power consumption) and could result in substantial improvement of performance and resource usage efficiency. An adaptive architecture is desirable where each layer of the protocol stack responds to the local variations as well as to the information from other layers. However, since there are many existing application layer protocols that use TCP, any modification of the transport layer of the fixed hosts is not feasible. Changes can be made only on MSs and mobile access points to ensure compatibility with existing applications. Such changes should be transparent to the application layer software that runs on top of the transport layer. Some of the approaches to improving the performance of TCP over wireless links are as follows:

End-to-End Solutions

End-to-end protocols attempt to make the TCP sender handle losses through the use of two techniques. First, they use some form of selective acknowledgments to allow the sender to recover from multiple packet losses in a window, without resorting to a coarse timeout. Second, they attempt to have the sender distinguish between congestion and other forms of losses using an explicit loss notification (ELN) mechanism.

- **TCP–SACK** [12.12]: Standard TCP uses a cumulative acknowledgment scheme, which does not provide the sender with sufficient information to recover quickly from multiple packet losses within a single transmission window. A selective acknowledgment (SACK) mechanism, combined with a selective repeat retransmission policy, can help to overcome these limitations. The receiving TCP sends back SACK packets to the sender, informing the sender of the data that have been received. The sender can then retransmit only the missing data segments. If the duplicate segment is received and is part of a larger block of noncontiguous data in the receiver's data queue, then the next SACK block should be used to specify this larger block.

- **Wireless wide-area transmission control protocol (WTCP)** [12.13]: WTCP protocol is a reliable transport layer protocol for a network with wireless links. WTCP runs on the BS that is involved in the TCP connection. In this protocol, the BS buffers data from the fixed host and uses separate flow and congestion control mechanisms for the link between itself and the MS. It temporarily hides the fact that a mobile link breakage has occurred by using local retransmissions of the data for which the MS has not sent an ACK. Once it has received an ACK from the MS, it sends this ACK to the fixed host, but only after changing the timestamp value in the ACK, so that the TCP's round-trip estimation at the fixed sender is not affected. This mechanism effectively hides the wireless link errors from the fixed sender.

- **Freeze-TCP protocol** [12.14]: The main idea behind freeze-TCP is to move the onus of signaling an impending disconnection to the client. A mobile node can certainly monitor signal strengths in wireless antennas and detect an impending handoff and, in certain cases, might even be able to predict a temporary disconnection. In such a case, it can advertise a zero window size, to force the sender into zero window probe mode and prevent it from dropping its congestion window.

- **Explicit bad state notification (EBSN)** [12.15]: Explicit bad state notification uses local retransmission from the BS to shield the wireless link errors and improve performance of TCP over the wireless link. However, while the BS is performing local recovery, the source could still timeout, causing unnecessary source retransmission. The EBSN approach avoids source timeout by using the EBSN message to the source during local recovery. The EBSN message causes the source to reset its timeout value. In this way, timeouts at the source during local recovery are eliminated.

- **Fast retransmission approach** [12.16]: The fast retransmission approach tries to reduce the effect of MS handoff. Regular TCP at the sender interprets the delay caused by a handoff process to be due to congestion. Therefore, whenever a timeout occurs, its TCP window size is reduced and these packets are retransmitted. The fast retransmission approach alleviates the retransmission problem by having the MS send a certain number of duplicate acknowledgments to the sender immediately after completing the handoff. This step causes TCP at the sender to reduce its window size immediately and retransmit packets starting from the first missing packet for which the duplicate acknowledgment has been sent, without waiting for the timeout period to expire.

Link Layer Protocols

There are two main classes of techniques employed for reliable link layer protocols:

1. Error correction using techniques such as FEC
2. Retransmission of lost packets in response to ARQ messages

- **Transport unaware link improvement protocol (TULIP)** [12.17]: TULIP provides a link layer that is transparent to the TCP, has no knowledge of the TCP's state, takes advantage of the TCP's generous timeouts, and makes efficient use of the bandwidth over the wireless link. TULIP provides reliability for only packets (frames) that require such service (service awareness), but it does not know any details of the particular protocol to which it provides reliable service for packets carrying TCP data traffic and unreliable service for other packet types, such as user datagram protocol (UDP) traffic. TULIP maintains local recovery of all lost packets at the wireless link in order to prevent unnecessary and delayed retransmission of packets over the entire path and a subsequent reduction in TCP's congestion window.

- **AIRMAIL protocol** [12.18]: AIRMAIL is the abbreviation of Asymmetric Reliable Mobile Access in Link Layer. This protocol employs a combination of FEC and ARQ techniques for loss recovery. The BS sends an entire window of data before the mobile receiver returns an acknowledgment. The rationale for this approach is not to waste bandwidth on ACKs and to limit the amount of work done by the mobile unit in order to conserve power.

- **Snoop protocol** [12.19]: In the snoop protocol, a transport layer aware agent, called a snoop agent, is introduced at the BS. The agent monitors the link interface for any TCP segment destined for the MS and caches it if buffer space is available. The BS also monitors the acknowledgments from the MS. A segment loss is detected by the arrival of duplicate acknowledgments from the MS or by a local timeout. The snoop agent retransmits the lost segment if it has been cached and suppresses the duplicate acknowledgments. The snoop agent essentially hides the link failures in the wireless link by using local retransmissions rather than allowing the TCP sender to invoke congestion avoidance mechanisms and the fast retransmission scheme.

Split TCP Approach

Split connection protocols split each TCP connection between a sender and receiver into two separate connections at the BS—one TCP connection between the sender and the BS, and the other between the BS and the MS. Over the wireless hop, a specialized protocol may be used that can tune into the wireless environment.

- **Indirect-TCP (I-TCP)** [12.20]: I-TCP is a split connection solution that uses standard TCP for its connection over the wireline link. The indirect protocol model for MSs suggests that any interaction from a MS to a fixed host should be split into two separate interactions—one between the MS and its mobile support router (MSR) over the wireless medium and another between the MSR and the fixed host over the fixed network. All the specialized support that is needed for the mobile applications and low speed and unreliable wireless medium can be built into the wireless side of the interaction while the fixed side is left unchanged at the transport layer. Handoff between two different MSRs is supported on the wireless side without having to reestablish the connection at the new MSR.

- **M-TCP protocol** [12.21]: In this approach, the BS relays ACKs back to the sender only when the receiver (MS) has acknowledged data; therefore, the end-to-end semantics is maintained though it also splits up the connection between a sender (fixed host) and a mobile receiver (MS) into two parts: one between fixed host and BS and another between BS and MS, which uses a customized wireless protocol. The receiver can make the sender enter the persist mode by advertising a zero window size in the presence of frequent disconnections. In this case, the sender freezes all packet retransmit timers and does not drop the congestion window so that the idle time during the slow start phase can be avoided. Whenever the BS detects a disconnection or packet loss, it sends back an ACK with a zero window size to force the sender into persist mode and to force it not to drop the congestion window.

● ● ● ● ● ● ● ● ● ● ● ● ● ● ● ● ●

12.5 Internet Protocol Version 6 (IPv6)

IPv6 [12.22] also known as IPng (Internet Protocol next generation) is proposed to address the unforeseen growth of the Internet and the limited address space provided by IPv4.

12.5.1 Transition from IPv4 to IPv6

IPv4 has extensively used for data communication in wired networks. We introduce this Internet protocol to understand its format. This is important, because a large number of IPv4-hosts and IPv4-routers have been installed and we need to maintain their compatibility. Figure 12.5 shows the IPv4 header format. The IPv4 uses a 32-bit address to provide unreliable and connectionless best effort delivery service. Datagrams (packets in the IP layer) may need to be fragmented into smaller datagrams due to the maximum packet size in some physical networks. It also depends on checksum to protect corruption during the transmission. However, the following are some disadvantages of IPv4:

1. Since the 32-bit address is not sufficient according to the rapidly increased size of Internet, more address space is needed.

2. Real-time audio and video transmissions are being used increasingly, and they require strategies to minimize transmission delay and resource reservation. Unfortunately, those features are neither provided nor supported by IPv4.

3. IPv4 does not have encryption or authentication.

Version (4 bits)	Header length (4 bits)	Type of service (8 bits)	Tota length (16 bits)	
Identification (16 bits)			Flags (3 bits)	Fragment offset (13 bits)
Time to live (8 bits)		Protocol (8 bits)	Header checksum (16 bits)	
Source address (32 bits)				
Destination address (32 bits)				
Options and padding (if any)				

Figure 12.5
IPv4 header format.

The transition from IPv4 to IPv6 is supposed to be simple and without any considerable (temporal) dependencies upon other measures. The IETF plans the following transition mechanisms:

- The basic principle should be Dual-IP-Stack (i.e., IPv4 hosts and IPv4 routers get an IPv6 stack in addition to their IPv4 stack). This coexistence ensures full compatibility between not yet updated systems, and already upgraded systems make it possible to employ IPv6 for communication right away.

- IPv6-in-IPv4 encapsulation: IPv6 datagrams can get encapsulated in IPv4 datagrams enabling IPv6 communication via pure IPv4 topologies. This so-called tunneling of IPv6 packets allows early worldwide employment of IPv6, although not all networks that are part of the communication path support IPv6. The tunnels between two routers must be manually configured, whereas tunnels between hosts and routers may be built up automatically. Tunneling of IPv6 datagrams can be removed as soon as all routers along the respective path have been upgraded with IPv6.

12.5.2   IPv6 Header Format

The format of IPv6 is shown in Figure 12.6 and Table 12.1.

12.5.3   Features of IPv6

IPv6 uses a 128-bit (16-byte) address to identify a host in the Internet. Some of the salient features of IPv6 are as follows:

- **Address space:** An IPv6 address is 128 bits long, which can effectively handle the problems created by a limited IPv4 address space.

- **Resource allocation:** IPv6 supports resource allocation by adding the mechanism of flow label. By using flow label, a sender can request special handling of the packet in the Internet.

- **Modified header format:** IPv6 separates options from the base header. This helps speed up the routing process since most of the options need not be checked by routers.

- **Support for security:** IPv6 supports encryption and decryption options, which provide authentication and integrity.

Version	Traffic class	Flow label	
Payload length		Next header	Hop limit
Source address			
Destination address			
Data			

Figure 12.6
Format of IPv6.

Table 12.1: ▶
Format of IPv6.

Name	Bits	Function
Version	4	IPv6 version number.
Traffic class	8	Internet traffic priority delivery value.
Flow label	20	Used for specifying special router handling from source to destination(s) for a sequence of packets.
Payload length	16, unsigned	Specifies the length of the data in the packet. When set to zero, the option is a hop-by-hop jumbo payload.
Next header	8	Specifies the next encapsulated protocol. The values are compatible with those specified for the IPv4 protocol field.
Hop limit	8, unsigned	For each router that forwards the packet, the hop limit is decremented by 1. When the hop limit field reaches zero, the packet is discarded. This replaces the time to live (TTL) field in the IPv4 header that was originally intended to be used as a time-based hop limit.
Source address	128	The IPv6 address of the sending node.
Destination address	128	The IPv6 address of the destination node.

12.5.4   Differences between IPv6 and IPv4

The main differences between IPv6 and IPv4 are as follows:

- **Expanded addressing capabilities**: In IPv6 the address space is increased from 32 to 128 bits. By doing so, more hierarchical address levels are possible and address prefix routing may be used more efficiently. Furthermore, the longer IPv6 addresses allow more devices and simplify address autoconfiguration. The multicast capabilities are improved and a new address type "anycast" is introduced for addressing the nearest interface out of a group of interfaces.

- **Simplified header format**: To optimize the speed of processing an IPv6 packet and to minimize its bandwidth requirements, some fields of the IPv4 header have been eliminated for IPv6 or made optional.

- **Improved support for options and extensions**: A new design concept for IPv6 is the extension header, which means that options and extensions can be more efficiently added, transmitted, and processed. The size of options is not so strictly limited as in IPv4 which facilitates flexibility for installing future options.

- **Flow labeling capabilities**: In IPv6, it is possible to label data flows which enables the sender to require a special treatment of packets (QoS) by routers on the way to the destination. This may be a nondefault QoS or a real-time service for multimedia applications such as audio or video. In particular, the capabilities of ATM can be used effectively.

- **Support for authentication and encryption**: IPv6 supports authentication of the sender (*i.e.*, a form of digital signature) and data encryption.

Furthermore, IPv6 supports mobility and auto configuration. MSs such as laptops are supposed to be reachable everywhere in the Internet with their home IP address, and a computer that is connected to a network is supposed to configure its correct address automatically.

12.6 Summary

In this chapter, basic mechanisms for providing successful transmission of information have been covered. Specific ways of extending these wireline techniques to wireless services have been discussed, and associated limitations have been pointed out. Some solutions to address these problems have also been suggested. The wireless world has been advancing at a fast pace, and a recent trend is to constitute a wireless connection among close-by devices. A specific class of such networks, called ad hoc and sensor networks, is discussed in Chapter 13.

12.7 References

[12.1] Darpa Internet Protocol Specification, "Internet Protocol," *RFC 791*, September 1981.

[12.2] J. Postel, "Internet Control Message Protocol," *RFC792*, 1981.

[12.3] W. Fenner, "Internet Group Management Protocol, Version 2," *RFC 2236*, November 1997.

[12.4] R. Droms, "Dynamic Host Configuration Protocol," *RFC 2131*, March 1997

[12.5] C. Hedrick, "Routing Information Protocol," *RFC 1058*, June 1988.

[12.6] J. Moy, "OSPF, Version 2," *RFC 1583*, March 1994.

[12.7] Y. Rekhter and T. Li, "A Border Gateway Protocol 4 (BGP-4)," *RFC 1771*, March 1995.

[12.8] G. Malkin, "RIP, Version 2," *RFC 1723*, November 1994.

[12.9] Darpa Internet Protocol Specification, "Transmission Control Protocol," *RFC 793*, September 1981.

[12.10] J. Postel, "Transmission Control Protocol," *RFC 793*, 1981.

[12.11] T. H. Cormen, C. E. Leiserson, R. L. Rivest, and C. Stein, *Introduction to Algorithms*, 2nd edition, The MIT Press, 2001.

[12.12] S. Floyd, J. Mahdavi, M. Mathis, and M. Podosky, "An Extension to the Selective Acknowledgement (SACK) Option for TCP," *RFC 2883*, July 2000.

[12.13] K. Ratnam and I. Matta, "WTCP: An Efficient Transmission Control Protocol for Networks with Wireless Links," *Technical Report NU-CCS-97-11*, Northeastern University, July 1997.

[12.14] T. Goff, J. Moronski, D. S. Phatak, and V. Gupta, "Freeze-TCP: A True End-to-End TCP Enhancement Mechanism for Mobile Environments," *Proccedings of IEEE 19th Infocom 2000*, pp. 1537–1545, Tel Aviv, Israel, 2000.

[12.15] B. S. Baksi, R. Krishna, N. H. Vaidya, and D. K. Pradhan, "Improving Performance of TCP Over Wireless Networks," *Proceedings of the 17th International Conference on Distributed Computing Systems*, Baltimore, MD, IEEE Computer Society Press, May 1997.

[12.16] M. Allman, V. Paxson, and W. R. Stevens, "TCP Congestion Control," *RFC 2581*, April 1999.

[12.17] N. H. Vaidya, M. Mehta, C. Perkins, and G. Montenegro, "Delayed Duplicate Acknowledgements: a TCP-Unaware Approach to Improve Performance of TCP Over Wireless," *TR-99-003*, TAMU, College Station, TX, 1999.

[12.18] E. Ayanoglu, S. Paul, T. F. Laporta, K. K. Sabnani, and R. D. Gitlin, "AIRMAIL: A Link Layer Protocol for Wireless Networks," *ACM Wireless Networks*, Vol. 1, No. 1, pp. 47–60, 1995.

[12.19] H. Balakrishnan, S. Seshan, E. Amir, and R. H. Katz, "Improving TCP/IP Performance Over Wireless Networks," *IEEE/ACM Transactions on Networking*, Vol. 5, No. 6, pp. 756–769, December 1997.

[12.20] A. Bakre and B. R. Badrinath, "I-TCP: Indirect TCP for Mobile Hosts," *Proccedings of 15th International Conference on Distributed Computing Systems*, pp. 136–146, Vancouver, BC, Canada, IEEE Computer Society Press, May 1995.

[12.21] K. Brown and S. Singh, "M-TCP: TCP for Mobile Cellular Networks," *ACM Computer Communications Review (CCR)*, Vol. 27, No. 5, 1997.

[12.22] S. Deering and R. Hinden, "Internet Protocol, Version 6 (IPv6) Specification," *RFC 2460*, December 1998.

● ● ● ● ● ● ● ● ● ● ● ● ● ● ● ● ●

12.8 Problems

P12.1. Describe the OSI model. In which layer(s) does CDPD operate?

P12.2. What are the differences between OSI and TCP/IP protocol models? Explain clearly.

P12.3. Look at your favorite Web site and find the difference between interior and exterior routing protocols.

P12.4. "Basic RIP supports single subnet masking for each IP network." Give a practical example where this becomes a critical issue.

P12.5. What are the differences between path-vector routing and shortest-path routing? Explain clearly.

P12.6. What is DHCP? How does DHCP support dynamic address allocation?

P12.7. With suitable examples, explain the differences between connection-oriented and connectionless protocols.

P12.8. What are the disadvantages of using wireline TCP over wireless networks?

P12.9. Explain the significance of initial sequence number in TCP.

P12.10. What are the inherent characteristics of wireless networks that require changes in existing TCP?

P12.11. What are particular advantages and disadvantages of using a split TCP approach for wireless networks?

P12.12. What are the problems faced by designers of wireless TCP stacks when using link layer protocols?

P12.13. What makes the fast-retransmission approach desirable in improving TCP performance over wireless networks?

P12.14. When is the reliable link layer useful in enhancing TCP performance?

P12.15. What is the operational difference between standard ACKs used in conventional TCP and SACKs used in wireless TCP? What improvement in performance does it provide for wireless networks?

P12.16. Both I-TCP and M-TCP are split TCP approaches to improving the performance of wireline TCP over wireless networks. What is the difference between these two approaches?

P12.17. Even though explicit bad state notification (EBSN) appears to be a very pragmatic approach for improving TCP performance over wireless networks, what is its most significant disadvantage?

P12.18. Can any of the methods (e.g., I-TCP, M-TCP, SACK, EBSN, etc.) be used to improve performance of TCP over wireless ad hoc networks? Suggest any ways by which this can be done.

P12.19. How many iterations are needed to calculate shortest path to all nodes from node 3? Determine the shortest distance to each node and the path used for each one of them.

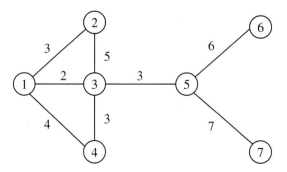

Figure 12.7
Figure for
Problem P12.19.

P12.20. Given the figure in Problem P12.19 as the connectivity graph of a network, you are allowed to go through only two steps of the Bellman-Ford algorithm at each node so that their complexity (and hence the time required) can be kept to a low value. What is the impact on shortest path calculations? Comment on the accuracy of the procedure?

P12.21. What kind of security measures are used in different layers of TCP/IP? Explain.

P12.22. What are the advantages of IPv6? Discuss whether an IPv6 network can support IPv4 packets and, if so, how?

P12.23. IPv6 supports resource allocation. Explain how this is achieved.

CHAPTER 13

Ad Hoc and Sensor Networks

• • • • • • • • • • • • • • • ⋯

13.1 Introduction

In Chapter 1, we briefly discussed mobile ad hoc networks (MANETs). In this chapter, we describe these networks in detail. A MANET consists of a number of mobile devices that come together to form a network as needed, without any support from any existing Internet infrastructure or any other kind of fixed stations. Formally, a MANET can be defined as an autonomous system of nodes or MSs (also serving as routers) connected by wireless links, the union of which forms a communication network modeled in the form of an arbitrary communication graph. This is in contrast to the well-known single-hop cellular network model that supports the needs of wireless communications by having BSs as access points. In these cellular networks, communication between two mobile nodes relies on the wired backbone and the fixed base stations. In a MANET, no such infrastructure exists and the network topology may change dynamically in an unpredictable manner since nodes are free to move and each node has limited transmitting power, restricting access to the node only in the neighboring range.

MANETs are basically peer-to-peer, multihop wireless networks in which information packets are transmitted in a store-and-forward manner from a source to an arbitrary destination, via intermediate nodes as illustrated in Figure 13.1. As nodes move, the connectivity may change based on relative locations of other nodes. The resulting change in the network topology known at the local level, must be passed on to the other nodes so that old topology information can be updated. For example, as MS2 in Figure 13.1 changes its point of attachment from MS3 to MS4, other nodes that are part of the network should use this new route to forward packets to MS2. Note that in Figure 13.1, and throughout this chapter, we assume that it is not possible to have all nodes within each other's radio range. In case all nodes are close by within each other's radio range, there are no routing issues to be addressed. In real-life scenarios, these networks will be made of vastly different types of devices, some more portable than the others. In such heterogeneous dynamic topologies, it is reasonable to assume that a node will not have enough transmitting power to reach all nodes in the network. In real situations, the power needed to obtain connectivity of all nodes

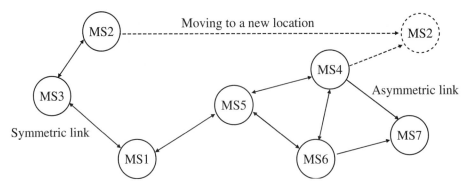

Figure 13.1
A mobile ad hoc
network (MANET).

in the network may be, at least, infeasible, and issues such as battery life come into play as well. Therefore, we are interested in scenarios in which only a few nodes are within each other's radio range. Figure 13.1 raises another issue, that of symmetric (bidirectional) and asymmetric (unidirectional) links. As we shall see, some of the protocols we discuss consider symmetric links with associative radio range; for example, if (in Figure 13.1) MS1 is within radio range of MS3, then MS3 is also within radio range of MS1. The communication links are symmetric. This assumption is not always valid because of differences in transmitting power levels and the terrain. Routing in such asymmetric networks is a relatively hard task. In certain cases, it is possible to find routes that exclude asymmetric links, since it is cumbersome to find the return path. Unless stated otherwise, throughout this text we consider symmetric links, with all nodes having identical capabilities and responsibilities.

The issue of efficient routing is one of the several challenges encountered in a MANET. The other issue is varying the mobility patterns of different nodes. Some nodes are highly mobile, while others are primarily stationary. It is difficult to predict a node's movement and direction of movement.

● ● ● ● ● ● ● ● ● ● ● ● ● ● ● ● ● ● ●
13.2 Characteristics of MANETs

Salient characteristics of ad hoc networks are as follows [13.1]:

1. **Dynamic topologies**: Nodes are free to move arbitrarily; thus, the network topology may change randomly and unpredictably and primarily consists of bidirectional links. In some cases, where the transmission power of two nodes is different, a unidirectional link may exist.

2. **Bandwidth-constrained and variable capacity links**: Wireless links continue to have significantly lower capacity than infrastructured networks. In addition, the realized throughput of wireless communications—after accounting for the effects of multiple access, fading, noise, interference conditions, and so on—is often much less than a radio's maximum transmission rate. One effect of relatively low to moderate link capacities is that congestion is typically the norm rather than the exception (i.e., aggregate application demand could likely approach or

exceed network capacity frequently). As a MANET is often simply an extension of the fixed network infrastructure, mobile ad hoc users would demand similar services.

3. **Energy-constrained operation**: Some or all of the MSs in a MANET may rely on batteries or other exhaustible means for their energy. For these nodes, the most important system design optimization criteria may be energy conservation.

4. **Limited physical security**: MANETs are generally more prone to physical security threats than wireline networks. The increased possibility of eavesdropping, spoofing, and denial of service (DoS) attacks should be carefully considered.

To reduce security threats, many existing link security techniques are often applied within wireless networks. As a side benefit, the decentralized nature of MANET control provides additional robustness against the single points of failure of centralized approaches. In addition, some envisioned networks (e.g., mobile military networks or highway networks) may be very large (e.g., tens or hundreds of nodes per routing area). Scalability is a serious concern in MANETs.

● ● ● ● ● ● ● ● ● ● ● ● ● ⋯⋯

13.3   Applications

Applications of wireless networks have been outlined in Chapter 1. Some specific applications of ad hoc networks include industrial and commercial applications involving cooperative mobile data exchange. There are many existing and future military networking requirements for robust, IP-compliant data services within mobile wireless communication networks, with many of these networks consisting of highly dynamic autonomous topology segments. Also, small electronic devices are being developed that could be worn on a human body and communicate with each other to deliver exciting services. Such developing technologies of "wearable" computing and communications provide innovative applications for MANETs. When properly combined with satellite-based information delivery, MANETs can provide an extremely flexible method for establishing communications for fire safety and rescue operations or other scenarios, requiring rapidly deployable communications with survivable and efficient dynamic networking. It is also likely that there are other applications for MANETs that are not presently realized or envisioned by researchers.

The technology of MANETs is somewhat equivalent to mobile packet radio networking (a term coined during early military research in the 1970s and 1980s); mobile mesh networking (a term that appeared in an article in *The Economist* regarding the structure of future military networks); and mobile, multihop, wireless networking (perhaps the most accurate term, although a bit cumbersome). Initially, the technology was developed keeping in mind the military applications of such a technology in areas such as the battlefield, where an infrastructured network is almost impossible to set up and maintain. In such situations, MANETs, with their self-organizing capability, can be used effectively where other technologies fail. Advanced features of MANETs, including data rates compatible with multimedia applications, global roaming capability, and coordination with other network structures, are enabling new applications.

1. **Defense applications**: Many defense applications require on-the-fly communications set up, and ad hoc/sensor networks are excellent candidates for their use in battle-field management. MANETs can be formed among soldiers on the ground or fighter planes in the air, while sensors could be deployed to monitor activities in the area of interest.

2. **Crisis-management applications**: These arise, for example, as a result of natural disasters in which the entire communication infrastructure is in disarray. Restoring communications quickly is essential. With wideband wireless mobile communications, limited and even total communication capability, including Internet and video services, could be set up in hours instead of days or even weeks required for restoration of wireline communications.

3. **Telemedicine**: The paramedic assisting the victim of a traffic accident in a remote location must access medical records (e.g., X-rays) and may need video conference assistance from a surgeon for an emergency intervention. In fact, the paramedic may need to instantaneously relay back to the hospital the victim's X-rays and other diagnostic tests from the site of the accident.

4. **Tele-geoprocessing applications**: The combination of geographical information systems (GIS), GPS, and high-capacity wireless mobile systems enables a new type of application referred to as tele-geoprocessing. Queries dependent on location information of several users, in addition to temporal aspects, have potential business applications. Another potential area that is being explored is the environmental monitoring [13.2] using a group of sensors.

5. **Virtual navigation**: A remote database contains the graphical representation of streets, buildings, and physical characteristics of a large metropolis. Blocks of this database are transmitted in rapid sequence to a vehicle, where a rendering program permits the occupants to visualize the needed environment ahead of time. They may also "virtually" see the internal layout of buildings, including an emergency rescue plan, or find possible points of interest.

6. **Education via the Internet**: Educational opportunities available on the Internet, both for K–12 students and individuals interested in continuing education, could be unavailable to people living in sparsely populated or remote areas because of the economic infeasibility of providing expensive last-mile wireline Internet access in these areas to all subscribers.

13.4 Routing

Routing in a MANET depends on many factors, including modeling of the topology, selection of routers, initiation of a route request, and specific underlying characteristics that could serve as heuristics in finding the path efficiently.

The low resource availability in MANETs necessitates efficient resource utilization; hence the motivation for optimal routing. Also, highly dynamic nature of these networks places severe restrictions on any routing protocol specifically designed for them. A network configuration is also called a network topology. There are three major goals when selecting a routing protocol:

1. Provide the maximum possible reliability by selecting alternative routes if a node connectivity fails.

2. Route network traffic through the path with least cost in the network by minimizing the actual length between the source and destination through the least number of intermediate nodes.

3. Give the nodes the best possible response time and throughput. This is especially important for interactive sessions between user applications.

In a MANET, each node is expected to serve as a router and each router is indistinguishable from another in the sense that all routers execute the same routing algorithm to compute routing paths through the entire network.

13.4.1   Need for Routing

MANET routing typically has the following goals:

1. Route computation must be distributed, because centralized routing in a dynamic network is impossible, even for fairly small networks.

2. Route computation should not involve maintenance of a global state, or even significant amounts of volatile nonlocal state. In particular, link state routing is not feasible due to the enormous state propagation overhead when the network topology changes.

3. As few nodes as possible, must be involved in route computation and state propagation, as this involves monitoring and updating at least some states in the network. On the other hand, every host must have quick access to the routes on demand.

4. Each node must care only about the routes to its destination and must not be involved in frequent topology updates for those portions of the network that have no traffic.

5. Stale routes must be either avoided or detected and eliminated quickly.

6. Broadcasts must be avoided as far as possible because broadcasts could be highly unreliable in MANETs.

7. If the topology stabilizes, then routes must converge to the optimal routes.

8. It is desirable to have a backup route when the primary route has become stale and is to be recomputed.

One of the major challenges in designing a routing protocol [13.3] for MANETs stems from the fact that, on the one hand, a node needs to know at least the reachability information to its neighbors for determining a packet route; on the other hand, in a MANET, the network topology can change very frequently. Furthermore, as the number of network nodes (MSs) can be large, the potential number of destinations is also large, requiring large and frequent exchanges of data (e.g., routes, route updates, or routing tables (RTs)) among the network nodes. Thus, the amount of update traffic can be high. This is in contradiction to minimized exchange of information as all updates travel over the air in a MANET.

13.4.2 Routing Classification

Existing routing protocols can be classified either as proactive or reactive [13.4]. Proactive protocols attempt to evaluate continuously the routes within the network, so that when a packet needs to be forwarded, the route is already known and can be immediately used. The family of distance vector protocols is an example of a proactive scheme. Reactive protocols, on the other hand, invoke a route determination procedure only on demand. Thus, when a route is needed, some sort of global search procedure is initiated. The family of classical flooding algorithms belongs to the reactive group. Examples of reactive (also called on-demand) ad hoc network routing protocols include ad hoc on-demand distance vector (AODV) [13.5] and temporally ordered routing algorithm (TORA) [13.6].

The advantage of the proactive schemes is that whenever a route is needed, there is negligible delay in determining the route. In reactive protocols, because route information may not be available at the time a datagram is received, the delay to determine a route can be significant. Furthermore, the global flood-search procedure of the reactive protocols incurs significant control traffic. Because of this long delay and excessive control traffic [13.7], pure reactive routing protocols may not be adequate for any real-time communication. However, pure proactive schemes are likewise not appropriate for the MANET environment, as they continuously use a large portion of the network capacity to keep the routing information current. Since the nodes in a MANET move fast and the changes may be more frequent than the route requests (RREQs), most of this routing information is never used. This is a waste of the wireless network capacity. The routing protocols may also be categorized as follows:

- Table-driven protocols
- Source initiated on-demand protocols

13.5 Table-Driven Routing Protocols

A comprehensive survey on different routing protocols for MANETs is given in [13.4], and here we summarize some of the important ones. These protocols are called table-driven because each node is required to maintain one or more tables to store routing information on every other node in the network. They are essentially proactive in nature so that the routing information is always consistent and up-to-date. The protocols respond to changes in network topology by propagating the updates throughout the network so that every node has a consistent view of the network. Some of the existing table-driven MANET routing protocols are discussed in the following subsections. They differ primarily in the number of necessary routing-related tables and the procedures to broadcast the network changes.

13.5.1   Destination-Sequenced Distance-Vector Routing

The destination-sequenced distance-vector (DSDV) [13.8] routing protocol is a table-driven routing protocol based on the classic Bellman-Ford routing algorithm discussed in Chapter 12. The algorithm works correctly, even in the presence of loops in the routing tables. As stated above, each mobile node maintains a routing table with a route to every possible destination in the network and the number of hops to the destination. Each such entry in the table is marked with a sequence number assigned by the destination node. The sequence numbers allow the mobile node to distinguish stale routes from new ones, and help avoid formation of routing loops.

A new route broadcast contains:

- The destination address.

- The number of hops required to reach the destination.

- The sequence number of the information received about the destination and a new sequence number unique to the broadcast.

If multiple routes are available for the same destination, the route with the most recent sequence number is used. If two updates have the same sequence number, the route with the smaller metric (e.g., hops) is used to optimize the routing. Further, if the routes fluctuate frequently, that may lead to large network traffic as broadcasts need to be sent each time a better route is discovered. To avoid such broadcasts, the mobile nodes keep track of the settling time of routes or the weighted average time before the route with best metric is discovered. The nodes can now delay the update broadcasts by settling time, during which a better route may be discovered, thus reducing network traffic.

Any updates in the routing tables are periodically broadcasted in the network to maintain table consistency. The amount of traffic generated by these updates can be huge. To alleviate this problem, the updates are done through two types of packets. The first is called a full dump [13.8]. A full dump packet carries all the available routing information and can require multiple network protocol data units (NPDUs). When there is only occasional movement, these packets are used rarely. Instead, smaller incremental packets are used to relay only the change in information since the last full dump. The incremental packets fit into a standard NPDU and hence decrease the amount of traffic generated. The nodes maintain a separate table in which they maintain all the information sent in the incremental routing information packets.

13.5.2   Cluster Head Gateway Switch Routing

The cluster head (CH) gateway switch routing (CGSR)protocol [13.9] is different from the previous protocol in the type of addressing and the network organization scheme employed. Instead of a flat network, CGSR uses CHs which control a group of ad hoc nodes and hence achieve a hierarchical framework for code separation among clusters, channel access, routing, and bandwidth allocation (Figure 13.2). Identification of appropriate clusters and selection of CHs is quite complex. Once clusters have been defined, it is desirable to use a distributed algorithm within the cluster to elect a node as the CH. The disadvantage of using a CH scheme is that

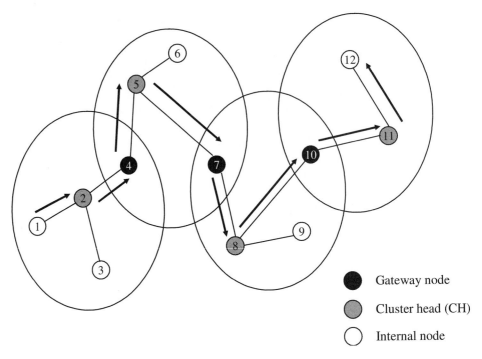

Figure 13.2
Routing in CGSR
from node 1 to
node 12.

● Gateway node

● Cluster head (CH)

○ Internal node

frequent changes adversely affect performance as nodes spend more time selecting a CH rather than relaying packets. Hence, the Least Cluster Change (LCC) clustering algorithm is used rather than CH selection every time the cluster membership changes. Using LCC, CHs only change when two CHs come into contact, or when a node moves out of contact with all other CHs.

CGSR uses DSDV as the underlying routing scheme and shares the overhead with the same. However, it modifies DSDV to use a hierarchical cluster-head-to-gateway routing approach. Gateway nodes are those within communication range of two or more CHs. A packet sent by a node is first transmitted to its CH. From there it is routed to the gateway node, then to another CH and so on till the packet reaches the CH of the destination. The packet is then transmitted to the destination as illustrated in Figure 13.2. To use this routing scheme, each node must maintain a cluster member table (CMT), which stores the destination CH for each node in the network. The CMTs are broadcast periodically by the nodes using the DSDV algorithm. When a node receives such a table from a neighbor, it can update its own information.

As expected, each node also maintains a routing table to determine the next hop required to reach any destination. While transmitting a packet, the node looks up the CMT and the routing table to determine the nearest CH along the route to the destination, and the next hop required to reach this CH. It then relays the packet to this node.

13.5.3   Wireless Routing Protocol

For the wireless routing protocol (WRP) [13.10], each node maintains four tables:

- Distance table
- Routing table
- Link-cost table
- Message retransmission list (MRL) table

The MRL records which updates in an update message should be retransmitted and which neighbors should acknowledge the retransmission. For this purpose, each entry in the MRL has a sequence number of the update message, a retransmission counter, an acknowledgment-required flag vector with one entry per neighbor, and a list of updates sent in the update message.

Nodes discover each other through hello messages. When a node receives a hello message from a new node, it adds the new node to its routing table and sends the new node a copy of its routing table. A node must send messages to its neighbors within a certain time to ensure connectivity. The messages sent by a node convey its existence to the neighbors, apart from the information contained in the message. In case, a node does not have any messages to send, it still must periodically send a hello message to ensure connectivity. Otherwise the neighboring nodes might interpret the absence of messages as the failure of the link connecting them and cause a false alarm.

Nodes inform each other of link changes through the use of update messages and contain a list of updates—the destination, the distance to the destination, and the predecessor of the destination. They also have a list of responses indicating which nodes would acknowledge the update. The update messages are sent after a node processes updates from its neighbors or detects a change in a link to a neighbor. In case a link between two nodes goes down, the nodes send update messages to their neighbors. The neighbors modify their table entries and explore new paths through other nodes. The new paths discovered are also relayed back to the original nodes.

A novel improvement in WRP is the method it uses to achieve freedom from routing loops. It belongs to the class of path-finding algorithms with an important distinction. In WRP, each node is forced to perform a consistency check on predecessor information reported by all its neighbors. Thus, WRP avoids the count-to-infinity problem, eliminates loops (although not instantaneously), and provides faster route convergence in case of link failures.

● ● ● ● ● ● ● ● ● ● ● ● ● ● ● ●

13.6   Source-Initiated On-Demand Routing

Source-initiated on-demand routing is essentially reactive in nature, unlike table-driven routing. The source-initiated approach generates routes only when a source demands it. In other words, when a source requires a route to a destination, the source initiates a route discovery process in the network. This process finishes when a route to the destination has been discovered or all possible routes have been examined without any success. The route thus discovered, is maintained by a route maintenance procedure, till the time it is no longer desired or the destination becomes inaccessible. Some of the popular source-initiated on-demand routing procedures are discussed subsequently.

13.6.1   Ad Hoc On-Demand Distance Vector Routing

Ad hoc on-demand distance vector (AODV) routing [13.11] is built over the DSDV algorithm described in Section 13.5.1. AODV is a significant improvement over DSDV. AODV is a pure on-demand route acquisition algorithm. The nodes that are not on a particular path do not maintain routing information, nor do they participate in the routing table exchanges. As a result, the number of broadcasts required to create the routes on demand via AODV is minimized rather than doing broadcasts to maintain complete route information in DSDV.

When a source needs to send a message to a destination and does not have a valid route to the latter, the source initiates a route discovery process. Source sends a route request (RREQ) packet to all its neighbors, the latter forward the request to all their neighbors, and so on, until either the destination or an intermediate node with "fresh enough" route to the destination is reached. Figure 13.3(a) illustrates the propagation of the broadcast RREQs across the network. As in DSDV, destination sequence numbers are used, to ensure that all routes are loop-free and contain the most recent route information. Each node has a unique sequence number and a broadcast ID which is incremented each time the node initiates a RREQ. The broadcast ID, together with the node's IP address, uniquely identifies every RREQ. The initiator node includes in the RREQ the following:

■ Its own sequence number

■ The broadcast ID

■ The most recent sequence number the initiator has for the destination

Intermediate nodes reply only if they have a route to the destination with a sequence number greater than or at least equal to that contained in the RREQ.

To optimize the route performance, intermediate nodes record the address of the neighbor from which they receive the first copy of the broadcast packet. This establishes the best reverse path. All subsequently received copies of the RREQ are discarded. Once the RREQ reaches the destination or an intermediate node with a fresh enough route to the destination, the intermediate/destination node sends a unicast route-reply (RREP) message back to the neighbor from which it received the first copy of the RREQ [Figure 13.3(b)]. As the RREP travels back on the reverse path, the nodes on this path set up their forward route entries to point to the node from which the RREP has just been received. These forward route entries indicate the active forward route. The RREP continues traveling back along the reverse path till it reaches the initiator of the route discovery. Thus, AODV can only support the use of symmetric links.

A route timer is associated with each route entry. This timer triggers the deletion of the route entry if it is not used within the specified lifetime. When a source node moves, it can reinitiate the route-discovery procedure to find new routes to the

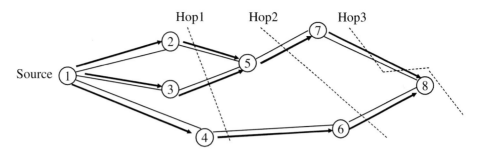

(a) Propagation of route request (RREQ) packet

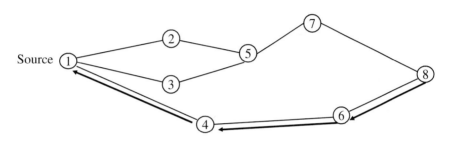

Figure 13.3
Route discovery in
the AODV protocol.

(b) Path taken by the route reply (RREP) packet

destination. If the nodes along the route move, their upstream neighbors (nodes just before them in route from source to destination) notice the movement and propagate a link failure notification to their own active upstream neighbors, and so on until the source node is reached. A link failure notification is essentially a RREP with infinite metric.

The source node can now choose to reinitiate the route discovery procedure if a route to that destination is still desired. Another protocol followed in route maintenance is the use of hello messages, periodic local broadcasts by a node to inform other nodes in its neighborhood of its presence. Hello messages ensure local connectivity. Nodes listen for retransmission of data packets to make certain that the next hop is still within reach. If such a retransmission is not heard, a variety of techniques may be used for recouping the path. One such method is the reception of hello messages to determine whether the next hop is within the communication range. The hello messages may also list other nodes from which a node has heard, thereby relaying more information about the network connectivity.

13.6.2 Dynamic Source Routing

Dynamic source routing (DSR) [13.12] is an on-demand routing protocol based on source routing. The mobile nodes maintain all source routes that they are aware of in cache. As the new routes are discovered, the cache is updated. The protocol works in two main phases: route discovery and route maintenance. When a mobile has a

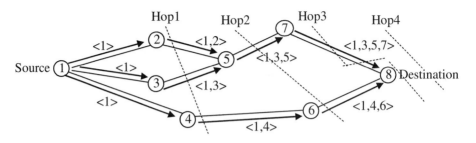

(a) Building record route during route discovery

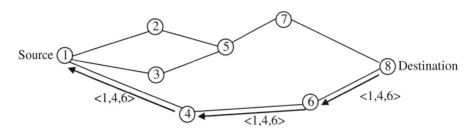

Figure 13.4
Creation of route
record in DSR using
symmetric links.

(b) Propagation of route reply with the route record

message to send, it consults the route cache to determine whether it has a route to the destination. If an active route to destination exists, it is used to send the message. Otherwise, the mobile initiates a route discovery by broadcasting a route-request packet. The route request contains the destination address, the source address, and a unique identification number. Each node that receives the route request checks whether it has a route to the destination. If it does not, it adds its own address to the route record of the packet and then rebroadcasts the packet on its outgoing links. To minimize the number of broadcasts, a node only rebroadcasts a packet if it has not seen the packet before and its own address was not already in the route record. Figure 13.4(a) illustrates the formation of route record as the route request propagates through the network.

When the route request reaches the destination or a node with a route to the destination, a route reply is generated. At this point, the route record indicates all the hops taken to reach the current node or destination. If the current node is the destination, it places the route record in the route request into the route reply. In case the responding node is an intermediate one, it appends the cached route (to the destination) to the route record and then places it into the route reply. The route reply packet is then sent back to the initiator. If the responding node has a route to the initiator in its cache, that route may be taken. Otherwise, if symmetric links are supported, a reverse path can be taken as in AODV. If symmetric links are not supported, the responding node must initiate its own route discovery and piggyback the route record on the new route request. A route reply with symmetric links is shown in Figure 13.4(b).

Route maintenance is carried by the use of route-error packets and acknowledgments. Route-error packets are generated at a node when the data link layer encounters a fatal transmission problem. On receiving a route-error packet, a node removes the hop in error from its route cache. It also truncates all routes containing the erroneous hop. In addition, acknowledgments are used to verify that route links are operating correctly. The acknowledgments may be passive in nature, when a node can hear the next hop retransmitting the data along the route.

13.6.3   Temporarily Ordered Routing Algorithm

The temporarily ordered routing algorithm (TORA) [13.6] is a loop-free and highly adaptive distributed routing algorithm based on the concept of link reversal. Due to the way it is designed, TORA minimizes the reaction due to topological changes. This is achieved by decoupling the generation of potentially far-reaching control messages from the rate of topological changes. The algorithm tries to localize such messages to a very small set of nodes in the neighborhood of the site of the change. It does not employ a dynamic, hierarchical routing mechanism like many other protocols which avoids the added complexity. This means that the route optimality suffers as the latter is given secondary importance. Longer routes are often used if the discovery of newer routes can be avoided.

TORA also exhibits multipath routing capability. The operation of TORA can be compared to that of water flowing downhill toward a sink node through a grid of tubes that model the routes in the real world network. The tube junctions represent the nodes, the tubes themselves represent the route links between the nodes, and the water in the tubes represents the packets flowing between nodes via the route links toward the destination as shown in Figure 13.5. Considering the data flow to be

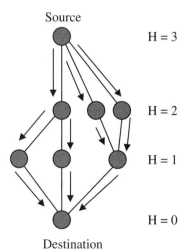

Figure 13.5
TORA height metric.

downhill, each node has a height with respect to the destination node. The analogy also makes it easy to correct routes in case of link failure or error. For example, if a tube between nodes A and B becomes blocked and water can no longer flow through it, the height of A is set to a level higher than any of its remaining neighbors. Now the water will flow back out of A and toward the other nodes (that may have been routing packets to the destination through A). Figure 13.5 illustrates the use of the height metric.

One of the main advantages of TORA is that it can operate smoothly in a highly dynamic mobile environment. It provides multiple routes for any source-destination pair. For this purpose, the mobile nodes must maintain routing information about their one-hop neighbors. The algorithm works in three main phases:

- Route creation

- Route maintenance

- Route erasure

A separate directed acyclic graph (DAG) is maintained by each node to every destination. When a route to a particular destination is required, the source node broadcasts a QUERY packet containing the destination address. The route query propagates through the network till it reaches either the destination or an intermediate node containing the route to the destination. This node then responds back with an UPDATE which contains its own height with respect to the destination (based on the path length it has to the destination). Each node that receives the UPDATE, in turn sets its height to a value greater than that of its neighbor from which the UPDATE has been received. This process creates a series of directed links from the originator of the query to the node that initially created the UPDATE.

When a node discovers that a route to a destination is no longer valid, it adjusts its height to be higher than its neighbors (local maximum) and then broadcasts an UPDATE packet. In case none of its neighbors has a finite height with respect to the destination, the source node initiates a new route search as described above.

When a node senses a network partition, it generates a CLEAR packet that resets the routing state and removes invalid routes from the network. TORA is placed above the Internet MANET encapsulation protocol—IMEP [13.13]. IMEP provides reliable, in-order delivery of all routing messages from a node to all its neighbors. It also notifies the routing protocol whenever a link to one of the neighbors is created or broken. IMEP attempts to reduce the overhead in this case by grouping together several TORA and IMEP control messages (called objects) into a single packet (as an object block) before transmission. Each block is identified by a unique sequence number. An object block also contains a response list of the other nodes from which an ACK has not been received. Only the latter nodes need to respond with an ACK on reception. Each block is retransmitted with a certain frequency. If needed, the retransmissions continue for a certain maximum total period. After this time, TORA is informed of the broken links due to nodes which have not yet sent an ACK. Furthermore, nodes periodically transmit a BEACON (or an equivalent) signal to sense the link status and maintain the neighbor list. Every node that hears the BEACON must respond back with a HELLO (or an equivalent) signal.

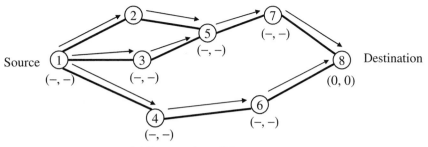

(a) Propagation of the query message

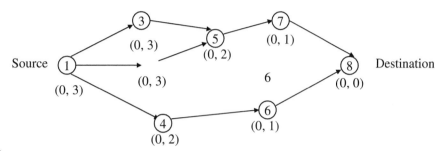

Figure 13.6
Height of the node
updated as a result
of the update message.

(b) Node's height updated as a result of the update message

In the route creation and maintenance phases, nodes use a height metric to establish a DAG rooted at the destination. Subsequently, the links are assigned an upstream or downstream direction according to the relative height metric of their neighboring nodes. This is illustrated in Figure 13.6(a). When a node moves, the DAG route is no longer valid. Hence, route maintenance must be performed to set up a DAG rooted at the same destination. As in Figure 13.6(b), when the last downstream link fails, the node generates a new reference level. The latter is propagated by the neighboring nodes and is vital in coordinating a structured reaction to the failure. In order to reflect the change in adapting to the new reference level, the links are reversed. This is essentially the same as reversing the direction of one or more links when a node has no downstream links.

The height metric in TORA depends on the logical time of a links failure. For this reason, timing becomes a crucial factor. The algorithm assumes that all nodes are synchronized with each other. This can be achieved by an external time source like the GPS. TORA has a 5-tuple metric which consists of:

- Logical time of link failure
- Unique ID of the node that defined the new reference level
- A reflection indicator bit
- A propagation ordering parameter
- Unique ID of the node

The first three elements together describe the reference level. Every time the last downstream link goes down, a new reference level must be defined. During the route erasure phase in TORA, a simple clear packet (CLR) is broadcasted throughout the network to obliterate invalid routes.

Finally, oscillation can occur while using TORA. This is especially likely when multiple sets of coordinating nodes are simultaneously detecting partitions, erasing routes or building routes based on each other. Since the nodes coordinate with each and share information, this problem of instability is similar to that of "count-to-infinity" in distance-vector routing protocols. However, these oscillations are only temporary and route eventually converges. In closure, an important point to note is that TORA is partially proactive and partially reactive. It is reactive since route creation is done on demand. On the other hand, it is proactive because multiple routing options are available in case of link failures.

13.6.4 Associativity-Based Routing

The associativity-based routing (ABR) [13.14] protocol is free from loops, deadlocks, and packet duplicates. A fundamental objective of ABR is to discover longer-lived routes. To this end, the protocol uses a new routing metric for MANETs. The metric is called the degree of association stability which is characterized by connection stability of one node with respect to another node over time and space. High association stability indicates a low state of node mobility. Conversely, a low degree of association stability may indicate high node mobility.

A new route is selected depending upon its degree of association stability. As in most other protocols, each node periodically transmits a beacon signal to broadcast its existence. The beacon signal causes the associativity ticks of the neighbors (those receiving the beacon) to be incremented. The associativity ticks are reset when a neighbors of a node or a node itself move out of proximity.

ABR operates in three phases:

- Route discovery
- Route reconstruction (RRC)
- Route deletion

The route discovery phase is facilitated by the use of broadcast query (BQ) await-reply (BQ-REPLY) cycle. All nodes, apart from the destination, that receive the BQ message append their addresses and the associativity ticks with their neighbors, along with the QoS information to the BQ message. The next such node in relay removes the associativity tick entries of the upstream neighbor. Only the entry concerned with the current node and its upstream neighbor is retained. In this manner, the packet arriving at the destination contains the associativity ticks of all the nodes along the route taken by the packet to reach the destination. The destination can now select the best route from all such packets received by examining the associativity ticks along the path. In case multiple paths with similar overall degree of association stability exist, the path with the minimum number of hops is selected. The destination now sends a REPLY packet back to the source along the selected path. Nodes propagating the REPLY mark their routes as active.

The RRC phase kicks in when there is movement of nodes along the path. When a source node moves, a BQ-REPLY process is initiated. A route notification (RN) message is used to erase route entries associated with the downstream nodes. When the destination moves, the immediate upstream node erases its route. It then checks if the destination is still reachable by a localized query (LQ [H]) process. Here [H] refers to the hop count from the upstream node to the destination. If the destination receives the LQ packet, it sends back a REPLY with the best partial route. Otherwise the initiating node times out and the process backtracks to the next upstream node. This is done by sending a RN[0] message to the next upstream node, which erases the invalid route and then invokes the LQ[H] process. If this process backtracks to more than halfway to the source, the LQ process is discontinued and a new BQ process is initiated at the source.

Finally, in case a route is no longer needed, the source node broadcasts a route delete (RD) message so that all the nodes along the route update their routing tables. The reason for using a full broadcast as opposed to a direct broadcast is that there might have been changes in the nodes along the route in RRC phase. The source may not be aware of these changes and must use a full broadcast.

13.6.5   Signal Stability-Based Routing

Signal stability-based routing (SSR) [13.15] is another on-demand routing protocol that selects routes depending on the signal strength between the nodes and a node's location stability. This mechanism selects routes that have "stronger" connectivity period [13.15] [13.16]. SSR can be divided into two cooperative protocols: the dynamic routing protocol (DRP) and the static routing protocol (SRP). The DRP is responsible for maintaining signal stability table (SST) and the routing table (RT). The SST is a record of the signal strengths of the neighboring nodes. The strength of a signal may be recorded as either a strong channel or a weak channel. All the transmissions are processed by the DRP. After the DRP updates the table entries, it passes a received packet to the SRP.

The SRP now processes the packet as follows: It passes the packet up the stack if it is the intended receiver; otherwise it looks up the destination in the RT and forwards the packet. If there is no entry for the destination in the RT, a route-search process must be initiated. These route requests are propagated throughout the network; however, they are only forwarded to the next hop if they were received over a strong channel and were not previously processed. The latter prevents looping in requests. The destination chooses the first route-search packet that it receives because it is highly probable that such a packet arrived via the shortest/least-congested route. The DRP now sends a route-reply message back to the initiator by the reverse route. The DRP of all the nodes along the reverse path update their RTs accordingly.

It is obvious that the route-search packets arriving at the destination have chosen paths of strong signal stability; otherwise they would have been dropped (when they arrive on a weak channel). There is a chance that no route exists with all string channels. For such a case, the source has a timeout associated with the route-search. When a link fails along a route, the intermediate node informs the source of the failure via an error message. The source sends an erase message to inform all the nodes of the broken link. The source now reinitiate a route-search process to find a new path to the destination.

● ● ● ● ● ● ● ● ● ● ● ● ● ● ● ● ●

13.7 **Hybrid Protocols**

Hybrid protocols attempt to take advantage of best of reactive and proactive schemes. The main idea behind such protocols is to initiate route-discovery ondemand but at a limited search cost. The subsections below discus some of the popular hybrid protocols in detail.

13.7.1 Zone Routing

The zone routing protocol (ZRP) [13.17] is a hybrid of proactive and reactive protocols. It tries to limit the scope of proactive search to the node's local neighborhood. At the same time, global search throughout the network can also be performed efficiently by querying selected nodes (and not all the nodes in the network). A node's local neighborhood is called a routing zone. Specifically, a node's routing zone is defined as the set of nodes whose minimum distance in hops from the node is no greater than the zone radius. A node maintains routes to all the destinations in the routing zone proactively. It also maintains its zone radius, and the overlap from the neighboring routing zones.

To construct a routing zone, the node must identify all its neighbors first which are one hop away and can be reached directly. The process of neighbor discovery is governed by the neighbor discovery protocol (NDP), a MAC-level scheme. ZRP maintains the routing zones via a proactive component called the intra-zone routing protocol (IARP) and is implemented as a modified distance vector scheme. Thus, IARP is responsible for maintaining routes within the routing zone. Another protocol called the inter-zone routing protocol (IERP) is responsible for discovering and maintaining the routes to nodes beyond the routing zone. This process uses a query-response mechanism on-demand basis. IERP is more efficient than standard flooding schemes.

When a source node has data to be sent to a destination which is not in the routing zone, the source initiates a route query packet. The latter is uniquely identified by the tuple - <source node ID, request number>. This request is then broadcasted to all the nodes in the source node's periphery. When a node receives this query, it adds its own ID to the query. Thus, the sequence of recorded nodes presents a route from the source to the current routing zone. Otherwise, if the destination is in the current node's routing zone, a route reply is sent back to the source along the reverse path from the accumulated record. A big advantage of this scheme is that a single route-request can result in multiple route replies. The source can determine the quality of these multiple routes based on such parameter(s) as hop count or traffic and choose the best route to be used.

13.7.2 Fisheye State Routing

The fisheye state routing (FSR) protocol [13.18] uses multilevel fisheye scopes to reduce the routing update overhead in large networks. The key idea is to exchange link-state entries with the neighbors with a frequency that depends on the distance to the destination. More effort is made in collecting topological data that is more likely to be required soon. With the basic assumption that nearby changes in network topology matter the most, FSR focuses its efforts on viewing the nearby changes with the highest resolution and very frequently. The changes at distant nodes are seen with a lower resolution and less frequently.

13.7.3 Landmark Routing (LANMAR) for MANET with Group Mobility

Landmark ad hoc routing (LANMAR) [13.19] combines the features of FSR and landmark routing. The major addition here is to use landmarks for each set of nodes that move together as a group (e.g., a company of soldiers in a battlefield). This reduces the overall routing update overhead. The nodes exchange the link-state information only with their neighbors, as in FSR. Routes within a fisheye scope are accurate, and the routes to remote groups of nodes called subnets are "handled" by the corresponding landmarks in the neighborhood. As the packet comes closer to the destination, it eventually switches to the accurate route provided by the fisheye.

A modified version of FSR is used for routing. The major difference between the two routing schemes is that in FSR the routing table contains all the nodes in the network. On the other hand, in LANMAR, the routing table contains only the nodes within the scope and the landmark nodes [13.20]. This reduces the routing table size and overhead of the update traffic and hence increases scalability of the scheme.

While relaying a packet, the logical subnet for the destination is looked up and the packet is routed toward the landmark node for that subnet. However, the packet need not pass through the landmark. For the updates in the routing table, LANMAR uses a scheme similar to that in FSR. Nodes periodically exchange the topological information with their immediate neighbors. In each update, a node sends entries within its fisheye scope. A distance vector with information about all the landmark nodes is also piggybacked onto this update.

13.7.4 Location-Aided Routing

The location-aided routing (LAR) [13.21] protocol uses location information of the nodes to limit the scope of route-request flood used in other protocols such as AODV and DSR. The location information may be obtained through GPS. The search for the route is limited to the request zone which is based on the expected location of the destination node at the time of route discovery.

Assume a node S needs to find a route to another node D. S also knows that D has been at location L at time t_0. The node S can speculate as to the expected zone of the node D at current time t_1 based on the priori knowledge. For example, if S knows that D travels with an average velocity v, the expected zone then becomes the

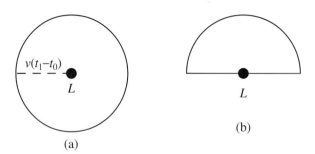

Figure 13.7
Examples of expected zone.

circular region of radius $v(t_1 - t_0)$ centered at L [Figure 13.7(a)]. An important note here is that the estimated zone is only an estimate of the current location of D. If the average speed of the node is more than v, the node can be outside the estimated circular region. If the node S does not have any information about prior location of node D, it cannot make a reasonable estimate toward its current location and the entire network becomes the potential expected zone. In general, more information regarding the prior location and mobility of a node can result in a smaller expected zone. Extending the example above, if S knows that D moves north in addition to the specifications above, the expected zone in Figure 13.7(a) can be reduced to that in Figure 13.7(b).

The next step is to determine a request zone based on the expected zone. When node S needs a route to node D, node S defines a request zone for the route request using the information about the expected zone of node D. The LAR algorithms now use flooding to find the route with one important modification. A node forwards the route request if and only if it belongs to the request zone. We can increase the probability of the route request reaching node D by including the entire expected zone within the request zone. The request zone can also include other regions around the expected zone.

The source node S uses the available information to determine the four corners of the request zone. These coordinates are included in the route request initiated by the source. When a node receives the route request, it discards the request if it is not inside the rectangle specified by the four coordinates. Otherwise it forwards the request to its neighbors. For example, in Figure 13.8, when node I receives a route request, node I forwards the request to its neighbors as it is within the rectangular request zone. On the other hand, node J is outside the request zone and discards the request. This algorithm is known as LAR scheme 1.

A similar scheme with a slight modification is called LAR scheme 2. Here, S knows the location (X_d, Y_d) of node D at some time t_0. S initiates a route request at time $t_1 \geq t_0$. Node S calculates its distance from the node D—the distance between points (X_s, Y_s) and (X_d, Y_d)—and includes this distance in the route request. The coordinates (X_d, Y_d) are also sent along with the route request. Given this information, a node J will only forward the request it receives from I (originated by S), if it is closer to the destination (X_d, Y_d) than I. This decreases the message overhead and improves the scalability of the algorithm.

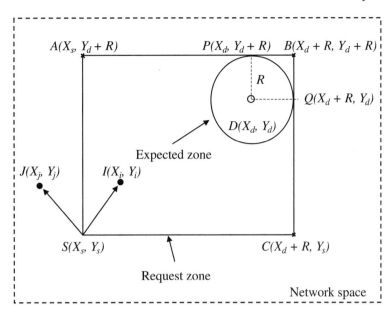

Figure 13.8
LAR scheme 1.

13.7.5 Distance Routing Effect Algorithm for Mobility

The distance routing effect algorithm for mobility (DREAM) [13.22] is built around two key ideas. The first is called the distance effect, which says that the farther the two nodes are from each other the slower they appear to be moving with respect to each other. This fact can be used to tune the rate of updates in the routing tables as a function of the distance between the nodes without compromising their accuracy. If two nodes are farther from each other, the updates in routing tables are needed less frequently than when the nodes are closer.

The second idea uses a similar frequency variance for updating the location information of a node. The location updates of a node are triggered by only one factor—the node's mobility rate. It is intuitive that routing information about a slowly moving node needs to be updated less frequently than a node that is moving fast. In this manner, each node can individually optimize the rate at which it sends updates to the rest of the network. The algorithm uses the routing tables and sends the message in the "recorded direction" of the destination node.

13.7.6 Relative Distance Microdiscovery Ad Hoc Routing

The relative distance microdiscovery ad hoc routing (RDMAR) [13.23] protocol is a highly adaptive, efficient and scalable routing protocol. The protocol is particularly suited for very large mobile networks where rate of topological change is moderate. The impact of link failures is localized to a very small region of the network and is achieved through the use of relative distance microdiscovery (RDM), a route discovery mechanism. The key concept is to limit the query floods by using the relative distance (RD) between two nodes. Every time a route search between two nodes is requested, an iterative algorithm computes an estimate of the RD between them, by using the average node mobility, previous RD, and the time elapsed since the last communication. The query flood is now limited to the region of the network that is centered at the source node and with a maximum propagation radius equal to the newly estimated RD between the source and the destination nodes. This localization of the query floods reduces the routing overhead and overall network congestion.

Each node maintains a routing table which lists all reachable destinations. For every destination, additional routing information is also stored. This includes the "default router" field, the "RD" field (in number of hops), the "time_last_update" (TLU) field, the "RT_timeout" field, and the "route flag" field.

RDMAR consists of two main algorithms:

- **Route Discovery**: When a source node s needs to send a message to a destination node d and no routes are known, node S initiates a route-discovery process. Node S can now either choose: To flood the entire network with route query or to limit the route discovery in a smaller region of the network.

- **Route Maintenance**: When an intermediate node S receives a data packet, it processes the routing header and then forwards the packet to the next hop. Furthermore, the node I sends an explicit message to determine whether a reverse link can be establishes with the previous node. Therefore, the nodes in RDMAR do not assume bidirectional links.

If the intermediate node I may not be able to forward the data packet correctly due to link or node failure, node I attempts additional retransmissions of the same data packet up to a maximum number of retries. If failure persists, a fresh route-discovery process is initiated.

13.7.7 Power Aware Routing

The power aware routing protocol uses power aware metrics [13.24][13.25] to determine routes in a MANET. Using such metrics can result in huge energy and cost savings for the entire network. For example, it has been shown that using these power aware metrics in a shortest-cost routing algorithm reduces the cost of routing by $5 \sim 30\%$ over shortest-hop routing. The energy consumption over the MAC layer protocol is also reduced by $40 \sim 70\%$.

An important point to note here is that the algorithm itself does not change. This means that although the mean time to node failure increases significantly; the packet delays and latencies do not increase. Another work [13.26] suggests using traffic characteristics and network congestion to select routes. Table 13.1 summarizes the main features of the protocols discussed so far.

13.7.8 Multipath Routing Protocols

Based on the route-discovery mechanism, routing protocols are classified as either reactive, proactive or hybrid protocols as discussed in previous sections. Similarly, based on the number of routes discovered between source and destination, protocols can be either unipath or multipath protocols. Multipath protocols aim at providing redundant paths to the destination. Availability of redundant paths to the same destination increases the reliability and robustness of the network. Providing

Table 13.1: ▶
Protocol Characteristics

Routing Protocol	Route Acquisition	Flood for Route Discovery	Delay for Route Discovery	Multipath Capability	Effect of Route Failure
DSDV	Computed a priori	No	No	No	Updates the routing tables of all nodes
WRP	Computed a priori	No	No	No	Ultimately, updates the routing tables of all nodes by exchanging MRL between neighbors
DSR	On demand, only when needed	Yes. Aggressive use of caching may reduce flood	Yes	Not explicitly. The technique of salvaging may quickly restore a route	Route error propagated up to the source to erase invalid path
AODV	On demand, only when needed	Yes. Controlled use of cache to reduce flood	Yes	No, although recent research indicates viability	Route error propagated up to the source to erase invalid path
TORA	On demand, only when needed	Basically one for initial route discovery	Yes. Once the DAG is constructed, multiple paths are found	Yes	Error is recovered locally
ZRP	Hybrid	Only outside a source's zone	Only if the destination is outside the source's zone	No	Hybrid of updating nodes' tables within a zone and propagating route error to the source
LAR	On demand, only when needed	Reduced by using location information	Yes	No	Route error propagated up to the source

multiple paths is beneficial, particularly in wireless ad hoc networks where routes are disconnected frequently due to mobility of the nodes and poor wireless link quality. However, multipath routing can lead to increased out-of-order delivery and resequencing of packets at the destination along with increased collision.

Multipath routing protocols can also aid in secure routing against denial of service attacks by providing multiple routes between the nodes. Nodes can switch over to an alternate route when the primary route has intermediate malicious nodes and appears to have been compromised. Various unipath protocols discussed in earlier sections can discover multiple paths between nodes. Diversity coding [13.27] takes advantage of multiple paths for fault-tolerant communication between nodes, where out of n paths available, m paths are used for transmitting data and remaining $n - m$ paths are used for transmitting redundant information. In this section we will review some of the proposed multipath routing protocols, few of which extend the idea of existing unipath protocols.

On-Demand Multipath Routing for Mobile Ad Hoc Networks

On-demand multipath routing [13.28] is an extension of the DSR protocol. It exploits multipath techniques in reducing the frequency of query floods used to discover new routes. It also improves performance by providing all intermediate nodes in the primary (shortest) route with alternate paths rather than providing only the source with alternate paths. Two multipath extensions for DSR have been proposed; in both, DSR starts route discovery by flooding the network using query messages. Each query message carries the sequence of hops it passed through in the message header. After receiving a query packet, the destination node replies with a reply packet that simply copies the route from the query packet and sends it back. Additionally, each node maintains a route cache, where complete routes to desired destinations are stored as learned from the reply packets. The destination node can receive many copies of the flooded query messages.

In the first multipath extension of DSR, the destination replies to a set of query packets that carry a source route that is link-wise disjoint from the primary source route. The primary source route is the route taken by first query reaching the destination node. The source caches all routes received in reply packets in its local route cache. When the primary route breaks, the remaining shortest route is used. The process continues till all the alternate routes are exhausted, and then a fresh route discovery is initiated. Alternate routes are therefore provided only to the source since reply packets sent by the destination node are addressed only to the source node. An intermediate link failure on the primary route results in a rote error packet being sent to the source, which will then use an alternate route. This leads to retransmissions of data packets already in transit from the broken link. To avoid these retransmission, in the second multipath extension of DSR all intermediate nodes are provided a disjoint alternate route so that in-transit data packets no longer face route loss. The destination node now replies to each intermediate node in the primary route with an alternate disjoint route to the destination. It is possible that not all intermediate nodes will get a different disjoint route (especially in sparse networks), and there still may be temporary route loss due to link failures, until an upstream node switches to an alternate route.

The advantage of this scheme can be understood by referring to Figure 13.9. Node n_1 (source node S) uses the primary route for sending data packets to node n_{k+1} (destination node D). When an intermediate link L_i is disrupted, the node i replaces the remaining portion of the route, $L_i - L_k$ in the packet header by the

Figure 13.9
Route construction
and maintenance
in the on-demand
multipath routing
protocol [13.28].
Credit: A. Masipuri
and S.R. Das, "On
Demand Multipath
Routing for Mobile
Ad Hoc Networks,"
*Proceedings of
Eighth International
Conference on
Computer
Communications
and Networks*,
pp. 64-70, Boston,
October 1999.

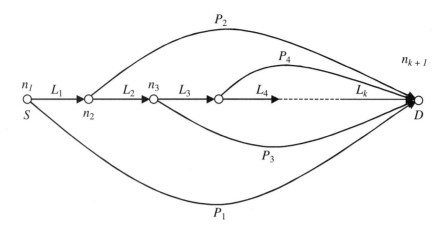

alternate route P_i. This continues till a link on P_i breaks, leading to transmission of an error packet backwards up to node n_{i-1}, which then switches all later packets to its own alternate route P_{i-1} by modifying the source route in the packet header. Thus, any intermediate node with an alternate path to the destination douses the error packet. This continues till the source gets an error packet and has no alternate route resulting in initiation of a new route discovery.

Ad Hoc On-Demand Distance Vector-Backup Routing

The ad hoc on-demand distance vector–backing routing (AODV–BR) [13.29] is a multipath routing protocol which constructs routes on demand and uses alternate paths only when the primary route is disrupted. This method utilizes a mesh arrangement to provide multiple alternate paths to existing on-demand routing protocols without extra control message overhead. Similar to its parent protocol AODV, this protocol also consists of two phases:

■ **Route Construction**: Source initiates route discovery by flooding a route request (RREQ) packet having a unique identifier so that intermediate nodes can detect and drop duplicate packets. Upon receiving a non-duplicate RREQ, the intermediate node stores the previous hop and the source node information in its route table. This process is also known as backward learning. It then broadcasts the RREQ packet or sends a route reply (RREP) packet, if it has a route to the destination. The destination node sends a RREP via the selected route when it receives the first RREQ packet or subsequent RREQs that have a better route than the previously replied route.

  The mesh construction and the alternate paths are established during the route reply phase. A node overhearing a RREP packet transmitted by a neighbor (on the primary route) but not directed to it, records that neighbor as the next hop to the destination in its alternate route table. A node may receive numerous RREPs for the same route if the node is within the radio range of more than one intermediate node of the primary route. The node then chooses the best route among them and inserts it into the alternate route table. When the RREP packet

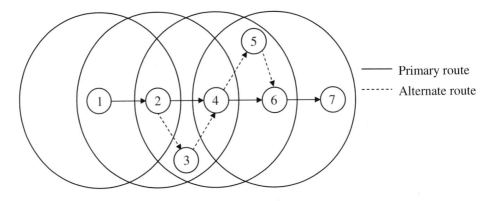

(a) Multiple routes from node 1 to node 7

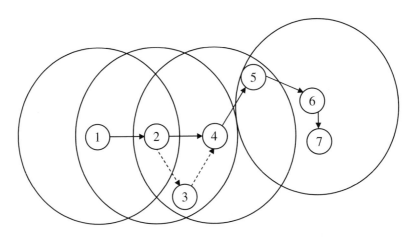

Figure 13.10
Multiple routes
in the AODV–BR
protocol.

(b) Alternate route used when primary disconnects

reaches the source, the primary route between the source and the destination is established and ready for use. Nodes that have an entry to the destination in their alternate route table become part of the mesh structure. The primary route and alternate routes together establish a mesh structure that looks like a fish bone, as shown in Figure 13.10(a).

■ **Route Maintenance and Mesh Routes**: Data packets are transmitted through the primary route unless there is a failure. If a node detects a route failure, it performs one hop data broadcast to its immediate neighbors specifying the detached link in the data header. Thus the packet is a candidate for "alternate routing." On receiving this packet, neighbor nodes that have an entry for the destination in their alternate route table unicast the packet to their next hop node. Packets are thus delivered through one or more alternate routes and are not dropped when route failure occurs, as shown in Figure 13.10(b). To prevent packets from going into a loop, these mesh nodes forward the data packet only if the packet is not received from their next hop to the destination and is not a duplicate packet. A node on the primary route also sends a route error (RERR) packet to the source if it detects a route failure, so that the route discovery can be initiated. Reconstruction of a new route instead of continuously using the alternate paths is done to ensure usage of a fresh and optimal route that reflects the current network topology.

Thus, the mesh connection is used only to "go around" the broken part of the link. Nodes that provide alternate paths overhear data packets and if the packet was transmitted by the next hop to the destination as indicated in their alternate route table, they update the path. If an alternate route is not updated during the timeout interval, the node deletes the path from the table.

Split Multipath Routing

Split multipath routing (SMR) [13.30] is an on-demand routing protocol that constructs maximally disjoint paths between a given source destination. Multiple routes are established, and data traffic is split into them to avoid congestion and facilitate efficient use of network resources. These routes may not be of equal lengths. SMR like other on-demand routing protocols builds multiple routes using request/reply cycles. The routing protocol consists primarily of two phases: route discovery and route maintenance.

■ **Route Discovery**: If a source node needs a route to a specific destination node and no route information is available, it broadcasts a RREQ packet. The packet header contains the source ID and a sequence number that identifies the packet uniquely. When a node other than the destination node receives a RREQ packet that is not a duplicate, it appends its ID and rebroadcasts the packet to the neighboring nodes. Instead of dropping all the duplicate packets, intermediate nodes forward duplicate packets that have arrived through a different incoming link (the link from which the first RREQ packet was received) and whose hop count is not greater than that of the first received RREQ packet. Besides, intermediate nodes do not send RREPs from their local route cache (as in DSR and AODV). This takes care of the problem of overlapped routes and helps in constructing disjoint paths.

When the destination node receives the first RREQ packet, it stores the entire path and sends a RREP packet to the source via this route. The RREP packet contains the entire path, and hence intermediate nodes can forward this packet using this information. The destination node waits for certain extra duration to receive more RREQs. It then selects another route that is maximally disjoint to the route already replied and generates another RREP packet to the source. Among many maximally disjoint routes, the destination node chooses the one with the shortest hop.

■ **Route Maintenance**: In the event of a node failing to deliver the data packet to the next hop of the route, it considers this as a link failure and sends a RERR packet to the upstream direction. The RERR message contains the route to the source and the immediate upstream and downstream nodes of the broken link. On receiving a RERR packet, the source cleans every entry in its route table that uses the broken link. If only one of the two routes of the session is invalidated, the source uses the remaining legitimate route to deliver data packets. The source can reinitiate the route discovery process when a particular route or both the routes of the session are broken. When the source receives a RREP packet, it uses the discovered route to transmit buffered data packets. If the source receives a second RREP packet, it has two routes to the target node and can split the data traffic into two routes.

Caching and Multipath Routing Protocol

The caching and multipath routing protocol (CHAMP) [13.31] makes use of temporal locality in dropped packets and targets at reducing packet loss due to a route breakdown. Every node maintains a small buffer for caching data packets that pass through it. When a downstream node discovers a error in forwarding, an upstream node with the relevant data in its buffer and a alternate route can retransmit the data.. This approach can be useful only if nodes maintain alternate routes to a destination. The main features of this protocol are therefore shortest multipath route discovery and cooperative packet caching.

Every node maintains a *route cache* and a *route request cache*. A route cache is a list containing forwarding information to every active destination. Each entry contains the destination identifier, distance to the destination, next hop nodes to the destination, the last time, and the number of times each successor node was used for forwarding. A route entry which has not been used for route life-time is deleted. The route request cache at a node is a list containing an entry for recent route request received and processed.

- **Route Discovery**: CHAMP operates on demand; a source node initiates a route discovery when it has data to send but has no available route. It then floods the network with a RREQ for the destination node. This establishes a DAG (direct acyclic graph) rooted at the source. When the destination node receives a RREQ, it sends back a RREP to an intermediate node through some nodes that are a subset of the DAG rooted at the source. Every RREQ from the source to destination has a forward count field, which is initialized to zero by the source and incremented by one every time the message is retransmitted by an intermediate node. The first time any intermediate node receives a RREQ from the source it initializes its hop count to the previous hop of the message. Every time it then receives a request from a path of the same length from the source, it includes the previous hop of the message in set of neighbors forwarding the same request. If it receives the same request via a shorter path, it resets its hop count and the previous hop of the message. The set of neighbors forwarding the same request can receive a corresponding RREP from the intermediate node, if it sends one.

  When a destination node receives a RREQ, it immediately sends back a RREP if the request is coming through the shortest path. Every RREP explicitly specifies the set of nodes that can accept the reply packet. The destination node initializes this field to the previous hop of the RREQ and hop count to zero. A node processes a RREP if it belongs to the set of nodes the RREP is intended for. It then accepts the route in the RREP if the route is shortest to the destination or its existing routes to destination have not been used for more than route fresh time and provided that the number of routes to destination is less than or equal to the maximum routes. It then also resets the set of next hop nodes to the destination table to contain the previous hop of the RREP. The node then computes its distance from the destination (which is equal to the hop count) and forwards the message to its upstream nodes by setting the set of nodes that can receive the RREP equal to the set of nodes that requested the route from this node to same the destination. It also increments the hop count by one if the corresponding request has not been replied yet. This process is repeated until the RREP reaches the source.

■ **Data Forwarding**: Data packets are identified by source identifier and a source-affixed sequence number. Each packet also includes the previous hop in its header. When a node has a data packet to forward, it chooses the least used next hop neighbor. It then saves a copy of the packet in its data cache, sets the previous hop field to its address and forwards the packet to the chosen next hop. If a node has no route to the destination and is the source of the packet, it saves the packet in its send buffer and performs a route discovery. However if it is not the source, it simply drops the packet and broadcasts a RERR containing the header information of the dropped packet. An upstream intermediate node on receiving the RERR packet will modify its set of next hop destination and will try to retransmit the data packet if it has a copy in its data cache and has an alternate route to the destination. If it does not have an alternate route or the packet in its data cache, it adds the data packet header information in a RERR packet and broadcasts it.

Neighbor-Table-Based Multipath Routing in Ad Hoc Networks

Neighbor-table-based multipath routing (NTBMR) [13.32] is a mixed multipath routing protocol which deals with regular topology changes in mobile ad hoc networks. In this scheme, multiple routes need not be disjoint as in SMR. Theoretical analysis has revealed that for error-prone wireless links, nondisjoint multipath routing has higher route dependability. In NTBMR every node maintains a neighbor table, which records its k-hop neighbor nodes. This scheme also consists of route discovery and route maintenance. The principal mechanism here is construction of a neighbor table and a route cache at every node. The routes in the neighbor table are used in the construction of route cache and are also used to establish the lifetime of wireless links to assist in route discovery.

■ **Establishment of Neighbor Table and Route Cache**: In the NTBMR protocol, all nodes in the network periodically transmit beacon packets. Using the time-to-live (TTL) field as a counter, these packets are transmitted only to two-hop neighbors. Each beacon packet has the following fields: packet type, source address, intermediate station address, unreachable station address, TTL, and sequence number. With the help of these beacon packets, a neighbor table is established based on the route information. The neighbor table can be time driven or data driven. With a time-driven mechanism, if a node receives the beacon packet along one particular route, it considers the route active and adds all the node IDs the packet has passed by to its neighbor table. This implies, one-hop neighbor can obtain one-hop route to the source node and that two-hop route neighbor obtains a two-hop route to the source node as well as a one-hop route to the intermediate relay station. However, if the station does not receive the beacon packet along the route within a predefined timeout period, it regards that route dormant and purges the corresponding stations along the route from the neighbor table.

One of the disadvantages of time-driven mechanisms is that a node cannot learn about changes in topology within the timeout period. To ease this, a data-driven mechanism is proposed whereby once a station detects that its one-hop neighbor is inaccessible; it will fill the address of the inaccessible station in the beacon packet and inform its other one-hop neighbors to revise their neighbor table. As soon as the other one-hop neighbors receive the beacon packet, they purge the "unreachable station" contained in the beacon packet from its two-hop neighbors in the neighbor table. The discovery of one-hop unreachable stations can be achieved by the link failure detection method of MAC layer or timeout of beacon packets.

Route discovery and maintenance is done using a route cache, which contains all the routes that the station is apprised of. If a neighbor table is updated at any time, it leads to changes in the route cache also. The route cache is kept up to date by monitoring route information contained in route-reply packets, route-error packets, route-request packets, and data packets. Priorities are given to routes based on the source they are obtained from. This process is known as *route extraction reason* and gives highest priority to routes learnt from reply packets and lowest priority to routes obtained from data packets. These priorities are also used to aid in route selection. Every node also computes the mean and variance of the wireless link lifetime and uses this to determine if a route is utilizable or not during route discovery.

- **Route Discovery**: A source tries to discover an effective route from its route cache. If many routes exist to the same destination node, it picks the route based on multiple parameters which include route setting up time, route distance, route extraction reason, and the like. If a node cannot find an appropriate route, the station will start the route-detection process, which is similar to DSR. After the node picks one route to the destination, it will fill the node addresses of the route in the corresponding fields of the data packet. Intermediate nodes can forward the packet based on these fields.

- **Route Maintenance**: If a route fails while a node is transmitting, alternate routes are used to overcome it. An intermediate node encountering a link malfunction will react differently based on two predefined transmission time threshold values indicated by T_1 and T_2 with $T_1 < T_2$. If an intermediate node receives a route error, it will make changes to its neighbor table and the route cache based on the route error information. The route error packet will then be transmitted to an upstream node. When the source receives a route error packet it will modify its neighbor table and the local route cache and then commence a route discovery process.

The NTBMR protocol has a good packet delivery ratio and lesser end-to-end delay as compared to DSR but incurs the additional overhead of beacon packets.

Table 13.2 summarizes the main characteristics of the multipath protocols discussed in this section.

Table 13.2: ▶
A Comparison of Different Multipath Protocols

Protocol	Types of routes	Number of routes	Routes used for transmission	Intermediate nodes have alternate routes?	Route caching?	Effect of single route failure
MDSR	Link-wise disjoint	No limit	Shortest route is used, alternate routes are kept as backup	Yes	Yes	Error packet is sent to the source. Intermediate node with alternate routes responds and shortest remaining alternate route is used.
AODV–BR	Not necessarily disjoint	No limit	Shortest route is used, alternate routes are kept as backup	Yes	No	Error packet broadcast to one-hop neighbors; neighbor with alternate route to destination responds and forwards data to destination. Route error packet sent to source to initiate route rediscovery
SMR	Maximally disjoint	Two	Shortest route is used, alternate route is used as backup	No	No	Error packet is sent to source and alternate route is used for further data communication.
CHAMP	Shortest multiple routes of equal lengths not necessarily disjoint	No limit	All routes are used in a round-robin fashion	Yes, every node must maintain at least two routes to every active destination for cooperative caching to be effective.	Yes	Node that detects link failure forwards data through alternate route if present, otherwise broadcasts error packet.

13.8 Wireless Sensor Networks

Wireless sensor networks are a new class of ad hoc networks that are expected to be deployed in the coming years, as they enable reliable monitoring and analysis of unknown and untested environments. A wireless sensor network is a collection of tiny disposable and low-power devices. A sensor node is a device that converts a physical attribute (e.g., temperature, vibrations) into a form understandable by users. Any of such devices may include a sensing module, a communication module (display or a medium to transmit data to the user), memory (to hold data until it can be used), and typically an exhaustable source of power like a small battery.

Wired sensor networks have been used for years for a number of applications. Some examples include distribution of thousands of sensors and wires over strategic locations in a structure such as an airplane, so that conditions can be constantly monitored both from the inside and the outside and a real-time warning can be issued as soon as a major problem is detected in the monitored structure.

The number of wired sensors can be made large to cover as much area as desirable. Each of these has a constant power supply and communicates with the end-user over a wired network. The organization of such a network should be planned to find strategic positions to place these nodes, and then the nodes should be installed appropriately. The failure of a single node might bring down the whole network or leave that region completely unmonitored. Sensor networks are usually unattended, and some degree of fault-tolerance needs to be incorporated so that the need for maintenance is minimized. This is especially desirable in those applications where the sensors may be embedded in the structures or places, which are inhospitable and inaccessible for service.

Advancement in technology has made it possible to have extremely small, low-power devices equipped with programmable computing, multiple-parameter-sensing, and wireless communication capabilities. Also, the low cost of sensors makes it possible to have a network of hundreds or thousands of them, thereby enhancing the reliability and accuracy of data and the area coverage. Also, it should be easy to deploy sensors since they require very low or no installation cost.

In short, the advantages of wireless sensor networks over wired ones are as follows:

1. **Ease of deployment**: These wireless sensors can be deployed (dropped from a plane or placed in a factory) at the site of interest without any preorganization, thus reducing the installation cost and increasing the flexibility of arrangement.

2. **Extended range**: One single huge wired sensor (macrosensor) can be replaced by many smaller wireless sensors for the same cost. One macrosensor can sense only a limited region, whereas a network of smaller sensors can be distributed over a wider region.

3. **Fault tolerance**: Since sensor networks are mostly unattended, they should be fault-tolerant. With macrosensors, the failure of one node makes that area completely unmonitored until it is replaced. In wireless sensors, failure of one node may not affect the network operation as there are other nodes collecting similar

data. At most, the accuracy of data collected may be reduced, but typically the entire area of interest is still covered.

4. **Mobility**: Since these wireless sensors are equipped with a battery, they can be mobile. Thus, if a region becomes unmonitored, we can have the nodes rearrange themselves to distribute evenly (i.e., these nodes can be made to move toward an area of interest). It should be noted that these nodes have limited mobility as compared to ad hoc networks.

The inherent limitations of wireless media, such as low-bandwidth, error-prone transmissions, and the need for collision-free channel access are also present in the sensor networks. In addition, since the wireless nodes are not connected in any way to a constant power supply, they derive energy from personal batteries. This limits the amount of energy available to the nodes, and since they are deployed in places where it is difficult to replace the nodes or their batteries, it is desirable to increase the lifetime of the network, and perferably all the nodes should die together so that all the nodes can be replaced simultaneously or new nodes can be put in the whole area. Finding individual dead nodes and then replacing those nodes selectively would require planned deployment and eliminate some of the advantages of these networks. Thus, the protocols designed for these networks must strategically distribute the usage of energy, which increases the average life of the overall system. In addition, environments in which these nodes operate and respond are very dynamic, with fast-changing physical parameters. Some of the parameters that might change depends on the application and can be defined as follows:

1. Power availability

2. Position (if the nodes are mobile)

3. Reachability

4. Type of task (i.e., attributes the nodes need to operate on)

These networks are fundamentally different from traditional wireless networks where data is exchanged between specific nodes. Sensor networks are based on "data centric" paradigms where, more than the specific nodes, the focus is on such attributes as temperature, motion, and region. Traditional routing protocols defined for MANETs are not well suited for wireless sensor networks. The application specific nature of these networks presents unique challenges in the design of generic protocols at different layers of the network architecture as follows:

- In traditional wired and wireless networks, each node is given a unique ID, used for routing. This cannot be used effectively in sensor networks; since these networks are data centric, routing to and from specific nodes is not required. Also, the large number of nodes in the network implies large IDs, which might be substantially larger than the actual data bits being transmitted.

- Adjacent nodes may have similar data. Therefore, rather than sending data separately from each node to the requesting node, it is desirable to aggregate similar data and then respond.

- The requirements of the network change with the application and hence are application specific. For example, in some applications the sensor nodes are fixed and not mobile where as others may need data based only on a single selected attribute (i.e., the attribute is fixed in the network).

Thus, sensor networks need protocols that are application specific, yet generic enough to be data centric, capable of data aggregation and minimizing energy consumption. An ideal sensor network should have the following additional features:

- The attribute-based addresses are composed of a series of attribute-value pairs that specify certain physical parameters to be sensed. For example, an attribute address may be (temperature $100°$C [$212°$F], location =?). Therefore, all nodes that sense a temperature greater than $100°$C ($212°$F) should respond with their location.

- Location awareness is another important issue. Since most data collection is based on location, it is desirable that the nodes know their position whenever needed.

- Another important requirement in some cases is that the sensors should react immediately to drastic changes in their environment (for example, in time-critical applications). The end-user should be made aware of any drastic deviation with minimum delay, while making efficient use of the limited wireless channel bandwidth and battery energy.

- Query handling is another additional feature. Users, using handheld wireless devices, should be able to request data from the network. Since these handheld devices are also energy constrained, the user should be able to query through the base station or through any of the sensor nodes, whichever is closer. Therefore, there should be a reliable mechanism to transmit the query to appropriate nodes that can respond to the query. The answer should then be rerouted to the user as quickly as possible. Since efficient query handling is a highly desirable feature, we explore it further in the following section.

13.8.1 Case Study

Consider the following scenario: Temperature sensors are placed around a factory (e.g., chemical, automotive). Each sensor has a sensing module, a communication module (wireless communication), a computing module, and memory.

Typical queries posed by the user include the following:

1. Report immediately if the temperature in the northeast quadrant goes below $5°$C ($41°$F).

2. Retrieve the average temperature in the southwestern quadrant.

3. For the next two hours report if the temperature goes beyond $100°$C ($212°$F).

4. Which areas had temperature between $5°$C ($41°$F) and $100°$C ($212°$ F) in the past two hours?

Such queries lead us to the following conclusions:

- Data from various nodes need to be aggregated, and typically aggregation of data from adjacent nodes is needed. This has the advantage of reducing traffic in the network.

- Queries that monitor the system are mostly duration-based queries.

- Time-critical queries should reach the user immediately.

- Some queries just require a snapshot view of the network at that instant.

In general, user queries can be broadly categorized into three types:

1. **Historical query**: This type of query is mainly used for analysis of historical data stored at the BS. For example, "What was the temperature two hours back in the northwest quadrant?"

2. **One-time query**: This type of query gives a snapshot view of the network. For example, "What is the current temperature in the northwest quadrant?"

3. **Persistent query**: This type of query is mainly used to monitor a network over a time interval with respect to some parameters. For example, "Report the temperature in the northwest quadrant for the next two hours."

In wireless sensor networks, where efficient usage of energy is very critical, larger latency for noncritical data is preferable for longer node lifetime. However, queries for time-critical data should not be delayed and need to be handled immediately. Some protocols try to use the energy intelligently by reducing unnecessary data transmission for noncritical data but transmitting time-critical data immediately, even if we have to keep the sensors on at all times. Periodic data are transmitted at longer intervals so that historical queries can be answered. All other data are retrieved from the system on demand.

Adapting to the Inherent Dynamic Nature of Wireless Sensor Networks

Some important objectives that need to be achieved are as follows: Exploit spatial diversity and density of sensor/actuator nodes to build an adaptive node sleep schedule; characterize the relationship between deployment density and network size; and explore the tradeoff between data redundancy and bandwidth consumption as follows:

- The nodes on deployment should spontaneously create and assemble a network, dynamically adapt to device failure and degradation, manage mobility of sensor nodes, and react to changes in task and sensor requirements.

- Some nodes may detect an event that triggers a big sensor, like a camera, generating heavy traffic. But when sensing activity is low, traffic should be light and hence able to adapt to changes in the traffic.

- It should allow finer control over an algorithm than simply turning it off or on. Nodes should be capable of dynamically trading precision for energy or scope for convergence time-based on incoming data.

The scalable coordination architectures for deeply distributed systems (SCADDS) project [13.33], also a part of DARPA SensIT program, focuses on adaptive fidelity, dynamically adjusting the overall fidelity of sensing in response to task dynamics (turn on more sensors when a threat is perceived). They use additional sensors (redundancy) to extend lifetime [13.34]. Neighboring nodes are free to talk to each other irrespective of their listen schedules; there is no clustering and no intercluster communication and interference. Adaptive self-configuring sensor network topologies (ASCENT) [13.35], which is a part of SCADDS, focuses on how to decide which nodes should join the routing infrastructure to adapt to a wide variety of environmental dynamics and terrain conditions, producing regions with nonuniform communication density. In ASCENT, each node assesses its connectivity and adapts its participation in its multihop network topology based on the measured operating region. A node signals and reduces its duty cycle when it detects high message loss, requesting additional nodes in the region to join the network in order to relay messages to it. It probes the local communication environment and does not join the multihop routing infrastructure until it is helpful to do so. It avoids transmitting dynamic state information repeatedly across the network.

13.8.2 DARPA Efforts toward Wireless Sensor Networks

The Defense Advanced Research Projects Agency (DARPA) has identified networked microsensor technology as a key application for the future. There are many interesting projects and experiments going on under the DARPA SensIT (Sensor Information Technology) program [13.36]. The SensIT program aims to develop the system framework for distributed microsensors. On the battlefield of the future, a huge networked system of smart, inexpensive microsensors, combining multiple sensor types, embedded processors, positioning ability, and wireless communication, will pervade the environment and provide commanders and soldiers with situation awareness. Therefore, software is needed to enable a variety of sensor nets, on the ground and in the air as well as on buildings and bodies, all functioning autonomously, operating with high reliability, and processing signals and information collaboratively in the network to provide useful information to soldiers in a timely manner.

● ● ● ● ● ● ● ● ● ● ● ● ● ● ● ⋯⋯

13.9 Fixed Wireless Sensor Networks

There is another aspect to employing and utilizing sensors in measuring some pre-specified physical parameters by placing them at fixed locations and linking them by a wireless network to perform distributed sensing tasks [13.37]. As these sensors are placed at predefined places, they are very useful for continuous and regular monitoring, such as facility and environmental sampling, security and surveillance, health care monitoring, and underwater measurements. Integration of sensing, signal processing, and wireless communication enables processing of events at the node, local neighborhood, and global levels. This requires multiple nodes to communicate for appropriate coordination and cooperation. The communication between sensors is

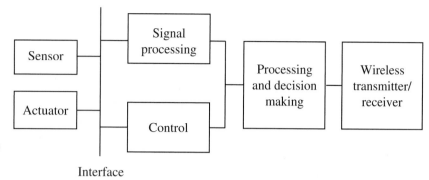

Figure 13.11
A general architecture
of a fixed sensor
node.

achieved by LOS infrared beam or conventional wireless radio communications like FDMA or TDMA.

The complexity of a sensor node depends on the expectations or functional requirements. A general architecture is shown in Figure 13.11, which is no different from that of a mobile sensor network. These stationary sensor nodes typically constitute an ad hoc LAN and possibly communicate to a base station or a wired backbone for further processing and decision making. The medium access using FDMA could use a CDMA model or fixed time slots in TDMA. Similar to mobile sensor nodes, the fixed nodes can measure different physical parameters and pass them on to the signal processor. This information, in turn, is passed on to other sensors as well as to a central controller, which can make an appropriate decision. In a similar way, the central controller can submit queries to the sensor nodes to find relevant information. The information can be retrieved from the fixed sensor nodes to a central controller, or the data sensed and collected by the sensors can be sent to the central controller. A lot more work is needed in this area, and the future of sensor nodes seems promising.

13.9.1   Classification of Sensor Networks

Looking at the various ways in which we can employ the network resources, sensor networks can be classified on the basis of their mode of operation or functionality and the type of target applications. Accordingly, sensor networks are classified into two types:

1. **Proactive networks**: The nodes in this network periodically switch on their sensors and transmitters, sense the environment, and transmit the data of interest. Thus, they provide a snapshot of the relevant parameters at regular intervals and are well suited for applications requiring periodic data monitoring.

2. **Reactive networks**: In this scheme the nodes react immediately to sudden and drastic changes in the value of a sensed attribute. As such, these are well suited for time-critical applications.

Once the type of network is decided, protocols that efficiently route data from the nodes to the users have to be designed, preferably using a suitable MAC sublayer protocol to avoid collisions. Attempts should be made to distribute energy dissipation evenly among all nodes in the network as we do not have specialized high-energy nodes in the network. There are some basic functionalities and characteristics expected from a protocol for proactive networks:

- **Report time**: This is the time period between successive reports sent by a node.
- **Attributes**: This is a set of physical parameters the user is interested in monitoring.

At every report time, the cluster members sense the parameters specified in the attributes and send the data to be aggregated on the requesting entity. This ensures that the user has a complete picture of the entire area covered by the network.

This scheme, however, has an important drawback. Because of the periodicity with which the data are sensed, it is possible that time-critical data may reach the user only after the report time, thus limiting its use to non-time-critical data sensing applications. We discuss both proactive and reactive protocols while emphasizing that the protocol to be chosen is directly related to the application requirements.

13.9.2 Fundamentals of MAC Protocol for Wireless Sensor Networks

The wireless medium is mostly a broadcast medium. All nodes within radio range of a node can hear its transmission. This can be used as a unicast medium by specifically addressing a particular node, and all other nodes drop the packet they receive. Accessing the medium must be coordinated as at any given time only one node needs to communicate in order to avoid collisions. There are two types of schemes available to allocate a single broadcast channel among competing nodes: static channel allocation and dynamic channel allocation.

- **Static channel allocation**: In this category of protocols, if there are N nodes, the bandwidth is divided into N equal portions either in frequency (FDMA), in time (TDMA), in code (CDMA), in space (SDMA: Space Division Multiple Access), or in Orthogonal Frequency Division Multiplexing (OFDM). Since each node is assigned a private portion, there is no interference between multiple users. These protocols work very well with efficient allocation mechanisms when there is only a small and fixed number of users, each of which has a buffered (heavy) load of data.

- **Dynamic channel allocation**: In this category of protocols, there is no fixed assignment of bandwidth. When the number of users changes dynamically and data is bursty at arbitrary nodes, it is advisable to use a dynamic channel-allocation scheme. These are contention-based schemes, wherein nodes contend for the channel when they have data while minimizing collisions with other nodes' transmissions. When there is a collision, the nodes are forced to retransmit data, thus leading to increased waste of energy on the nodes and unbounded delay. Example protocols include CSMA (persistent and nonpersistent) [13.38] and MACAW (multiple access collision avoidance protocol for wireless LANs) [13.39], IEEE 802.11 [13.40].

Information retrieval in sensor networks can be done assuming either a flat topology or a hierarchical model. In a hierarchical clustering model, once clusters have been formed, the number of nodes in the cluster is fixed and is also not large. Therefore, with such a scenario, it is better to use one of the static channel-allocation schemes. Studies [13.41], [13.42] have pointed out the uses of TDMA for wireless sensor networks. In this scheme all the nodes transmit data in their slot to the CH, and at all other times the radio can be switched off, thereby saving valuable energy. When it is not possible to use TDMA, the nodes can use nonpersistent CSMA since the data packets are of fixed size.

TDMA is suitable for either type of network. In proactive networks, since we have the nodes transmitting periodically, we can assign each node a slot and thus avoid collisions. In reactive networks, since adjacent nodes have similar data, when a sudden change takes place in some attribute being sensed, all the nodes will respond immediately. This will lead to collisions, and it is possible that the data never reaches the user in time. For this reason, TDMA is employed so that each node is given a slot and transmits, only in that slot. Even though this increases the delay and many slots might be empty, it is better than the delay and energy consumption incurred by dynamic channel-allocation schemes.

CDMA is used to avoid intercluster collisions. Though this means that more data need to be transmitted per bit, it allows for multiple transmissions using the same frequency. A number of advantages have been pointed out for using a TDMA/CDMA combination to avoid intra-/inter-cluster collision in ad hoc and sensor networks [13.43].

13.9.3 Flat Routing in Sensor Networks

Routing in wireless sensor networks is very different from the traditional wired or wireless networks. Sensor networks are data centric, requesting information satisfying certain attributes, and thus do not require routing of data between specific nodes. Also, since adjacent nodes have almost similar data and might almost always satisfy the same attributes, rather than sending data separately from each node to the requesting node, it is desirable to aggregate similar data in a certain region before sending it. This aggregation is also known as "data fusion" [13.44][13.45].

Many protocols have been proposed that collect data based on the queries injected by the user or that collect data always so that the network is ready to answer any query the user has. These protocols are based on the same concept as ad hoc networks, wherein a route is set up only when needed (on-demand routing) or there is a route from each node to every other node so that when it is needed, it is immediately available (proactive). We now look into protocols that collect data to answer queries injected by the user.

Directed Diffusion

Directed diffusion [13.42] was one of the first data dissemination protocols developed for sensor networks. The query is disseminated (flooded) throughout the network, and gradients are set up to draw data satisfying the query toward the requesting node.

Events (data) start flowing toward the requesting node from multiple paths. A small number of paths can be reinforced to prevent further flooding.

This type of information retrieval is well suited only for persistent queries where requesting nodes are expecting data that satisfy a query for some duration. This makes it unsuitable for historical or one-time queries as it is not worth setting up gradients for queries that employ the path only once. Also, this type of data collection does not fully exploit the fact that adjacent nodes have similar data, as it uses a flat topology. At most, in this protocol, data can be aggregated at the intermediate nodes.

SPIN

A family of adaptive protocols called sensor protocols for information via negotiation (SPIN) [13.46] disseminates all the information at each node to every node in the network. This enables a user to query any node and get the required information immediately. These protocols make use of the property that nearby nodes have similar data and thus distribute only the data which other nodes do not have. These protocols work proactively and distribute the information all over the network, even when a user does not request any data.

COUGAR

Distributed query processing may result in several orders of magnitude fewer messages than a centralized query-processing scheme. References [13.47] and [13.48] discuss the two approaches for processing sensor queries: **warehousing** and **distributed**. In the warehousing approach, data is extracted in a predefined manner and stored in a central database (BS). Subsequently, query processing takes place on the BS. In the distributed approach, only relevant data is extracted from the sensor network, whenever the data is needed.

A model for sensor database systems known as COUGAR provides user representation and internal representation of queries. The format of the sensor queries is also important to aggregate the data and to combine two or more queries. These protocols use a flat topology that is not suitable for wireless sensor networks because one sensor node cannot aggregate data from a number of nearby nodes in this topology and cannot take full advantage of the specific feature in sensor nodes. It is shown that a hierarchical clustering scheme is the most suitable for wireless sensor networks. COUGAR has a three-tier architecture:

- **Query proxy**: A small database component running on the sensor nodes to interpret and execute queries.

- **Front-end component**: A powerful query proxy that allows the sensor network to connect to the outside world. Each front end includes a full fledged database server.

- **Graphical user interface (GUI)**: Through the GUI, users can pose ad hoc and long-running queries on the sensor network. A map component allows the user to query by region and visualize the topology of sensors in the network.

Queries are formulated regardless of the physical structure or the organization of the sensor network. Sensor data is different from traditional relational data since it is not stored in a database server and varies over time. Aggregate queries or correlation queries that give a bird's-eye view of the environment also focus on a particular region of interest. Each long-running query defines a persistent view, which it maintains during a given time interval. In addition, a sensor database should account for sensor and communication failures. Sensor data is measured with an associated uncertainty, and it is desirable to establish and run a distributed query execution plan without assuming global knowledge of the sensor network. In summary, the protocols we have seen so far use a flat topology that is not suitable for some applications of wireless sensor networks since through this topology we cannot aggregate data from a number of nearby nodes and do not take full advantage of the specific features in sensor networks. There are a number of clustering algorithms in literature, and we discuss some of them in the following section. It is always important to keep in mind that different algorithms, whether viewing the topology as flat or hierarchical, are best suited for different application environments.

Hierarchical Routing in Sensor Networks

Some authors suggest that a hierarchical clustering scheme is the most suitable for wireless sensor networks, as this model enables us to take advantage of all the features that are specific to sensor networks. The network is assumed to consist of a BS, away from the nodes, through which the end user can access data from the sensor network. All the nodes in the network are homogeneous and begin with the same initial energy. The BS, however, has a constant power supply and so has no energy constraints. It can transmit with high power to all the nodes and there is no need for routing from the BS to any specific node. However, the nodes cannot always reply to the BS directly due to their power constraints, resulting in asymmetric communication. BS can also be used as a database to hold data. Consider the partial network structure shown in Figure 13.12. Each cluster has a CH that collects data from its cluster members, aggregates it, and sends it to the BS or an upper-level CH. For example, nodes 1.1.1, 1.1.2, 1.1.3, 1.1.4, 1.1.5, and 1.1 form a cluster with node 1.1 as the CH. Similarly, there exist other CHs, such as 1.2. These CHs, in turn, form a cluster with node 1 as their CH. Therefore, node 1 becomes a second-level CH as well. This pattern is repeated to form a hierarchy of clusters, with the uppermost level cluster nodes reporting directly to the BS. The BS forms the root of this hierarchy and supervises the entire network. The main features of this architecture are as follows:

- All the nodes transmit only to their immediate CH, thus saving energy.

- Only the CH needs to perform additional computations on the data, such as aggregation. Therefore, energy is again conserved.

- The cluster members are mostly adjacent to each other and have similar data. Since the CHs aggregate similar data, aggregation is said to be more effective.

- CHs at increasing levels in the hierarchy need to transmit data over relatively longer distances. As they need to perform extra computations, they end up consuming energy faster than the other lower level nodes. In order to distribute

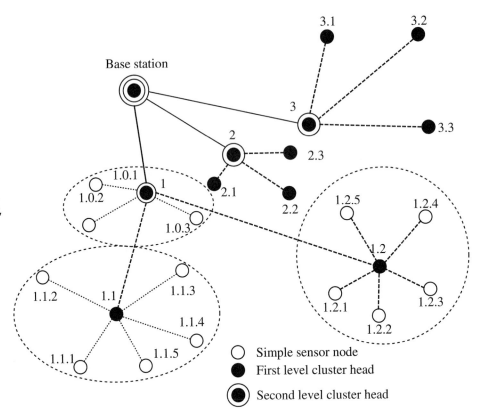

Figure 13.12
Hierarchical clustering.
Credit: A. Manjeshwar,
Q.A. Zeng, and
D.P. Agrawal, "An
Analytical Model
for Information
Retrieval in Wireless
Sensor Networks
using Enhanced
APTEEN Protocol,"
*IEEE Transaction
on Parallel and
Distributed Systems*,
Vol. 13, No. 12,
pp. 1290-1302,
December 2002.

this consumption evenly, all the nodes take turns, becoming the CH for a time interval T, called the cluster period.

■ Since only the CHs need to know how to route the data toward their own CH or BS, complexity in data routing is reduced.

For applications that need to collect data for analysis of the situation or circumstances, it is adequate if we get data when the sensors are able to send data. But in applications that need to get data when something critical happens, such as the "temperature going beyond $100°C$ ($212°F$)," "more than 20 tanks passing by a region," but do not really care what happens in the network at other times, it is not desirable to waste sensors' energy transmitting all the data they have collected. Ideally, it would be better if we could have flexibility in the network so that the user could decide how the network should behave based on the requirements.

Cluster-Based Routing Protocol

A cluster-based routing protocol (CBRP) has been proposed in [13.49] for sensor networks. It divides the network nodes into a number of overlapping or disjoint two-hop-diameter clusters in a distributed manner. Here, the cluster members just send the data to the CH and the CH routes the data to the destination. But this protocol is not suitable for wireless sensor networks as, due to high mobility, it requires a lot of "hello messages" to maintain the clusters. The sensor nodes do not have as much mobility, and two-hop-diameter clusters are not adequate to exploit the underlying feature of "adjacent nodes have similar data" in sensor networks.

Scalable Coordination

In [13.50], a hierarchical clustering method is discussed, with emphasis on localized behavior and the need for asymmetric communication and energy conservation in sensor networks. In this method (no experimental results are provided) the cluster formation appears to require a considerable amount of energy. Periodic advertisements are needed to form the hierarchy. Also, any changes in the network conditions or sensor energy level result in reclustering, which is not always acceptable as some parameters tend to change dynamically.

Low-Energy Adaptive Clustering Hierarchy (LEACH)

The low-energy adaptive clustering hierarchy (LEACH) is actually a family of protocols [13.43] that suggests both distributed and centralized schemes; they have minimal setup time and are very energy efficient. One important feature of LEACH is that it utilizes randomized rotation of local cluster heads (CHs) to distribute the energy load evenly among the sensors in the network. LEACH also makes use of a TDMA/CDMA MAC to reduce inter-cluster and intra-cluster collisions. LEACH is a good approximation of a proactive network protocol, with some minor differences. Once the clusters are formed, the CHs broadcast a TDMA schedule giving the order in which the cluster members can transmit their data. Every node in the cluster is assigned a slot in the frame, during which it transmits data to the CH. When the last node in the schedule has transmitted its data, the schedule is repeated. The report time is equivalent to the frame time in LEACH. The frame time is not broadcast by the CH but is derived from the TDMA schedule. However, it is not under user control. Also, the attributes are predetermined and are not changed after initial installation. This network can be used to monitor machinery for fault detection and diagnosis. It can also be used to collect data about temperature (or pressure or moisture) change patterns over a particular area. But data collection is centralized and done periodically. Therefore, it is most appropriate only for constant monitoring of networks. In most cases, the user does not always need all that data (immediately). Therefore, periodic data transmissions are unnecessary. Repeated transmissions result in increased energy usage at each sensor. This approach is similar to the warehousing approach.

Threshold-Sensitive Energy Efficient Network (TEEN)

In this subsection, a reactive network protocol called TEEN (threshold sensitive energy efficient sensor network) Protocol [13.51] is discussed whose timeline is depicted in Figure 13.13. In this scheme, at every cluster change time, in addition to the attributes, the CH broadcasts the following messages to its members:

Figure 13.13
Timeline for TEEN.
Credit: A. Manjeshwar,
and D.P. Agrawal,
"TEEN: A Protocol
for Enhanced Efficiency
in Wireless Sensor
Networks," *Proceedings
of the 1st International
Workshop on Parallel
and Distributed
Computing Issues in
Wireless Networks
and Mobile Computing,*
in conjunction with
2001 IPDPS, April
23-27, 2001.

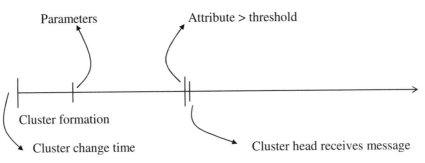

- **Hard threshold (HT)**: This is a threshold value for the sensed attribute developed for reactive networks. It is the absolute value of the attribute beyond which the node sensing this value must switch on its transmitter and report to its CH.

- **Soft threshold (ST)**: This is a small change in the value of the sensed attribute that triggers the node to switch on its transmitter and transmit.

The nodes sense their environment continuously. The first time a parameter from the attribute set reaches its hard threshold value, the node switches on its transmitter and sends the sensed data. The sensed value is also stored in an internal variable in the node, called the sensed value (SV). The nodes will next transmit data in the current cluster period, only when both the following conditions are true:

- The current value of the sensed attribute is greater than the hard threshold.

- The current value of the sensed attribute differs from SV by an amount equal to or greater than the soft threshold.

Whenever a node transmits data, SV is set equal to the current value of the sensed attribute. Thus, the hard threshold tries to reduce the number of transmissions by allowing the nodes to transmit only when the sensed attribute is in the range of interest. The soft threshold further reduces the number of transmissions by eliminating all the transmissions that might have otherwise occurred when there is little or no change in the sensed attribute once the hard threshold is reached. The main features of this scheme are as follows:

- Time-critical data reach the user almost instantaneously. Therefore, this scheme is eminently suited for time-critical data sensing applications.

- Message transmission consumes much more energy than data sensing. Therefore, even though the nodes sense continuously, the energy consumption in this scheme can be much less than in proactive networks, because data transmission is done less frequently.

- The soft threshold can be varied, depending on the criticality of the sensed attribute and the target application.

- A smaller value of the soft threshold gives a more accurate picture of the network, at the expense of increased energy consumption. Thus, the user can control the tradeoff between energy efficiency and accuracy.

- At every cluster change time, the parameters are broadcast afresh; thus, the user can change them as required.

The main drawback of this scheme is that if the thresholds are not reached, the nodes will never communicate, the user will not get any data from the network at all, and the user will never be able to know even if all the nodes have died. Thus, this scheme is not well suited for applications where the user needs to get data on a regular basis. Another possible problem with this scheme is that a practical implementation would have to ensure that there are no collisions in the cluster. TDMA scheduling of the nodes can be used to avoid this problem. This will, however, introduce a delay in reporting of time-critical data. CDMA is another possible solution to this problem. This protocol is best suited for time-critical applications such as intrusion and explosion detection.

Adaptive Periodic Threshold-Sensitive Energy-Efficient Sensor Network Protocol (APTEEN)

There are applications in which the user wants time-critical data and also wants to query the network for analysis of conditions other than collecting time-critical data. In other words, the user might need a network that reacts immediately to time-critical situations and gives an overall picture of the network at periodic intervals, so that it is able to answer analysis queries. None of the aforementioned sensor networks can do both jobs satisfactorily since they have their own limitations.

Adaptive periodic threshold-sensitive energy-efficient sensor network protocol (APTEEN) [13.52][13.53] is able to combine the best features of proactive and reactive networks while minimizing their limitations to create a new type of network called a hybrid network. In this network, the nodes not only send data periodically, they also respond to sudden changes in attribute values. This uses the same model as the TEEN protocols with the following changes. In APTEEN, once the CHs are decided, the following events take place in each cluster period. The CH first broadcasts the following parameters:

- **Attributes**: This is a set of physical parameters which the user is interested in.

- **Thresholds**: This parameter consists of a HT and a ST. HT is a value of an attribute beyond which a node can be triggered to transmit data. ST is a small change in the value of an attribute that can trigger a node to transmit.

- **Schedule**: This is a TDMA schedule similar to the one used in [13.43], assigning a slot to each node.

- **Count time (CT)**: Count time is the maximum time period between two successive reports sent by a node. It can be a multiple of the TDMA schedule length, and it introduces the proactive component in the protocol.

Figure 13.14
Timeline for APTEEN.
Credit: A. Manjeshwar,
and D.P. Agrawal,
"APTEEN: A Hybrid
Protocol for Efficient
Routing and
Comprehensive
Information Retrieval
in Wireless Sensor
Networks," *Proceedings
of the 2nd International
Workshop on Parallel
and Distributed
Computing Issues in
Wireless Networks
and Mobile
Computing*, April
2002.

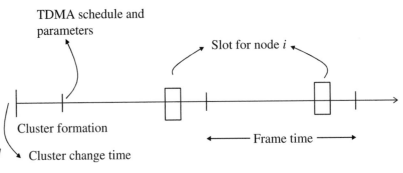

The nodes sense their environment continuously. However, only those nodes that sense a data value at or beyond the hard threshold transmit. Furthermore, once a node senses a value beyond HT, it next transmits data only when the value of that attribute changes by an amount equal to or greater than the soft threshold ST. The exception to this rule is that if a node does not send data for a time period equal to the count time, it is forced to sense and transmit the data, irrespective of the sensed value of the attribute. Since nodes near each other may fall in the same cluster and sense similar data, they may try sending their data simultaneously, leading to collisions between their messages. Hence, a TDMA schedule is used and each node in the cluster is assigned a transmission slot, as shown in Figure 13.14. In the sections to follow, data values exceeding the threshold value are referred to as critical data. The main features of this scheme are as follows:

- It combines both proactive and reactive policies. By sending periodic data, it gives the user a complete picture of the network, like a proactive scheme. It also senses data continuously and responds immediately to drastic changes, making it responsive to time-critical situations. Thus it behaves as a reactive network.

- It offers a lot of flexibility by allowing the user to set the count time interval (CT) and the threshold values for the attributes.

- Changing the count time as well as the threshold values can control energy consumption and can support both proactive and reactive behavior in a sensor network.

The main drawback of this scheme is the additional complexity required to implement the threshold functions and the count time. However, this is a reasonable tradeoff.

Table 13.3 illustrates the characteristics of hierarchical and flat topologies for the sensor networks.

Table 13.3: ►
Hierarchical Versus Flat Topologies for Sensor Networks

Hierarchical	Flat
Reservation-based scheduling	Contention-based scheduling
Collisions avoided	Collision overhead present
Reduced duty cycle due to periodic sleeping	Variable duty cycle by controlling sleep time of nodes
Data aggregation by cluster head	Node on multihop path aggregates incoming data from neighbors
Simple but less than optimal routing	Routing is complex but optimal
Requires global and local synchronization	Links formed on the fly, without synchronization
Overhead of cluster formation throughout the network	Routes formed only in regions that have data for transmission
Lower latency as multihop network formed by cluster heads is always available	Latency in waking up intermediate nodes and setting up the multihop path
Energy dissipation is uniform	Energy dissipation depends on traffic patterns
Energy dissipation cannot be controlled	Energy dissipation adapts to traffic pattern
Fair channel-allocation	Fairness not guaranteed

13.10 Summary

This chapter presents an overview of ad hoc and sensor networks that enable close-by nodes to communicate with each other. As the topology is not known and is changing dynamically due to mobility, search for a communication path from a source to an arbitrary destination is somewhat cumbersome. Such peer-to-peer routing makes routing in sensor networks challenging due to the portable nature of sensors. Efficient access to the medium is important both from a performance and a reliability viewpoint. There are many open issues, and enhancing their manageability will greatly increase the usefulness of ad hoc and sensor networks.

13.11 References

[13.1] R. Duggirala, "A Novel Route Maintenance Technique for Ad Hoc Routing Protocols," M.S. Thesis, University of Cincinnati, November 2000.

[13.2] D. P. Agrawal, M. Lu, T. C. Keener, M. Dong, and V. Kumar, "Exploiting the Use of Wireless Sensor Networks for Environment Monitoring,"*Journal of the Enviromental Management*, pp. 35–41, August 2004.

[13.3] J. Jubin and T. Truong, "Distributed Algorithm for Efficient and Interference-free Broadcasting in Radio Networks," *Proceedings of IEEE INFOCOM'87*, Vol. 3, No. 4, pp. 21–32, January 1987.

[13.4] E. Royer and C.K. Toh, "A Review of Current Routing Protocols for Ad Hoc Mobile Wireless Networks," *IEEE Personal Communications*, Vol. 7, No. 4, pp. 46–55, April 1999.

[13.5] C. Perkins and E. Royer, "Ad Hoc On-Demand Distance Vector Routing," *IEEE Workshop on Mobile Computing Systems and Applications.* Vol. 3, No. 4, pp. 90–100, February 1999.

[13.6] V. D. Park and M. S. Corson, "A Highly Adaptive Distributed Routing Algorithm for Mobile and Wireless Networks," *Proceeding of IEEE INFOCOM'97*, pp. 103–112, April 1997.

[13.7] P. Jacquet and L. Viennot, "Overhead in Mobile Ad Hoc Network Protocols," *INRIA Research Report RR-3965*, available at *http://www.inria.fr/Equipes/HIPERCOM-eng.html*, July 2000.

[13.8] C. E Perkins and P. Bhagwat, "Highly Dynamic Destination-Sequenced Distance-Vector Routing (DSDV) for Mobile Computers," *Computer Communications Review*, pp. 234–244, October 1994.

[13.9] C. C. Chiang, "Routing in Clustered Multihop, Mobile Wireless Networks with Fading Channel," *Proceedings of IEEE SICON*, pp. 197–211, April 1997.

[13.10] S. Murthy and J. J. Garcia-Luna-Aceves, "An Efficient Routing Protocol for Wireless Networks," *ACM Mobile Networks and Applications Journal*, pp. 183–197, October 1996.

[13.11] R. Castenada and S. Das, "Query Localization Techniques for On-Demand Routing Protocols in Ad Hoc Networks," *Mobile Computing and Communications Conference*, Vol. 3, No. 4, pp. 113–120, August 1999.

[13.12] D. Johnson and D. Maltz, "Dynamic Source Routing in Ad Hoc Wireless Networks," *Mobile Computing*, Chapter 5, Kluwer Academic, pp. 153–181, 1996.

[13.13] M. S. Corson and V. D. Park, "An Internet MANET Encapsulation Protocol (IMEP) Specification," *Internet-Draft*, November 1997.

[13.14] C.-K. Toh, "Associativity-Based Routing for Ad-Hoc Networks," *Wireless Personal Communications Journal, Special Issue on Mobile Networking and Computing Systems*, Vol. 4, No. 2, pp. 103–139, March 1997.

[13.15] R. Dube, "Signal Stability Based Adaptive Routing for Ad Hoc Mobile Networks," *Proceedings of IEEE Personal Communications*, pp. 36–45, February 1997.

[13.16] L. Chlamtac and A. Lerner, "Link Allocation in Mobile Radio Networks with Noisy Channel," *Proceedings of the IEEE INFOCOM*, pp. 1243–1257, 1986.

[13.17] Z. Haas and M. Pearlman, "The Performance of Query Control Schemed for the Zone Routing Protocol," *Proceedings of ACM SIGCOMM '98*, pp. 360–368, June 1998

[13.18] A. Iwata, C. C. Chiang, G. Pei, M. Gerla, and T. W. Chen, "Scalable Routing Strategies for Ad Hoc Wireless Networks," *IEEE Journal on Selected Areas of Communications*, pp. 1369–1379, August 1999.

[13.19] G. Pei, M. Gerla, and X. Hong, "LANMAR: Landmark Routing for Large Scale Wireless Ad Hoc Networks with Group Mobility," *ACM MobiHoc*, Boston, MA, August 2000.

[13.20] P.F. Tsuchiya, "The Landmark Hierarchy: A New Hierarchy for Routing in Very Large Networks," *Computer Communication Review*, Vol. 18, No. 4, pp. 35–42, August 1988.

[13.21] Y. B. Ko and N. H. Vaidya, "Location-Aided Routing (LAR) in Mobile Ad Hoc Networks," *Proceedings of MOBICOM '98*, pp. 66–75, 1998.

[13.22] S. Basagni, I. Chlamtac, V. R. Syrotiuk, and B. A. Woodward, "A Distance Routing Effect Algorithm for Mobility (DREAM)," *ACM/IEEE International Conference on Mobile Computing and Networking*, pp. 76–84, October 1998.

[13.23] G. Aggelou and R. Tafazolli, "RDMAR: A Bandwidth-Efficient Routing Protocol for Mobile Ad Hoc Networks," *ACM International Workshop on Wireless Mobile Multimedia(WoWMoM)*, August 1999.

[13.24] S. Singh, M. Woo, and C. S. Raghavendra, "Power-Aware Routing in Mobile Ad Hoc Networks," *Proceedings, ACM/IEEE Mobicom*, pp. 181–190, 1998.

[13.25] K. T. Jin and D. H. Cho, "A MAC Algorithm for Energy-Limited Ad Hoc Networks," *Proceedings of IEEE VTC 2000, Fall*, pp. 219–222, Sept. 24–28, 2000.

[13.26] S. H. Lee and D. H. Cho, "A New Adaptive Routing Scheme Based on the Traffic Characteristics in Mobile Ad Hoc Networks," *Proceedings of Fall VTC 2000*, pp. 2911–2914, September 2000.

[13.27] E. Ayanoglu, I. Chih-Lin, R. D. Gitlin, and J. E. Mazo, "Diversity Coding for Transparent Self-Healing and Fault-Tolerant Communication Networks," *IEEE Transactions on Communications*, Vol. 41, pp. 1677–1686, November 1993.

[13.28] A. Nasipuri and S. R. Das, "On-Demand Multipath Routing for Mobile Ad Hoc Networks," *Proceedings of Eight International Conference on Computer Communications and Networks*, pp. 64–70, Boston, October 1999.

[13.29] S. J. Lee and M. Gerla, "AODV-BR: Backup Routing in Ad Hoc Networks," *Proceeding of the IEEE, Wireless Communications and Networking Conference 2000 (WCNC 2000)*, Vol. 3, pp. 1311–1316, Chicago, September 2000.

[13.30] S. J. Lee and M. Gerla, "Split Multipath Routing with Maximally Disjoint Paths in Ad Hoc Networks," *Proceeding of the IEEE International Conference on Communications 2001 (ICC 2001)*, Vol. 10, pp. 3201–3205, Helsinki, June 2001.

[13.31] A. Valera, W. K. G. Seah, and S. V. Rao, "Cooperative Packet Caching and Shortest Multipath Routing in Mobile Ad Hoc Networks," *Proceedings of the IEEE INFOCOM 2003*, pp. 260–269, March 2003.

[13.32] Z. Yao, Z. Ma, and Z. Cao, "A Multipath Routing Scheme Combating with Frequent Topology Changes in Wireless Ad Hoc Networks," *Proceedings of International Conference on Communication Technology 2003 (ICCT 2003)*, pp. 1250–1253, April, 2003.

[13.33] SCADDS Project, *http://www.isi.edu/scadds/*.

[13.34] D. Hall, *Mathematical techniques in multisensor data fusion*, Artech House, 1992.

[13.35] A. Cerpa and D. Estrin, "Adaptive Self-Configuring Sensor Networks Topologies," *UCLA CS Department Tech. Report UCLA/CSD-TR-01-0009*, May 2001.

[13.36] DARPA SensIT Program, *http://www.darpa.mil/ito/research/sensit*.

[13.37] K. Sohrabi, J. Gao, V. Ailawadhi, and G. J. Pottie, "Protocols for Self-Organization of a Wireless Sensor Network," *IEEE Personal Communications*, pp. 16–27, October 2000.

[13.38] A. Tanenbaum, *Computer Networks*, Prentice Hall PTR, Upper Saddle River, NJ, 1996.

[13.39] V. Bhargavan, A. Demers, S. Shenker, and L. Zhang, "MACAW: A Media Access Protocol for Wireless LANs," *Proceedings of 1994 SIGCOMM Conference*, pp. 215–225, 1994.

[13.40] B. P. Crow, I. Wadjaja, J. G. Kim, and P. T. Sakai, "IEEE 802.11 Wireless Local Area Networks," *IEEE Communications Magazine*, pp. 116–126, September 1997.

[13.41] W. B. Heinzelman, "Application-Specific Protocol Architectures for Wireless Networks," Ph.D. Thesis, Massachusetts Institute of Technology, June 2000.

[13.42] C. Intanagonwiwat, R. Govindan, and D. Estrin, "Directed Diffusion: A Scalable and Robust Communication Paradigm for Sensor Networks," *Proceedings of the 6th annual ACM/IEEE Conference on Mobile Computing and Networking (MOBICOM)*, pp. 56–67, August 2000.

[13.43] W. Heinzelman, A. Chandrakasan, and H. Balakrishnan, "Energy-Efficient Communication Protocols for Wireless Microsensor Networks," *Proceedings of Hawaiian International Conference on Systems Science*, January 2000.

[13.44] R. Brooks and S. Iyengar, *Multi-Sensor Fusion*, Prentice Hall, Upper Saddle River, NJ, 1998.

[13.45] P. Varshney, *Distributed detection and data fusion*, Springer-Verlag, 1997.

[13.46] W. Heinzelman, J. Kulik, and H. Balakrishnan, "Adaptive Protocols for Information Dissemination in Wireless Sensor Networks," *Proceedings of 5th ACM/IEEE Mobicom Conference (MobiCom'99)*, August 1999.

[13.47] P. Bonnet, J. Gehrke, and P. Seshadri. "Querying the Physical World," *IEEE Personal Communications, Special Issue on Smart Spaces and Environments*, October 2000.

[13.48] P. Bonnet, J. Gehrke, and P. Seshadri, "Towards Sensor Database Systems," *2nd International Conference on Mobile Data Management*, January 2001.

[13.49] M. Jiang, J. Li, and Y. Tay, "Cluster Based Routing Protocol (CBRP) Functional Specification," *Internet Draft*, 1998.

[13.50] D. Estrin, et al., "Next Century Challenges: Scalable Coordination in Sensor Networks," *ACM Mobicom*, 1999.

[13.51] A. Manjeshwar and D. P. Agrawal, "TEEN: A Protocol for Enhanced Efficiency in Wireless Sensor Networks," *Proceedings of the 1st International Workshop on Parallel and Distributed Computing Issues in Wireless Networks and Mobile Computing, in conjunction with 2001 IPDPS*, April 23–27, 2001.

[13.52] A. Manjeshwar and D. P. Agrawal, "APTEEN: A Hybrid Protocol for Efficient Routing and Comprehensive Information Retrieval in Wireless Sensor Networks," *Proceedings of the 2nd International Workshop on Parallel and Distributed Computing Issues in Wireless Networks and Mobile Computing*, April 2002.

[13.53] A. Manjeshwar, Q-A. Zeng, and D. P. Agrawal, "An Analytical Model for Information Retrieval in Wireless Sensor Networks using Enhanced APTEEN Protocol," *IEEE Transactions on Parallel and Distributed Systems*, Vol. 13, No. 12, pp. 1290–1302, December 2002.

● ● ● ● ● ● ● ● ● ● ● ● ● ● ●

13.12 Problems

P13.1. What are the differences between cellular and ad hoc networks?

P13.2. Why is it not possible to use circuit switching in ad hoc networks?

P13.3. A given ad hoc network consists of 100 nodes, and the mobility of the nodes is such that every one second, two existing radio connections are broken, while two new radio links are established. Assuming each node is connected to exactly four adjacent nodes, find the total number of communications links in the network.

P13.4. In Problem P13.3, if the updated message is sent every 5 seconds, what is the upper limit on the number of messages initiated periodically if a table-driven routing protocol is to be used? Explain clearly.

P13.5. In Problem P13.4, if the destination node is located at 5 hops apart from a given source node, what is the maximum possible value of:
(a) Number of alternate paths of length of 5 hops?
(b) Alternate disjoint paths of length 5 hops?

P13.6. Repeat Problem P13.5, if the distance is changed to 8 hops.

P13.7. A snapshot of an ad hoc network is shown in the figure.

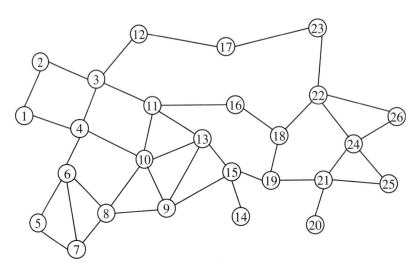

Figure 13.15
Figure for
Problem P13.7.

Describe briefly the process taken to do the following:

(a) You need to create a route from the source node 6 to the destination node 23 using the DSR algorithm.

(b) Repeat part (a) using TORA routing.

(c) What changes would you make in part (a) if you use the AODV protocol?

P13.8. How does signal stability affect the route in Problem P13.7?

P13.9. Assuming that the location of the destination node 23 is known to be located in the northeast direction, what changes do you need to make in determining a route in Problem P13.7? Explain clearly.

P13.10. In ad hoc networks, it is sometimes desirable to transmit packets of a single message using multiple paths.

(a) Can you think of any specific reasons for this?

(b) If you need to employ two alternate paths, how would you determine that in Problem P13.7 using DSR?

P13.11. Consider an ad hoc network where communication (message or packet transfer) is to take place from node X to node Y. The route has already been established and a data packet is to be transferred over n hops. To transfer the packet, the kth node uses the following medium-access protocol:

■ It waits for time $t(k)$ after which the channel becomes free. $t(k) = k\alpha$ time units.

■ It transfers the data packet to the next hop. This takes α time units.

■ It receives an acknowledgment. This takes another $\alpha/2$ time units.

The time $t(k)$ before the kth node actually transmits the data packet is given by $t(k) = k\alpha$ time units.

(a) Find an expression for time taken for the data to cover n hops (i.e., from node 1 to node $n + 1$).

(b) If the time taken to travers n hops is $T = 2n\sqrt{n\alpha}$, what is the value of n?

P13.12. Illustrate how multipath routing can be done between nodes 1 and 26 of Problem P13.7. Which multipath routing technique is beneficial and why? Explan clearly.

P13.13. What is meant by piggybacking and what are the advantages? Explain clearly.

P13.14. What are the implications of using CDMA in an ad hoc network? Explain in detail.

P13.15. What are the advantages and disadvantages of reactive and proactive protocols? Which one would you prefer and why? Explain with specific conditions.

P13.16. What are the similarities and differences between ad hoc networks and sensor networks? Explain clearly.

P13.17. In a sensor network, the energy consumed by different functions by a sensor is as follows:

Mode	Energy Consumed (in nJ/bit)
Sleeping mode	0
Sensing or idle mode	0.5
Aggregation	5
Communication to cluster head	100
Cluster head to BS	1000

Assume the total number of nodes as P, the number of non-cluster nodes as n, the number of cluster heads to be m, and the frame size to be B bits.

(a) Find the power consumption, during a frame time period if sensing and communication is done during every frame, assuming the other half of the nodes are sleeping at that time.

(b) Find the power consumption in the idle frame when sensing and communication to CH is done in every alternate frame. Remember that power is consumed even in sleeping mode of the cycles, when sensing is not carried out.

(c) Find the total power consumption in different frames if sensing is done every alternate cycle, while transmission to CH is done every fourth frame.

(d) Repeat part (b) if there are 10 clusters, with each cluster consisting of 8 sensor nodes and aggregation done by CH every 8 frames while CH to base station communication takes place every 16 frames.

P13.18. Assume that CDMA/TDMA is used for each cluster of Problem P13.17(d). Can you come up with a time-slot schedule for each cluster of the sensor network when the TEEN protocol is to be used? Assume that two levels of clustering is present. Remember that CHs need to communicate with the base station as well, using a different CDMA code.

P13.19. What changes you need to make in Problem P13.18 if the APTEEN protocol is to be used?

P13.20. A wireless sensor has a transmitter/receiver range of 2 m, and many such sensors need to be installed in a nuclear plant building of size 50 m × 50 m with the height of 25 m. Can you think of an efficient arrangement of the sensor arrays? Explain clearly.

P13.21. What will be the impact in Problem P13.20 if the sensor range could be increased to 10 m?

P13.22. Repeat Problem P13.20, if it is to be done for a 56 m long airplane, whose cross-section is represented as shown in the figure below.

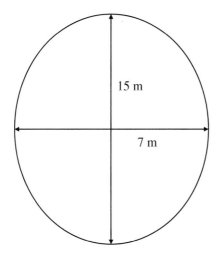

15 m

7 m

Figure 13.16
Figure for
Problem P13.22.

P13.23. Repeat Problem P13.19, if it is to be done for a lake of size 250 m length × 50 m width × 5 m deep, if biosensors are to be installed to monitor pollutant level, and if the range of each sensor is 0.5 m.

P13.24. Why do you use a "data-centric" approach in a sensor network?

P13.25. What are the advantages and limitations of a "directed-diffusion" approach in a sensor network? Explain clearly.

P13.26. A clustering approach has been suggested to locally collect and "aggregate" information in a sensor network. What kind of aggregation is desirable?

P13.27. Can the past response location of a query be helpful in limiting the flooding area? Explain clearly.

P13.28. From your favorite Web site, find what is meant by "gossiping-based routing." What are the advantages and limitations of such an approach? Expain clearly.

P13.29. In a sensor network, energy consumption is one of the major constraints. Keeping this in mind, what factors would one consider when designing a security scheme for such networks? Explain.

P13.30. How can you provide security in an ad hoc network? What are some possible schemes and their relative advantages?

CHAPTER 14

Wireless MANs, LANs, and PANs

● ● ● ● ● ● ● ● ● ● ● ● ● ● ●

14.1 Introduction

During the past 25 years, several different wireless technologies have found success in bringing innovative and versatile services to the market. This revolution has been made possible by the development of new networking technologies and paradigms, such as wireless metropolitan area networks (WMANs), wireless local area networks (WLANs), and wireless personal area networks (WPANs). The incredible penetration of the IEEE 802.11b WLAN standard, popularly known as Wi-Fi (wireless-fidelity), has shown the economic feasibility of such solutions. Wi-Fi hotspots have sprung up at varied places such as Starbucks cafes, McDonald's, malls, beaches, hotels, community halls, and convention centers. People have started talking about deployment and uses of WiMAX and WMAN technologies covering a larger area.

WMAN, WLAN, and WPAN—all aim to provide wireless data connectivity, but with different characteristics and expectations and therefore different market segments. A WMAN is meant to cover an entire metropolitan area, a WLAN provides similar services but covers a much smaller area (e.g., a building, an office campus, lounges). A WPAN is an extremely short–range network, formed around the personal operating space of a user. Typically, WPANs are used to replace cables between a computer and its peripheral devices, but very often they can be used for transmitting images, digitized music, and other data.

Of the three types of networks, it is the WLAN that has garnered a lot of attention, primarily because of the unprecedented popularity and commercial success of the IEEE 802.11b, stemming from its cost effectiveness and ease of deployment. Other standards include HiperLAN2 (from ETSI) [14.1] and the newer IEEE 802.11a and IEEE 802.11g [14.2][14.3]. Bluetooth [14.4] (which is also the IEEE 802.15.1) is the most visible face of the WPAN, but the IEEE 802.15.3 and IEEE 802.15.4 are also being developed. The WMAN has been slow to catch on commercially. The only deployment is the Ricochet [14.5] based on a proprietary solution. The IEEE has recently begun standardization work in the form of the IEEE 802.16 Working Group.

Figure 14.1 clearly shows the operating space of the various IEEE 802 wireless standards and activities that are still in progress.

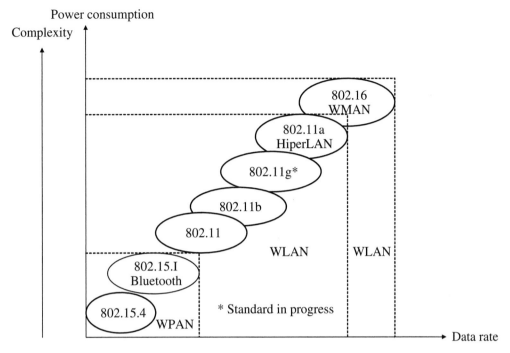

Figure 14.1   The scope of various WMAN, WLAN, and WPAN standards.

WLAN is becoming increasingly important for people within work environments like a warehouse, or for students and faculty members moving around the campus. It is extensively used for data transfer, while voice communication has yet to be accepted.

● ● ● ● ● ● ● ● ● ● ● ● ● ● ● ● ●

14.2 Wireless Metropolitan Area Networks (WMANs)

14.2.1   IEEE 802.16

The IEEE 802.16 standard has been designed to evolve as a set of air interfaces based on a common MAC protocol but with physical layer specifications approved in 2001; it addresses frequencies from 10 to 66 GHz. A new project, currently in the balloting stage, expects to complete an amendment denoted IEEE 802.16a [14.6], also known as WiMAX. This document will extend the air interface support to lower frequencies in the 2–11 GHz band, including both licensed and license-exempt spectra. Compared to the higher frequencies, such spectra offer a less expensive opportunity to reach many more customers, although at generally lower data rates.

MAC Layer

The IEEE 802.16 MAC protocol supports point-to-multipoint broadband wireless access. It allows very high bit rates, in both the forward and reverse links, at the same time allowing hundreds of terminals per channel that may potentially be shared by multiple end-users. The versatile services required by these MSs include legacy TDM voice and data, IP connectivity, and packetized voice over IP (VoIP). The IEEE 802.16 MAC must therefore be able to accommodate both continuous and bursty traffic. Additionally, these services are expected to be assigned QoS in keeping with the traffic types. The IEEE 802.16 MAC provides a wide range of service types analogous to the classic ATM service categories as well as newer categories such as guaranteed frame rate (GFR) [14.7].

The IEEE 802.16 MAC protocol must also support a variety of backhaul requirements, including both ATM and packet-based protocols. Convergence sublayers are used to map the transport-layer–specific traffic to a MAC and offers features such as payload header suppression, packing, and fragmentation; the convergence sublayers and MAC work together in a form that is often more efficient than the original transport mechanism.

Issues of transport efficiency are also addressed at the interface between the MAC and the PHY layer. For example, modulation and coding schemes are specified in a burst profile that may be adjusted to each subscriber station adaptively for each burst. The MAC can make use of bandwidth-efficient burst profiles under favorable link conditions but shift to more reliable, though less efficient, alternatives as are required to support the planned 99.999 percent link availability.

The request-grant mechanism is designed to be scalable, efficient, and self-correcting. The IEEE 802.16 access system does not loose efficiency when presented with multiple connections per terminal, multiple QoS levels per terminal, and a large number of statistically multiplexed users. It takes advantage of a wide variety of request mechanisms, balancing the stability of contentionless access with the efficiency of contention-oriented access.

Along with the fundamental task of allocating bandwidth and transporting data, the MAC includes a privacy sublayer; this provides authentication of network access, thereby avoiding theft of service and providing key exchange and encryption for data privacy. To accommodate more demanding physical environments and different service requirements of the frequencies between 2 and 11 GHz, the IEEE 802.16a project is providing a MAC to support automatic repeat request (ARQ) for mesh network architectures.

MAC Layer Details

The MAC includes service-specific convergence sublayers that interface to higher layers to carry out the key MAC functions. The privacy sublayer is located below the common part sublayer.

– Service Specific Convergence Sublayers

The IEEE 802.16 defines two general service-specific convergence sublayers for mapping services to and from the IEEE 802.16 MAC connections. The ATM convergence sublayer is defined for ATM services, and the packet convergence sublayer is defined for mapping packet services such as IPv4, IPv6, Ethernet, and virtual local area network (VLAN). The primary task of the sublayer is to classify service data units (SDUs) to the proper MAC connection, preserve QoS, and enable bandwidth allocation. The mapping takes various forms depending on the type of service. In addition to these basic functions, the convergence sublayers can also perform more sophisticated functions such as payload header suppression and reconstruction to enhance airlink efficiency.

– Common Part Sublayer

Introduction and General Architecture—The IEEE 802.16 MAC is designed to support a point-to-multipoint architecture with a central BS handling multiple independent sectors simultaneously. On the downlink (DL) (forward channel), data to the subscriber stations (SSs—essentially the MSs) are multiplexed in TDM fashion. The uplink (UL) (reverse channel) is shared between SSs in TDMA fashion.

The IEEE 802.16 MAC is connection oriented. All services, including inherently connectionless services, are mapped to a connection. This provides a mechanism for requesting bandwidth, associating QoS and traffic parameters, transporting and routing data to the appropriate convergence sublayer, and all other actions associated with the contractual terms of the service. Connections are referenced with 16-bit connection identifiers (CIDs) and may require continuous availability of bandwidth or bandwidth on demand.

Each SS has a standard 48-bit MAC address, but this serves mainly as an equipment identifier, since the primary addresses used during operation are the CIDs. Upon entering the network, the SS is assigned three management connections in each direction. These three connections reflect the three different QoS requirements used by different management levels. The first of these is the basic connection, which is used for the transfer of short, time-critical MAC and radio link control (RLC) messages. The primary management connection is used to transfer longer, more delay-tolerant messages such as those used for authentication and connection setup. The secondary management connection is used for the transfer of standard-based management messages such as dynamic host configuration protocol (DHCP), trivial file transfer protocol (TFTP), and simple network management protocol (SNMP).

The MAC reserves several connections for other purposes. One connection is reserved for contention-based initial access. Another is reserved for broadcast transmissions in the forward channel as well as for signaling broadcast contention-based polling of SS bandwidth needs. Additional connections are reserved for multicast, rather than broadcast, contention-based polling. SSs may be instructed to join multicast polling groups associated with these multicast polling connections.

MAC PDU Formats—The MAC PDU (protocol data unit) is the data unit exchanged between the MAC layers of the BS and its SSs. A MAC PDU consists of a fixed-length MAC header, a variable-length payload, and an optional cyclic redundancy check (CRC). Two header formats, distinguished by the HT field, are defined: the generic header (see Figure 14.2) and the bandwidth request header. Except for bandwidth containing no payload, MAC PDUs have either MAC management messages or convergence sublayer data.

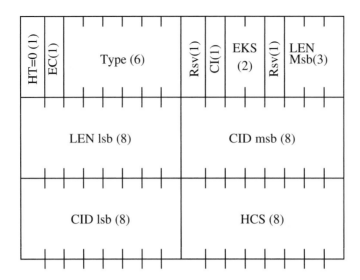

Figure 14.2
Generic header for
MAC PDU.

Three types of MAC subheader may be present. A grant management subheader is used by the SS to convey bandwidth management needs to its BS. A fragmentation subheader indicates the presence and orientation within the payload of any fragments of the SDUs. The packing subheader is used to indicate packing of multiple SDUs into a single PDU. The generic header follows a grant management, and fragmentation subheaders may be inserted into MAC PDUs. The packing subheader may be inserted before each MAC SDU if shown by the type field.

Transmission of MAC PDUs—The IEEE 802.16 MAC supports various higher-layer protocols such as ATM or IP. Incoming MAC SDUs from corresponding convergence sublayers are formatted according to the MAC PDU format, possibly with fragmentation and/or packing, before being conveyed over one or more connections in accordance with the MAC protocol. After traversing the airlink, MAC PDUs are reconstructed back into the original MAC SDUs so that the format modifications performed by the MAC layer protocol are transparent to the receiving entity.

The IEEE 802.16 takes advantage of packing and fragmentation processes, and their effectiveness, flexibility, and efficiency are maximized by appropriate bandwidth allocation. Fragmentation is the process in which a MAC SDU is divided into one or more MAC SDU fragments. Packing is the process in which multiple MAC SDUs are packed into a single MAC PDU payload. Both processes may be initiated by either a BS for a DL or for a SS for an UL connection. The IEEE 802.16 allows simultaneous fragmentation and packing for efficient use of the bandwidth.

PHY Support and Frame Structure—The IEEE 802.16 MAC supports both TDD and FDD. In FDD, both continuous and burst DLs are possible. Continuous DLs allow for certain robustness enhancement techniques, such as interleaving. Burst DLs (either FDD or TDD) allow the use of more advanced robustness and capacity enhancement techniques, such as subscriber-level adaptive burst profiling and advanced antenna systems.

The MAC builds the DL subframe starting with a frame control section containing the DL-MAP (downlink MAP) and UL-MAP (uplink map) messages. These indicate PHY transitions on the DL as well as bandwidth allocations and burst profiles on the UP. The DL-MAP is always applicable to the current frame and is always at least two FEC blocks long. To allow adequate processing time, the first PHY transition is expressed in the first FEC block. In both TDD and FDD systems, the UL-MAP provides allocations starting no later than the next DL frame. The UL-MAP can, however, start allocating in the current frame, as long as processing times and round-trip delays are observed.

Radio Link Control—The advanced technology of the IEEE 802.16 PHY requires equally advanced RLC, particularly a capability of the PHY to change from one burst profile to another. The RLC must control this capability as well as the traditional RLC functions of power control and ranging. RLC begins with periodic BS broadcast of the burst profiles that have been chosen for the UL and DL. Among the several burst profiles used on a channel, one particular is chosen based on a number of factors, such as rain region and equipment capabilities. Burst profiles for the DL are each tagged with a DL interval usage code (DIUC) and those for the UL are tagged with an UL interval usage code (UIUC).

During initial access, the SS performs initial power leveling and ranging using ranging request (RNG-REQ) messages transmitted in initial maintenance windows. The adjustments to the SS's transmit time advance, as well as power adjustments, and are returned to the SS in ranging response (RNG-RSP) messages. For ongoing ranging and power adjustments, the BS may transmit unsolicited RNG-RSP messages instructing the SS to adjust its power or timing. During initial ranging, the SS can also request service in the DL via a particular burst profile by transmitting its choice of DIUC to the BS. The selection is based on received DL signal-quality measurements performed by the SS before and during initial ranging. The BS may confirm or reject the choice in the ranging response. Similarly, the BS monitors the quality of the UL signal it receives from the SS. The BS commands the SS to use a particular UL burst profile simply by including the appropriate burst profile UIUC with the SS's grants in UL-MAP messages.

After initial determination of UL and DL burst profiles between the BS and a particular SS, RLC continues to monitor and control the burst profiles. Harsher environmental conditions, such as rain fades, can force the SS to request a more robust burst profile. Alternatively, exceptionally good weather may allow an SS to temporarily operate with a more efficient burst profile. The RLC continues to adapt the SS's current UL and DL burst profiles, ever striving to achieve a balance between robustness and efficiency. Because the BS is in control and directly monitors the UL signal quality, the protocol for changing the UL burst profile for an SS is simple: the BS merely specifies the profile's associated UIUC whenever granting the SS bandwidth in a frame. This eliminates the need for an acknowledgment, since the SS will always receive either both the UIUC and the grant or neither. Hence, there exists no chance of UL burst profile mismatch between the BS and the SS.

In the DL, the SS is the entity that monitors the quality of the receive signal and therefore knows when its DL burst profile should change. The BS, however, is the entity in control of the change. There are two methods available to the SS to request a change in DL burst profile, depending on whether the SS operates in the grant per connection (GPC) or grant per SS (GPSS) mode. The first method would typically apply (based on the discretion of the BS scheduling algorithm) only to GPC SSs. In this case, the BS may periodically allocate a station maintenance interval to the SS. The SS can use the RNG-REQ message to request a change in DL burst profile. The preferred method is for the SS to transmit a DL burst profile change request (DBPC-REQ). In this case, which is always an option for GPSS SSs and can be an option for GPC SSs, the BS responds with a DL burst profile change response (DBPC-RSP) message confirming or denying the change.

Because messages may be lost due to irrecoverable bit errors, the protocols for changing SS's DL burst profile must be carefully structured. The order of the burst profile change actions is different when transitioning to a more robust burst profile than when transitioning to a less robust one. The standard takes advantage of the fact that any SS is always required to listen to more robust portions of the DL as well as the profile that has been negotiated.

Channel Acquisition—The MAC protocol includes an initialization procedure designed to eliminate the need for manual configuration. Upon installation, SS begins scanning its frequency list to find an operating channel. It may be programmed to register with one specific BS, referring to a programmable BS ID broadcast by each. This feature is useful in dense deployments where the SS might hear a secondary BS due to selective fading or when the SS picks up a side-lobe of a nearby BS antenna.

After deciding on which channel or channel pair to start communicating, the SS tries to synchronize to the DL transmission by detecting the periodic frame preambles. Once the physical layer is synchronized, the SS looks for periodic DCD (DL channel descriptor) and UCD (UL channel descriptor) broadcast messages that enable the SS to learn the modulation and FEC schemes used on the carrier.

IP Connectivity—After registration, the SS acquires an IP address via the DHCP and establishes the time of day via the Internet time protocol. The DHCP server also provides the address of the TFTP server from which the SS can request a configuration file. This file provides a standard interface for providing vendor-specific configuration information.

Physical Layer [14.7]

10–66 GHz: For the deployment of single-carrier modulation in the air interface "WirelessMAN-SC" (WMAN–SC), a precondition is that line-of-sight (LOS) conditions should exist. This is provided in the design of the PHY specification for 10–66 GHz. The point-to-point communication is enabled through a TDM scheme whereby a BS transmits the signal sequentially to each MS in its allocated slot. Access in the UL direction is by TDMA. The burst design selected allows coexistence of both TDD and FDD forms of communication. In the TDD scheme, both the UL and DL are possible over the same channel but not at the same time. In FDD, the uplink and downlink occur over separate channels and could occur together. At the cost of increasing hardware complexity, half-duplex FDD support was added, and this resulted in making the technology cheaper by a small margin. In order that modulation and coding can be programmed dynamically, both TDD and FDD alternatives support adaptive burst profiles.

2–11 GHz: The standards for both licensed and license-exempt in the 2–11 GHz bands are being formulated, and the final draft has no yet been completed [14.7]. IEEE project 802.16a addresses these issues, and three air interfaces are defined

Table 14.1: ▶
Three 2–11 GHz Air Interface of the IEEE 802.16a Draft 3 Specifications

Air Interface	Specification
WMAN–SC2	A single-carrier modulation is used.
WMAN–OFDM	License-exempt bands necessarily use this TDMA access interface. OFDM is present with a 256-point transform.
WMAN–OFDMA	Each receiver is assigned a set of multiple carriers to enable multiple access. OFDM is present with a 2048-point transform.

in Table 14.1. One of these has to be implemented by each system compliant with 802.16a. All the three interfaces can provide interoperability. It is envisaged that outdoor application, especially in urban areas could involve nonlight-of-sight (NLOS) links between a BS and the user. Owing to the expected multipath propagation, the design of the 2–11 GHz physical layer is driven by the need for NLOS. The hardware expense and installation costs involved in outdoor-mounted antennas are other factors that need further consideration.

It is important to note that the IEEE 802.16a amendment has not yet been completed and hence could exhibit significant changes. The propagation requirements necessitate the use of advanced antenna systems. Notwithstanding the reasonably stable draft that has been achieved, modes could be added or deleted and hence the specifications could be changed through the ballot.

Physical Layer Details

In the PHY specification, burst single-carrier modulation with adaptive burst profiling is used for the 10–66 GHz frequency band. The channel bandwidths are 20, 25 MHz (typical U.S. allocation) or 28 MHz (typical European allocation). The systems use Nyquist square-root raised cosine pulse shaping with a roll-off factor of 0.25. By using this adaptive burst profiling, each SS may adjust the transmission parameters like the modulation and coding schemes, individually frame-by-frame. Both the TDD variant and the burst FDD variant are defined in this specification.

The data bits are randomized to minimize the possibility of transmission of an unmodulated carrier and to ensure adequate numbers of bit transitions to support clock recovery. The data is also FEC coded using Reed-Solomon GF (256) which allow variable block size and has appropriate error correction capabilities. An inner block convolutional code is used to robustly transmit critical data like frame control and initial accesses. The FEC coded data is mapped to a QPSK, 16-state QAM (16-Quadrature Amplitude Modulation) or 64-state QAM (64-Quadrature Amplitude Modulation) to form burst profiles with varying robustness and efficiency. The block may be shortened if the last FEC block is not filled.

The frame size can be 0.5, 1, or 2 ms. There are UL subframes and DL subframes in each frame. A frame is divided into physical slots, and the physical slot is the unit for bandwidth allocation and identification of PHY transitions. A physical slot has 4QAM symbols. For the TDD variant and the FDD variant, different framings are defined. In the TDD variant, a frame starts with a DL subframe followed by a UL subframe. In the FDD variant, UL and DL are using different frequencies. The BS controls the UL and DL in the UL-MAP and DL-MAP. In the DL-MAP, the first part is a frame control section which contains control information for all SSs. Following the frame control section is the TDM portion. A negotiated burst profile is used to provide synchronization with the DL. For the FDD variant, a TDMA segment is used to transmit data to half-duplex SSs. This permits some SSs to transmit data earlier than they were scheduled. The synchronization with the DL may get lost because of the half-duplex nature. However, the TDMA preamble provides a way to get synchronization back. Because the bandwidth requirements may vary from time to time, the mixture and duration of burst profiles and the presence or absence of the TDMA portion may vary from frame to frame. The recipient SS is included in the MAC headers not in the DL-MAP; therefore, all of the DL subframes are listened to by all SSs for the potential reception. For full-duplex SSs, this means they receive all burst profiles of equal or greater robustness than they would have by negotiating with the BS. Unlike the DL, specific SSs are granted bandwidth by UL-MAP. Now the SSs start transmitting, using the burst profile specified by the UL interval usage code (UIUC) in UL-MAP entry, in their assigned allocations, thus granting them bandwidth. Contention-based allocations are also provided in the UL subframe for initial system access and broadcast or multicast bandwidth requests. Properly sized access opportunities for initial system access are allowed extra guard time for these SSs, which are not yet resolved with the transmit time advances necessary to offset the round-trip delay to the BS.

The transmission convergence (TC) sublayer resides in between the PHY and MAC layers. This layer delivers the transformation of variable length MAC PDUs into the fixed length FEC blocks (with possibly a shortened block at the end) of each burst. A sized PDU contained in the TC layer fits in the FEC block currently being filled. As shown in Figure 14.3, the pointer indicates to the next MAC PDU header that starts within the FEC block. The TC PDU format allows resynchronization to the next MAC PDU in the event that the previous FEC block had irrecoverable errors. In the absence of the TC layer, a receiving SS or BS would potentially lose the entire remainder of a burst with the occurrence of an irrecoverable bit error.

WMAN has been envisioned to be a data network that covers an entire city. Network access is provided by WMAN to buildings through exterior antennas, communicating with central radio BSs (base stations). It further offers an option to cabled access networks, such as fiber optic links, coaxial systems using cable modems, and DSL links. The nomadic access is explicitly handled by its fundamental design. The Ricochet [14.5] network can be thought of as the only pure WMAN commercial

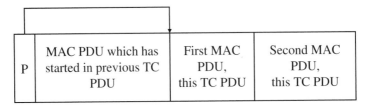

Figure 14.3
TC PDU format.

service. The air interfaces for WMANs, WirlessMAN IEEE 802.16, were published on April 8, 2002 [14.6]. In the following two sections we will look at these two technologies in more detail.

14.2.2   Ricochet

Ricochet provides secure mobile access outside the office and is more of a WMAN service than a WLAN, as the typical coverage is of an entire city. The Ricochet service was introduced by Metricom, a commercial Internet service provider (ISP), and was available primarily at airports and some selected areas. On July 2, 2001, Metricom filed for bankruptcy, and the Ricochet service was turned off. By September 2002, it was back again in Denver, this time promoted by Aerie Networks Inc. and San Diego was recently added to the Richochet map.

   The Ricochet access service allows a link to the Internet without phone lines. It is a wide area wireless system using spread spectrum, packet switching data technology, and Metricom's patented frequency hopping, checker architecture. The network operates in the license-free 902–928 MHz ISM band. The Ricochet wireless micro-cellular data network (MCDN), shown in Figure 14.4, consists of shoebox-sized radio

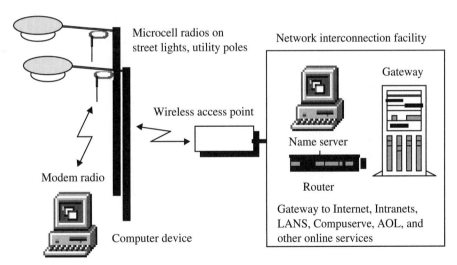

Figure 14.4
Ricochet wireless microcellular data network.

transceivers, also called microcell radios, which are typically mounted on streetlights or utility poles. The microcells require only a small amount of power from the streetlight itself (connected with a special adapter). They are strategically placed every quarter to half mile in a mesh pattern. Each microcell radio employs 162 frequency hopping channels and uses a randomly selected hopping sequence. Installation of each microcell radio takes less than five minutes. The Ricochet network has a main system called name server, which provides service validation and path information.

The original Ricochet modem weighs 13 ounces, has the general dimensions of a small paperback book but is less than half an inch thick, and plugs directly into a desktop, laptop, or PDA standard serial port within a coverage area. The company's recent modem is extremely small and is in the form of a personal computer memory card international association (PCMCIA) card. RF to phone line connections required for Ricochet's TMA sevice are made using specially designed Ricochet modems. When a Ricochet modem is configured to operate in bridge mode, it translates signals from other Ricochet modems into signals that can be received by a wired modem.

The Ricochet wireless network is based on frequency-hopping, spread-spectrum packet radio technology, with transmissions randomly hopping every two-fifths of a second over 162 channels. The RF signals are passed from radios onto the Ricochet wired backbone network. Radios are configured to send their incoming packets through specific wired access points, thereby reducing the number of "hops" an RF packet might take to reach the Metricom backbone network. A comparison between WMAN standards as given in Table 14.2.

Table 14.2: ▶
Comparison of WMAN Standards [14.8]
Credit: R.L. Ashok, and D.P. Agrawal, "Next Generation Wearable Networks," *IEEE Computer*, November 2003, Vol. 36, No. 11, pp. 31-39.

Technology	Wireless MAN	
	IEEE 802.16	Ricochet
Operational spectrum	10–66 GHz, LOS required, 20/25/28 MHz channels	900 MHz
Physical layer	TDMA-based uplink, QPSK, 16-QAM, 64-QAM	FHSS
Channel access	TDD and FDD variants	CSMA
Nominal data rate possible	120/134.4 Mbps for 25/28 MHz channel	176 kbps
Coverage	Typically a large city	As of September, 2002 only Denver, CO
Power level issues	Complicated power control algorithms for different burst profiles	Low power modem compatible with laptops and hand-helds
Interference	Present but limited	Present
Price complexity	Not available	Medium
Security	High. Defines an extra privacy sublayer for authentication.	High (Patented security system)

14.3 Wireless Local Area Networks (WLANs)

WLANs have gained immense popularity during the past two years. They are now standard equipment on most laptops and several high-end PDAs. The low cost, ease of installation, and almost no maintenance have resulted in several businesses looking at the WLAN as a convenient corporate solution. The IEEE, ETSI,and HomeRF WG have been involved in developing standards for the WLAN. These include the IEEE 802.11x, HiperLANx, and HomeRF. Of these, the IEEE 802.11 family of protocols have clearly become the dominant standard for WLAN in the world. HiperLAN has some market share, especially in Europe, and HomeRF has no share at all. In fact the HomeRF WG officially ceased to exist as of January 1, 2003 [14.9]. Its promoter companies (including Intel and Proxim) switched to the more popular IEEE 802.11 standard. It is however an innovative and interesting technology and is still available to researchers in universities and labs. In the following sections we will look at the IEEE 802.11, HiperLAN, and HomeRF.

14.3.1   IEEE 802.11

The IEEE group that proposed the standards for indoor LANs (e.g., Ethernet) in the early 1980s published a standard for WLANs and named it the IEEE 802.11 [14.2][14.3] (now known as IEEE 802.11a). This physical layer PHY (physical layer) and MAC standard specifies carrier frequencies in the 2.4 GHz range bandwidths with data rate of 1 or 2 Mbps, protocols, power levels, modulation schemes, and so on [14.10]. These are just the standards for which a compatible product can be manufactured. It does not address the difficulties in manufacturing a terminal unit to that specification.

User demand for higher bit rates and international availability of the 2.4 GHz ISM band has resulted in development of a high-speed standard in the same carrier frequency range. This standard, called the IEEE 802.11b (popularly known as Wi-Fi), specifies a PHY layer providing a basic rate of 11 Mbps and a fall-back rate of 5.5 Mbps. Products supporting this higher data rate have been released and are being used extensively in the market.

Wireless technology is improving at a fast pace. Future products could operate at higher frequencies and provide higher bit rates. To meet such demands, the IEEE 802.11 group has added another layer in the 5.2 GHz band, utilizing OFDM to provide data rates up to 54 Mbps. This standard, known as the IEEE 802.11a, became the first to use OFDM in packet-based communication. OFDM is also used in HiperLAN2.

The IEEE 802.11 and IEEE 802.11b standards could be used to provide communication between a number of terminals as an ad hoc network (Figure 14.5) using peer-to-peer mode, or as a client/server wireless configuration (Figure 14.6), or a fairly complicated distributed network (Figure 14.7). The key behind all these networks is the wireless cards and WLAN access point (AP) (simply known as AP). There are many companies (Lucent, Roxim, Netgear, among others.) that manufacture and support these devices. The IEEE standards allow two types of transmissions:

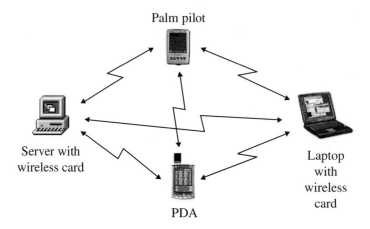

Figure 14.5
Peer-to-peer wireless ad hoc mode.

Palm pilot

Server with wireless card

PDA

Laptop with wireless card

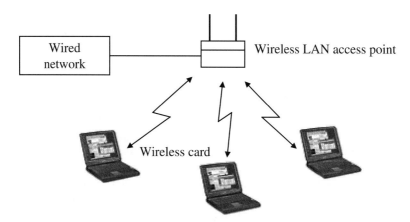

Figure 14.6
Client/server wireless configuration.

Wired network

Wireless LAN access point

Wireless card

frequency hopping spread spectrum (FHSS) and direct sequence spread spectrum (DSSS). FHSS is primarily used for low-power, low-range applications, and DSSS is popular in providing Ethernet-like data rates.

In the ad hoc network mode, as there is no central controller, the wireless access cards use the CSMA/CA protocol to resolve shared access of the channel. In the client/server configuration, many PCs and laptops, physically close to each other (20 to 500 meters), can be linked to a central hub (known as access point [AP]) that serves as a bridge between them and the wired network. The wireless access cards provide the interface between the PCs and the antenna, while the AP serves as the WLAN hub. The AP is usually placed at the ceiling or high on the wall and supports a number of (115 to 250) users receiving, buffering, and transmitting data between the WLAN and the wired network. The AP can also be programmed to select one of the hopping sequences, and the WLAN cards tune in to the corresponding sequence.

A larger area can be covered by installing several APs in the building and as with a cellular structure, there can be overlapped access areas. The access points track movement of users within a coverage area and make decisions on whether

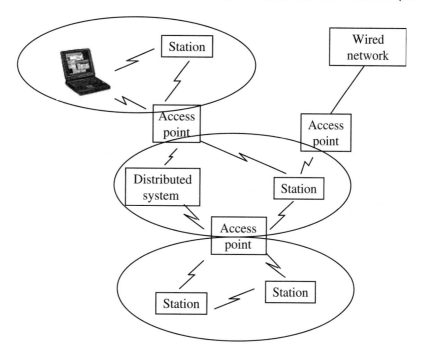

Figure 14.7
Distributed wireless network.

to allow users to communicate through them. An elaborate wireless distributed configuration, shown in Figure 14.7, allows several LANs to be interconnected using APs. In all these schemes, handoff and roaming can be easily supported across different APs. Encryption can also be provided using the optional shared-key RC4 (Ron's Code 4, alternatively known as Rivest's Cipher 4) algorithm. The WLAN cards could be operated in continuous aware mode (radio always on) and power-saving polling mode (radio in sleep state to extend battery life). In the latter mode, the AP keeps data in its buffer for the users and sends a signal to wake them up.

Wi-Fi (Wireless Fidelity)

In August 1999, a group of industry leaders formed a nonprofit organization called the wireless Ethernet compatibility alliance (WECA) to promote the IEEE 802.11 high-rate standard (which eventually became the IEEE 802.11b) as a commercial standard to assure the interoperability of different vendors' products. WECA selected an independent test lab to test and certify the interoperability of the IEEE 802.11b products. In March 2000, the first round of products passed the test and bore the Wi-Fi (wireless-fidelity) tag. WECA was later renamed the Wi-Fi alliance and certifies all the IEEE 802.11 high-rate standards (which include the IEEE 802.11b, IEEE 802.11a, and IEEE 802.11g) products. Almost all companies selling the IEEE 802.11 equipment are members of the Wi-Fi alliance. Currently, they are working on a security certification ("Wi-Fi protected access") which is based on the IEEE 802.11i draft.

MIT Roofnet

Roofnet [14.11], an experimental multi-hop IEEE 802.11b mesh network, consists of about 50 nodes in apartments of Cambridge, Massachussetts. Each node is in radio range of a subset of the other nodes, and can communicate with the rest of the nodes via multi-hop forwarding. A few of the nodes act as gateways to the wired Internet. The network requires no preconfiguration, and users can connect to it on the fly. A new user can turn on a new node and start using it for Internet connectivity with no configuration beyond installing the hardware. The new user need not allocate an IP address, aim a directional antenna, or ask existing users to perform any special actions to add the new node. Roofnet uses a new routing protocol called SrcRR, which is inspired by the DSR protocol. The typical maximum useful radio range is about 100 meters.

14.3.2 ETSI HiperLAN

HiperLAN [14.12] stands for high-performance LAN. While all of the previously discussed technologies have been designed specifically for an ad hoc environment, HiperLAN is derived from traditional LAN environments and can support multimedia data and asynchronous data effectively at high rates (23.5 Mbps). Also, a LAN extension via access points can be implemented using standard features of the HiperLAN/1 specification. However, HiperLAN does not necessarily require any type of access point infrastructure for its operation. HiperLAN started in 1992, and standards were published in 1995. It employs the 5.15 GHz and 17.1 GHz frequency bands and has a data rate of 23.5 Mbps with a coverage of 50 m and mobility < 10 m/s. It supports a packet-oriented structure, which can be used for networks with or without a central control (BS–MS and ad hoc). It supports 25 audio connections at 32 kbps with a maximum latency of 10 ms, 1 video connection of 2 Mbps with 100 ms latency, and a data rate of 13.4 Mbps.

HiperLAN/1 [14.1] is specifically designed to support ad hoc computing for multimedia systems, where there is no requirement to deploy centralized infrastructure. It effectively supports MPEG or other state-of-the-art real-time digital audio and video standards. The HiperLAN/1 MAC is compatible with the standard MAC service interface, enabling support for existing applications to remain unchanged. The HiperLAN describes the standards for service and protocols of the two lowest layers of the OSI model. HiperLAN type 2 has been specifically developed to have a wired infrastructure, providing short-range wireless access to wired networks such as IP and ATM. The two main differences between HiperLAN types 1 and 2 are as follows:

- Type 1 has a distributed MAC with QoS provisions, whereas type 2 has a centralized scheduled MAC.

- Type 1 is based on Gaussian minimum shift keying (GMSK), whereas type 2 is based on OFDM.

The mobile terminals communicate with one AP at a time over an air interface. HiperLAN/2 automatically performs handoff to the nearest AP. The AP is basically a radio BS that covers an area of about 30 to 150 meters, depending on the environment. MANETs can also be created easily. The goals of HiperLAN are as follows:

- QoS (to build multiservice networks)

- Strong security

- Handoff when moving between local area and wide areas

- Increased throughput

- Ease of use, deployment, and maintenance

- Affordability

- Scalability

One of the primary features of HiperLAN/2 is its high-speed transmission rates (up to 54 Mbps). It uses a modulation method called OFDM to transmit analog signals. This can, however, be dynamically adjusted to a lower rate by using different modulation schemes. It is connection-oriented, and traffic is transmitted on bidirectional links for unicast traffic, and unidirectional links toward the MSs for multicast and broadcast traffic. This connection-oriented approach makes support for QoS easy, which in turn depends on how the HiperLAN/2 network interoperates with the fixed network, using Ethernet, ATM, or IP.

HiperLAN/2 supports automatic frequency allocation, eliminating the need for manual frequency planning as in cellular networks. The APs in HiperLAN/2 have built-in support for automatically selecting an appropriate radio channel for transmission within the coverage area. Security is provided by key negotiation, authentication (network access identifier [NAI] or X.509), and encryption using DES or 30DES. A mobile terminal will automatically initiate a handoff if it moves out of signal range from an AP.

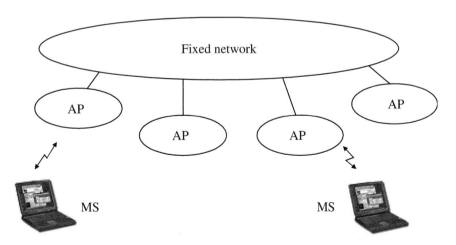

Figure 14.8
A simple HiperLAN
system.

The HiperLAN/2 architecture shown in Figure 14.8 allows for interoperation with virtually any type of fixed network, making the technology both network and application independent. Interoperation with Ethernet networks and support for ATM, PPP (point-to-point protocol), Firewire, and IP are integrated into Hiper-LAN/2. A MS may at any time request the AP and enter a low-power state for a sleep period. At the end of this negotiated sleep period, the MS searches for the presence of any wake-up indication. If there is no wake-up indication, the MS goes back to its low-power state for another sleep period. The channel spacing is 20 MHz, allowing high bit rates per channel.

Control is centralized at the AP, which informs the MS to transmit data using time division duplex and dynamic TDMA and adapts according to the request for the resources from the MS. The basic MAC frame structure comprises transport channels for broadcast control, frame control, access control, forward link and reverse link data transmission, and random access. Selective repeat ARQ is an error control mechanism used to increase reliability over the radio link. Packets are delivered in sequence by assigning a sequence number per connection.

The radio link control (RLC) protocol provides the following services:

- Association control with feature negotiation

- Encryption algorithms and convergence layers, authentication, key negotiation, and convergence layer negotiation

- Radio resource control to support handoff capability, to perform radio measurements in assisting the APs in selecting an appropriate radio channel, and to run the power-saving algorithm

- Connection control for the establishment and release of user connections

A HiperLAN/2 network can be used between the MSs and the network/LAN. The HiperLAN/2 network supports mobility within the same LAN/subnet, and the rest of the issues are handled by the upper layers. Therefore, the possibility of a single node working as a node in an ad hoc network and reverting to its role as a part of a LAN can be done easily. HiperLAN/2 networks can be deployed at "hot spot" areas such as airports and hotels, as an easy way of offering remote access and Internet services. An access server to which the HiperLAN/2 network is connected can route a connection request for a PPP or for Internet access. HiperLAN/2 can also be used as an alternative access technology to third generation networks.

14.3.3   HomeRF

It is estimated that 43 million U.S. homes now contain more than one personal computer. Approximately 13 million households in the United States contain a home business and need reliable and fast networking solutions.

After considering networks with few nodes and less than 10 meters range, we consider MANETs, which span an enclosed area such as a home or an office building or a warehouse floor in a workshop. These are broadly divided into the two categories of home (HomeRF[14.13]) and business workspace (HiperLAN). This difference is deemed necessary because a considerable amount of traffic in home is voice and there are devices with a lot of different demands on the network. On the other hand, in a business workspace, the traffic tends to be of only one kind, in most cases, data. Also, data rates need to be very high for business.

A home network typically consists of one high-speed Internet access port providing data to multiple networked nodes (PCs, handheld devices, or smart appliances). Home networking allows all computers in a home to simultaneously utilize the same high-speed ISP account.

Home networking provides two options: wired solution and wireless solution. Ethernet is based on the IEEE 802.3 standard with a data rate of 10 Mbps. Each PC is connected to a special device called an Ethernet hub to control communication in the whole home network. A 56 Kb analog, ISDN, cable, or ADSL (asymmetric digital subscriber line) modem provides connection to the Internet. The Ethernet network uses CSMA/CD for media access.

Wireless networks use high-frequency electromagnetic waves, either infrared (IR) or radio frequency, to transmit information from one point to another without relying on physical connections. Data and voice traffic are superimposed, or modulated, onto the radio waves, or carriers, and extracted at the receiving end. Multiple radio carriers can exist in the same space at the same time without interfering with each other by transmitting at different frequencies. To extract data, a receiver tunes in or selects one radio frequency while filtering out others. A wireless network at home offers advantages of mobility and flexibility, is simple, economical, and secure, and is based on industry standards.

One PC is the main access port, transmitting and receiving from other PCs on the network, with the master PC providing network addressing and routing between the home and the Internet. This solution addresses the PC-related network elements in the home, such as file and printer sharing, multiuser game playing, and a single shared ISP account. It leaves other elements, such as voice communications and control and monitoring applications, without a solution.

HomeRF Technology

Imagine switching on a coffee machine in kitchen, increasing the volume of the living-room stereo, and running hot water in bathtub, all from your bed! The requirements for such a system are different. A typical home needs a network inside the house for access to a public network telephone (isochronous multimedia) and Internet (data), entertainment networks (cable television, digital audio and video with the IEEE 1394), transfer and sharing of data and resources (printer, Internet connection), and home control and automation. The devices should be able to self-configure and maintain connectivity with the network. The devices need to be plug-and-play–enabled so that they are available to all other clients on the network as soon as they are switched on, which requires automatic device discovery and identification in the

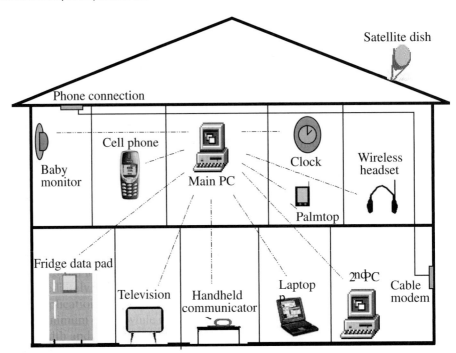

Figure 14.9
Architecture of
HomeRF system.

system. Home networking technology should also be able to accommodate any and all lookup services, such as Jini. HomeRF [14.13] products allow you to simultaneously share a single Internet connection with all of your computers—without the hassle of new wires, cables, or jacks.

HomeRF [14.9] visualizes a home network as shown in Figure 14.9. A network consists of resource providers, which are gateways to different resources like phone lines, cable modem, satellite dish, and so on, and the devices connected to them such as cordless phone, printers, fileservers, and TV. The goal of HomeRF is to integrate all of these into a single network suitable for all applications and to remove all wires and utilize RF links in the network suitable for all applications. This includes sharing PC, printer, fileserver, phone, Internet connection, and so on, enabling multiplayer gaming using different PCs and consoles inside the home, and providing complete control on all devices from a single mobile controller. With HomeRF, a cordless phone can connect to PSTN, but can also connect through a PC for enhanced services. HomeRF makes an assumption that simultaneous support for both voice and data is needed. Table 14.3 compares WLAN technologies regarding some relevant parameters.

Table 14.3: ▶
Comparison of WLAN Standards [14.8]
Credit: R.L. Ashok and D.P. Agrawal, "Next Generation Wearable Networks," *IEEE Computer*, November 2003, Vol. 36, No. 11, pp. 31-39.

Technology	Wireless LAN		
	802.11b (Wi-Fi)	HomeRF	HiperLAN2
Operational spectrum	2.4 GHz	2.4 GHz	5 GHz
Physical layer	DSSS	FHSS with FSK	OFDM with QAM
Channel access	CSMA–CA	CSMA–CA and TDMA	Central resource control/TDMA/TDD
Nominal data rate possible	22 Mbps	10 Mbps	32–54 Mbps
Coverage	100 m	>50 m	30–150 m
Power level issues	< 350 mA current drain	<300 mA peak current	Uses low power states like sleep
Interference	Present	Present	Minimal
Price/complexity	Medium (<\$ 100)	Medium	High (>\$ 100)
Security	Low	High	High

14.4 Wireless Personal Area Networks (WPANs)

14.4.1 Introduction

Bluetooth is the only WPAN technology to be commercially available. Even though it was developed in the mid-nineties, it is only since 2002 that its presence has become visible in a gamut of devices ranging from laptops to wireless mouse to cameras and cell phones. The IEEE has now taken a significant interest in WPANs and has initiated the development of the IEEE 802.15.x protocols to address the needs of WPANs with varied data rates. Bluetooth has been adopted as the IEEE 802.15.1 (medium rate) while the IEEE 802.15.3 (high rate) and 802.15.4 (low rate) are in their final stages of development.

The IEEE 802.15 working group is formed by four task groups (TGs)[14.14]:

■ The IEEE 802.15 WPAN/Bluetooth TG1: The TG1 has been established to support applications which require medium-rate WPANs (such as Bluetooth). These WPANs will handle a variety of tasks ranging from cell phones to PDA communications and will have a QoS suitable for voice applications.

■ The IEEE 802.15 Coexistence TG2: Several wireless standards, such as Bluetooth and the IEEE 802.11b, and appliances, such as microwaves, operate in the unlicensed 2.4 GHz ISM frequency band. TG2 (the IEEE 802.15.2) is developing recommended practices to facilitate coexistence of WPANs (the IEEE 802.15) and WLANs (the IEEE 802.11).

- The IEEE 802.15 WPAN/High Rate TG3: The TG3 for WPANs is chartered to draft a new standard for high-rate (20 Mbps or greater) WPANs. Besides a high data rate, the new standard provides low-power and low-cost solutions, addressing the needs of portable consumer digital imaging and multimedia applications.

- The IEEE 802.15 WPAN/Low Rate TG4: The goal of the TG4 is to provide a standard for ultra-low complexity, cost, and power for low-data-rate (200 kbps or less) wireless connectivity among inexpensive fixed, portable, and moving devices. Location awareness is being considered as a unique capability of the standard. The scope of the TG4 is to define the physical and medium access control layer specifications. Potential applications are sensors, interactive toys, smart badges, remote controls, and home automation.

One key issue in WPANs is the interworking of wireless technologies to create heterogeneous wireless networks. For instance, WPANs and WLANs will enable an extension of devices without direct cellular access to 3G cellular systems (i.e., UMTS, W-CDMA, and cdma2000). Moreover, devices interconnected in a WPAN should be able to utilize a combination of both 3G access and WLAN by selecting the access mechanism that is best suited at a given time. In such networks, 3G, WLAN, and WPAN technologies do not compete against each other but enable the user to select the best connectivity for intended purposes.

14.4.2 IEEE 802.15.1 (Bluetooth)

Bluetooth [14.15] is named after the King of Denmark, who unified different factions in Christianity throughout Denmark. If you are in a building not wired, and suddenly an email is sent to your notebook and if your cellular phone is in your briefcase, you will be unable to respond the email. If Bluetooth is present with your cellular phone (basically, Bluetooth is a wireless wire), as illustrated in Figure 14.10, you can easily reply.

Bluetooth has been designed to allow low-bandwidth wireless connections to become so simple to use that they seamlessly integrate into your daily life [14.16].

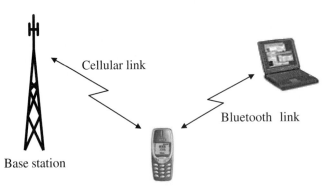

Figure 14.10
Use of Bluetooth to connect notebook.

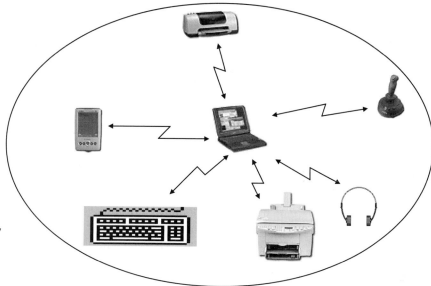

Figure 14.11
Bluetooth connecting printers, PDAs, desktops, FAX machines, keyboards, joysticks, and virtually any other digital device.

Ericsson, Intel, IBM, Nokia, and Toshiba started this in 1998 by establishing a Bluetooth special-interest group. In December 1999, many companies, including 3COM, Lucent, Motorola, and Microsoft, joined in an attempt to evolve a reliable universal link for short-range RF communication. It is widely recognized that as time progresses, the number of short wires connecting computer peripherals has been increasing day by day. Low-cost, low-power, radio-based wireless links eliminate the need for short cables. An infrared link can easily provide a link with speeds up to 10 Mbps at very low cost and ease of installation, but requires line of sight and offers only a point-to-point link. Hence, the concept of Bluetooth evolved to provide a universal standard for short-range RF communication of both voice and data.

Bluetooth [14.17] offers many options to the user by replacing the cable used to connect a laptop to a cellular phone. printers, desktops, FAX machines, keyboards, joysticks, and virtually any other digital device can be networked by the Bluetooth system (Figure 14.11). Bluetooth also provides a universal bridge to existing data networks (Figure 14.12) and a mechanism to form small private MANETs (Figure 14.13).

A simple example of a Bluetooth application is updating the phone directory of your PC from a mobile telephone. With Bluetooth, entering numbers of all your contacts between your phone and your PC could happen automatically and without any user involvement. Of course, you can easily expand to include your calendar, to-do list, memos, email, and so on. It is reasonable to assume that it would be feasible to find the price of all sale items automatically on your cell phone or PDA.

The ultimate goal is to make computers (PCs/laptops) have only one wire attached to them, which is the power cord, and make a portable computer truly portable. In the case of a PDA, the power cord is also eliminated. Communication protocols between two computers in a conference room environment do exist for

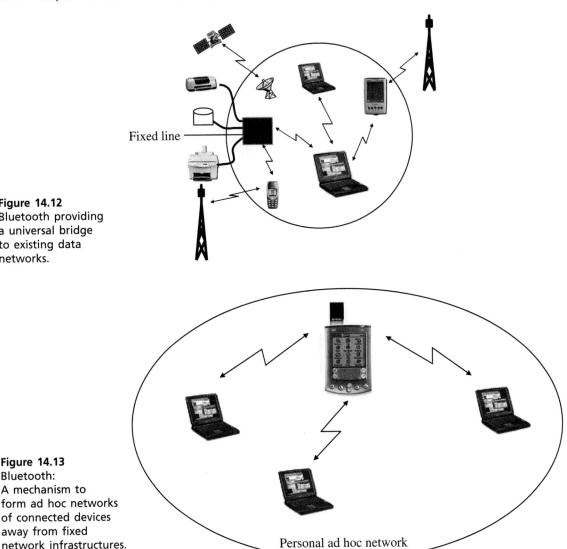

Figure 14.12
Bluetooth providing
a universal bridge
to existing data
networks.

Fixed line

Figure 14.13
Bluetooth:
A mechanism to
form ad hoc networks
of connected devices
away from fixed
network infrastructures.

Personal ad hoc network

Bluetooth. However, the demands placed on the network by the voice and data traffic are different; although multimedia traffic is likely to use most of the asynchronous real-time interactive data. These packets consume nearly 1/3 of bandwidth in traditional peer-to-peer networks and much more in connections involving peripherals. Any of the existing transport protocols cannot be used in this scenario and efficient protocols to handle this general situation need to be developed.

Bluetooth utilizes the unlicensed ISM band at 2.4 GHz. A typical Bluetooth device has a range of about 10 meters. The communication channel supports data (asynchronous) and voice (synchronous) with a total bandwidth of 1 Mbps. The synchronous voice channels are provided using circuit switching (slot reservation at fixed intervals). The asynchronous data channels are provided using packet switching utilizing a polling access scheme. A combined data-voice packet is also defined to provide 64 kbps voice and 64 kbps data in each direction. The time slots can be reserved for synchronous packets with a frequency hop for each transmitted packet. A packet usually covers a single time slot but can be extended to cover up to five slots. The Bluetooth specification defines two power levels: a low power level that covers a small personal area within a room, and a high power level that can cover a medium range, such as an area within a home. Software controls and identity coding built into each microchip ensure that only those units preset by their owners can communicate with the following characteristics:

- Fast frequency hopping to minimize interference
- Adaptive output power to minimize interference
- Short data packets to maximize capacity
- Fast acknowledgments allowing low coding overhead for links
- CVSD (continuous variable slope delta) modulation voice coding, which can withstand high bit-error rates
- Flexible packet types that support a wide application range
- Transmission/reception interface tailored to minimize power consumption

Architecture of the Bluetooth System

Bluetooth devices can interact with other Bluetooth devices in several ways (Figure 14.14). In the simplest scheme, one of the devices acts as the master and (up to) seven others as slaves and it is known as a piconet. A single channel (and bandwidth) is shared among all devices in the piconet. Each of the active slaves has an assigned 3-bit active member address. Many other slaves can remain synchronized to the master though remaining inactive slaves, referred to as parked nodes. The master regulates channel access for all active nodes and parked nodes. If two piconets are close to each other, they have overlapping coverage areas. This scenario, in which nodes of two piconets intermingle, is called a scatternet. Slaves in one piconet can participate in another piconet as either a master or slave through time division multiplexing. In a scatternet, the two (or more) piconets are not synchronized in either time or frequency. Each of the piconets operates in its own frequency hopping channel while any devices in multiple piconets participate at the appropriate time via time division multiplexing. Before any connections in a piconet are created, all devices are in STANDBY mode, where unconnected units periodically "listen" for messages every 1.28 seconds. Each time a device wakes up, it tunes on the set of 32 hop frequencies defined for that unit.

Piconet supports both point-to-point and point-to-multipoint connections, and details of Bluetooth technological characteristics are shown in Table 14.4.

The connection procedure for a piconet is initiated by any of the devices, which then becomes master of the created piconet. A connection is made by sending a PAGE message if the address is already known, or by an INQUIRY message followed by a subsequent PAGE message if the address is unknown. In the PAGE state, the master unit sends a train of 16 identical messages using 16 different hop frequencies

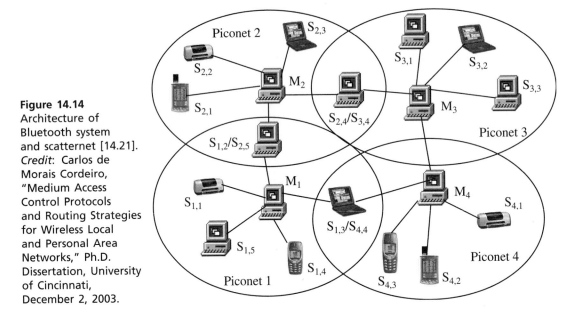

Figure 14.14
Architecture of Bluetooth system and scatternet [14.21]. *Credit*: Carlos de Morais Cordeiro, "Medium Access Control Protocols and Routing Strategies for Wireless Local and Personal Area Networks," Ph.D. Dissertation, University of Cincinnati, December 2, 2003.

Table 14.4: ▶
Bluetooth Technological Characteristics

Frequency band	2.4 GHz (unlicensed ISM band)
Technology	Spread spectrum
Transmission method	Hybrid direct sequence and frequency hopping
Transmission power	1 milliwatt (0 dBm)
Range	10 meters (40 feet)
Number of devices	8 per piconet, 10 piconets per coverage area
Data speed	Asymmetric link: 721+57.6 kbps Symmetric link: 432.6 kbps
Maximum voice channels	3 per piconet
Maximum data channels	7 per piconet
Security	Link layer with fast frequency hopping (1600 hops/s)
Power consumption	30 μA sleep, 60 μA hold, 300 μA standby, 800 μA max transmit
Module size	3 square cm (0.5 square inches)
Price	Expected to fall to $5 in the next few years
C/I cochannel	11 dB (0.1% BER)
C/I 1 MHz	−8 dB (0.1% BER)
C/I 2 MHz	−40 dB (0.1% BER)
Channel switching time	220 μs

defined for the device to be paged (slave unit). If it does not get any response, the master transmits a train on the remaining 16 hop frequencies. The maximum delay before the master reaches the slave is twice the wake-up period (0.64 seconds). A power-saving mode can be used for units in a piconet if there are no data to be transmitted. The master unit can put slave units into HOLD mode, where only an internal timer is running. Slave units can also demand to be put into HOLD mode. Data transfer restarts instantly when units move out of HOLD mode. The HOLD is used when connecting several piconets or managing a low-power device such as a temperature sensor. In the SNIFF mode, a slave device listens to the piconet at a reduced rate, reducing its duty cycle. The SNIFF interval is programmable and depends on the application. In the PARK mode, a device is still synchronized to the piconet but does not participate in the traffic.

The Bluetooth core protocols are shown in Figure 14.15; the rest of the protocols are used only as needed. Service discovery protocol (SDP) provides a means for applications to discover which services are provided by or are available through a Bluetooth device. Logical link control and adaptation layer protocol (L2CAP) supports higher-level protocol multiplexing, packet segmentation, and reassembly, and the conveying of quality of service information. Link manager protocol (LMP) is used by the link managers (on either side) for link setup and control. The baseband and link control layer enables the physical RF link between Bluetooth units forming a piconet. It provides two different kinds of physical links with their corresponding baseband packets, SCO and ACL, which can be transmitted in a multiplexing manner on the same RF link.

Each link type supports up to 16 different packet types. Four of these are control packets which are common for both SCO and ACL links. Both link types use a TDD scheme for full-duplex transmissions. The SCO link is symmetric and typically supports time-bounded voice traffic. SCO packets are transmitted over reserved intervals. Once the connection is established, both master and slave units may send SCO packet types and allow both voice and data transmissions—with only the data portion being retransmitted when corrupted.

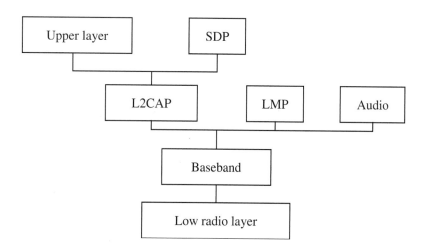

Figure 14.15
Bluetooth core
protocols.

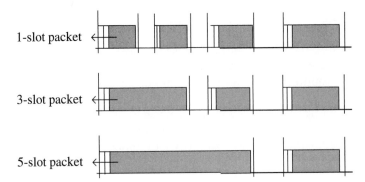

Figure 14.16
Packet transmission
in Bluetooth.

The ACL link is packet oriented and supports both symmetric and asymmetric traffic. The master unit controls the link bandwidth and decides how much piconet bandwidth is given to each slave and the symmetry of the traffic. Slaves must be polled before they can transmit data. The ACL link also supports broadcast messages from the master to all slaves in the piconet. There are three error-correction schemes defined for Bluetooth baseband controllers:

- 1/3 rate FEC
- 2/3 rate FEC
- ARQ scheme for data

There are three and five slot packets as depicted in Figure 14.16. A TDD scheme divides the channel into 625 μs slots at a 1 Mb/s symbol rate. As a result, at most 625 bits can be transmitted in a single slot. However, to change the Bluetooth device from transmit state to receive state and tune to the next frequency hop, a 259 μs turn around time is kept at the end of the last slot. This results in reduction of effective bandwidth available for data transfer. Table 14.5 summarizes the

Table 14.5: ▶
Bluetooth Packet
Types [14.16]
Credit: IEEE 802.15
Working Group for
WPANS, HYPERLINK
"http://grouper.ieee.
org/groups/802/15"

Type	User Payload (bytes)	FEC	Symmetric (kbps)	Asymmetric (kbps)	
DM1	0–17	Yes	108.0	108.8	108.8
DH1	0–27	No	172.8	172.8	172.8
DM3	0–121	Yes	256.0	384.0	54.4
DH3	0–183	No	384.0	576.0	86.4
DM5	0–224	Yes	286.7	477.8	36.3
DH5	0–339	No	432.6	721.0	57.6
HV1	0–10	Yes	64.0		
HV2	0–20	Yes	128.0		
HV3	0–30	No	192.0		

available packet types and their characteristics [14.4]. Bluetooth employs HVx (high-quality voice) packets for SCO transmissions and DMx (data medium-rate) or DHx (data high-rate) packets for ACL data transmissions, where $x = 1$, 3 or 5. In case of DMx and DHx, x represents the number of slots a packet occupies as shown in Figure 14.16, while in the case of HVx, it represents the level of forward error correction (FEC). The purpose of the FEC scheme on the data payload is to reduce the number of retransmissions. In the ARQ scheme, data transmitted in one slot are directly acknowledged by the recipient in the next slot, performing both the header error check and the cyclic redundancy check.

14.4.3   IEEE 802.15.3

The IEEE 802.15.3 Group is developing an ad hoc MAC layer suitable for multimedia WPAN applications and a PHY capable of data rates in excess of 20 Mbps. The current draft of the IEEE 802.15.3 standard (being dubbed WiMedia) specifies data ratesof up to 55 Mbps in the 2.4 GHz unlicensed band. The technology employs an ad hoc PAN topology not entirely dissimilar to Bluetooth, with roles for "master" and "slave" devices. The draft standard calls for drop-off data rates from 55 Mbps to 44 Mbps, 33 Mbps, 22 Mbps, and 11 Mbps. The IEEE 802.15.3 is not compatible with either Bluetooth or the IEEE 802.11 family of protocols though it reuses elements associated with both.

IEEE 802.15.3 MAC and PHY Layer Details

The IEEE 802.15.3 MAC layer specification is designed from the ground up to support ad hoc networking, multimedia QoS provisions, and power management. In an ad hoc network, devices can assume either master or slave functionality based on existing network conditions. Devices in an ad hoc network can join or leave an existing network without complicated setup procedures. The IEEE 802.15.3 MAC specification provides provisions for supporting multimedia QoS. Figure 14.17 illustrates the MAC superframe structure that consists of a network beacon interval, and a contention access period (CAP) reserves for guaranteed time slots (GTSs). The boundary between the CAP and GTS periods is dynamically adjustable.

A network beacon is transmitted at the beginning of each superframe, carrying WPAN-specific parameters, including power management, and information for new devices to join the ad hoc network. The CAP period is reserved for transmitting non-QoS data frames such as short bursty data or channel access requests made by the devices in the network. The medium access mechanisms during the CAP period is CSMA/CA. The remaining duration of the superframe is reserved for GTS to carry data frames with specific QoS provisions. The type of data transmitted in the GTS can range from bulky image or music files to high-quality audio or high-definition video streams. Finally, power management is one of the key features of the IEEE 802.15.3 MAC protocol, which is designed to significantly lower the current drain while being connected to a WPAN. In the power saving mode, the QoS provisions are also maintained.

The IEEE 802.15.3 PHY layer operates in the unlicensed frequency band between 2.4 GHz and 2.4835 GHz, and is designed to achieve data rates of 11–55 Mb/s

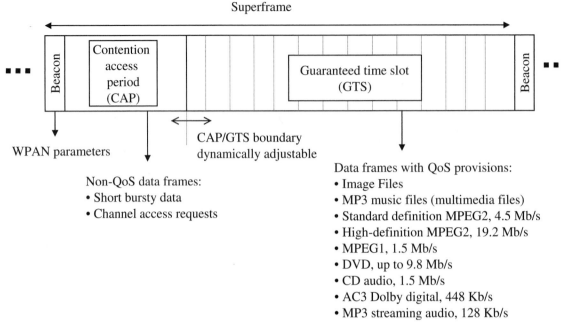

Figure 14.17 IEEE 802.15.3 MAC superframe [14.22].
Credit: J. Karaoguz, "High-rate Wireless Personal Area Networks," *IEEE Communications Magazine*, Vol. 39, No. 12, pp. 96-102, December 2001.

that are commensurate with the distribution of high-definition video and high-fidelity audio. The IEEE 802.15.3 systems employ the same symbol rate, 11 Mbaud, as used in the IEEE 802.11b systems. Operating at this symbol rate, five distinct modulation formats are specified, namely, uncoded QPSK modulation at 22 Mb/s and trellis coded QPSK, 16/32/64-QAM at 11, 33, 44, 55 Mb/s, respectively (Trellis Coded Modulation-TCM). The base modulation format is QPSK (differentially encoded). Depending on the capabilities of devices at both ends, the higher data rates of 33–55 Mb/s are achieved by using 16, 32, 64-QAM schemes with 8-state 2D trellis coding. Finally, the specification includes a more robust 11 Mb/s QPSK TCM transmission as a dropback mode to alleviate the well-known hidden node problem. The IEEE 802.15.3 signals occupy a bandwidth of 15 MHz, which allows for up to four fixed channels in the unlicensed 2.4 GHz band. The transmit power level complies with the FCC rules with a target value of 0 dBm.

The RF and baseband processors used in the IEEE 802.15.3 PHY layer implementations are optimized for short-range transmission limited to 10 m, enabling low-cost and small-form-factor MAC and PHY implementations for integration in consumer devices. The total system solution is expected to fit easily in a compact flash card. The PHY layer also requires low current drain (less than 80 mA) while actively transmitting or receiving data at minimal current drain in the power saving mode.

From an ad hoc networking point of view, it is important that devices have the ability to connect to an existing network with a short connection time. The

IEEE 802.15.3 MAC protocol targets connection times much less than 1 sec. Reviewing the regulatory requirements, it should be noted that the operation of WPAN devices in the 2.4 GHz band is highly advantageous since these devices cannot be used outdoors in Japan while operating in the 5 GHz band. The outdoor use of most portable WPAN devices prohibits the use of 5 GHz band for worldwide WPAN applications.

14.4.4   IEEE 802.15.4

The IEEE 802.15.4 defines a specification for low-rate, low-power WPANs (LR-WPANs) [14.18]. It is extremely well suited to those home networking applications where the key motivations are reduced installation cost and low power consumption. The home network has varying requirements. There are some applications that require high data rates such as shared Internet access, distributed home entertainment, and networked gaming. However, there is an even bigger market for home automation, security, and energy conservation applications, which typically do not require the high bandwidths associated with the former category of applications. Instead, the focus of this standard is to provide a simple solution for networking wireless, low-data-rate, inexpensive, fixed, portable and moving devices. Application areas include industrial control; agricultural, vehicular, and medical sensors; and actuators that have relaxed data-rate requirements.

Inside the home, there are several areas where such technology can be applied effectively: PC peripherals including keyboards, wireless mice, low-end PDAs, and joysticks; consumer electronics including radios, TVs, DVD players, and remote controls; home automation including heating, ventilation, air conditioning, security, lighting, and control of windows, curtains, doors, locks; health monitors, and diagnostics. These typically need less than 10 kbps, while the PC peripherals require a maximum of 115.2 kbps. Maximum acceptable latencies vary from 10 ms for the PC peripherals to 100 ms to home automation.

As we have seen, the IEEE 802.15.1 and 802.15.3 are meant for medium- and high-data-rate WPANs respectively [14.19]. The IEEE 802.15.4 effort is geared towards those applications that do not fall in the above two categories, which have low bandwidth requirements and very low power consumption and are extremely inexpensive to build and deploy. These are referred to as LR–PANs. In 2000, two standards groups, the Zigbee alliance (a HomeRF spinoff) and the IEEE 802 working group came together to specify the interfaces and the working of the LR–PAN. In this coalition, the IEEE group is largely responsible for defining the MAC and the PHY layers, while the Zigbee alliance which includes Philips, Honeywell and Invensys Metering Systems, among others, is responsible for defining and maintaining higher layers above the MAC. The alliance is also developing application profiles, certification programs, logos and a marketing strategy. The specification is based on the initial work done mostly by Philips and Motorola for Zigbee [14.20]—previously known as PURLnet, FireFly and HomeRF Lite.

The IEEE 802.15.4 standard—like all other IEEE 802 standards—specifies those layers up to and including portions of the data link layer. The choice of higher-level protocols is left to the application, depending on specific requirements. The important criteria would be energy conservation and the network topology. The draft, as such, supports networks in both the star and peer-to-peer topology. Multiple address types—both physical (64 bit) and network assigned (8 bit)—are allowed. Network layers are also expected to be self-organizing and self-maintaining to minimize cost to the customer.

Currently, the PHY and the data link layer (DLL) have been more or less clearly defined. The focus now is on the upper layers, and this effort is largely led by the Zigbee Alliance [14.20], which aims to bring this innovative and cheap technology to the market by 2003. In the following sections the MAC and PHY layer issues of the IEEE 802.15.4 are described.

IEEE 802.15.4 Data Link Layer (DLL) Details

The DLL is split into two sublayers—the MAC and the logical link control (LLC). The LLC is standardized in the IEEE 802 family, while the MAC varies depending on the hardware requirements. Figure 14.18 shows the correspondence of the IEEE 802.15.4 to the ISO–OSI reference model.

The IEEE 802.15.4 MAC provides services to an IEEE 802.2 type I LLC through the service-specific convergence sublayer (SSCS). A proprietary LLC can access the MAC layer directly without going through the SSCS. The SSCS ensures compatibility between different LLC sublayers and allows the MAC to be accessed through a single set of access points. The MAC protocol allows association and disassociation, acknowledged frame delivery, channel-access mechanism, frame validation, guaranteed time slot management, and beacon management. The MAC sublayer provides the MAC data service through the MAC common part sub layer (MCPS–SAP),

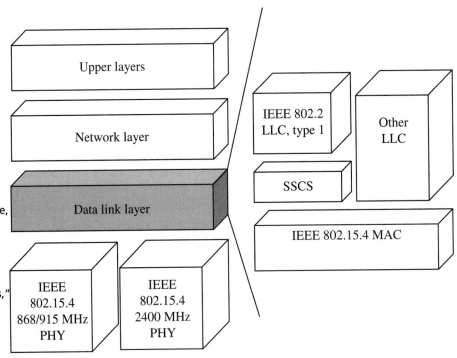

Figure 14.18
IEEE 802.15.4 in the ISO–OSI layered network model [14.23].
Credit: E. Callaway, P. Gorday, L. Hester, J.A. Gutierrez, M. Naeve, B. Heile, V. Bahl, "Home Networking with IEEE 802.15.4: A developing standard for low-rate wireless personal area networks," *IEEE Communications Magazine*, Vol. 40, No. 8, pp. 70-77, Aug. 2002.

and the MAC management services through the MAC layer management entity (MLME–SAP). These provide the interfaces between the SSCS (or another LLC) and the PHY layer. MAC management service has only 26 primitives as compared to the IEEE 802.15.1, which has 131 primitives and 32 events.

The MAC frame structure has been designed in a flexible manner, so that it can adapt to a wide range of applications, while maintaining the simplicity of the protocol. There are four types of frames: beacon, data, acknowledgment, and command frames. The overview of the frame structure is illustrated in Figure 14.19.

The MAC protocol data unit (MPDU), or the MAC frame, consists of the MAC header (MHR), MAC service data unit (MSDU), and MAC footer (MFR). The MHR consists of a 2-byte frame control field that specifies the frame type and the address format and controls the acknowledgment; the 1-byte sequence number, which matches the acknowledgment frame with the previous transmission; and a variable-sized address field (0–20 bytes). This allows either only the source address, possibly in a beacon signal, or both source and destination address as in normal data frames, or no address at all as in an acknowledgment frame. The payload field is variable in length, but the maximum possible size of an MPDU is 127 bytes. The beacon and the data frames originate at the higher layers and actually contain some data, while the acknowledgment and the command frame originate in the MAC layer and are used to simply control the link at a peer-to-peer level. The MFR completes the MPDU and consists of a frame check sequence (FCS) field, which is basically a 16-bit CRC code.

The IEEE 802.15.4 provides dedicated bandwidth and low latencies to certain types of applications by operating in a superframe mode. One of the devices—usually one that is less power constrained than the others—acts as the PAN coordinator, transmitting superframe beacons at predetermined intervals that range from 15 ms

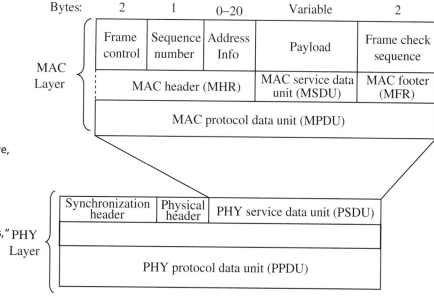

Figure 14.19
The general MAC frame format [14.23]. *Credit*: E. Callaway, P. Gorday, L. Hester, J.A. Gutierrez, M. Naeve, B. Heile, V. Bahl, "Home Networking with IEEE 802.15.4: A developing standard for low-rate wireless personal area networks," *IEEE Communications Magazine*, Vol. 40, No. 8, pp. 70-77, Aug. 2002.

to 245 ms. The time between the beacons is divided into 16 equal time slots independent of the superframe duration. The device may transmit at any slot, but must complete its transmission before the end of the superframe. Channel access is usually contention based, though the PAN may assign time slots to a single device. This is known as a guaranteed time slot (GTS) and introduces a contention-free period located immediately before the next beacon. In a beacon-enabled superframe network, a slotted CSMA/CA is employed, while in nonbeacon networks, the unslotted or standard CSMA/CA is used.

An important function of MAC is to confirm successful reception of frames. Valid data and command frames are acknowledged; otherwise, it is simply ignored. The frame control field indicates whether a particular frame has to be acknowledged or not. The IEEE 802.15.4 provides three levels of security: no security, access control lists, and symmetric key security using AES-128. To keep the protocol simple and the cost minimum, key distribution is not specified but may be included in the upper layers.

IEEE 802.15.4 PHY Layer Details

The IEEE 802.15.4 offers two PHY layer choices based on the DSSS technique and shares the same basic packet structure for low–duty cycle, low-power operation. The difference lies in the frequency band of operation. One specification is for the 2.4 GHz ISM band, to be available worldwide, and the other is for the 868/915 MHz for Europe and the United States, respectively. These offer an alternative to the growing congestion in the ISM band due to a large-scale proliferation of microwave ovens and the like. They also differ with respect to the data rates supported. The ISM band PHY layer offers a transmission rate of 250 kbps, while the 868/915 MHz layer offers 20 and 40 kbps. The lower rate can be translated into better sensitivity and larger coverage area, while the higher rate of the 2.4 GHz band can be used to attain lower duty cycle, higher throughput, and lower latencies.

The range of LR–WPAN is dependant on the sensitivity of the receiver, which is −85 dB for the 2.4 GHz PHY and −92 dB for the 868/915 MHz PHY. Each device should be able to transmit at least 1 mW but actual transmission power depends on the application. Typical devices (1 mW) are expected to cover a range of 10–20 m, but with good sensitivity and a moderate increase in power, it is possible to cover the home in a star network topology.

The 868/915 MHz PHY supports a single channel between 868.0 and 868.6 MHz and 10 channels between 902.0 and 928.0 MHz. Since these are regional in nature it is unlikely that all 11 channels ought to be supported on the same network. It uses a simple DSSS in which each bit is represented by a 15-chip maximal length sequence (m-sequence). Encoding is done by multiplying the m-sequence with +1 or −1, and the resulting sequence is modulated by the carrier signal using BPSK.

The 2.4 GHz PHY supports 16 channels between 2.4 GHz and 2.4835 GHz with 5 MHz channel spacing for easy transmit and receive filter requirements. It employs a 16-ary quasi-orthogonal modulation technique based on DSSS. Binary data is grouped into 4-bit symbols, each specifying one of 16 nearly orthogonal 32-bit chip pseudo-noise (PN) sequences for transmission. PN sequences for successive data symbols are concatenated, and the aggregate chip is modulated onto the carrier using minimum shift keying (MSK). The use of "nearly orthogonal" symbol sets simplifies the implementation, but incurs a minor performance degradation (< 0.5 dB). In terms of energy conservation, orthogonal signaling performs better than differential BPSK. However, in terms of receiver sensitivity, the 868/915 MHz has a 6–8 dB advantage.

The two PHY layers though different, maintain a common interface to the MAC layer (i.e., they share a single packet structure as shown in Figure 14.20).

The packet or PHY protocol data unit (PPDU) consists of the synchronization header, a PHY header for the packet length, and the payload itself, which is also referred to as the PHY service data unit (PSDU). The synchronization header is made up of a 32-bit preamble, used for acquisition of symbol and chip timing and possible coarse frequency adjustment, and an 8-bit start-of-packet delimiter, signifying the end of the preamble. Of the eight bits in the PHY header, seven are used to specify the length of the PSDU, which can range from 0 to 127 bytes. Channel equalization is not required for either PHY layer because of the small coverage area and the relatively low chip rates. Typical packet sizes for monitoring and control applications are expected to be in the order of 30–60 bytes.

Since the IEEE 802.15.4 standard specifies working in the ISM band, it is important to consider the effects of the interference that is bound to occur. The applications envisioned by this protocol have few or no QoS requirements. Consequently, data that does not go through on the first attempt will be retransmitted, and higher latencies are tolerable. Too many transmissions also increases the duty cycle and therefore affects the consumption of power. Once again, the application areas are such that transmissions will be infrequent, with the devices in a passive mode of operation for most of the time. Table 14.6 gives a comprehensive comparison of WPAN solutions.

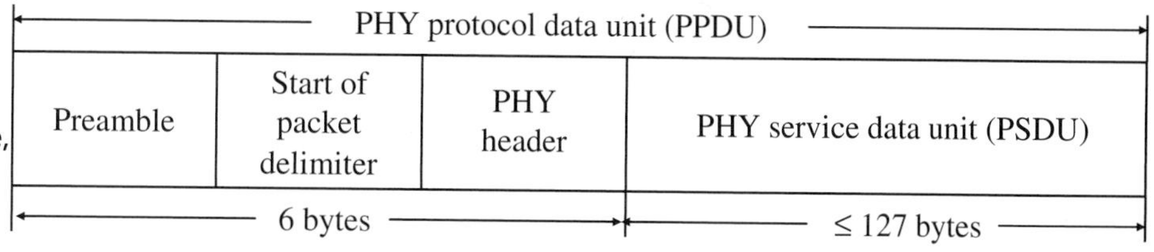

Figure 14.20
IEEE 802.15.4 PHY layer packet structure [14.23].
Credit: E. Callaway, P. Gorday, L. Hester, J.A. Gutierrez, M. Naeve, B. Heile, V. Bahl, "Home Networking with IEEE 802.15.4: A developing standard for low-rate wireless personal area networks," *IEEE Communications Magazine*, Vol. 40, No. 8, pp. 70-77, Aug. 2002.

	PHY protocol data unit (PPDU)			
Preamble	Start of packet delimiter	PHY header	PHY service data unit (PSDU)	
	6 bytes		≤ 127 bytes	

PHY packet fields:
· Preamble (32 bits) synchronization
· Start of packet delimiter (8 bits) signify end of preamble
· PHY header (8 bits) specify length of PSDU
· PSDU (≤ 127 bytes) PHY layer payload

Table 14.6: ▶
Comparison of WPAN Systems[14.8]
Credit: R.L. Ashok and D.P. Agrawal, "Next Generation Wearable Networks," *IEEE Computer*, November 2003, Vol. 36, No. 11, pp. 31-39.

Technology	Bluetooth (802.15.1)	802.15.3	802.15.4
Operational spectrum	2.4 GHz ISM band	2.402–2.480 GHz ISM band	2.4 GHz and 868/915 MHz
Physical layer details	FHSS, 1600 hops per second	Uncoded QPSK trellis coded QPSK or 16/32/64-QAM scheme	DSSS with BPSK or MSK (O–QPSK)
Channel access	Master slave polling, time division duplex (TDD)	CSMA–CA, and guaranteed time slots (GTS) in a superframe structure	CSMA–CA, and guaranteed time slots (GTS) in a superframe structure
Maximum data rate	Up to 1 Mbps	11–55 Mbps	868 MHz–20, 915 MHz–40, 2.4GHz–250 kbps
Coverage	<10 m	<10 m	<20 m
Power level issues	1 mA–60 mA	<80 mA	Very low current drain (20–50 μA)
Interference	Present	Present	Present
Price	Low (<$ 10)	Medium	Very low

• • • • • • • • • • • • • • •

14.5 Summary

In this chapter we have looked at WMANs, WLANs, and WPANs—all wireless connectivity solutions primarily distinguished by the range they cover and therefore, to some extent, the services they provide. Even though there are several types of protocols within each category, only one from each of these has been able to have a measure of commercial success. The WLAN world is completely dominated by the IEEE 802.11. Most laptops today come with built-in 802.11b cards. The only WPAN standard that has reached the market as a mass consumer technology is Bluetooth. WiMedia™ [14.14] (the IEEE802.15.3) also seems to be promising.

• • • • • • • • • • • • • • •

14.6 References

[14.1] ETSI, *"High Performance Radio Local Area Network (HIPERLAN) Type 1; Functional Specification,"* 105 pages, *http://webapp.etsi.org/pda/home.asp?wki id=6956.*

[14.2] B. P. Crow, I. Wadjaja, J. G. Kim, and P. T. Sakai, "IEEE 802.11 Wireless Local Area Networks," *IEEE Communications Magazine*, pp. 116–126, London, UK, September 1997.

[14.3] IEEE "Wireless LAN Medium Access Control (MAC) and Physical Layer (PHY) Specification: Higher Speed Physical Layer (PHY) Extension in the 2.4 GHz Band," 65 pages, 1999.

[14.4] *http://www.bluetooth.com*.

[14.5] *http://www.ricochet.com*.

[14.6] IEEE P802.16a/D3-2001: "Draft Amendment to IEEE Standard for Local and Metropolitan Area Networks—Part 16: Air Interface for Fixed Wireless Access Systems—Medium Access Control Modifications and Additional Physical Layers Specifications for 2–11 GHz," March 25, 2002.

[14.7] C. Eklund, R. B. Marks, K. L. Stanwood, and S. Wang, "IEEE Standard 802.16: A Technical Overview of the WirelessMANT Air Interface for Broadband Wireless Access," *IEEE Communications Magazine*, June 2002.

[14.8] R. L. Ashok and D. P. Agrawal, "Next Generation Wearable Networks," *IEEE Computer*, November 2003 Vol. 36, No. 11, pp. 31–39, November 2003.

[14.9] R. Shim, "HomeRF Working Group Disbands," 7 January 2003, CNET News.com; *news.com.com/2100-1033-979611.html*.

[14.10] R. T. Valadas, A. R. Tarares, A. M. D. O. Burate, A.C. Moreira, and C. T. Lomba, "The Infrared Physical Layer of the IEEE 802.11 Standard for Wireless Local Area Networks," *IEEE Communications Magazine*, pp. 107–112, December 1998.

[14.11] D. Aguayo, J. Bicket, S. Biswas, D. S. J. De Couto, and R. Morris, "MIT Roofnet Implementation," *http://www.pdos.lcs.mit.edu/roofnet/*.

[14.12] M. Johnson, *"HiperLAN/2-The Broadband Radio Transmission Technology Operating in the 5 GHz Frequency Band,"* 21 pages, *http://www.hiperlan2.com/ site/specific/whitepaper.exe*.

[14.13] K. Negus, A. Stephens, and J. Lansford, "HomeRF: Wireless Networking for the Connected Home," *IEEE Personal Communications*, pp. 20–27, February 2000.

[14.14] IEEE 802.15 Working Group for WPANs, *http://grouper.ieee.org/groups/802/15/*.

[14.15] J. Haartsen, "The Bluetooth Radio System," *IEEE Personal Communications*, pp. 28–36, February 2000.

[14.16] The Bluetooth Special Interest Group, "Baseband Specifications," *http://www.bluetooth.com*.

[14.17] J. Bray and C. F. Sturman, *Bluetooth: Connect without Cables*, Prentice Hall PTR; 1st edition, December 15, 2000.

[14.18] J. A. Gutierrez, M. Naeve, E. Callaway, M. Bourgeois, V. Mitter, and B. Heile, "IEEE 802.15.4: A Developing Standard for Low-Power Low-Cost Wireless Personal Area Networks," IEEE Network, Vol. 15, No. 5, pp. 12–19, September/October 2001.

[14.19] J. Karaoguz, "High Rate Wireless Personal Area Networks," *IEEE Communications Magazine*, December 2001.

[14.20] *http://www.zigbee.org*.

[14.21] C. M. Cordeiro, "Medium Access Control Protocols and Routing Stratgies for Wireless Local and Personal Area Networks," Ph.D. Dissertation, University of Cincinati, December 2, 2003.

[14.22] J. Karaoguz, "High Rate Wireless Personal Area Networks," *IEEE Communications Magazine*, Vol. 39,No. 12, pp. 96–102, December 2001.

[14.23] E. Callaway, et-al., "Home Networking with IEEE 802.15.4: A Developing Standard for Low-Rate Wireless Personal Area Networks," *IEEE Comminications Magazine* Vol. 40, No. 8, pp. 70–77, August 2002

14.7 Problems

P14.1. What happens if you use two household cordless phones at the same time? Explain with appropriate reasons.

P14.2. You might have observed the following:

(a) When you open your garage door, your next-door neighbor's door might also open.

(b) Your neighbor complains sometimes that the TV channel is changing automatically.

Can you think of ways to avoid such phenomena? Explain clearly.

P14.3. A set of small robots needs to be equipped with wireless devices. Consider the usefulness of the following devices if used in a laboratory environment:

(a) Infrared (IR)

(b) Diffused infrared

Obtain information about infrared communication from your favorite Web site.

P14.4. Repeat Problem P14.3, if the robots are employed for a field application.

P14.5. A set of small robots needs to be equipped with wireless devices. Consider the usefulness of the following devices if used in a laboratory environment:

(a) WMAN

(b) WLAN

(c) WPAN

P14.6. Do Bluetooth devices and household microwave ovens interfere? Explain.

P14.7. What impact will Bluetooth devices connected to mobile units have on the piconet?

P14.8. In a hypothetical wireless system, five adjacent frequency bands (f_1, f_2, f_3, f_4, f_5) are allowed for frequency hopping sequences. Enumerate how many different hopping sequences are possible and prove their correctness.

P14.9. In Problem P14.8, it was decided to add five additional channels, (f_6, f_7, f_8, f_9, f_{10}) while keeping the frequency hopping sequence to five bands. Is it advisable to maintain frequency hopping within each of the channels (f_1, f_2, f_3, f_4, f_5) and (f_6, f_7, f_8, f_9, f_{10}) or it is better to select five channels among the bands (f_1, f_2, f_3, f_4, f_5, f_6, f_7, f_8, f_9, f_{10})? Explain your answer with some quantitative measures.

P14.10. A conference organizer decided to have eight separate groups of panels, A, B, C, D, E, F, G, and H to make decisions on eight parallel tracks for a professional meeting. To facilitate communication between six members of each group, a piconet is formed using Bluetooth-enabled laptops. The following hopping sequence is followed by a piconet of each group.

Group	Allocated Frequency Hopping Sequence							
A	f_1	f_5	f_9	f_{13}	f_{17}	f_{21}	f_{25}	f_{29}
B	f_2	f_6	f_{10}	f_{14}	f_{18}	f_{22}	f_{26}	f_{30}
C	f_3	f_7	f_{11}	f_{15}	f_{19}	f_{23}	f_{27}	f_{31}
D	f_4	f_8	f_{12}	f_{16}	f_{20}	f_{24}	f_{28}	f_{32}
E	f_{13}	f_{17}	f_{21}	f_{25}	f_{29}	f_1	f_5	f_9
F	f_{14}	f_{18}	f_{22}	f_{26}	f_{30}	f_2	f_6	f_{10}
G	f_{15}	f_{19}	f_{23}	f_{27}	f_{31}	f_3	f_7	f_{11}
H	f_{16}	f_{20}	f_{24}	f_{28}	f_{32}	f_4	f_8	f_{12}

If there is a collision, quantify the fraction of time during which such an interference may be present.

P14.11. Assuming that channel $f_i \neq f_j$ for $i \neq j$, find if there can be a collision and an interference in Problem P14.10 if

(a) Group (A, B, C, and D) or Group (E, F, G, and H) are simultaneously operating due to overlapping of memberships between these groups.

(b) All eight groups are communicating simultaneously.

(c) Any six groups are operational at a time due to limitations on the availability of laptops.

P14.12. In Problem P14.11(a), Group A may need to communicate with Group B about a submission, overlapping between the two subject areas. How is it possible to establish such an interaction? Consider all possible feasibilities and explain with appropriate justifications.

P14.13. The group of reviewers is to be redistributed in the afternoon as follows: Group (A, B, E, F) and Group (C, D, G, H). Repeat Problem P14.11 for this new distribution.

P14.14. What are the advantages and disadvantages of using Bluetooth-based devices as a sensor network? Explain your answer from a possible feasibility point of view.

P14.15. Forming a cluster of Bluetooth devices into a piconet is important. Can you think of any strategy to define members of a piconet? Justify your answer.

P14.16. A bridge node provides access between two adjacent piconets. How can you schedule from one piconet to bridge to the second piconet so that information can be transferred? Explain.

P14.17. Can you apply different ad hoc network routing protocols to a scatternet? Explain clearly with suitable examples.

P14.18. What is the rationale behind using different slot sizes in Bluetooth? Explain clearly.

P14.19. How do you ensure that two adjacent piconets do not use the same frequency hopping sequence? Explain.

P14.20. Can you possibly use "orthogonal latic squares" to avoid the problem indicated in Problem P14.19? Find out details on "orthogonal latic squares" from Web search. Explain clearly.

P14.21. Compare HyperLAN 2 and Bluetooth.

P14.22. Compare the usefulness and limitations of WMANs, WLANs, and WPANs.

P14.23. What is the fundamental difference between the Ricochet solution and IEEE 802.16?

P14.24. Describe the wireless solutions that you would recommend for the following situations. Some situations may need multiple standards. Explain clearly.
(a) A person carries a PDA, laptop, bio-sensors, and wrist watch with applications that are collaborative in nature and communicate with the Internet.
(b) A salesman on the road needs to keep track of product inventories.
(c) A group of executives meet in a conference room and want to digitally exchange their business cards.
(d) A group of conference organizations needs to take "conflict of interest" into account while discussing conference submissions and making acceptance decisions.

CHAPTER 15

Recent Advances

• • • • • • • • • • • • • • •
15.1 Introduction

Wireless and mobile technology has been advancing at an unparalleled rate and its impact is being observed in many facets of our daily life. Recent advances and future directions are being explored for home, industrial systems, commercial, and military environments. In a house, a central access point (AP) is expected to communicate with various appliances and control them using wireless mode, even from a remote location. HomeRF, Bluetooth, and Jini projects, which are being pursued by a consortium of companies, seem to be a significant step in this direction. A system like this could support a bracelet, which would constantly monitor various body functions and parameters and indicate abnormalities. However, a lot more effort is needed before such a system can be realized.

In commercial applications, the issues are the range of the system, the number of APs, and the number of users for each AP. For example, in a department store, each floor may have one AP, while in a factory, several uniformly spaced APs per floor may be needed. The communication could be either voice or data packets, or a combination of both. In defense applications, effective communication could be achieved using either an infrastructure system, or could be supported by a decentralized, peer-to-peer, MANET formed with close-by mobile users. In all these systems, security both in terms of authentication and encryption is critical. It is important to optimize power usage and routing table size and sustain a path during a transmission session in MANETs.

A wireless system, in general, is expected to provide "anytime, anywhere" service. This feature is essential only for military, defense, and a few critical areas such as nuclear power, aviation, and medical emergencies. For most applications, "many times, many where" attributes may be adequate. Attempts are being made to move intelligence to the user side as much as possible, and usage charges based on service time and not purely on connection time are being considered. Emphasis is on a scalable communication paradigm. Different kinds of mobility, are being characterized, and the corresponding effect on handoff in various layers needs to be examined. To minimize handoff, the use of a macrocellular infrastructure and multilevel overlapped schemes being investigated are for users with different mobility characteristics. However, it may be better to have a large number of small cells, rather than a few larger cells. On the other hand, small cells cause frequent handoffs, especially for highly mobile users. Therefore, there is a tradeoff, and an optimum solution may depend on service requirements and mobility characteristics.

In second-generation wireless systems, the emphasis was on voice communication, and data loss was not considered. Now, there is a need to provide seamless Internet access, and ways to handle integrated voice and data traffic need to be examined carefully. Third-generation systems must support real-time data communication, while maintaining compatibility with existing second-generation systems. Also, the kind of language support needed to provide a seamless Web access in the sky needs to be examined carefully. The future direction in the wireless and mobile systems area was summarized in a recent National Science Foundation–sponsored workshop [15.1].

A recent FCC approval of additional frequency bands has encouraged the use of ultra-wideband (UWB) communication technology, as multimedia applications demand a lot of bandwidth. This also necessitates the use of a unified model to represent voice and data over mobile IP. One approach being explored is to classify packets as real-time and non–real-time and control the bandwidth by assigning priority to both handoff and real-time calls. This could be considered as an attempt to satisfy QoS for different applications, including protocols for multimedia service as applied to laptops, PDAs, Palm Pilots, and cell phones. The multimedia traffic often needs to be multicasted to a group of subscribers, and each type may need a slightly different type of support.

Another class of networks, described as MANETs, is being explored for numerous applications, and it is important to look at how routes in these networks could be maintained for successful transmission of information between two arbitrary MSs. In this respect, Bluetooth , Wi-Fi, and WiMAX devices are adding another dimension by augmenting existing wireless capabilities. A system level adaptation of wireless devices employing a minimum level of interaction is desirable, and the impact of software portability and language constraints should be examined carefully. Security issues are critical in all such systems. Many of these are discussed here as an indication of future research endeavors. In this chapter, we consider some of the research areas being pursued in wireless systems.

● ● ● ● ● ● ● ● ● ● ● ● ● ● ● ● ●

15.2 Ultra-Wideband Technology

Ultra-wideband (UWB) technology, also known as impulse or zero-carrier radio technology, appears to be one of the most promising wireless radio communication technologies of our time. Unlike conventional radio systems, which operate within a relatively narrow bandwidth, the UWB radio system operates across a wide range of the frequency spectrum by transmitting a series of extremely narrow (10–1000 per second) and low-power pulses [15.2]. The low-power signaling is accomplished by reusing previously allocated RF bands, by hiding the signals under the noise floor of the spectrum [15.3]. When properly implemented, UWB systems can share this spectrum with other traditional radio systems without causing noticeable interference and provide a highly desirable way of easing the bottleneck due to the scarcity of the radio spectrum [15.3].

This technology is not an entirely new concept. Some early pioneers—Heinrich Hertz [15.4] and others—used spark gaps to generate UWB signals even before sinusoidal carriers were introduced at the beginning of the last century. However, only recently, it has been possible to generate and control UWB signals and apply modulation, coding, and multiple-access techniques to make UWB attractive for wireless communication applications [15.4]. Early UWB systems were developed mainly as a military surveillance tool because they could "see through" trees and walls and below ground surfaces. Now, UWB technology is focused on consumer electronics device communications as well.

15.2.1 UWB System Characteristics

The UWB signal is defined as a signal with bandwidth greater than 25% [15.5][15.6] of the center frequency or with bandwidth greater than 1 GHz. This wide bandwidth makes it possible to share the spectrum with other users. Recent results reveal that UWB signals are naturally suited for location determination applications. There are several methods of generating these UWB signals. Two of the popular methods are low duty cycle impulse UWB implemented as time modulated—UWB (TM–UWB), and high duty cycle direct sequence phase coded UWB (DSC–UWB) [15.5]. Wide spectra are generated in these two methods. The propagation characteristics and application capabilities vary considerably in these two methods [15.5].

- *TM–UWB technology*: The basic element in TM–UWB technology is the monocycle wavelet. Typically, wavelet pulse widths are between 0.2 and 1.5 nanoseconds, corresponding to center frequencies between 600 MHz and 5 GHz. The pulse-to-pulse intervals are between 25 and 1000 nanoseconds [15.5]. In TM–UWB, the system uses a modulation technique called pulse position modulation [15.5]. The TM–UWB transmitter emits ultra-short monocycle wavelets with tightly controlled pulse-to-pulse intervals, which are varied on a pulse-by-pulse basis in accordance with an information signal and a channel code. The modulation makes the signal less detectable as the signal spectrum is made smoother by the modulation [15.5]. A pulse generator generates the transmitted pulse at the required power. The transmitter also has a picosecond precision timer that enables precise time modulation, pseudonoise (PN) encoding, and distance determination. The TM–UWB receiver directly converts the received RF signal into a baseband digital or analog output signal with the help of a front-end cross correlator. There is no intermediate frequency stage, which reduces the complexity of the transmitter and the receiver design [15.5]. Generally, multiple monocycles carry a single bit of information and at the receiver these pulses are combined to recover the transmitted information. The precise pulse timing inherently enables accurate positioning and location capability in a TM–UWB system [15.5].

- *DSC–UWB technology*: A second method of generating useful UWB signals is the [15.5] DSC–UWB approach. Here, the signal is spread by direct sequence modulating a wavelet pulse trains at duty cycles approaching that of a sine wave carrier [15.5]. The spectrum spreading, channelization, and modulation is provided by a PN (pseudonoise) sequence, and the chipping rate is maintained as some fraction of carrier center frequency.

15.2.2   UWB Signal Propagation

Fundamentally, UWB impulse wavelets propagate by the free space law. The coherent interaction of signals arriving by many paths causes the Rayleigh or multipath fading in RF communications. Inside buildings, when continuous sine waves are transmitted wherein the channels exhibit multipath differential delays in the nanosecond range, the multipath fading occurs naturally [15.5]. This issue cannot be resolved by relatively narrowband channels, and hence a significant Rayleigh fading effect must be contended with in systems like IS-95.

Properly designed UWB systems can have bandwidths exceeding 1 GHz and are capable of resolving multipath components with differential delays of a fraction of a nanosecond. For example, when a monocycle arrives at the receiver using two different paths [15.5], the receiver can lock on to either pulse and receive a strong signal. More than one correlator can be used to lock on to different signals and energy from the signals can be added, thereby increasing the received S/N. It is natural and possible that a given pulse may interfere with another late-arriving reflection from the previous pulse in a train of transmitted pulses. However, these interfering pulses can be ignored, as each individual pulse is subject to PN time modulation and more than one pulse carries the bit energy. Consequently, this multipath interference may not cause any loss at an UWB receiver and instead in an in-building environment the UWB system architecture can improve the performance (S/N) by 6 to 10 dB [15.5].

15.2.3   Current Status and Applications of UWB Technology

The application of UWB technology was once restricted to military, police, and firefighter systems. However, in early 2002, the FCC cleared the way to use UWB technology for commercial wireless applications. Concerns about interference with frequencies currently in use by radio, TV, and mobile phone carriers prompted the FCC to put restrictions on which frequencies UWB could be operated in, taking special care to avoid interference with those used by the military and GPS services. Various companies and research organizations around the world have been involved in developing prototype applications to study the feasibility of commercial use of UWB technology. TM–UWB has the potential to create more bandwidth in the increasingly crowded radio spectrum. This technology has three distinct application capabilities [15.5]: communications, advanced radar sensing, and precision location and tracking.

The noiselike spectral characteristics of UWB signals enable secure communication with a greatly reduced probability of detection. Short pulse wavelets of TM–UWB that are relatively immune to multipath interference are suitable for robust in-building communications, especially in urban areas [15.5]. The precision timing of pulses (in TM–UWB) has enabled the development of through-the-wall radar with detection, ranging, and motion sensing of personnel and objects with centimeter precision [15.5]. Another more accurate radar—ground penetrating and vehicle anti-collision radar—is also feasible [15.5].

Precision timing also enables applications involving accurate location and tracking capabilities as well as unmanned vehicle applications [15.7]. DSC–UWB is suitable for most data communication applications [15.5]. UWB technology is appropriate for the high-performance wireless home network, which mandates support for large bit rate (50 Mbps), high-speed, affordable connectivity between devices, simultaneous data transmission from multiple devices, and full-motion video capability [15.8].

Since the UWB signal can provide undetectable interference with other signals, UWB can coexist with other technology (Bluetooth, 802.11a/b/g) without mutual disruption. UWB technology can also be useful for a lot of WPAN applications, such as enabling high-speed wireless universal serial bus (WUSB), wireless PC peripheral connectivity, and replacing cables in next-generation Bluetooth technology devices, such as 3G cell phones, high speed and low power MANET devices etc. [15.9].

15.2.4 Difference Between UWB and Spread Spectrum Techniques

As stated before, UWB technology differs from conventional narrowband RF and spread spectrum technologies. UWB uses an extremely wide band of RF spectrum to transmit more data in a given period of time than the other traditional technologies.

The spread spectrum techniques include direct sequence spread spectrum and frequency hopping spread spectrum, and applications such as Bluetooth technology, IEEE 802.11a/g. In such spread spectrum systems, the spread spectrum signal is modulated by a carrier with a PN pseudorandom code or hopping pattern to move the already spread signal to the most suitable band for transmission. UWB is a time-domain concept and there is no carrier modulation. Actually, the spread bandwidth for a UWB waveform is generated by time hopping modulation, and this modulation process is limited to a very short duration pulse. Since individual transmission bits are subdivided into biphase-modulated chipping intervals or distinct frequency changes in spread spectrum systems, the carrier of such systems always has 100 percent duty cycle. While in UWB, pulse durations are very short compared with its pulse interval durations; therefore, the duty cycle is extremely small percent (about 0.5 percent) and such a low-duty cycle leads to a large peak-to-average ratio and low power consumption.

15.2.5 UWB Technology Advantages

The combination of larger spectrum, lower power, and pulsed data means that UWB causes less interference than narrowband radio designs while yielding low probability of detection and excellent multipath immunity. This wide spectrum signature provides UWB with even greater advantages, like very precise range information, which could be used for security purposes in a WLAN/WPAN environment, as well as a strong capability for overcoming very high levels of interference from other narrowband devices [15.2]. In addition, UWB systems are much less complex, allowing for significantly lower cost and smaller size, since they do not use any radio frequency/intermediate frequency (RF/IF) conversion stages, local oscillators, mixers, and other expensive surface acoustic wave (SAW) filters common to traditional radio technologies [15.2]. Broad consumer adoption of wireless networking technology can finally become a reality [15.2].

15.2.6   UWB Technology Drawbacks

UWB is a disruptive technology for wireless networking applications [15.2], and its use would not be appropriate for a WAN deployment such as wireless broadband access. UWB devices are power limited because they must coexist on a noninterfering basis with other licensed and unlicensed users across several frequency bands. Furthermore, antenna gain cannot be increased to operate at greater range since power limits on UWB devices are angle independent. An implementation in low-voltage CMOS (Complementary Metal Oxide Semiconductor) is not possible as some UWB systems might exhibit a high peak-to-average ratio (PAR) [15.2]. For UWB systems using PPM as their modulation technique, limited jitter requirements could be an issue.

15.2.7   Challenges for UWB Technology

To make the highly promising new UWB technology a popular scheme for commercial wireless applications, the following challenges need to be addressed:

- UWB system designers need to provide an extremely accurate pulse design that produces emissions with flat and wide power spectral densities [15.10][15.11].
- Harmful interference effects of UWB signals to narrowband receivers and those of narrowband transmitters to UWB receivers must be understood completely [15.10][15.11].
- Requirements for the PHY and MAC functions of wireless devices based on UWB-radio technology (UWB-RT) [15.4] must be understood.
- Since UWB radio devices are suitable for communications and location tracking applications and services, there is a need to determine under what conditions and in what way the functions of communication and location tracking can or should be combined [15.10][15.11].
- If it is necessary to ensure that implementation of UWB technology does not cause interference to systems operated in the radio spectrum used for aeronautical safety, public safety, emergency and medical, military, and other consumer and business product services.
- New measurement techniques are needed to measure the characteristics of noise-like UWB signals having transient behavior [15.10][15.11].
- If it is necessary to identify and standardize the requirements and characteristics for a wireless home network with a variety of devices connected to each other [15.8].

15.2.8   Future Directions

UWB technology has several unique characteristics, including high capacity, low probability of multipath fading, interference immunity, low probability of detection, and frequency diversity, which allow for a simpler and more cost-efficient radio design. UWB is suitable for a broad variety of applications and, when implemented efficiently, has the potential to address the "spectrum drought."

● ● ● ● ● ● ● ● ● ● ● ● ● ● ● ● ●
15.3   Multimedia Services Requirements

Two general trends can distinguish multimedia requirements with respect to telecommunications: an increasing demand for bandwidth and a need for transparent support for user mobility. Existing telecommunication networks offering high bandwidth are mainly wired networks, whereas existing mobile communication systems mainly offer relatively low data rates. The third-generation mobile communication systems (IMT-2000) offer higher data rates than the current wireless mobile systems, and high data rates could be achived if mobility is very restricted.

With respect to high-speed wireless access systems, wireless ATM appears to be a promising technique. Wireless ATM offers transparent wireless connectivity to ATM terminals and LANs, customer premises network, local loop, and peer-to-peer setups. Different research projects in this field are being pursued [15.12]. On today's Internet, 90% of the traffic uses TCP (75% of which is for the WWW [World Wide Web]), while 80% of the networking is done over the IP network. Multimedia streaming on IP has become a main issue.

QoS for a network is defined in different parameters, such as bandwidth, latency, jitter, packet loss, and packet delay. For voice applications, QoS is based on bandwidth, whereas for voice-over IP (VoIP) it is based on latency (i.e., end-to-end delay, which should not be more than 200 ms). Classes of services (CoS) are used to manage each type of traffic in a particular way. ETSI has introduced four CoSs. Class 1 is a best effort service, whereas Class 4 is QoS guaranteed. QoS can be linked to network level or application level. On the network level, QoS depends on the network policy (i.e., mechanisms such as filters, rerouting in the core of the network, and control access at the corners of the network)—for example, intelligence in the routers (OSPF, RIP, SNMP (Simple Network Management Protocol), BGP, etc.). The current Internet uses integrated services (IntServ), differentiated services (DiffServ), multiprotocol label switching (MPLS), and IPv6 for guaranteeing QoS.

In order to form complete end-to-end systems for streaming and communication over a wireless network, standards need to be defined in the following areas:

- Media codecs
- Transport protocols
- Media control protocols
- File formats

- Capability exchange

- Metadata (media description, menus, etc.)

Specific standards from the standards bodies can be found in [15.13].

15.3.1 Media Codecs

MPEG-4 Visual/MPEG-Audio

MPEG-1 and MPEG-2 are limited to audio and video compression, whereas MPEG-4 [15.14][15.15] describes the coded representation of natural and synthetic multimedia objects. These objects may include images, video, audio, text, graphics, and animation. The objects are usually included in "scenes" by describing their relative positions in space. Built-in interactivity is one of the main features of MPEG-4, enabling the viewer to change object locations or remove them from the display.

Several different types of audio coding, including natural and synthetic sounds, speech and music coding, and virtual-reality content, are integrated in MPEG-4 audio. The MPEG-4 audio capabilities include the following tools:

- **Speech tools**: Used for compression of synthetic and natural speech.

- **Audio tools**: Used for compression of recorded music and other audio sound tracks.

- **Synthesis tools**: Used for very low bit rate description, transmission, and synthesis of synthetic music and other sounds.

- **Composition tools**: Used for object-based coding, interactive functionality, and audiovisual synchronization.

- **Scalability tools**: Used for the creation of bit streams that can be transmitted, without recoding, at several different bit rates.

15.3.2 File Formats

Two file formats are defined for MPEG-4 format: one is based on the Apple Quick-Time format, and the other is based on the Microsoft ASF (Advanced Streaming Format) [15.16] format. The MPEG committee has decided to use the scheme based on the QuickTime format and calls it MPEG-4 Intermedia.

15.3.3 HTTP

Hypertext transfer protocol (HTTP) is a very simple and widely used way to stream media files, since it works with regular Web servers and does not require a special media server. The file format on the wire is the same as the file format for storage since HTTP can be used to stream multiplexed files directly. In MPEG-4 systems, the transport of streams is divided into four layers [15.16] as follows:

- **Compression layer**: Includes elementary (raw) media streams (audio, video, etc.).

- **Synchronization layer**: Adds a header to each access unit of an elementary stream, which includes timestamps, reference to a clock elementary stream, and identification of key frames (random AP). This is similar to the task of the real-time transport protocol (RTP) [15.17] [15.18][15.19][15.20] in IP networks.

- **Flexmux layer**: Groups elementary streams according to common attributes, such as QoS requirements.

- **Transmux layer**: This is the actual transport protocol, like RTP/UDP in MPEG-2. MPEG-4 does not define its own transport protocol but assumes that the application relies on an existing transport protocol.

15.3.4   Media Control Protocols

To enable full streaming systems, a media control protocol needs to be defined to support the following features:

1. Seeking (forward/rewind/skip)
2. Bandwidth scalability
3. Live streaming

The real-time streaming protocol (RTSP) [15.21] establishes and controls time-synchronized streams of continuous media such as audio and video and acts as a "network remote control" for multimedia servers.

15.3.5   SIP

The session initiation protocol (SIP) [15.22] is an application-layer control (signaling) protocol for creating, modifying, and terminating sessions with one or more participants. In addition to multimedia over cellular networks, IP broadband networks have caught researchers' attention. The multimedia vision for broadband networks is shown in [15.23]. Providing fixed mainstream transport for broadband multimedia services brings wireless into the mainstream of communications networking. The need for multimedia in broadband Internet is given in [15.23].

15.3.6   Multimedia Messaging Service

The multimedia messaging service (MMS)[15.24] is an open industry specification developed by the WAP forum for the third-generation partnership program (3GPP). This service is a significant enhancement to the current SMS service, which allows only text. MMS has been designed to allow rich text, color, icons and logos, sound clips, photographs, animated graphics, and video clips and works over the broadband wireless channels in 2.5G and 3G networks. MMS and SMS are similar in the sense that both are store-and-forward services where the message is first sent to the network, which then delivers it to the final destination. But unlike SMS, which can only be sent to another phone, the MMS service can be used to send messages to a phone or may be delivered as an email.

The main components of MMS architecture are

- MMS Relay: Transcodes and delivers messages to mobile subscribers.
- MMS Server: Provides the store in the store-and-forward MMS architecture.
- MMS User Agent: An application server giving users the ability to view, create, send, edit, delete and manage their multimedia messages.
- MMS User Databases: Contains records of user profiles, subscription data and the like.

The content of MMS messages is defined by the MMS Conformance Specification Version 2.0.0, which specifies SMIL 2.0 Basic profile for the presentation format.

Although MMS is targeted toward 3G networks, carriers all over the world have been deploying MMS on networks such as 2.5G using WAP. This is a way of generating revenue from older networks.

Some of the possible application scenarios are as follows:

- Next generation voicemail: It is now possible to leave text, pictures, and even video mail.
- Immediate messaging: MMS features "push" capability. That is to say, as long as the receiving terminal is on, the message is delivered instantly rather than having to be "collected" from the server. With the prospect of "always-on" terminals, this opens up exciting possibilities of multimedia chat in real time.
- Choosing how, when, and where to view the messages: Not everything has to be instant. With MMS, users have an unprecedented range of choices about how their mail should be managed. They can predetermine what categories of messages are to be delivered instantly, stored for later collection, redirected to their PC, or deleted. What is more, they also have dynamic control with the ability to make ad hoc decisions about whether to open, delete, file, or transfer messages as they arrive.
- Mobile fax: Any fax machine can be used to print out an MMS message.
- Sending multimedia postcards: A clip of holiday video could be captured through the user's handset's integral video cam or uploaded via Bluetooth from a standard camcorder and then combined with voice or text messages and mailed instantly to family and friends.

15.3.7   Multimedia Transmission in MANETs

Multimedia transmission usually requires higher bandwidth due to time-sensitive nature of the data. Providing multimedia communication in MANETS implies supporting communication among a group of users having unpredictable size and membership, and integrated objects of multimedia data could be delivered by establishing multiple virtual channels between the server and the end stations. These channels are also of varying bandwidth capacity and different end-to-end delay and path lifetime, which introduce real challenges in realizing desired level of QoS. In MANETs, node mobility may lead to frequent disconnection of wireless links and dynamically change the network topology. The major factors affecting the ad hoc channel performance are large route discovery delays in the event of route change and throughput performance deterioration with increase in the number of hops. A multimedia document server generates multiple objects in the data stream that renders different-sized data. Such variable bandwidth requirements are also present when different users retrieve data at random intervals of time. For a network to deliver QoS guarantees, it must reserve and appropriately control resources.

The major limitation in transmitting real-time video information over MANET is the issue of link reliability. To improve the quality of the video reception, a cross-layer feedback control mechanism has been proposed [15.25] that can allow the application layer to adapt itself to a dynamically changing network topology. The AODV [15.26] routing protocol used in conjunction with IEEE 802.11b MAC layer, relies on positive acknowledgment. If the transmitter does not receive any acknowledgment after a given number of retransmission attempts, a link breakage is triggered. The application layer resumes its communication as soon as a new route is established. From video transmission point of view, video frame can be switched to the I-frame mode (Intraframe) to encode the incoming frame as soon as the new route is established. I-frame coding, though having a higher bit rate (compared with interframe prediction, P-frame), has the advantage of being independent of previously coded frames and can therefore speed up the re-synchronization of the video at the decoder. Switching to P-frame can also be done, as shown in Figure 15.1. If redundant packets are present in a frame, the identical part is removed, reducing the packet size and re-encapsulating into a single RTP (real-time protocol) packet. If a link breakage occurs, redundant packets can also be used to replace the missing frame.

In MANETs, QoS routing can be supported by employing a dynamic channel assignment. The two most widely used algorithms are minimum-blocking channel assignment and bandwidth-reallocation channel assignment (MBCA/BRCA) [15.27]. These methods can increase a system's aggregate traffic by 2.8 kbps as compared to a

Figure 15.1
Effect of a route change on a consecutive coded video frame and packet transmission strategy in a MANET [15.25].
Credit: H. Gharavi and K. Ban, "Dynamic Adjustment Packet Control for Video Communications over Ad-hoc Networks," *IEEE ICC* 2004, pp. 3086-3090, 2004.

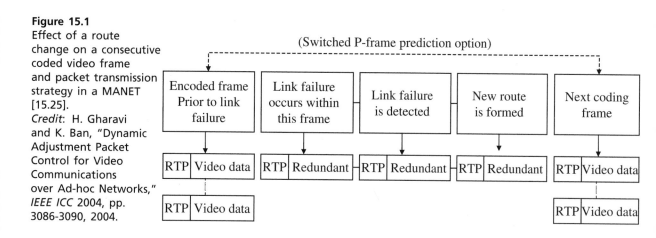

conventional scheme. MBCA is a novel method for slot reservation that also leads to enhanced channel usage. When a MS requests a new connection with QoS bounds, the network admission controller checks whether the link has sufficient available resources to satisfy the minimum bandwidth required (LBW). Once the admission controller at each radio along the multihop path accepts the call, it passes the admission decision and the requirement profile of the MS to the packet-sorter module, which classifies traffic packets of admitted flows. During the reverse pass, the network reserves and allocates resources. Bandwidth, buffer space, and schedulability are the main reserved resources. If the MANET employs clustering of MSs, then TDMA is chosen as the channel access scheme within a cluster for real-time traffic. CDMA could be overlaid on top of the TDMA infrastructure to minimize interference while enhancing usage of limited bandwidth.

15.4 Push-to-Talk (PTT) Technology

Push-to-talk (PTT) is a "walkie-talkie-type" service implemented over cellular networks. It is also abbreviated as P2T, PoC (PTT over Cellular). PTT terminals have a PTT button that a user presses to start a conversation. The conversation can be a person-to-person conversation or one of various types of group conversations. It is an instant, half-duplex communication medium that allows callers to connect rapidly with each other. Nextel, the U.S. operator of PTT, first introduced the service on its integrated digital enhanced network (iDEN) almost ten years ago.

PTT is a quick, short, and spontaneous communication from the users' perspective. The users of PTT pay only for the resources consumed, which is measured in number of bits transferred carrying talk burst rather than the period of connection. Once a PTT call has been established, the participants can communicate immediately. PTT can be implemented over packet networks, and users can be reached for traditional circuit switched household telephones.

From a mobile systems point of view, PTT is a new type of service with distinctive features. It is an add-on feature to normal cell phones, and its underlying streaming characteristics makes it very suitable for packet networks, such as always connected. However, it can significantly increase the GPRS traffic in today's networks. It is also considered as a front-runner in peer-to-peer services over IP, as the IMS (IP multimedia subsystem) architecture provides the capabilities and foundation [15.28].

15.4.1 PTT Network Technology

PTT does not broadcast to all the radios within range. A PTT handset unicasts to the nearest BS. From the BS, the call enters the network, where switching and IP call routing take place. The call is only retransmitted over the air from the BS where the receiving-party is located [15.29]. There are mainly two kinds of PTT. One is Motorola and Nextel's iDEN cellular networks and the other is PTT in non-iDEN cellular networks. In the autumn of 2003, Ericsson, Motorola, and Siemens submitted their jointly defined PoC specifications to OMA (open mobile alliance) to facilitate multi-vendor interoperability for PTT products [15.30].

15.4.2   PTT in iDEN Cellular Networks

Nextel phones offer the service "direct connect," which has the PTT features. It has an entirely separate special cellular network that has its own frequencies and equipment in addition to the normal cell network shared with other providers. This network is based on Motorola's iDEN and makes direct connect possible. iDEN was first introduced in 1994 by Motorola. It uses the 800 MHz portion of the radio spectrum assigned to specialized mobile radio (SMR) service. iDEN uses TDMA technology to split a 25 KHz frequency into six separate time slots. Using a combination of half-duplex and full-duplex signals. iDEN can provide the following services: normal cell phone voice communications; messaging (pager, email); digital two-way radio (one-to-one and group); and data services (wireless Web and private networks) [15.31]. The digital two-way radio service uses a half-duplex signal as a direct connect call uses only a single frequency. PTT requires the person speaking to press a button while talking and then release it when he or she is done. The listener then presses his or her button to respond. This way the system knows which direction the signal should be traveling in. To enable direct connect, Nextel configures your phone to use the dispatch call service to reach the person or persons you specify. This person (or group) must use Nextel's service also.

15.4.3   PTT in Non-iDEN Cellular Networks: PoC

PoC uses the GPRS network to send packetized speech between participants of a PoC session. The signaling architecture of PoC is based on SIP, which is used to establish and maintain sessions between users. Within UMTS, PoC forms one of the services supported by the IMS [15.32].

In the OMA specifications, the architecture of PoC is defined as shown in Figure 15.2. This architecture is based on the requirements listed for the system in the PoC requirements document [15.33] and includes the functional entities, interfaces, system concepts, and high-level procedures of the PoC services. The access network used by the PoC architecture includes both the radio access as well as the other nodes required to gain IP connectivity and IP mobility. PoC utilizes the SIP/IP core based on capabilities from IMS as specified in 3GPP (3GPP TS 23.228)[15.34] and 3GPP2 (3GPP2 X P0013.2) [15.35]. PoC functional entities include PoC client, PoC server, and group and list management server (GLMS).

- **PoC Client**: The PoC client resides on the MS and is used to access PoC service [15.33].

- **PoC Server**: The PoC server implements the application level network functionality for the PoC service [15.33].

- **Group and List Management Server (GLMS)**: PoC users use the GLMS to manage groups and lists (e.g., contact and access lists) that are needed for the PoC service [15.33].

External entities providing services to the PoC system include SIP/IP core, charging entity among others.

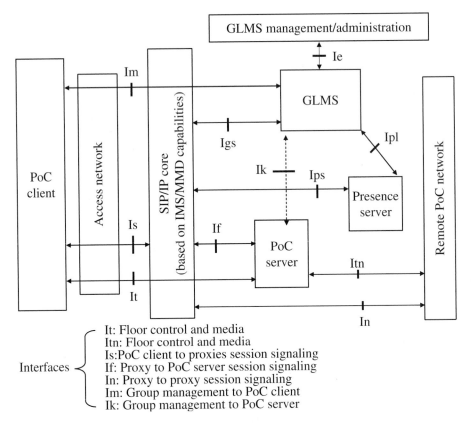

Figure 15.2
PoC architecture
[15.33].

Interfaces
It: Floor control and media
Itn: Floor control and media
Is: PoC client to proxies session signaling
If: Proxy to PoC server session signaling
In: Proxy to proxy session signaling
Im: Group management to PoC client
Ik: Group management to PoC server

15.4.4 Limitations of Current Services

PTT attracts users from many different segments due to its characteristics inherited from other popular services (telephony, messaging, and walkie-talkie) and the simplicity of using and understanding the service [15.28]. As more companies start offering this service, their service interaction compatibility with each other becomes increasingly important. Standards for PTT are being established so that customers would get better services. GPRS, WCDMA, and CDMA2000 can all meet PoC's technical requirements.

However, there still remain many challenges. PTT suffers from latency: pauses of a second or more between pushing the phone's walkie-talkie button and receiving a call, are typical of handling voice calls on a data network. If you want to initiate PTT talk, you just press the button and are ready to talk immediately. You just need to wait for the other side to reply. Also if both ends press the button at the same time, either of the two subscribers can hear each other. Also customers have to purchase a new handset, and widespread acceptance of PTT will hinge on roaming and/or interoperability agreements among carriers. The quality of PTT service still needs to be improved.

● ● ● ● ● ● ● ● ● ● ● ● ● ● ● ●

15.5 Mobility and Resource Management for Integrated Systems

Numerous companies and service providers are pursuing a fully integrated service solution for wireless mobile networks. Current voice, fax, and paging services will combine with data transfer, video conferencing, and other mobile multimedia services. The next generation of wireless mobile networks, such as the IMT-2000 network and the UMTS, have been designed to support a true combination of both real-time service and non–real-time service and then form a global personal communication network [15.36]. In order to support such a wide range of traffic efficiently, appropriate resource management schemes are crucial.

15.5.1   Mobility Management

First of all, mobility management has to be taken into consideration while designing the infrastructure itself for wireless mobile networks. Effective and efficient handoff is the key factor enabling the mobile user to move seamlessly from one cell to another cell, from one service area to another, and so on.

Mobility management features two tasks—location management and handoff management—that enable mobile networks to locate roaming MSs for call delivery and maintain connections as the MSs are moving around. Location management enables the wireless network to discover the current point of attachment of the MS and deliver calls. The first stage of location management is the location registration (or location update). In this stage, the MS periodically notifies the network of its new AP, allowing the network to authenticate the user and revise the user's location profile. The second stage is the call delivery, in which the wireless mobile network is queried for the MS location profile and the current position of the MS is found. Handoff primarily represents a process of changing some of the parameters of a channel (frequency, time slot, spreading code, or a combination of them) associated with the current connection in progress. The handoff process usually consists of two phases: the handoff initialization phase and the handoff-enabling phase. In the handoff initialization phase, the quality of the current communication channel is monitored in order to decide when to trigger the handoff process. In the handoff execution phase, the allocation of new resources by a new BS is initiated and processed. Poorly designed handoff schemes tend to generate very heavy signaling traffic, and thereby result in a dramatic decrease in quality of integrated service in the wireless network.

Mobility management requests are often initiated either by an MS's movement by crossing a cell boundary, or by a deteriorated quality of signal received on a currently employed channel. With the anticipated increased penetration of wireless services, the next generation of wireless mobile networks will provide an architectural basis to support a drastic increase in traffic bandwidth. According to the IMT-2000 outline of ITU, a simultaneous operation of high capacity picocells, urban terrestrial microcells and macrocells, and large satellite cells will be exploited in IMT-2000.

Much more frequent handoffs will occur when the size of the cell becomes smaller or there is a drastic change in the propagation condition of the signal. Therefore, mobility management should be given more careful consideration in next-generation wireless mobile networks.

Various handoff initiating criteria have been proposed recently. In order to decide when to trigger the handoff, the quality of the current communication channel is monitored. Handoff is a very rigorous process; therefore, unnecessary handoffs should be avoided. If the handoff criteria are not chosen carefully, the call might be handed back and forth several times between two neighboring BSs's, especially when the MS is moving around the overlapping region between the two BSs coverage area boundaries. If the criteria are too conservative, then the call may be lost before the handoff can take place. Based on the link status, the measurement process determines the need for handoff and the new target cell for transfer. Since the propagation condition between the BS and the MS is made up of the direct radio propagation paths (direct, reflection, refraction), the following types of handoff-initiating criteria have been proposed [15.37][15.38][15.39]:

- **Word error indicator**: A metric that indicates whether the current burst was demodulated properly in the MS.

- **Received signal strength indication**: A measure of the received signal strength that indicates useful dynamic range, typically between 80 and 100 dB.

- **Quality indicator**: An estimate of the "eye opening" for a radio signal, which is related to the signal to interference and noise ratio, including the effects of dispersion. The quality indicator has a narrow range (relating to the range of SIR from 5 dB to 25 dB).

In the design of a good handoff scheme, it is desirable that the blocking probability for calls originated in a cell be minimized as much as possible. However, from the user's point of view, how to handle a handoff request is more important. If new resources cannot be allocated in a timely fashion, the ongoing call has to face forced termination, which is much more disastrous than the blocking of a new call. In addition, attempts should be made to decrease the transmission delay of non–real-time service calls as well as increase channel utilization in a fair manner. Therefore, the handoff strategy for integrated service in next-generation wireless networks needs to take different features of these services into account (i.e., the ideal handoff processes have to be service dependent). For example, transmission of real-time service is very sensitive to interruptions. On the other hand, transmission delay of non–real-time service does not have any significant impact on the performance of service (i.e., non–real-time service is delay insensitive). Therefore, a successful handoff without interruption is very important for real-time services, but not so critical for non–real-time services. In order to provide better service for a MS with limited frequency spectrum, a wireless system must manage radio resources efficiently.

15.5.2   Resource Management

With the rapid increase in the size of the wireless mobile communication and its demands for high-speed multimedia communications, the spectrum resources have become very limited. Therefore, managing radio resources efficiently is very important. A BS can only serve the MSs within its coverage area if the transmission conditions are good enough to maintain the connections with acceptable QoS. Links have to be established both ways, one from the BS to the MS (downlink or forward link) and another from the MS to the BS (uplink or reverse link). The bandwidth needs to be managed carefully so that service can be provided to as many users as possible. Moreover, since each type of service has distinct characteristics and QoS requirements, a thorough understanding of the user requirements (i.e., the required QoS and the traffic characteristics) is useful in supporting multiclass service effectively.

1. (a) **Complete sharing (CS) and complete partitioning (CP)**: Two extreme resource-allocation strategies are complete sharing (CS) and complete partitioning (CP) [15.40]. As the names suggest, all traffic classes share the entire bandwidth in CS while in CP, bandwidth is divided into distinct portions with each portion corresponding to a particular traffic class. CS does not provide any priority differentiation among service classes and a temporary overload of one traffic class results in degrading the connection quality of all other classes. CP is wasteful of bandwidth if the predicted bandwidth demand for a particular traffic class is greater than the actual bandwidth demand. Strategies in between are generally referred to as hybrid strategies.

   (b) **Guard channels**: The use of guard channels for handoff has been commonly employed by voice cellular networks. The guard channel handoff scheme is similar to the nonprioritized handoff scheme except that a number of channels in each BS are exclusively reserved for handoff request calls. Therefore, the total number of channels is divided into two groups: the normal channels, which serve both originating calls and handoff request calls, and the reserved channels, which serve handoff request calls exclusively. In this way, there is a built-in priority for a handoff request call over an originating call as long as guard channels are still available. System performance is better than with the nonprioritized handoff scheme.

   (c) **Queuing scheme**: The queuing scheme for priority handoff is based on the fact that there are overlapped areas between adjacent cells in a wireless mobile network. This area is called the handoff area, where a call can be handled by either BS of adjacent cells. The time that a MS spends in the handoff area is referred to as the handoff area dwell time. In the queuing-based priority handoff scheme, each BS has one or more queuing buffers for all incoming calls. When a call arrives at the BS, it checks whether a channel is available. A call can be serviced immediately if there is an available channel in the BS. However, even if no channel is available when a call arrives, the call will not be blocked or dropped as long as there is free space in the queue for this kind of service. The incoming call is kept in the queue to wait for the next channel available. Whenever a channel is released, the BS first checks whether there are any waiting calls in the queue. If there are, then the released channel is assigned to a waiting call in the queue, and this is usually done on a FIFO basis. In a queuing priority handoff scheme, there are two issues that are of major concern: How many queues should the BS have and what kind of service call should be included in the queue? If a handoff request call can be queued, the queued handoff request calls can keep communicating with the old BS as long as it is still in the handoff area, so that the forced termination probability can be decreased. If originating calls can be queued, the blocking probability of originating calls can be decreased.

(d) **Priority reservation handoff**: One way of giving priority reservation to real-time service handoff requests [15.41] is to reserve a number of channels for real-time service handoff requests. Queues are allowed for real-time service handoff requests and non–real-time service handoff requests. Moreover, a non–real-time service handoff request in the queue can be transferred to another queue in an adjacent cell when the MS moves out of the cell before getting a channel.

(e) **Priority reservation with preemptive priority**: A service-dependent priority handoff scheme for integrated wireless mobile networks has been proposed [15.42][15.43]. Calls are divided into four different service types: originating real-time service calls; originating non–real-time service calls; real-time service handoff request calls; and non–real-time service handoff request calls. Correspondingly, the channels in each cell are divided into three groups. One is for real-time service calls (including originating and handoff request calls), the second is for non–real-time service calls (including originating and handoff request calls), and the last is for overflowed handoff requests from the previous two groups. Of the three groups, some channels are reserved exclusively for real-time service handoff requests. Therefore, the real-time service handoff requests have priority over non–real-time service handoff requests, and all the handoff requests have priority over originating calls. To give the real-time service handoff requests higher priority over non–real-time service handoff requests, a preemptive priority procedure can be introduced, in which the real-time service handoff request has the right to preempt the non–real-time service call when it finds no channel available on its arrival. Individual queues are added for both real-time and non–real-time service handoff requests, so that the interrupted non–real-time service call can return back to the last position of its own queue and wait for the next available channel. The non real-time service handoff requests waiting in the queue can be transferred from the current BS to one of the target BSs when the MS moves out of the current cell before it gets service.

15.5.3 Recent Advances in Resource Management

There are many open issues in the area of resource management as resources are expected to efficiently support multi-class service, with each type of services having distinct characteristics and QoS requirements. An efficient algorithm for near optimal channel allocation is presented in [15.44][15.45] to be applied when different types of service are to be provided. A preemptive priority scheme for an integrated wireless and mobile network is proposed by first dividing channels into three independent groups and classifying traffic into four different types and modeling the system by a multidimensional Markov chain model. A set of relations are obtained that correlate performances with various system parameters. A novel recursive algorithm is developed to determine the minimal number of channels in each channel group that would be necessary to satisfy the QoS requirements.

Existing schemes employ either complete partitioning of channels for each type of service or complete sharing with guard channels wherein overall throughput in terms of channel unitization drops or degrades the QoS of services for traffic with lower priorities. A new cutoff priority scheme is proposed in [15.46], which is quite general, provides much better QoS for higher-priority services than common sharing schemes and improves overall throughput.

Resource management for future-generation MANETs ought to support multimedia services. Voice, video telephone, or video conference require different bandwidth, and these services can be degraded in case of congestion as long as they are still within the prespecified tolerable range. On the flip side, this flexibility may penalize classes having higher bandwidth requirement at heavy traffic load. How the admission-control and bandwidth-allocation scheme can achieve fairness among each class and management of adaptive multimedia applications is becoming a new research topic in wireless and mobile networks. There should be new QoS metrics to evaluate such schemes, such as the degradation ratio of each class, the QoS fluctuation frequency, and the fairness among different classes, which have not been addressed in the existing proposals.

An adaptive admission-control and bandwidth-allocation scheme is proposed in [15.47] which can support multi-class traffic by dynamically adjusting the priority of each class. It is also important to determine whether new coming calls can be accepted and how much bandwidth should be provided to them based on the current system state. A comprehensive set of QoS metrics is described and an analytical model is developed to give closed form expressions for these QoS performance parameters.

● ● ● ● ● ● ● ● ● ● ● ● ● ● ● ●
15.6 Enhancement for IEEE 802.11 WLANs

The IEEE 802.11 standard defines detailed MAC and PHY specifications for WLANs [15.48]. WLANs are growing in popularity because they can provide mobility and flexibility. Furthermore, existing Ethernet-based LANs can be easily extended to support WLAN by using the services of WLAN AP. The basic topology of an 802.11 WLAN consists of two or more wireless nodes, or stations (MSs), which have recognized each other and have established communications. MSs communicate directly with each other in a peer-to-peer fashion. This type of network, covered in Chapter 13, is often formed on the fly and is referred to as a MANET.

The basic access method for 802.11 is the DCF which is based on CSMA/CA, and considerable work has been done in evaluating the performance of this protocol. However, most analytical work is confined to saturation performance of single-hop ad hoc networks. In [15.49][15.50], a linear feedback model is employed to evaluate the performance for CSMA/CA according to the Poisson distributed traffic in both single-hop and multi-hop ad hoc networks. The model consists of a finite population of MSs. An embedded Markov chain is used to analyze the throughput and delay performance. The results show that although RTS/CTS (i.e., request to send/clear to send) do add overhead to the system, they become essential when either the hidden terminal problem is dominant, the traffic is heavy, or the packet length is very large. The results also show that performance degrades dramatically in multi-hop ad hoc networks when the number of competing MSs increases, which implies that scalability is still a major problem in ad hoc networks. It is observed that in multi-hop ad hoc networks, the hidden terminal problem still exists even when employing RTS/CTS. This happens when neighbors of sender/receiver do not receive RTS/CTS correctly.

The IEEE 802.11e working group has developed enhanced DCF (EDCF) to improve the access mechanism of IEEE 802.11 so that the differentiated service could be provided [15.51]. The basic idea is to introduce traffic categories (TC) and provide different priorities to different TCs. The IEEE 802.11e architecture can support multiple queues (up to eight) for different priorities. EDCF has two priority schemes, one of which is the interframe space (IFS) priority scheme, the other is the contention window (CW) priority scheme. In the IFS priority scheme, an arbitration interframe space (AIFS) is used, and a station can send a data packet or start to decrease its backoff counter after it detects the channel being idle for an AIFS. The AIFS is at least as large as the DIFS and can be adjusted for each TC according to the corresponding priority. Thus, the stations with shorter AIFS have a higher priority to access the channel than the stations with longer AIFS. The CW priority scheme implements service differentiation by using a different CW_{min} (i.e., the minimum contention window) between TCs. Since CW is used to determine the waiting time before a station is allowed to transmit its packet, smaller CW implies higher priority.

A very limited analysis of EDCF exists in the literature as it is a new protocol, and most related work is often confined to simulation. In [15.52], the performance of EDCF is evaluated by dividing the traffic into two categories: real-time packets and non–real-time packets. An analytical model is proposed to quantify the performance of both IFS priority and CW priority in the EDCF. In the IFS priority scheme, the AIFS for real-time packets is DIFS, and the AIFS for non–real-time packets is DIFS+SLOT. This priority scheme is evaluated with the average delays for real-time packets and non–real-time packets with the assumption that each station always has a packet available for transmission. Suppose at each station, the fraction of the real-time packets is R_r and that of the non–real-time packets is $(1 - R_r)$. Since the AIFS for real-time packets is one slot shorter than the AIFS for non–real-time packets, according to CSMA/CA, real-time packets can decrease the backoff counter after an idle duration of DIFS when a transmission is finished, while non–real-time packets have to wait for an idle duration of DIFS+SLOT to decrease the backoff counter. Taking the original scheme (i.e., no IFS priority and no backoff priority) be the base, the results show that the IFS priority scheme works better when the number of competing stations is large and it can improve up to 50% for the real-time packets delay when R_r is 0.5. In the CW priority scheme, non–real-time packets use

CW_{min} and real-time packets use $R_r \cdot CW_{min}$ as the minimum contention window. The improvement in backoff priority scheme is observed to be about 33% for the real-time packet delay, no matter how many stations are present in the system. A new priority scheme is also proposed, which allows the user to continuously send real-time packets. To get a good balance between the fairness and the priority, the maximum number of real-time packets (i.e., fairIndex) that a user can continuously send is defined. The proposed scheme is shown to provide much better results than IEEE 802.11e EDCF. It works best and can improve up to 80% for the real-time packets delay when fairIndex equals 2. Furthermore, since the proposed priority scheme can greatly reduce the number of collisions, it can even improve about 30% for overall system performance. Although IEEE 802.11e EDCF can improve the performance for higher-priority traffic, it cannot guarantee all the QoS requirements since contention for the channel still exists. The 802.11e EDCF still needs further investigation before it can become a standard.

The purpose of the contention window is to reduce the collisions. When traffic is light, there are almost no collisions in the system. Therefore, it does not seem so important to optimize the CW for this case. When the traffic is heavy, the collision probability strongly depends on the contention window for CSMA/CA. Then, how to choose the CW becomes essential. The 802.11 standard adopts an exponentially increasing CW. In [15.53], it is shown that an exponential CW cannot provide optimal performance. Based on the analytical model to evaluate the performance of the 802.11 MAC protocol, the optimal CW is obtained. It is observed that the optimal CW scheme outperforms the exponential CW scheme greatly. From the research results, we can conclude that it is highly desirable to look at methods for improving the performance of the MAC layer protocol in the IEEE 802.11.

15.6.1   Issues in MAC Protocols

WLANs are facing the challenges of 802.11-related security as well as the support of multicast and location management. More work is also necessary to address WLAN scalability before WLANs are widely employed. 802.11i is the security standard for Wi-Fi networks (i.e., WLANs) that upgrades the former wireless security standard, wired equivalent privacy (WEP). WEP can easily be cracked by those with the right tools. The industry consortium, Wi-Fi Alliance, introduced Wi-Fi protected access (WPA). It is a subset of the abilities of 802.11i, which include encryption with temporal key integrity protocol (TKIP), setup using a pre-shared key, and RADIUS-based 802.1X authentication of users. 802.11i has all the abilities of WPA and adds the requirement to use the advanced encryption standard (AES) for encryption of data.

Many mobile applications such as distance education, interactive games, and military command and control require support for group communication. Wireless multicast is the most efficient way supporting group communication, as it allows transmission and routing of packets to multiple destinations using fewer network resources. It can update membership information for network traffic routing when

MSs move to different locations or leave the group, or new users join the group. However, how to ensure reliability, privacy, quality of service, and low delay in WLANs are major technical challenges due to the characteristics of WLANs.

Location-based services that personalize the user's experience attract more MSs to use WLANs. These services include location-based billing, information services such as providing listings of local restaurants or movie theaters, emergency services, and tracking services such as vehicle tracking.

Scalability is a major concern to WLANs. The large-scale deployment of WLANs presents technical as well as economic challenges. When multiple network access providers set up more WLANs in hot spots, interference between WLANs exists and QoS cannot be guaranteed. It is necessary that some type of coordination is needed to limit the number of different WLANs in the same area. That is why the Wi-Fi Alliance has certified numerous "Wi-Fi zones" where wireless providers need to meet the strict deployment and service requirements.

In addition to multiple 802.11 standards, there are other standards for WLANs such as the European Hiper-LAN2 [15.54][15.55], which stands for high performance radio local area network. Hiper-LAN2 is a wireless LAN standard developed by the broadband radio access networks (BRAN) division of the European Telecommunications Standards Institute (ETSI). It defines a very efficient, high-speed wireless LAN technology that meets the requirements of Europe's spectrum regulatory bodies. Similar to IEEE 802.11a, HiperLAN/2 operates in the 5 GHz frequency band using OFDM and offers data rates of up to 54 Mbps. In fact, the physical layer of HiperLAN/2 is very similar to 802.11a. However, the MAC layer is much different between 802.11a and HiperLAN/2. 802.11a uses CSMA/CA to transmit packets, while HiperLAN/2 uses TDMA.

With CSMA/CA, all the 802.11 stations share the same radio channel and contend for access. If a MS happens to be transmitting, all other MSs will wait until the channel is free. A problem with CSMA/CA is that it causes stations to wait for an indefinite period of time. As a result, there's no guarantee of when a particular MS can send a packet. The lack of regular access to the medium is not desirable when supporting real-time data such as voice and video information. HiperLAN/2, however, offers a regular time relationship for network access by using TDMA. This TDMA system is a centralized scheduling system, which dynamically assigns each MS a time slot based on the station's demand for the radio channel. The MSs then transmit at regular intervals during their respective time slots, so that they can more efficiently use the medium and improve the support of voice and video applications. HiperLAN/2 is designed to interface with other high-speed networks, including 3G cellular, asynchronous transfer mode (ATM), and other Internet protocol–based networks. This can be a real advantage when integrating wireless LANs with cellular systems and wide area networks. About HiperLAN/2, it can be said that it is defining the future rather than developing a standard for today only. HiperLAN/2 deployment will be one of the best technologies to accommodate the growing requirements of corporate WLANs, along with being able to support next-generation WLAN deployments.

● ● ● ● ● ● ● ● ● ● ● ● ● ● ●

15.7 Multicast in Wireless Networks

15.7.1 Recent Advances in Multicast over Mobile IP

As discussed in Chapter 9, IETF has proposed two methods to support multicast over Mobile IP: remote subscription and bidirectional tunneling (BT) [15.56]. A brief description of these schemes and the MoM protocol has already been provided in Chapter 9. In this section we focus on some of the recently proposed protocols for providing multicast over Mobile IP and also provide some directions for future research.

An enhancement of MoM, called range-based MoM (RBMoM) [15.57], provides a tradeoff between the shortest delivery path and the frequency of the multicast tree reconfiguration. It selects a router, called the multicast home agent (MHA), which is responsible for tunneling multicast packets to the MS's currently subscribed FA. MHA serves MSs, which are roaming around the foreign networks and are within its service range. If a MS is out of service range, then a MHA handoff will occur. Initially, the MHA of a MS is its HA. Every MS can have only one MHA, which dynamically changes based on the location of the MS, whereas a HA of a MS never changes. The protocol requires that each MHA must be a multicast group member.

Multicast for Mobility Protocols (MMP)

MMP [15.58] provides a fast and efficient handoff for MSs in foreign networks and also enables location-independent addressing. MMP combines the concepts of Mobile IP and core-based trees (CBT), where the former controls communication up to the foreign network, whereas the latter manages movement of the MSs inside them. Here the foreign domain forms a hierarchy of multicast supporting routers. Just like FA, BSs acting as multicast routers transmit periodic beacons, which include multicast care-of-address (CoA). Once having acquired the CoA, the MS sends a registration message to the BS, which triggers a multicast tree join and transmits a CBT join request toward the core. The core relays the registration request to the HA of the MS by replacing the CoA with its own address, thus hiding the multicast part of the protocol and acting as a sole foreign agent. In case the domain consists of a hierarchy of multicast routers, the border router can be selected as the core of the network. The drawback of this scheme is that it assumes a large-scale deployment of multicast-capable routers in each domain. Also, it is not protocol independent.

Mobicast

Mobicast [15.59] is designed for an internetwork environment with small wireless cells. It assumes that a set of cells are grouped together and are served by a domain foreign agent (DFA). DFAs serve as multicast forwarding agents and are meant to isolate the mobility of the mobile host from the main multicast delivery tree. This hierarchical mobility management approach tries to isolate the mobility of the FAs from the main multicast delivery tree.

An approach to address the case in which a MS is both source and recipient of a multicast session is proposed in [15.60]. Unlike previous approaches, which mainly handle recipient mobility, this approach also considers the case when a MS is also a source. In general, the effect of receiver movement on the multicast tree is local, whereas it may be global for a source movement and may affect the complete multicast delivery tree. This scheme uses a MS initiated approach for multicast handoff and tries to make the effect of the MS movement local, irrespective of whether it is working as a source or as a recipient of a multicast group.

Mobicast is based on an IETF proposed method to support multicast over Mobile-IP. As mentioned earlier, to handle the case when a MS is both the source and recipient of a multicast session, one needs to minimize the possibility of rebuilding the complete multicast tree at each foreign domain the MS visits. Although bidirectional tunneling looks like an obvious solution here, it makes the reception of multicast packets to the MS very inefficient. Here the scheme proposes to use only a reverse tunnel from the MS's current point of attachment to its HA to forward multicast packets. To receive multicast packets, it uses a remote subscription method. To illustrate, consider a source-based multicast tree as shown in Figure 15.3, with S as one of the multicast sources. Assume S now moves to a foreign domain, which is not presently a member of the multicast group subscribed by S. Hence S sends a multicast join message in its foreign domain and a notify message to its HA. As shown in the figure in steps 2 and 3, a bidirectional tunnel is created between HA and FA. In the meantime, S also initiates the multicast tree join (Step 4). Hence, the existence of the bidirectional tunnel in this scheme is temporary and it is kept till the multicast agent at the foreign domain starts receiving packets from its tree-joining request. Once the FA starts receiving packets (Step 5), it verifies the packet sequence number from that tunneled to it from the bi-directional tunnel. In case there is a

— — Remote subscription (receiving packets)

— · — Reverse tunnel (sending packets)

Figure 15.3
Handling multicast source movement [15.60].
Credit: H. Gossain, S. Kamat, and D.P. Agrawal, "A Framework for Handling Multicast Source Movement over Mobile IP," *IEEE International Conference on Communications*, May 2002.

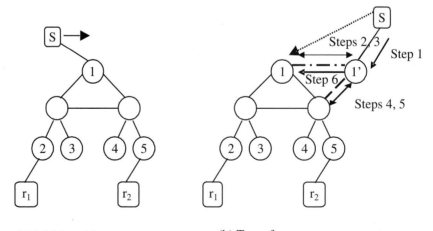

(a) Initial multicast tree   (b) Tree after source movement

missing packet, it waits till the packet is forwarded from the HA. Consequently, it requests to discontinue the forward tunnel from the HA to FA (Step 6) and keeps only the reverse tunnel. By this way, the FA localizes the effect of the MS movement, even if it is a source of the multicast group.

It is to be noted that the service disruption period due to handoff is caused by three main entities:

(a) Duration during which MH has no network connectivity

(b) MS registration duration in the foreign domain

(c) Multicast tree join delay in the foreign domain.

For a given handoff, any or all of the above entities may contribute to multicast service disruption. Here a smooth handoff technique helps to reduce the service disruption for the multicast session during the handoff.

With the help of extensive simulation work, authors show that the proposed approach is not only better than some of the existing approaches but also scalable with the number of MSs in a domain and host moving probability [15.61].

The adoption of wired multicast protocols to a MANET [15.62], which completely lacks any infrastructure, appears less promising. These protocols have been designed for infrastructured wireless networks and may fail to keep up with node movements and frequent topology changes due to host mobility. Also the protocol overheads may increase substantially. Rather, new protocols which operate on demand are being proposed and investigated. Studies indicate that tree-based on-demand protocols (AMRIS, MAODV, LAM, LGT) are not necessarily the best choice for all applications. When the network topology changes very frequently, mesh-based protocols (ODMRP, CAMP, FGMP) seem to outperform tree-based schemes, mainly because of the availability of alternative paths, which allows multicast datagrams to be delivered to all or most of the multicast receivers even if some link fails. Between tree-based and mesh-based approaches, there are hybrid schemes (AMRoute, MCEDAR) which take advantage of both tree-and mesh-based schemes and are suitable for medium mobility networks. Finally, stateless multicast schemes (like DDM) are designed to support multiple small groups.

15.7.2   Reliable Wireless Multicast Protocols

Here we discuss two different multicast protocols that are designed for reliable delivery of messages.

RMDP Protocol

The reliable multicast data distribution protocol (RMDP) presented in [15.63] is meant to be implemented for use on the MBONE. It relies on the use of FEC and ARQ information to provide reliable multicast service. Here redundant information is inserted into the FEC, which helps a receiver to reconstruct the original packet. In case such information is not enough, an ARQ is sent to the multicast source, which, in turn, retransmits the multicast packet to all the receivers. In RMDP, a data object is a file identified by a unique name—for example, its uniform resource locator (URL). Each file is assumed to have a finite size and is split into packets of s bytes each.

RMDP uses an (n, k) encoder with $n \gg k$ to generate packets for transmission, and assumes the underlying multicast network provides unreliable but efficient delivery of data packets.

RMDP offers scalability and efficiency when used in a reliable media. However, one of the main drawbacks of RMDP is that data encoding and decoding is done through software, resulting in a processing overhead and, therefore, results in performance degradation. For resource-limited receivers, decoding cost is of major concern. In addition, in highly unreliable wireless media, errors typically occur in bursts, causing the protocol to generate a large amount of ARQ packets, which triggers substantial amount of packet retransmissions. Hence, RMDP's retransmission scheme based on ARQ packets does not help in conserving network resources. On the other hand, this problem could be minimized if a hierarchical scheme is employed instead.

RM2 Protocol

Reliable mobile multicast (RM2) [15.64] is a reliable multicast protocol and is used for both wired and wireless environments. RM2 guarantees sequential packet delivery to all its multicast members without any packet loss. RM2 relies on the Internet group management protocol (IGMP) to manage multicast group membership, and on the IETF's Mobile IP to support user mobility. RM2 is a hierarchical protocol that divides a multicast tree into subtrees, whereby subcasting within these smaller regions is applied using a tree of retransmission servers (RSs). Each RS has a retransmission subcast address shared by its members, which may be dynamically configured using IETF's Multicast address dynamic client allocation protocol (MADCAP) [15.65]. In order to guarantee an end-to-end reliability, the receivers are required to send NACKs. In other words, RM2 implements selective packet retransmission.

Table 15.1 compares various mobile IP-based wireless multicast routing protocols.

Table 15.1: ▶
Mobile IP-Based Wireless Multicast Routing Protocols [15.56]
Credit: H. Gossain, C.M. Cordeiro, and D.P. Agrawal, "Multicast: Wired to Wireless," *IEEE Communications Magazine*, pp. 116-123, June 2002.

Parameter	Optimal routing	Reliability	Packet redundancy	Multicast routing protocol	Join and graft delays
Remote subscription	Yes	No	No	Independent	Yes
Bidirectional tunneling	No	No	Yes	Independent	No
MoM	No	No	Minimal	Independent	No
MMP	No	No	Minimal	CBT	No
Mobicast	Yes	No	Minimal	Independent	Yes
RMDP	No	Yes	Yes	Independent	Yes
RM2	Yes	Yes	No	Independent	Yes

15.7.3   Future Directions

As must be evident from the forgoing discussion, multicast is a field in which there is no one-size-fits-all protocol that can optimally serve the needs of all types of multicast applications. Hence, both wired and wireless multicast proposals have been designed to cope with specific application needs, which often lead to unexpected behavior when applied to unfamiliar environments. As the impact of multicast spans numerous domains, analyzing and indicating the suitability of a protocol is very hard, especially when considering both wired and wireless multicast. In this section, we gave an overview of current research in wireless multicast. This area is rapidly evolving, and there are still many challenges that need to be addressed. Questions that remain to be addressed are whether multicast ought to be application oriented, and how to integrate the wireless multicast infrastructure into Mobile IP. Furthermore, a detailed investigation is still needed for both unreliable and reliable environments.

15.8   Directional and Smart Antennas

In Chapter 13, we discussed wireless technologies for MANETs. However, recent research results have shown that wireless MANETs have limited throughput due to poor spatial reuse imposed by omnidirectional antennas at each MS. Here a MS can transmit only in one direction and all other MSs lying in the silent zones of the communicating nodes have to sit idle for the duration of communication. However, a MS equipped with a directional antenna can form a beam in the direction of the communication, thereby reducing the coverage of silence zones (refer to Figure 15.4)

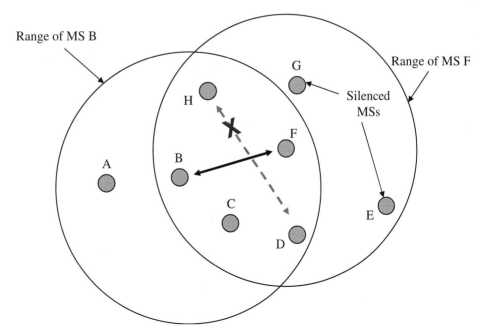

Figure 15.4
Communication
with omnidirectional
antennas.

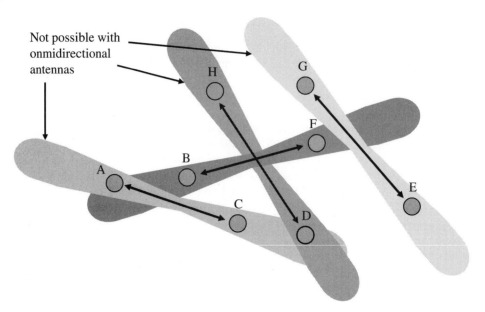

Not possible with onmidirectional antennas

Figure 15.5
Communication with directional antennas.

and allowing neighboring nodes to communicate simultaneously, thereby enhancing system throughput (Figure 15.5).

15.8.1 Types of Antennas

An antenna is the medium for releasing electromagnetic energy into the air interface and also the means for trapping the RF energy in reception, thus being the essential facilitator in wireless communication. The spatial distribution of power radiated from an antenna determines its radiation (power) pattern [15.66]. The type of radiation pattern of an antenna is used to broadly classify the antenna type. An antenna that transmits power equally in all directions is called an omnidirectional antenna, and an antenna that concentrates the radiated power in a particular directed zone in space is called a directional antenna. A special type of directional antenna that has the inbuilt intelligence to form a beam in a particular direction to transmit or receive based on a certain criterion is called a smart antenna. A smart antenna that can form more than one beam (i.e., one each for receiving signal from a different direction) is called a multiple-beam smart antenna. The above-mentioned classification of antenna type has an important impact on the method of medium access. This is especially true in a MANET, where there is no centralized coordinator for medium access. The antenna type must be taken into account while defining the medium access control scheme, so that the hidden terminal and exposed terminal problems are adequately solved [15.67].

15.8.2 Smart Antennas and Beamforming

Simultaneous transmission (or simultaneous reception) by a MS requires smart antennas equipped with spatial multiplexing and demultiplexing capability. Beamforming is a technique whereby the gain pattern of an adaptive array is steered to a desired direction through either beam-steering or null-steering signal-processing algorithms allowing the antenna system to focus the maxima of the antenna pattern toward the desired user while minimizing the impact of noise, interference, and other effects from undesired users that can degrade signal quality. Smart antennas are implemented as an array of omnidirectional antenna elements, each of which is fed with the signal, with an appropriate change in its gain and phase. This array of complex quantities constitutes a steering vector, and allows the resultant beam to form the main lobe and nulls in certain directions. With an L-element array, it is possible to specify $(L-1)$ maximas and minimas (nulls) in desired directions by using constrained optimization techniques when determining the beamforming weights. This flexibility of an L-element array to be able to fix the pattern at $(L-1)$ places is known as the degree of freedom of the array [15.68]. Smart antennas can be classified into two groups, both systems using an array of (omnidirectional) antenna elements: switched beam and adaptive beamforming antenna systems.

Switched Beam

A switched-beam system consists of a set of predefined beams, of which the one that best receives the signal from a particular desired user is selected. The beams have a narrow main lobe and small sidelobes so signals arriving from directions other than that of the desired main lobe direction are significantly attenuated. A linear RF network called a fixed beamforming network (FBN) is used that combines M antenna elements to form up to M directional beams.

Adaptive Beam

Adaptive antenna arrays, on the other hand, rely on beamforming algorithms to steer the main lobe of the beam in the direction of the desired user and simultaneously place nulls in the direction of the interfering users' signals. An adaptive antenna array has the ability to change antenna pattern dynamically to adjust to noise, interference, and multipath. It consists of several antenna elements (array) whose signals are processed adaptively by a combining network; the signals received at different antenna elements are multiplied with complex weights and then summed to create a steerable radiation pattern. Popular beamforming algorithms (e.g., the recursive least squares (RLS) algorithm) use a training sequence to obtain the desired beam pattern, while blind beamforming methods such as the constant modulus algorithm (CMA) do not impose such a requirement [15.69].

15.8.3   Smart Antennas and SDMA

Space division multiplexing access (SDMA) is simultaneous multiple reception (or transmission) of data at the base station (BS) using smart antennas equipped with spatial multiplexers and demultiplexers. Efficient use of directional antennas implies forwarding of packets to other nodes based on the knowledge of location. Ko, Shankarkumar, and Vaidya [15.70] have proposed a MAC layer protocol for directional antennas that exploits location information. Another MAC layer protocol using directional antennas has been proposed in [15.71]. Location information may be made available through the global positioning system (GPS) [15.72] installed at nodes. GPS uses triangulation of beams received from satellites to determine location. The magnitude of throughput enhancement that may be achieved by using directional RTS (DRTS) and directional CTS (DCTS) messages over spatial sub-channels (instead of omnidirectional RTS/CTS) in wireless ad hoc networks has been explored in [15.73]. Another MAC layer protocol for SDMA in wireless ad hoc networks proposed in [15.74] is based on the use of the ready-to-receive (RTR) concept and is illustrated in Figure 15.6.

A MS that wants to initiate reception sends out an omnidirectional RTR packet to poll all neighboring nodes simultaneously for data. The RTR packet contains the unique training sequence assigned to the receiver MS R, and transmitter MSs (A, B, C) use this training sequence to form directional beams in the direction of the receiver. The receiver MS also advertises the maximum size of the data packet (a network parameter) that it shall accept in the RTR packet. The potential transmitter MSs that have packets for MS R reply to the RTR message, each with their RTS requests, after forming directional beams in the direction of R. Each of these MSs also transmits its training sequence, allowing MS R to simultaneously form beams toward them. They also inform the receiver of the size of the data packet that they intend to transmit a parameter not greater than the size advertised by the receiver. After this, the receiver informs each of the potential transmitters of the negotiated packet size, which is the maximum of all the packet sizes requested by

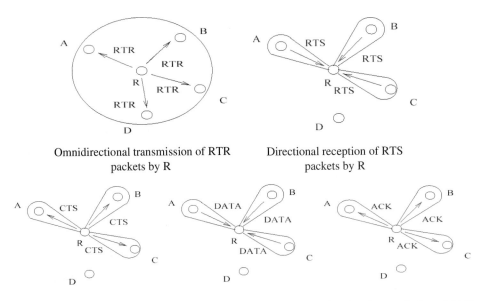

Omnidirectional transmission of RTR
packets by R

Directional reception of RTS
packets by R

Directional transmission of CTS   Directional reception of DATA   Directional transmission of ACK
packets by R   packets by R   packets by R

Figure 15.6
The basic function
of smart antenna
[15.137].

the transmitters. This is done in a CTS packet, which is transmitted directionally toward the intended transmitters. All transmitters pad their DATA packet size up to the negotiated value, after marking the logical "End of Packet." The DATA packets are transmitted directionally toward the receiver. A possible optimization to save transmission power is that transmitters need not perform bit stuffing if they actually calculate the expected time of ACK arrival based on the negotiated packet size, which is obtained from the CTS packet. After receiving the DATA frames simultaneously, the receiver replies with simultaneous directional ACKs to each of the transmitters. In this manner, synchronization is achieved for all received signals to a MS. An assumption in the above mechanism is that the MSs have low mobility, so that beams can be formed on the basis of training sequences in the RTR and RTS packets, and the same directional beams can be used for transmission or reception (which are reciprocal processes) for the entire duration of the DATA and ACK exchange. Also, MSs that are not attempting to receive data from others are listening for any RTRs that they may receive.

Each MS records the control information that it overhears from other ongoing transmissions, and uses this information to modify its radiation pattern by placing nulls in appropriate directions. This state information is maintained in a spatial null angle table (which is analogous to the network allocation vector of 802.11), that lists the transmitting MS, its radial direction relative to the MS maintaining the angle table, the time this entry was made, and the time after which the entry must be purged.

● ● ● ● ● ● ● ● ● ● ● ● ● ● ●

15.9   Design Issues in Sensor Networks

In Chapter 13, we covered two possible architectural designs of sensor networks—hierarchical and flat networks. A hierarchical organization of the sensor network typically uses a cluster-based routing protocol such as LEACH, while the flat organization of the network makes use of the directed diffusion paradigm for routing data. Hierarchical topology is better suited for applications where most of the area covered by the sensor network is to be diagnosed, or when the traffic is light. Time-critical applications that cannot tolerate the initial latency in setting up routes for data gathering or applications that demand fair allocation of bandwidth typically use the hierarchically clustered networking approach. On the other hand, flat topology is used in applications where traffic conditions change frequently in a random fashion and routes need to be adapted dynamically to these conditions, depending on the energy level of sensor nodes. Also, in cases where the user wants better network performance for a query that has a higher priority, the network can dynamically allocate more network resources to such queries, routing them through lower latency paths. A reactive protocol TEEN has been introduced [15.75], based on hierarchical clustering, where nodes transmit data only when the value of the sensed attribute changes beyond a threshold value. This reduces unnecessary data transmissions, while the sensors are busy monitoring their environment to pass on time-critical data almost instantaneously. Therefore, this scheme is eminently suited for time-critical data sensing applications. Even though the nodes sense continuously, the energy consumption in this scheme can be much smaller than in the proactive network, because data transmission is done less frequently. APTEEN [15.76] [15.77] is an improvement over TEEN, as it can emulate a combination of both proactive and reactive network characteristics. Data transmission can be triggered by a change in the value of attributes beyond a threshold value similar to TEEN. On the other hand, after a specified time, a node is forced to sense and transmit the data, irrespective of the sensed value of the attribute. This provides the user with a hybrid network that reacts immediately to time-critical situations and gives an overall picture of the network. A third way of characterizing protocols have also been proposed that provide the user the flexibility to request either past, present, or future data from the network in the form of historical, one-time, and persistent queries, respectively. The delay incurred in handling various types of queries has also been analytically determined. These three protocols offer versatility to the users while consuming energy very efficiently by minimizing noncritical data transmissions. The performance of these protocols has been evaluated for a simple temperature sensing application with a Poisson arrival rate for queries. In terms of energy efficiency, these protocols have been observed to outperform existing conventional warehousing sensor network protocols.

Current research is focused on developing schemes for time-critical information retrieval in a sensor network with a flat topology using directed diffusion for routing. Flat networks have higher initial latency in establishing a multihop path but can better adapt to variable traffic conditions, by rerouting data through alternative paths, and hence are robust to topology variations due to dying or mobile sensor nodes or their mobility. At the MAC layer, local and global time synchronization is not required, unlike TDMA scheduling used in hierarchically clustered sensor networks. A priority can be associated with every query injected in the network in terms of the accuracy and speed with which the response is expected. Reducing initial latency in setting up the route from the user to the desired regions in the network is being pursued and caters to time-critical applications besides carrying out efficient periodic monitoring.

15.9.1   Sensor Databases

Work is being done in the area of sensor databases. Researchers at Cornell are developing a model for sensor database systems known as COUGAR [15.78] to run a distributed query execution plan without assuming global knowledge of the sensor network. Recently, a Web database system has been developed that determines the appropriate number, placement, and content of multiple, redundant data caches throughout the network in order to minimize a composite cost function based on data criticality requirements and power consumption. This innovative software offers users the flexibility to adapt to new missions, situations, capabilities, and usage without sacrificing high efficiency and reliability.

15.9.2   Collaborative Information Processing

Collaborative processing is another challenging area in sensor networks. Nodes need to collaborate and aggregate the data they gather periodically, requiring efficient localized beamforming algorithms. ECCS Dynamic Declarative Network Configuration, Massachusetts Institute of Technology Lincoln Laboratory (part of DARPA SensIT), focuses on demonstrating the value of collaborative processing through the development of cost and performance models and analyzes traditional unattended ground sensors versus ad hoc networked sensors. The algorithms and methods [15.79] employed for collaborative sensing strongly influence the overall performance of ad hoc networked sensors.

15.9.3   Operating System Design

TinyOS architechture [15.80] developed by the researchers at Berkeley is an ultra-low-power sensor platform, including hardware and software, that enables low-cost deployment of sensor networks. It is a system-level bridge that combines advances in low-power RF technology with micro-electro mechanical systems (MEMS) transducer technology. MagnetOS [15.81], being developed at the Cornell University, is a single system image (SSI) operating system. The entire MANET looks like a single Java virtual machine. MagnetOS partitions applications into mobile components that communicate via remote procedure calls (RPCs) to find a good placement of components on the nodes in a MANET.

15.9.4   Multipath Routing in Sensor Networks

The goal of multiple path routing in sensor networks is to distribute the routing of data packets generated by the multiple queries between a given source and a sink on as many nodes as possible, so that excessive energy depletion of just a few nodes along the single selected route (usually the shortest) could be avoided.

As the user may inject a query at a random location in the network, most of the time, it is impossible to predict the traffic pattern. When traffic is heavy, a large disparity is introduced in the energy level of the nodes lying on the direct path connecting the source to the sink with respect to the rest of the network. These nodes lying on the shortest path have a relatively lower energy level compared to the other nodes in the network and hence become the bottleneck nodes. Therefore, one motivation behind multipath routing is to improve the capacity of the network by attempting to maintain the same rate of energy depletion for every node in the network.

Current energy aware protocols are designed in sensor networks based on an assumption that queries injected in different parts of the network are equally critical and are routed along the shortest paths. This may not be a realistic assumption for many large-scale applications such as battlefield surveillance where besides the periodic monitoring of attributes such as temperature, and tracking of vehicles, there are a few user-initiated, real-time queries that need a quick response time or there are alarm signals to warn the soldiers of some impending danger or unusual change in the value of the sensed attributes that needs immediate attention. In order to support a large range of applications, sensor networks must be able to fulfill application-specific demands without sacrificing the broad objectives of a robust network with adequate longevity. User-specific requirements may conflict with the system-specific characteristics like energy efficiency that are essential to provide a better service to all user queries.

The maintenance overhead of a multiple-path scheme is measured by the energy required to maintain these alternate paths using periodic keep-alive beacons. This suggests that a multiple-path scheme would be preferable when either the density of simultaneously active sources in the network is high and their location is random or there are few data sources with very high traffic intensity such that traffic in the network is unevenly distributed among the network nodes. Multipath routing is cost-effective for a heavy load scenario, while a single-path routing scheme with a lower complexity may otherwise be more desirable.

Classical multiple-path routing has been extensively studied and used in all kinds of existing communication networks such as the Internet, high-speed networks [15.82], and ATM networks [15.83]. In multiple-path routing each source discovers and maintains the set of routes that can be used to reach its destination; the possible routes can be discovered by applying a source-routing algorithm. The advantage of using multiple paths is two-fold. First, it provides an even distribution of the traffic load or energy consumption over the network. Second, in spite of a route disruption, the source is able to send data to the destination by using an alternative functioning route.

To combat the inherent unreliability of these networks, the Split Multipath Routing scheme has been proposed by Lee et al. [15.84] that uses multiple paths simultaneously by splitting the information among the multitude of paths, so that the probability that any essential portions of the information will be received at the destination can be increased without incurring excessive delay.

A substantial amount of work has been reported on single-path routing as compared to multiple-path routing in wireless ad hoc networks. Some applications of multiple-path routing for ad hoc networks have been considered by [15.85] [15.86][15.87]. TORA is a source-initiated routing protocol for ad hoc networks that creates multiple-paths on demand. There is a need to adapt multiple-path routing to overcome the design constraints of a sensor network. Important design considerations that drive the design of sensor networks are energy efficiency and scalability [15.88] of the routing protocol. Discovery of all possible paths between a source and a sink might be computationally exhaustive. Besides, updating the source about the availability of these paths at any given time might involve considerable communication overhead. The routing algorithm must depend only on the local information [15.89] or the information piggy-backed with data packets, as global exchange of information is too energy consuming due to the large number of nodes. Multipath routing specifically for sensor networks has been explored by [15.90][15.91] [15.92][15.93].

Assuming each node to have a limited lifetime, Chang and Tassiulas have proved [15.90] the overall lifetime of the network can be improved if the routing protocol minimizes the disparity in the residual energy of every node, rather than minimizing the total energy consumed in routing. Ganesan et al. [15.91] have proposed a multiple-path scheme to achieve high resilience to node failure with low maintenance overhead. In their scheme, in order to keep the available paths alive, the source periodically floods low-rate data over each alternate path. The frequency of these low-rate data events determines how quickly their mechanism recovers from failures of the primary path.

Shah et al. [15.94] have modified the directed diffusion protocol to improve the overall network lifetime. Instead of reinforcing a single optimal, shortest path for routing, alternate good paths discovered during the route discovery phase of the directed diffusion are also cached, and one of them is chosen for routing in a probabilistic fashion.

Servetto et al. [15.92] have also implemented multiple-path routing using random walks between a source and sink, and thereby avoiding the overhead of caching paths. They assume the nodes to be powered by a renewable source of energys, hence node failure is temporary.

Jain et al. [15.95] propose a distributed and scalable traffic scheduling algorithm that splits the traffic generated at the data source among multiple-paths constructed between the source and the sink in proportion to their residual energy. The multiple-paths are constructed with low communication overhead and spread over a large symmetrical area bounded by the source and sink. They further introduce [15.96] priority-based treatment of data packets by routing time critical packets through shorter paths and the non–real-time data over longer paths using load shedding and QoS–based classification of available paths.

15.9.5   Service Differentiation

Service requirements could be diverse in a network infrastructure. Some queries are useful only when they are delivered within a given time frame. Service differentiation is popularly used to split the traffic into different classes based on QoS desired by each class.

Chen et al. [15.97] describe two-fold goals of QoS routing: (a) selecting network paths that have sufficient resources to meet the QoS requirements of all admitted connections and (b) achieving global efficiency in resource utilization. Arbitrary placement of nodes causes large disparity between geographical distances separating the nodes. In MANETs, static provisioning is not enough because MS's mobility necessitates dynamic allocation of resources. In sensor networks although user mobility is practically absent, dynamic changes in the network topology may be present because of MS's loss due to battery outage. Hence, multihop ad hoc routing protocols must be able to adapt to the variation in the route length and its signal quality while providing the desired QoS. It is difficult to design provisioning algorithms that achieve simultaneously good service quality as well as high resource utilization. Since the network does not know in advance where packets will go, it will have to provision enough resources to all possible destinations to provide high service assurance. This results in a severe under utilization of resources.

Bhatnagar et al. [15.98] discusses the implications of adapting these service differentiation paradigms from wired networks, to sensor networks. They suggest the use of adaptive approaches, the sensor nodes learn the network state using eavesdropping or by explicit state dissemination packets. The nodes use this information to aid their forwarding decisions—for example, low-priority packets could take a longer route to make way for higher-priority packets through shorter routes. The second implication of their analysis is that the applications should be capable of adapting their behavior at run time based on the current allocation, which must be given as a feedback from the network to the application.

● ● ● ● ● ● ● ● ● ● ● ● ● ● ● ●

15.10   **Bluetooth Networks**

In the past quarter of a century, we have seen the three generations of wireless cellular systems, attracting end users by providing efficient mobile communications. Wireless technology has become an important component in providing networking infrastructure for localized data delivery as well. This later revolution was made possible by the introduction of new networking technologies and paradigms, such as WLAN, WPAN, and various possible options covered in Chapter 14.

15.10.1   Interference on Bluetooth Networks

WLAN and WPAN technologies are complementary rather than competitive technologies. They are likely to operate in proximity so that more advanced services can be obtained, such as Internet access by Bluetooth devices through existing WLAN infrastructure. Since these technologies operate in the same ISM unlicensed frequency band, mutual interference may be present. Despite using a frequency hopping scheme, Bluetooth devices experience a perceptible drop in throughput due to interference when multiple piconets operate in the vicinity, as well as in the presence of the network using IEEE 802.11 protocol.

Based on an analytical and simulation study, a mathematical model has been derived [15.99] for the packet success probability due to interference, which takes into account factors such as the network load, propagation loss law and shadowing, and showed that interference is a dominant source of drop in Bluetooth piconet throughput. Furthermore, with increased transceiver sensitivity, it is likely that the number of piconets that may interfere with each other will increase dramatically. This kind of interference on nearby piconets is defined as intermittent interference. Such a definition is due to the frequency hopping nature of the Bluetooth radio that generates interference in an intermittent fashion.

Analytical models are very important because they open various possibilities for new research. In the specific case of the study performed in [15.99], it is possible to propose mechanisms to cope with these negative performance implications resulting from interference. According to the IEEE 802.15.2 WG (working group), two classes of coexistence mechanisms are possible: collaborative and noncollaborative mechanisms. If it is possible for the WPAN and the WLAN to exchange information between one another, then it is possible to develop a collaborative coexistence mechanism where the two wireless networks negotiate to minimize mutual interference. If there is no method to exchange information between the two wireless networks, then you can use a noncollaborative coexistence mechanism. Proposals range from packet segmentation algorithms to the application of adaptive frequency hopping.

Interference-aware Bluetooth dynamic segmentation (IBLUES) [15.100] is one of such noncollaborative coexistence algorithms, which employ dynamic packet segmentation to mitigate interference effects on Bluetooth devices. The Bluetooth standard defines different packet types to adjust to different application requirements. These range from a single unprotected 1-slot packet to a FEC-encoded 5-slot packets. Ultimately, the application is responsible for selecting the packet type that it thinks best suits its needs. In case of a real-time application, FEC-encoded packet should be chosen, whereas in a non–real-time application, unprotected packets are the best option since they provide higher effective throughput. In Bluetooth, the adaptation layer is responsible for receiving messages from upper layers and segmenting them into small pieces of data large enough to fit into a Bluetooth standard packet for best utilization. Ideally, the adaptation layer should choose the best suitable packet for transmission, based both on the application requirements and on the wireless channel condition. Furthermore, this choice cannot be static for the entire message due to the dynamic error-rate nature of a wireless channel. IBLUES takes advantage of this infrastructure and defines a dynamic mechanism to switch between different packet types according to current wireless channel conditions, while it builds on the analytical model devised in [15.99] to achieve optimal performance. Moreover, interference originated from multiple piconets and the IEEE 802.11 are both taken into account in the development of the protocol.

The impact of interference on the Bluetooth packet throughput is given in [15.99]. This clearly indicates that different packet types should be employed according to interference levels. An aggregate throughput for an increasing number of piconets shows [15.100] that after a certain number of piconets the overall throughput cannot get any better due to increased interference. Results show that IBLUES can effectively cope up with the time-varying characteristics of the Bluetooth wireless channel by employing dynamic segmentation.

15.10.2 Bluetooth Dynamic Slot Assignment

So far, we can envision three waves of Bluetooth-based applications. Initially, Bluetooth was designed to enable a wide range of devices such as laptops, PDAs, mobile phones, and headsets, to form ad hoc networks in a semi-autonomous fashion [15.101]. The second wave of applications was the development of APs (with functionality similar to the IEEE 802.11 APs) enabling hundreds of Bluetooth units to access the wired network in places such as theaters, stadiums, conferences, pavilions, and so on [15.102] [15.103]. However, in the third wave of applications, the low cost, effortless and instant connection provided by Bluetooth technology has also become attractive for automatically forming a MANET of a large number of low-power sensor nodes [15.104][15.105]. These sensor network applications are characterized by thousands of nodes embedded in the physical world, and a Bluetooth RF link to enable them to form a network by bridging only these sensor nodes within radio range [15.106][15.107]. These call for solutions to be developed that are applicable to both small-scale and large-scale Bluetooth networks and that aim at keeping interference at minimum levels [15.99][15.100] due to the envisioned large number of piconets.

By virtue of the master/slave communication model, the Bluetooth medium access provides for simplicity, low power (as compared to other standards), and low cost, these being the major forces driving the usefulness of the technology. However, this design choice also brings in major shortcomings, such as the inability for slaves to communicate directly with each other since their packets must be forwarded through the master device [15.101]. Moreover, in Bluetooth there is no built-in support for the many applications that require group communication [15.108][15.109]; while this can be achieved only by multiple unicast packets or by a piconet-wide broadcast. As a result of these limitations, the packet forwarding among slaves in Bluetooth becomes sub-optimal, bandwidth is wasted by forwarding through the master, end-to-end packet delay increases, and power consumption is significantly increased at the master unit due to its frequent medium access for both transmission and reception. Therefore, the adoption of the master/slave paradigm in its present form does not seem to be the most appropriate solution.

To overcome these issues and address the shortcomings of the current Bluetooth master/slave communication model, a novel dynamic slot assignment (DSA) scheme to be coordinated by the master of a piconet is proposed in [15.109] [15.110]. Based on the piconet traffic patterns, the master device dynamically allocates slots for direct communication between slaves. This way, packets do not need to be forwarded by the master while it periodically reevaluates slot assignments and changes them accordingly. This novel communication architecture not only enables direct slave-to-slave communication but also serves as multi-slave communication, hence emulating a group (i.e., multicast-like) communication within the piconet. Extensive simulations of DSA have been carried out, and a drastic enhancement in current Bluetooth performance has been observed. In unicast scenarios, piconet throughput increases by up to 300%, delay is reduced to one-third, and overhead is halved, whereas in multicast-like communication throughput boosts up to 500%, delay decreases to approximately one-thirtieth, and overhead is merely one-seventh of existing Bluetooth implementations. Additionally, we have also shown that power consumption at the master is dramatically reduced due to the smaller number of transmissions/receptions, and in certain scenarios the reduction achieved is up to 80%.

15.10.3 BlueStar: Enabling Efficient Integration Between Bluetooth WLANs and WPANs

As previous studies have pointed out [15.99] [15.100][15.102] [15.111][15.112] it is much likely that Bluetooth devices and IEEE 802.11 WLAN [15.48] stations (here, we use the terms WLAN and IEEE 802.11/802.11b interchangeably) operating in the 2.4 GHz industrial-scientific-medical (ISM) frequency band should be able to coexist as well as cooperate with each other and access each other's resources. These technologies are complementary, and such an integrated environment is envisioned that Bluetooth devices will obtain information through the WLAN, and ultimately the Internet. These cooperative requirements have led to an intuitive architecture called BlueStar [15.113], whereby a few selected Bluetooth devices, called Bluetooth wireless gateways (BWGs), are also members of a WLAN, empowering low-cost, short-range devices to access the global Internet infrastructure through the use of WLAN-based high-powered transmitters. It is also possible that Bluetooth devices might access the WAN through a 3G cellular infrastructure like UMTS and cdma2000. However, from the point of view of cost and performance, it is advantageous for Bluetooth devices to have access to the WAN through a WLAN system where a WLAN infrastructure is readily available. In these scenarios, Bluetooth (or WPAN) devices would only make use of the cellular network infrastructure where WLAN coverage is not provided.

An important challenge in defining the BlueStar architecture is that both Bluetooth and WLANs employ the same 2.4 GHz ISM band and can possibly impact performance [15.100][15.114][15.115]. We refer to the interference generated by WLAN devices over the Bluetooth channel as persistent interference [15.113], while the presence of multiple piconets in the vicinity creates interference [15.99][15.100] [15.102][15.111]referred to as intermittent interference [15.116][15.117].

To combat both of these interference sources and provide effective coexistence, BlueStar employs a unique hybrid approach of adaptive frequency hopping (AFH) [15.118] and a new mechanism called Bluetooth carrier sense (BCS) in BlueStar. AFH seeks to mitigate persistent interference by scanning the channels during a monitoring period and labeling them as "good" or "bad," based on whether the packet error rate (PER) of the channel is below or above a given threshold. BCS takes care of the intermittent interference by mandating that before any Bluetooth packet transmission, the transmitter has to sense the channel to determine the presence of any ongoing activity. This channel sensing is performed during the turnaround time of the current slot, and it does not require any changes to the current Bluetooth slot structure. According to the IEEE 802.15 Coexistence Task Group 2 terminology [15.119], BlueStar would be classified as a noncollaborative solution in the sense that the Bluetooth and the WLAN system operate independently [15.120], with no exchange of information (as a matter of fact, this is required by the Federal Communications Commission regulations for the license-free bands).

The industry has also been making efforts toward integrating Bluetooth and WLAN [15.121][15.122][15.123]. However, most recent solutions do not tackle the issue of simultaneous operation of Bluetooth and WLANs—that is, either Bluetooth or WLANs, but not both, can access (i.e., be active on) the wireless medium at a time, as only a single card is available. Moreover, this implies that additional integrated cards have to be acquired. BluetStar enables simultaneous operation by using existing WLAN hardware infrastructure, while relying on the availability of Bluetooth interfaces.

15.10.4   Traffic Engineering over Bluetooth MANETs

With respect to Bluetooth-enabled networks, we identify three major shortcomings related to the physical, the data link, and the network layer, referred to as inefficiency, limited capacity and random topology. When there are a large number of connections, either the delay drastically increases or the new incoming traffic is simply blocked. Therefore, there is a lack of traffic engineering techniques in the current Bluetooth. As we know, traffic engineering has been shown to be extremely useful for the Internet [15.116][15.124], by efficiently transferring information from a source to an arbitrary destination with controlled routing function and steering traffic through the network. Its systematic application helps in enhancing the QoS delivered to end users. Traffic engineering suggests both demand side and supply side policies for minimizing congestion and improving QoS. Demand side policies restrict access to congested resources, dynamically regulate the demand to alleviate the overloaded condition, or control the way the data is routed in the network. Supply side policies augment network capacity to better accommodate the traffic.

In [15.125], traffic engineering is incorporated into Bluetooth by employing the demand side and supply side policies [15.117][15.124][15.126] in the form of pseudo-role-switching (PRS) and pseudo-partitioning (PPR) schemes respectively. PRS maximizes the bandwidth utilization and minimizes latency within a piconet, while PPR breaks the boundaries by dynamically partitioning piconets as traffic demand exceeds Bluetooth capacity. Extensive simulation shows up to 50% reduction in the network overhead and up to 200% increase in the aggregate throughput [15.125].

15.10.5 Distributed Topology Construction

The ubiquitous use of information-intensive consumer devices such as cell phones, PDAs, and laptop computers makes necessary a new networking paradigm for their interconnection. The goal is to create a WPAN that accommodates seamless information transfer between different devices with varying capacity in an ad hoc manner without the need for manual configuration, cables, or wired infrastructure. Bluetooth was motivated in part by the need for suitable link-layer WPAN technologies. The Bluetooth communication substrate consisting of radio, baseband, link controller, and link manager layers specifies mechanisms for establishing connection with nearby devices in an ad hoc manner. Unlike traditional WLANs, which rely on distributed contention resolution mechanisms, Bluetooth is based on a centralized master/slave polling scheme known as TDD (time division duplex).

The Bluetooth specification alludes to the concept of internetworking multiple piconets, known as a scatternet, but does not specify how it is to be formed. Before this concept becomes a reality, it is important to solve a number of challenging problems. We can identify three main challenges to the development of self-organizing Bluetooth scatternets as [15.127]

■ Topology formation

■ Link scheduling

■ Packet routing

In broadcast-based WLANs such as IEEE 802.11, the physical distance between nodes determines the network topology. In Bluetooth, an explicit topology formation process is required since nearby devices need to discover each other and explicitly establish a point-to-point link. During the link-formation process, the two Bluetooth nodes synchronize the frequency hopping sequence and gather necessary clock information. The essential ad hoc discovery process could be lengthy, and clever solutions are required to quickly form a network topology that spans across all nodes within the transmission proximity.

15.11 Low-Power Design

The world today is moving from bulky computers to wearable devices. This shift in paradigm brings with it the need to think about the power conservation of the wireless devices, because limited energy resources characterize most wireless devices. Ad hoc and sensor networks today are in a nascent stage of development; however, when their use is commercialized, power consumption is expected to be a major hurdle in the smooth functioning of wireless nodes. For example, consider wireless sensors deployed in forests to detect the spread of wildfires. In this case, these sensors might be air dropped and might have to last for months [15.128]. Another example is sea exploration to gather data about currents, tides, flash floods, and so on. In these cases, it is desirable that the devices do not run out of power at the crucial stage because once deployed, replacement of their batteries is difficult and the only choice may be to replenish the whole sensor system. However, battery technology is progressing slowly, whereas computation and communication demands are increasing rapidly. To compensate for this, the scientific community is coming up with innovative methods to conserve battery power.

The traditional approach to saving power is to use power-down features to minimize the power consumption of unused hardware. For portable computers this means turning off the hard disk, processor, screen, modem sound, and so on. For the MS, it means switching off the display power while it is not in use. Another approach commonly used to respond to low-power requirements is to reduce the supply voltage of the computer chips. For example, lowering the supply voltage from the standard 5.0 V to 3.3 V reduces the power consumption by 56%. However, lowering the supply voltage requires all components to operate at a low voltage. To improve upon this, a new method has been designed whereby the input voltage of specific parts can be lowered. Since in an ad hoc and sensor network, a major portion (30–50%) of the power is consumed by the processor itself, research has been oriented in the direction of conserving the processor power consumption. Intel Corporation set the initial building block. In 1995, Intel introduced the first x86 processor that operated at a lower voltage (2.9 V) than the PC motherboard (3.3 V) [15.129].

From all the basic advances in the field of processor power conservation, came the concept of a dynamic voltage scaling (DVS) mechanism to reduce CPU power requirements without significant performance degradation. With today's processors' speed reaching gigahertz levels, an inherent power dissipation level on the order of tens of watts becomes an important concern in digital design. To give a fair idea of how it works, we take the example of the dynamic power dissipation P_{dynamic} in CMOS circuits. The equation has a quadratic dependency on the supply voltage V_{dd} ($P_{\text{dynamic}} \propto C V_{dd}^2 f$), where C is the collective switching capacitance and f is the supply frequency. From the preceding equation, it can be inferred that dynamically varying the voltage and the frequency can save us a lot of CPU power. Since the proportionality depends on the square of the voltage, reducing the voltage will save us more power [15.130]. Experiments have shown that the energy per instruction at minimal speed (59 MHz) and low voltage is 1/5 of the energy required at full speed

(251 MHz) and high voltage [15.129]. Transmeta TM5400, or the "Crusoe," is one of the few processors that actually supports voltage scaling.

The main philosophy of changing voltage is attributed to the fact that wireless devices are often in idle state or doing some very trivial work. During those idle hours, their power consumption far exceeds what is required. Therefore, if the supply voltage is reduced during those periods, then a lot of energy can be saved and the device's battery will last longer. For instance, in a sensor network, when the sensing parameter does not vary too frequently with time, the device can be governed to sense data at certain intervals of time and the processor can turn to sleep mode and save precious power for future use. Some researchers argue that varying the voltage may increase the response time of the devices as it takes time for components to switch from low power to high power. However, since most of the wireless applications (e.g., sensors in a forest) are not real-time applications, the delay in response time is acceptable.

Mobile computers are being used for video processing (e.g., in sending pictures of rare aquatic life from deep within the sea). Video processing is a key component of multimedia information exchange. Since the battery power is limited, conserving energy plays a significant role [15.131]. As mentioned previously, variable voltage techniques and variable clock speed processors are being seriously considered for multimedia applications to prolong the lifetime of the battery.

A new direction is being added to the field of power saving. Until now the emphasis has been on providing hardware solutions to power saving, like turning off the display during the periods of inactivity (implemented in BIOS [basic input/output system] or screen savers) and slowing down the CPU according to work load. However, now research is being made in the direction of software level techniques for conserving power—for example, modeling the various protocol layers (TCP) [15.132]. However, the research is still in its initial stage and directed primarily to applications. A lot of work is needed in this direction and a general hardware solution like DVS should be implemented.

• • • • • • • • • • • • • • •

15.12 XML

15.12.1 HTML Versus Markup Language

Hypertext markup language (HTML) is the simplest and the most popular markup language. In 1989 Tim Berners-Lee, with an intention to enable scientists to share information from any location, developed hypertext documents that could be linked to and read anywhere. An HTML file is a text file containing small markup tags that instruct the Web browser how to display the page. HTML's phenomenal success has been partly due to its simplicity. However, HTML was not designed to be interactive between the server and client.

HTML has been extremely popular on the WWW for its usefulness and flexibility and can be used in wireless applications as well. However, in a wireless environment, bandwidth is limited and a direct use of HTML is too expensive and prohibitive from a delay point of view. Therefore, researchers have to find ways to minimize the amount of data transfer without sacrificing the quality significantly so that the Web can be accessed by wireless devices like Palm Pilots™ and cell phones. This has encouraged the adoption of eXtensible Markup Language (XML) for the wireless world.

Markup languages are static and do not process information. A document with markup can do nothing by itself. However, a programming language can easily process the information presented in markup format. Essentially a markup language contains identical information and brings intelligence to a document so that applications can be read and processed effectively.

Technically, XML is a metalanguage—that is, it can create its own markup language. It is extensible, which means that it can create its own elements. XML describes a class of data objects called XML documents, which are stored on computers, and partially describes the behavior of programs that process these objects. Documents can be customized according to the kind of information that needs processing. Hence, if the discipline is wireless, a document type might be created by marking elements like <Sender>, <Receiver>, <RTS>, <CTS>. XM is a subset or a restricted form of SGML, the standard generalized markup language (ISO 8879). XML is extensible, with precise and deep structures.

XML has a low-level syntax for representing structured data. A simple syntax can be used to support a wide variety of formats. The number of XML applications is growing rapidly, and the growth pattern is likely to continue. There are many areas—for example, the health care industry, the wireless devices for government and finance—where XML applications are used to store and process data. The use of XML leads to a simple method for data representation and organization, and problems of data incompatibility and tedious manual rekeying can become manageable. One such application of XML in the wireless area is wireless markup language (WML).

15.12.2   WML: XML Application for Wireless Handheld Devices

WML is a markup language based on XML, with its specification developed and maintained by the WAP Forum, an industry-wide consortium founded by Nokia, Phone.com, Motorola and Ericsson. This specification defines the syntax, variables, and elements used in a valid WML file. The actual WML 1.1 document type definition (DTD) is available for those familiar with XML at "http://www.wapforum.org/DTD/wml_1.1.xml." A valid WML document must correspond to this DTD; otherwise it cannot be processed. If a phone or other communications device is said to be WAP capable, this means that it has a piece of software loaded onto it (known as a microbrowser) that fully understands how to handle all entities in the WML 1.1 DTD.

WML was designed for low-bandwidth and small-display devices. As part of this design, the concept of a deck of cards has been utilized. A single WML document (i.e., the elements contained within the <WML> document element) is known as a deck. A single interaction between an agent and a user is known as a card. The beauty of this design is that multiple screens can be downloaded to the client in a single retrieval. Using WML Script, user selections or entries can be handled and routed to already loaded cards, thereby eliminating excessive transaction transmissions with remote servers. Of course, with limited client capabilities comes another tradeoff. Depending on client memory restrictions, it may be necessary to split cards into multiple decks to prevent a single deck from becoming too large. WML predefines a set of elements that can be combined to create a WML document.

15.13 Threats and Security Issues

The use of wireless communication technology continues to grow at an exponential rate as it supports robust and efficient operation by incorporating automatic routing functionality in a way transparent to all mobile subscribers. However, recent concerns indicate that larger security challenges are present in wireless networks than the conventional wireline networks. Security becomes mandatory in such a mobile network due to the existence of hackers, viruses, intruders, Internet-based attacks, disgruntled current and ex-employees, and industrial espionage.

15.13.1 Security Threats to Wireless Networks

Wireless networks are vulnerable to the following security risks [15.133] [15.134]:

- **Accidental attack**: Exposure due to failure of components.

- **Passive attack**: In passive attacks the goal of the intruder is to obtain information that is being transmitted. These are classified based on the nature of eavesdropping on transmission or monitoring of information transfer. The two types of attacks involved are releasing contents of a message and, obviously, traffic characteristics. Passive attacks are very difficult to detect because they do not involve any alteration of data. Hence, passive attacks need to be prevented and not necessarily detected.

- **Active attack**: This happens when data modification or false data transmission takes place. It can be divided into four categories:

(a) **Masquerade**: This is said to be present when one entity pretends to be a different entity.

(b) **Replay**: Implies capturing of the information and then retransmitting to cause unexpected consequences.

(c) **Modification of a message**: A portion of the actual message is reordered, delayed, or altered to produce unexpected effects.

(d) **Denial of service (DoS)**: Implies temporary prevention of communication facilities for their normal use. It may have some prespecified target. Another form of DoS is the disruption of an entire network or flooding the network with a large number of messages to an extent that the performance of the system is drastically degraded.

- **Unauthorized usage**: Another security hole is the growing use of the Internet. If users from inside can get onto the Internet, then users from outside can get into a network if proper precautions are not taken. This can leave the network vulnerable to hackers, viruses, and intruders. Firewall products offer packet filtering, and proxy servers can give protection against this threat. User authentication is one of the best ways to secure the network from these threats.

- **Broadcast-based**: An eavesdropper is able to tap into the communication on the wireless communication channels, by positioning itself within transmission range.

- **Infrastructure**: These attacks are based on some weakness in the system, such as a software bug or hardware failure. The protection is to keep the possible damages as small as possible.

- **Device vulnerability**: Mobile devices can be hijacked easily and if the secret IDs and codes are embedded in the device, hijackers may get access to private information stored on it. They could then access other resources of the network.

- **Heterogeneity**: Mobile nodes have to adjust to potentially different physical communication protocols as they move to different places.

- **Limited computational ability**: This may exclude techniques such as public key cryptography during normal operations to conserve power.

15.13.2 Why Existing Wired Solutions Are Not Applicable to Wireless Networks

The dissimilarities between the wired Internet and wireless networks make it difficult to import contemporary intrusion detection systems from wired networks to wireless networks. The most significant difference is that data in a wireless network is transmitted over the broadcast medium, which can be received by all the nodes in the vicinity. Apart from this, another major difference lies in the communication pattern in a wireless network with limitations like slower link, battery power constraints [15.135][15.136], limited bandwidth, and higher cost. Therefore, disconnected operations are very common in wireless networks.

15.13.3 Current Approaches

IEEE 802.11 defines two authentication schemes: open system authentication and shared key authentication. In the open system authentication scheme, which is the default scheme, a terminal announces that it wishes to associate with an AP, and typically the AP allows the association. The shared key method requires that the wired equivalent privacy (WEP) [15.137] algorithm be implemented on both the wireless terminal and the AP. Mobiles that are allowed to connect to the network use the same shared key, so this authentication method is only able to verify whether the particular mobile user belongs to the group allowed to connect to the network, but there is no way to distinguish one mobile from another. Also, there are no means available to authenticate the network. The IEEE 802.11 does not define any key management functions.

The IEEE 802.11 defines an optional WEP mechanism to implement the confidentiality and integrity of the traffic in the network. WEP is used at the station-to-station level and does not offer any end-to-end security.

Secure key generation and distribution: Secure key generation and distribution capability is required by any system [15.138] that contains cryptographic authentication, confidentiality, and identification. Mobile systems have constraints such as minimal computational capabilities and authentication (while setting up the protocol, the user and the carrier never meet face to face).

The following approaches are being used for key generation:

1. Key generation by the telephone manufacturer and distribution to the service provider (SP) via a backbone network

2. Generation by the SP using either a manual entry by the customer or electronic distribution at the point of sale

3. Over-the-air phone activation with key exchange

An approach introduced in [15.139] provides interrouter authentication while incurring minimal overhead and handling the replay problems in wireless ad hoc networks. Reference [15.139] has proposed a scheme that assumes mutual trust among the network nodes. Nodes authenticate route reply messages to prevent replay attack. The scheme also implements message authentication code (MAC) to ensure integrity of route request packets.

In MANETs, [15.140] proposed to identify some internal attacks specific to the AODV routing protocol. To mitigate the identified attacks, the intrusion detection mechanism (IDM), which identifies misbehaving nodes using the knowledge base, and the response model (RM), which isolates these nodes from the network, were proposed. The existing security techniques for wired networks are not directly applicable to wireless networks. Security techniques should be present in all the networking layers. Therefore, new techniques must be developed for the wireless network. Current techniques for MANETs do not address internal attacks. Securing a wireless network is a nontrivial task, and only partial solutions are presently feasible.

15.13.4 Intrusion Detection in MANETs

MANETs provide infrastructure-free communication over the shared wireless channel. Without the support of a fixed infrastructure, it opens a new wireless networking transmission paradigm by employing multi-hops for information transfer. The underlying self-organizing property enables many powerful applications, such as instant conference in battlefield, emergency, and disaster situations. However, the great flexibility of wireless ad hoc networks also brings many research challenges, and one important issue usually ignored is the security problem.

Several mechanisms for securing MANETs have been proposed and can be classified into two types: intrusion prevention and intrusion detection. Intrusion prevention implies developing new secured protocols for MANETs or modifying the logic of existing protocols to enhance their security. Most of the current work using encryption or authentication mechanisms belongs to this group. While the intrusion prevention approach, as a first line of defense, can provide a certain level of security to wireless ad hoc networks, it is far from being sufficient. Intrusion detection, as a complementary mechanism, is designed to protect the availability, confidentiality, and integrity of critical networked information systems. The key idea of intrusion detection is to discover a set of system features that characterize system's normal behavior and use them to recognize anomalies and known intrusions, based on the assumption that normal system behavior and an attack are distinct and can be separated.

The concept of intrusion detection is not new as it has been widely researched in wired networks [15.141] [15.142]. However, the fundamental characteristics of MANETs present more stringent requirements for intrusion detection. First, the self-organizing property without the support of a fixed infrastructure requires a distributed intrusion detection system. Second, the thin difference between normal and anomalous patterns in wireless ad hoc networks requires a high detection accuracy and sensitivity of the intrusion detection system. Third, in the real network environment, a purely labeled training dataset may not be available, which requires the intrusion detection system be tolerant to the noise buried in the training dataset. Fourth, the constrained battery power requires highly efficient intrusion detection algorithms. Therefore, a practical distributed intrusion detection system with high detection accuracy, good noise tolerance, and low computational overhead is needed to secure MANETs.

Intrusion Detection Models

We discuss two system models: a distributed hierarchical system model and a completely distributed system model (shown in Figure 15.7) [15.143].

Both these system detection models are distributed in nature. The advantage of the distributed hierarchical model is that the data collected by a cluster head

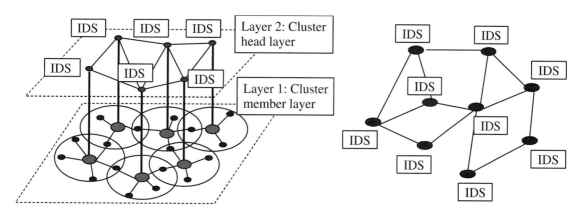

(a) A distributed hierarchical detection model.   (b) A completely distributed detection model.

Figure 15.7 Two intrusion detection models [15.143].
Credit: H. Deng, Q.A. Zeng, and D.P. Agrawal, "SVM-based Intrusion Detection System for Wireless Ad Hoc Networks," *Proceedings of IEEE Vehicular Technology Conference Fall 2003*, Orlando, October 6-9, 2003.

(CH) may be more comprehensive, which enhances the reliability of detection results. However, it is based on a hierarchical clustering scheme, and how effective selection of a CH in a dynamically changing environment poses another problem. The distributed hierarchical system model is good for ad hoc networks with lower mobility, such as wireless sensor networks. A completely distributed system model is more suitable for MANETs with high mobility, but more false alarms are anticipated to be present since only an incomplete data set is available and used. It should be remembered that both system models are based on the assumption that the number of malicious nodes are few as compared to the network size; otherwise, the scheme fails.

SVM–Based Intrusion Detection System

A comprehensive intrusion detection system [15.143] consists of four components (see Figure 15.8): local data collection module (DCM), support vector machine-based intrusion detection module (SVMDM), local response module (LRM), and global response module (GRM). The DCM gathers streams of audit data from various network sources and passes it to the SVMDM. The SVMDM analyzes the gathered local data traces using the SVM classification algorithm and identifies misbehaving nodes in the network. In the SVMDM, two types of SVM [15.144] based detection methods are present, depending on whether the attack data are available or not. One-class SVM classifier based intrusion detection (1-SVMDM) is used whenever no attack data are available, while conventional two-class SVM based intrusion detection (2-SVMDM) is applied in the situation when attack data are available. In practice, the 1-SVMDM can be used in the early stage of intrusion detection to find possible network-intrusive behaviors. After collecting some attack instances, 2-SVMDM can be used. The LRM is responsible for sending out the local detection results based on the locally collected data set. The GRM collects the local detection results from the LRM and makes a global response. Whenever any misbehaving node is detected, the GRM sends out alarm messages to the whole network to isolate the misbehaving node.

Figure 15.8
Proposed SVM–based intrusion detection system [15.143].
Credit: H. Deng, Q.A. Zeng, and D.P. Agrawal, "SVM-based Intrusion Detection System for Wireless Ad Hoc Networks," *Proceedings of IEEE Vehicular Technology Conference Fall 2003*, Orlando, October 6-9, 2003.

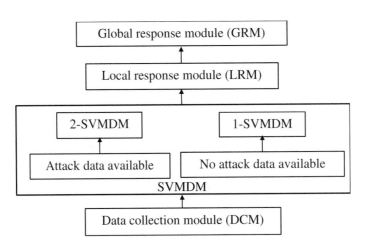

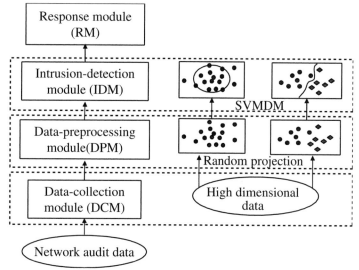

Figure 15.9
Random projection
technique for intrusion
detection [15.146].

Random Projection for Network Intrusion Detection Systems

Considering the constrained capabilities of wireless nodes, a new and more practical intrusion detection system is proposed using random projection technique [15.145] [15.146], which takes a labeled or unlabeled very high dimensional noisy dataset as input, and can be used in real-time network intrusion detection (see Figure 15.9).

The main idea of this approach is to first project a high-dimensional dataset to a lower dimensional space using random projection technique and then do the intrusion detection in the projected lower dimension by SVM classifier. The thrust of our proposed method lies in the fact that if the projected lower dimensional dataset can provide a comparable detection performance with the original high-dimensional dataset, then the complexity of the detecting algorithm will decrease drastically. Moreover, low-dimensional data can be stored and transmitted efficiently, thereby saving system resources. In addition, this approach can detect intrusions on an unlabeled dataset, without the requirement of purely labeled training dataset, which makes which the intrusion-detection system more practical.

● ● ● ● ● ● ● ● ● ● ● ● ● ● ● ● ●

15.14 Summary

The field of wireless devices and technology is rapidly changing, and research findings are becoming obsolete fairly quickly. We have attempted to provide an overview of research being carried out from the computing point of view. Issues like characterizing human movements and mobility modeling of MSs are critical not only in resource allocation, but also in ascertaining degree of QoS. The effect of power control and tradeoffs among power, space, and speed ought to be established. Usefulness of coding and automatic retry on minimization of errors and corresponding processing complexity may be worth exploring. Minimization of various handshaking signals needs to be considered carefully, and new technologies need to be developed that could support continuous media streaming without interruption. In brief, the future of wireless and mobile systems seems to be very promising, and it is hoped that many new applications will emerge as potential users of this exciting technology.

● ● ● ● ● ● ● ● ● ● ● ● ● ● ● ● ●

15.15 References

[15.1] D. P. Agrawal, "Future Directions in Mobile Computing and Networking Systems," *Report on NSF Sponsored Workshop held at the University of Cincinnati*, June 13–14, 1999, *Mobile Computing and Communications Review*, Vol. 3, No. 4, pp. 13–18, October 1999, and revised version at *http://www.ececs.uc.edu/~dpa*.

[15.2] Xtremespectrum: *http://www.xtremespectrum.com/products/faq.html*.

[15.3] K. Siwiak, P. Withington, and S. Phelan, "Ultra-Wide Band Radio: The Emergence of an Important New Technology," *Time DomainCorp.*, *http://www.time-domain.com*.

[15.4] G. R. Aiello, M. Ho, and J. Lovette, "Ultra-Wideband: An Emerging Technology for Wireless Communications," *Fantasma Networks, Inc.* (*http://www.fantasma.net*).

[15.5] K. Siwiak, "Ultra-Wide Band Radio: Introducing a New Technology," *Time Domain Corp.*, *http://www.timedomain.com*.

[15.6] J. Foerster, E. Green, S. Somayazulu, and D. Leeper, "Ultra-Wideband Technology for Short or Medium Range Wireless Communications," *Intel Corp.*, *http://www.Intel.com*.

[15.7] R. J Fontana, J. F. Larric, and J. E. Cade, "An Ultra Wideband Communications Link for Unmanned Vehicle Applications," *MultiSptectral Solutions, Inc.* (*http://www.his.com/~mssi*).

[15.8] "High Performance Wireless Home Networks: An Ultra-Wide Band Solution," *Fantasma Networks, Inc.* (*www.fantasma.net*).

[15.9] "Ultra-Wideband (UWB) Technology Enabling High-Speed Wireless Personal Area Networks," (white paper), Intel, *http://www.intel.com/technology/ultrawideband/downloads/ Ultra-Wideband.pdf*.

[15.10] W. Hirt, "UWB Radio Technology (UWB-RT) Short Range Communication and Location Tracking," *IBM Research, Zurich Research Laboratory*, Switzerland, *http://www.ibm.com*.

[15.11] "FCC, Notice of Proposed Rules Making (NPRM), Revision of Commission's Rules Regrading UWB Transmission Systems," *ET Docket 98–153*, May 11, 2000 (released), *http://www.fcc.com*.

[15.12] S. Rudd, "Data Only Networking," *http://www.interop.com*.

[15.13] D. Gill, "Standards for Multimedia Streaming and Communication over Wireless Networks," *http://www.emblazer.com/tech_mpeg4_2.shtml.*

[15.14] "Information Technology—Coding of Audio-Visual Objects—Part 1: Systems," *MPEG Committee Document N2501 (MPEG-4 Systems Version 1), http://www.emblazer.com/teeh_mpeg4_1.shtml.*

[15.15] D. Singer and W. Belknap, "Text for ISO/IEC 14496-1/PDAM1 (MPEG-4 Version 2 Intermedia Format—MP4)," *MPEG Committee Document N2801 Subpart 4*, July 1999.

[15.16] "Advanced Streaming Format (ASF) Specification, Public Specification Version 1.0," *Microsoft Corporation*, February 26 1998.

[15.17] K. Tanigawa, T. Hoshi, and K. Tsukada, "Simple RTP Multiplexing Transfer Methods for VoIP," *IETF Draft*, work in progress, *http://alternic.net/draft-t-u/draft-tanigawa-rtp-multiplex-01.txt*, or *http://www.ietf.org/proceedings/98/doc/slides/draft-ietf-avt-germ-00.txt.*

[15.18] J. Rosenberg and H. Schulzrinne, "An RTP Payload Format for User Multiplexing," *IETF Draft*, work in progress, *http://www.ietf.org/proceedings/99mar/I-D/draft-ietf-avt-aggregation-00.txt.*

[15.19] B. Subbiah and S. Sengodan, "User Multiplexing in RTP Payload between IP Telephony Gateways," *IETF Draft*, work in progress, *http://www.ietf.org/proceedings/99mar/I-D/draft-ietf-avt-mux-rtp-00.txt.*

[15.20] M. Handley, "GeRM: Generic RTP Multiplexing," *IETF Draft*, work in progress, *http://www.ietf.org/proceedings/99mar/I-D/draft-ietf-avt-germ-00.txt.*

[15.21] H. Schulzrinne, A. Rao, and R. Lanphier, "Real Time Streaming Protocol (RTSP)," *RFC 2326*, April 1998, *http://www.ietf.org/rfc/rfc2326.txt.*

[15.22] M. Handley, H. Schulzrinne, E. Schooler, and J. Rosenberg, "Session Initiation Protocol (SIP)," *RFC 2543*, March 1999, *http://www.ietf.org/rfc/rfc2543.txt.*

[15.23] M. Oelsner and C. Ciotti, "Wireless ATM and Wireless LAN—An Overview of Research, Standards and Systems," *ACTS Mobile Communications Summit*, Aalborg, Denmark, October 1997.

[15.24] *http://www.symbian.com/technology/mms.html.*

[15.25] H. Gharavi and K. Ban, "Dynamic Adjustment Packet Control for Video Communications over Ad-hoc Networks," *IEEE ICC 2004*, pp. 3086–3090, 2004.

[15.26] C. E. Perkins and E. M. Royer, "Ad Hoc On-Demand Distance Vector Routing," Proceedings of the *2nd IEEE Workshop on Mobile Computing Systems and Applications*, New Orleans, LA, pp. 90–100, February 1999.

[15.27] G. Aggelou, "On the Performance Analysis of the Minimum-Blocking and Bandwidth-Reallocation Channel-Assignment (MBCA/BRCA) Methods for Quality-of-Service Routing Support in Mobile Multimedia Ad Hoc Networks," *IEEE Transactions on Vehicular Technology*, Vol. 53, No. 3, May 2004.

[15.28] *http://www.northstream.se.*

[15.29] *http://www.kerton.com.*

[15.30] *http://www.technewsworld.com/story/33151.html.*

[15.31] *http://electronics.howstuffworks.com/question530.htm.*

[15.32] *http://www.mpirical.com/companion/mpirical_companion.html#http://*
www.mpirical.com/companion/Multi_Tech/
POC_-_Push_to_talk_Over_Cellular.htm.

[15.33] *http://member.openmobilealliance.org/ftp/public_documents/POC/.*

[15.34] *http://www.3ggp.org/.*

[15.35] *http://www.3gpp2.org/.*

[15.36] I. F. Akyildiz, J. McNair, et al., "Mobility Management in Next Generation Wireless System," *Proceedings of the IEEE*, Vol. 87, No. 8, pp. 1347–1384, August 1999.

[15.37] Y.-B. Lin and I. Chlamtac, *Wireless and Mobile Network Architecture*, John Wiley and Sons, Hoboken, NJ, 2001.

[15.38] M. D. Austin and G. L. Stuber, "Direction Biased Handoff Algorithms for Urban Microcells," *Proceedings of the IEEE VTC-94*, pp. 101–105, June 1994.

[15.39] A. Murase, I. C. Symington, and E. Green, "Handover Criterion for Macro and Microcellular Systems," *Proceedings of the IEEE VTC-91*, pp. 524–530, June 1991.

[15.40] B. Epstein and M. Schwartz, "Reservation Strategies for Multimedia Traffic in A Wireless Environment," *Proceedings of the IEEE VTC*, pp. 165–169, July 1995.

[15.41] Q-A. Zeng and D. P. Agrawal, "Performance Analysis of a Handoff Scheme in Integrated Voice/Data Wireless Networks," *Proceedings of the IEEE VTC-00*, pp 845–851, September 2000.

[15.42] J. Wang, Q-A. Zeng, and D. P. Agrawal, "Performance Analysis of Integrated Wireless Mobile Network with Queuing Handoff Scheme," *Proceedings of the IEEE Radio and Wireless Conference 2001*, pp. 69–72, August 2001.

[15.43] J. Wang, Q-A. Zeng, and D. P. Agrawal, "Performance Analysis of Preemptive Handoff Scheme for Integrated Wireless Mobile Networks," *Proceedings of the IEEE Globecom 2001*, November 2001.

[15.44] H. Chen, Q-A. Zeng, and D. P. Agrawal, "A Novel Optimal Channel Partitioning Algorithm for Integrated Wireless and Mobile Networks," *Journal of Mobile Communication, Computation and Information.* October 2004, pp. 507–517.

[15.45] H. Chen, Q-A. Zeng, and Dharma P. Agrawal, "A Novel Analytical Modeling for Optimal Channel Partitioning in the Next Generation Integrated Wireless and Mobile Networks," *Proceedings of MSWiM Workshop 2002*, in conjunction with *Mobicom 2002*, September 28, 2002.

[15.46] H. Chen, Q-A. Zeng, and Dharma P. Agrawal, "A Novel Channel Allocation Scheme in Integrated Wireless and Mobile Networks," *Proceedings of 2003 Workshop on Mobile and Wireless Networks*, May 19–22, 2003.

[15.47] H. Chen, Q-A. Zeng, and D. P. Agrawal, "An Adaptive Call Admission and Bandwidth Allocation Scheme for Future Wireless and Mobile Networks," *IEEE VTC Fall 2003*, Orlando, October 2003.

[15.48] IEEE Standard for Wireless LAN, "Medium Access Control (MAC) and Physical Layer (PHY) Specification," P802.11, November 1999.

[15.49] Y. Chen, Q-A Zeng, and D. P. Agrawal, "Performance of MAC Protocol in Ad Hoc Networks," *Proceedings of Communication Networks and Distributed Systems Modeling and Simulation Conference*, pp. 55–61, January 2003.

[15.50] F. Tobagi and V. Hunt. "Performance Analysis of Carrier Sense Multiple Access with Collision Detection," *Computer Networks*, Vol. 4, pp. 245–259, 1980.

[15.51] M. Benveniste, G. Chesson, M. Hoeben, A. Singla, H. Teunissen, and M. Wentink, "EDCF Proposed Draft Text," *IEEE working document 802.11-01/131r1*, March 2001.

[15.52] Y. Chen, Q-A Zeng, and D. P. Agrawal, "Performance Analysis of IEEE 802.11e Enhanced Distributed Coordination Function," *Proceedings of IEEE International Conference on Networking*, September 2003.

[15.53] Y. Chen, Q-A Zeng, and D. P. Agrawal, "Performance Analysis and Enhancement of IEEE 802.11 MAC Protocol," *Proceedings of IEEE International Conference on Telecommunications*, Feburary 2003.

[15.54] ETSI, "Broadband Radio Access Networks (BRAN); HIPERLAN Type 2 Technical Specification; Physical (PHY) Layer," August 1999.

[15.55] ETSI, "Broadband Radio Access Networks (BRAN); HIPERLAN Type 2; Data Link Control (DLC) Layer; Part 1: Basic Transport Functions," December 1999.

[15.56] H. Gossain, C. M. Cordeiro, and D. P. Agrawal, "Multicast: Wired to Wireless," *IEEE Communications Magazine*, pp. 116–123, June 2002.

[15.57] C. R. Lin and K-M. Wang, "Mobile Multicast Support in IP Networks," *IEEE INFOCOM'00*, pp. 1664–1672, March 2000.

[15.58] A. Mihailovic, M.Shabeer, and A. H. Aghvami, "Sparse Mode Multicast as a Mobility Solution for Internet Campus Networks," *Proceedings of PIMRC'99*, Osaka, Japan, September 1999.

[15.59] C. L. Tan and S. Pink, "Mobicast: A multicast Scheme for Wireless Networks," *Mobile Networks*. pp. 259–271, Appl. 5, 4 Dec. 2000.

[15.60] H. Gossain, S. Kamat, and D. P. Agrawal, "A Framework for Handling Multicast Source Movement over Mobile IP," *IEEE International Conference on Communications*, May 2002.

[15.61] H. Gossain, S. Kamat, D. P. Agrawal, "A Framework for Handling Multicast Source Movement over Mobile-IP," *IEEE International Conference on Communications,* May 2002.

[15.62] C. M. Cordeiro, H. Gossain and D. P. Agrawal, "Multicast over Wireless Mobile Ad Hoc Networks: Present and Future Directions," *IEEE Network*, *Special Issue on Multicasting: An Enabling Technology*, January/February 2003.

[15.63] L. Rizzo and L. Vicisano, "RMDP: A FEC-Based Reliable Multicast Protocol for Wireless Environments," *Mobile Computing and Communications Review*, Vol. 2, No. 2, pp. 23–32, April 1998.

[15.64] D. H. Sadok, C. M. Cordeiro, and J. Kelner, "A Reliable Subcasting Protocol for Wireless Environments," *The 2nd International Conference on Mobile and Wireless Communication Networks*, Paris, France, May 2000.

[15.65] MADCAP Protocol, *http://www.ietf.org/internet-drafts/ draft-ietf-malloc-madcap-07.txt*.

[15.66] C. A. Balanis, *Antenna Theory: Analysis and Design*, Wiley, New York, 1997.

[15.67] S. Keshav, *An Engineering Approach to Computer Networking*, Addison Wesley, Boston, MA, 1997.

[15.68] L. C. Godara, "Application of Antenna Arrays to Mobile Communications, Part I: Performance Improvement, Feasibility and System Considerations," *Proceedings of the IEEE*, July 1997.

[15.69] J. C. Liberti and T. S. Rappaport, *Smart Antennas for Wireless Communications: IS-95 and Third Generation CDMA Applications.* Prentice Hall, Upper Saddle River, NJ, 1999.

[15.70] Y. B. Ko, V. Shankarkumar, and N. H. Vaidya, "Medium Access Control Protocols Using Directional Antennas in Ad Hoc Networks," *Proceedings of IEEE INFOCOM'2000*, March 2000.

[15.71] A. Nasipuri, S. Ye, J. You, and R. E. Hiromoto, "A MAC Protocol for Mobile Ad Hoc Networks Using Directional Antennas," *Proceedings of the IEEE Wireless Communications and Networking Conference (WCNC)*, September 2000.

[15.72] "All About GPS," available on the world wide web at *http://www.trimble.com/gps/*.

[15.73] D. Lal, R. Gupta, and D. P. Agrawal, "Throughput Enhancement in Wireless Ad Hoc Networks with Spatial Channels—A MAC Layer Perspective," *Proceedings of Seventh IEEE International Symposium on Computers and Communications*, July 2002.

[15.74] D. Lal, R. Toshniwal, R. Radhakrishnan, J. Caffery, and D. P. Agrawal, "A Novel MAC Layer Protocol for Space Division Multiple Access in Wireless Ad Hoc Networks," *Proceedings of IEEE International Conference on Computer Communications and Networking (ICCCN)*, October 2002.

[15.75] A. Manjeshwar and D. P. Agrawal, "TEEN: A Routing Protocol for Enhanced Efficiency in Wireless Sensor Networks," *Proceedings of the 1st International Workshop on Parallel and Distributed Computing Issues in Wireless Networks and Mobile Computing*, April 2001.

[15.76] A. Manjeshwar and D. P. Agrawal, "An Efficient Sensor Network Routing Protocol (APTEEN) with Comprehensive Information Retrieval," *Proceedings of the 2nd International Workshop on Parallel and Distributed Computing Issues in Wireless Networks and Mobile Computing*, April 2002.

[15.77] A. Manjeshwar, Q-A. Zeng, and D. P. Agrawal, "An Analytical Model for Information Retrieval in Wireless Sensor Networks Using Enhanced APTEEN Protocol," *IEEE Transactions on Parallel and Distributed Systems, PDS Special Issue on Mobile Computing and Wireless Networks*, Vol. 13, No. 12, pp. 1290–1302, December 2002.

[15.78] A. Faradjian, J. E. Gehrke, and P. Bonnet, "GADT: A Probability Space ADT for Representing and Querying the Physical World," *Proceedings of the 18th International Conference on Data Engineering (ICDE 2002)*, San Jose, February 2002.

[15.79] "Collaborative Processing," *http://dss.ll.mit.edu/dss.web/SensIT.html*.

[15.80] J. Hill, R. Szewczyk, A. Woo, S. Hollar, D. Culler, and K. Pister, "System Architecture Directions for Network Sensors," *Proceedings of the Conference of Architecture Support for Programming Languages and Operating Systems, 2000 (ASPLOS 2000)*, 2000.

[15.81] R. Barr, J. C. Bicket, D. S. Dantas, B. Du, T. W. D. Kim, B. Zhou, and E. G. Sirer, "On the Need for System-Level Support for Ad Hoc and Sensor Networks," *Operating Systems Review, ACM*, Vol. 36(2): pp. 1–5, April 2002.

[15.82] N. F. Maxemchuk, "Dispersity Routing in High-speed Networks," *Computer Networks and ISDN System 25*, pp. 645–661, 1993.

[15.83] H. Suzuki and F. A. Tobagi, "Fast Bandwidth Reservation Scheme with Multi-link & Multi-path routing in ATM networks," *Proceedings of IEEE INFOCOM*, 1992.

[15.84] S. Lee and M. Gerla, "Split Multipath Routing with Maximally Disjoint Paths in Ad Hoc Networks," *Proceedings of the IEEE ICC'01*, pp. 3201–3205, 2001.

[15.85] A. Nasipuri and S. Das, " On-Demand Multipath Routing for Mobile Ad Hoc Networks," *Proceedings of the 8th Annual IEEE International Conference on Computer Communications and Networks (ICCCN)*, pp. 64–70, October 1999.

[15.86] M. R. Pearlman, Z. J. Hass, P. Sholander, and S. S.Tabrizi, "On the Impact of Alternate Path Routing for Load Balancing in Mobile Ad Hoc Networks" *Proceedings of IEEE/ACM MobiHoc*, 2000.

[15.87] V. D. Park and M. S. Corson, "A Highly Distributed Routing Algorithm for Mobile Wireless Networks," *Proceedings of IEEE INFOCOM*, pp. 1405–1413, 1997.

[15.88] C. Intanagonwiwat, R. Govindan, and D. Estrin, "Directed Diffusion: A Scalable and Robust Communication Paradigm for Sensor Networks," *Proceedings of the 6th Annual ACM/IEEE International Conference on Mobile Computing and Networking (MOBICOM)*, pp. 56–67, August 2000.

[15.89] W. Heinzelman, A. Chandrakasan, and H. Balakrishnan, "Energy-Efficient Communication Protocols for Wireless Microsensor Networks," *Proceedings Hawaian International Conference on Systems Science*, January 2000.

[15.90] J. Chang and L. Tassiulas, "Maximum Lifetime Routing in Wireless Sensor Networks," *Proceedings of Advanced Telecommunications and Information Distribution Research Program*, 2000.

[15.91] D. Ganesan, R. Govindan, S. Shenker, and D. Estrin, "Highly Resilient, Energy Efficient Multipath Routing in Wireless Sensor Networks," *Mobile Computing and Communications Review (MC2R)*, Vol. 1, No. 2, 2002.

[15.92] S. D. Servetto and G. Barrenechea, "Constrained Random Walks on Random Graphs: Routing Algorithms for Large Scale Wireless Sensor Networks,"

Proceedings of the 1st ACM International Workshop on Wireless Sensor Networks and Applications, pp. 12–21, September 2002.

[15.93] C. Schurgers and M. B. Srivastava, "Energy Efficient Routing in Wireless Sensor Networks," *MILCOM'01*, October 2001.

[15.94] R. C. Shah and J. Rabaey, "Energy Aware Routing for Low Energy Ad Hoc Sensor Networks," *Proceedings of IEEE Wireless Communications and Networking Conference (WCNC)*, March 2002.

[15.95] N. Jain, D. K. Madathil, and D. P. Agrawal, "Energy Aware Multi-Path Routing for Uniform Resource Utilization in Sensor Networks," *Proceedings of the IPSN'03 International Workshop on Information Processing in Sensor Networks*, Palo Alto, CA, April 22, 2003.

[15.96] N. Jain, D. K. Madathil, and D. P. Agrawal, "Exploiting Multi-Path Routing to achieve Service Differentiation in Sensor Networks," *Proceedings of the 11th IEEE International Conference on Networks (ICON 2003)*, Sydney, Australia, October 2003.

[15.97] S. Chen and K. Nahrstedt, "Distributed Quality-of-Service Routing in Ad-Hoc Networks," *IEEE Journal on Special Areas in Communications*, Vol. 17, No. 8, August 1999.

[15.98] S. Bhatnagar, B. Deb, and B. Nath, "Service Differentiation in Sensor Networks," *Proceedings of Fourth International Symposium on Wireless Personal Multimedia Communications*, 2001.

[15.99] C. Cordeiro, D. Agrawal, and D. Sadok, "Piconet Interference Modeling and Performance Evaluation of Bluetooth MAC Protocol," *IEEE Transactions on Wireless Communications*, accepted for publication.

[15.100] C. Cordeiro and D. Agrawal, "Employing Dynamic Segmentation for Effective Co-located Coexistence between Bluetooth and IEEE 802.11 WLANs," *Proceedings of IEEE GLOBECOM*, Taiwan, November 2002.

[15.101] Bluetooth SIG, "Bluetooth Specification," *http://www.bluetooth.com*.

[15.102] Y. Lim, S. Min, and J. Ma, "Performance Evaluation of the Bluetooth-based Public Internet Access Point," *Proceedings of the 15th ICOIN*, pp. 643–648, 2001.

[15.103] N. Rouhana and E. Horlait, "BWIG: Bluetooth Web Internet Gateway," *Proceeding of IEEE Symposium on Computer and Communication.*, July 2002.

[15.104] O. Kasten and M. Langheinrich, "First Experience with Bluetooth in the Smart-Its Distributed Sensor Network," *Proceedings of the Workshop in Ubiquitous Computing and Communications*, October 2001.

[15.105] F. Siegemund and M. Rohs, "Rendezvous Layer Protocols for Bluetooth-enabled Smart Devices," *Proceedings of International Conference on Architecture of Computing Systems*, April 2002.

[15.106] D. Estrin, R. Govindan, and J. Heidmanm, "New Century Challenges: Scalable Cordination in Sensor Networks," *Proceedings of ACM Mobicom*, pp. 263–270, 1999.

[15.107] J. Kahn, R. Katz, K. Pister, "New Century Challenges: Mobile Networking for Smart Dust," *Proceedings of ACM Mobicom*, pp. 271–278, 1999.

[15.108] C. Cordeiro, H. Gossain, and D. Agrawal, "Multicast over Wireless Mobile Ad Hoc Networks: Present and Future Directions," *IEEE Network, Special Issue on Multicasting: An Enabling Technology*, January/February 2003.

[15.109] C. Cordeiro, S. Abhyankar, and D. P. Agrawal, "A Dynamic Slot Assignment Scheme for Slave-to-Slave and Multicast-like Communication in Bluetooth Personal Area Networks," *Proceedings of IEEE Globecom*, San Francisco, December 2003.

[15.110] C. Cordeiro, S. Abhyankar, and D. P. Agrawal, "A Novel Energy Efficient Communication Architecture for Bluetooth Ad Hoc Networks," *ACM/IFIP Personal Wireless Communications (PWC) Conference*, Venice, Italy, September 2003.

[15.111] S. Zurbes, W. Stahl, K. Matheus, and J. Harrtsen, "Radio Network Performance of Bluetooth," *Proceedings of IEEE ICC 2000*, Vol. 3, pp. 1536–1567, 2000.

[15.112] D. Famolari, "Link Performance of an Embedded Bluetooth Personal Area Network," *Proceedings of IEEE ICC 2001*, Helsinki, June 2001.

[15.113] C. M. Cordeiro, S. Abhyankar, R. Toshiwal, and D. P.Agrawal, "BlueStar: Enabling Efficient Integration between Bluetooth WPANs and IEEE 802.11 WLANs," *ACM/Kluwer Mobile Networks and Applications (MONET) Journal, Special Issue on Integration of Heterogeneous Wireless Technologies*, to appear, Fall 2003.

[15.114] N. Golmie and F. Mouveaux, "Interference in the 2.4 GHz ISM Band: Impact on the Bluetooth Access Control Performance," *Proceedings of the IEEE ICC'01*, Helsinki, Finland, June 2001.

[15.115] M. Fainberg, and D. Goodman, "Analysis of the Interference Between IEEE 802.11b and Bluetooth Systems," *Proceedings of the IEEE VTC Fall 2001*, October 2001.

[15.116] W. Lai, B. Christian, R. Tibbs, and S.Berghe, "A Framework for Internet Traffic Engineering Measurement," *Internet Draft, draft-ietf-tewg-measure-02.txt*.

[15.117] G. Ash, "Traffic Engineering & QoS Methods for IP, ATM, & TDM-Based Multiservice Network," *Internet Draft, draft-ietf-tewg-qos-routing-04.txt*.

[15.118] H. Gan and B. Treister, "Adaptive Frequency Hopping Implementation Proposals for IEEE 802.15.1/2 WPAN," *IEEE 802.15-00/367r0*, 2000.

[15.119] IEEE 802.15 Coexistence Task Group 2, *http://www.ieee802.org/15/pub/TG2.html*.

[15.120] M. Albrecht, M. Frank, P. Martini, M. Schetelig, A. Vilavaara, A. Wenzel, "IP services over Bluetooth: Leading the Way to A New Mobility," *Proceedings of LCN'99*, pp. 2–11, 1999.

[15.121] Mobilian, Mobilian True Radio, *http://www.mobilian.com*.

[15.122] Possio, Possion PX20, *http://www.possio.com*.

[15.123] Red-M, Genos Wirelessware, *http://www.red-m.com*.

[15.124] D. Awduche, A. Chiu, A. Elwalid, I Widjaja, and X. Xiao, "Overview and Principles of Internet Traffic Engineering," *Internet Draft draft-ieft-tewg-principles-02.txt*.

[15.125] S. Abhyankar, R. Toshiwal, C. M. Cordeiro, and D. P. Agrawal, "On the Application of Traffic Engineering over Bluetooth Ad Hoc Networks," *Proceedings of ACM International Workshop on Modeling, Analysis and Simulation of Wireless and Mobile Systems (MSWiM), in conjunction with ACM Mobicom*, San Diego, CA, September 2003.

[15.126] K. Owens and V. Sharma, "Network Survivability Considerations for Traffic Engineered IP Networks," *Internet Draft, draft-owens-te-network-survivability-01.txt.*

[15.127] Q. Wang and D. P. Agrawal, "A Dichotomized Rendezvous Algorithm for Mesh Bluetooth Scatternets," *An International Journal of Ad Hoc & Sensor Wireless Networks,* Old City Publishing, Vol. 1, January 2005, pp. 1–24, also see details of simulator at the web site http://www.ececs.uc.edu/~cdmc/ucbt/.

[15.128] *http://nesl.ee.ucla.edu/research.htm.*

[15.129] J. Pouwelse, K. Langendoen, and H. Sips, "Dynamic Voltage Scaling on a Low-Power Microprocessor," *Technical Report*, Delft University of Technology, 2000.

[15.130] A. Azevedo, I. Issenin, R. Gupta, N. Dutt, A. Veidenbaum, and A. Nicolau, "Profile-Based Dynamic Voltage Scheduling Using Program Checkpoints in the COPPER Framework," *Design Automation and Test in Europe*, March 2002.

[15.131] P. Agrawal, J-C. Chen, S. Kishore, P. Ramanathan, and K. Sivalingam, "Battery Power Sensitive Video Processing in Wireless Networks," *Proceedings of the IEEE PIMRC'98*, Boston, MA, September 1998.

[15.132] R. Kravets and P. Krishnan, "Application-Driven Power Management for Mobile Communication," *Proceedings of the Fourth Annual ACM/IEEE International Conference on Mobile Computing and Networking (MobiCom)*, Dallas, TX, pp. 263–277, October 1998.

[15.133] "Network Security in a Wireless LAN," *BREEZECOM Wireless Communications*, Inc., Feb 1999, *http://www.summitonline.com/security/papers/breeze1.html.*

[15.134] S. Uskela, "Security in Wireless Local Area Networks," *http://www.tml.hut.fi/Opinnot/Tik-110.501/1997/wireless_lan.html.*

[15.135] M. Flinn and M. Satyanarayanan, "Energy-Aware Adaptation for Mobile Application," *ACM SOSP*, pp. 48–63, December 1999.

[15.136] L. Feeney and M. Nilsson, "Investigating the Energy Consumption of a Wireless Network Interface in an Ad Hoc Networking Environment," *Proceeding of the IEEE INFOCOM 2001*, 2001.

[15.137] M. Dave and F.-W. Areth, "Wired on Wireless: A New Class of 802.11 Devices Go the Distance," March 1999, *http://www.networkcomputing.com/1006/1006r2.html.*

[15.138] C. Carroll, Y. Frankel, and Y. Tsiounis, "Efficient Key Distribution for Slow Computing Devices," *IEEE Symposium on Security and Privacy*, pp. 66–76, May 1998.

[15.139] L. Venkatraman and D. P. Agrawal, "A Security Scheme for Routing in Ad Hoc Networks," *Proceedings of the 13th International Conference on Wireless Communications*, Calgary, Canada, pp. 129–146, July 2001.

[15.140] S. Bhargava and D. P. Agrawal, "Security Enhancements in AODV Protocol for Wireless Ad Hoc Networks," *Proceeding of the Vehicular Technology Conference, Fall 2001*, Atlantic City, NJ, October 2001.

[15.141] D. Denning, "An intrusion detection model," *IEEE Transactions on Software Engineering*, 12(2): pp. 222–232, 1987.

[15.142] C. Warrender, S. Forrest, and B. Pearlmutter, "Detecting Intrusions using System Calls: Alternative Data Models," *Proceedings of the 1999 IEEE Symposium on Security and Privacy*, pp. 133–145, 1999.

[15.143] H. Deng, Q-A. Zeng, and D. P. Agrawal, "SVM-based Intrusion Detection System for Wireless Ad Hoc Networks," *Proceedings of IEEE Vehicular Technology Conference Fall 2003*, Orlando, October 6–9, 2003.

[15.144] V. Vapnik, *Statistical Learning Theory*, John Wiley & Sons Inc., New York, 1998.

[15.145] H. Deng, Q-A. Zeng, and D. P. Agrawal, "Network Intrusion Detection System using Random Projection Technique," *Proceedings of the 2003 International Conference on Security and Management (SAM'03)*, Las Vegas, NV, pp. 10–16, June 23–26, 2003.

[15.146] H. Deng, Q-A. Zeng, and D. P. Agrawal, "An Unsupervised Network Anomaly Detection System Using Random Projection Technique," *Proceedings of the 2003 International Workshop on Cryptology and Network Security (CANS03)*, Florida, September 24–26, 2003.

15.16 Problems

P15.1. How is UWB different from frequency hopping used in Bluetooth? Explain.

P15.2. Multimedia services have two components of video and voice data. Can you characterize them as non–real-time and real-time traffic? Explain clearly.

P15.3. Can you use PTT technology to process multimedia traffic? Explain clearly.

P15.4. Assume that traffic is assigned four different priority levels taking into account real-time and handoff traffic. How can you handle such traffic while supporting mobility?

P15.5. What are the pros and cons of employing satellite communication for multicasting? Explain.

P15.6. Can you do multicasting in a MANET? If yes, how and if no, why not? Explain clearly.

P15.7. If preemption is allowed in Problem P15.4, how would you do the scheduling and what are the relative advantages and disadvantages?

P15.8. If you are given a choice of using either a reactive or a proactive routing protocol, which one would you prefer for Problem P15.7 with MANET and why?

P15.9. Using your favorite MANET simulator, create an arbitrary MANET and find disjoint paths between an arbitrary source-destination pair. Assume all needed parameters.

P15.10. Compare the time requirements in Problem P15.9, if
  (a) The number of MSs is doubled.
  (b) The radio coverage area is doubled.
  (c) The node connectivity is doubled.
  (d) Both parts (a) and (b) are done.

P15.11. Name the applications for which direct-diffusion–based flat architectures or cluster-based sensor networks could be considered useful? Explain in detail.

P15.12. What are the uses of different types of queries in sensor networks? Explain clearly.

P15.13. Using your favorite MANET simulator, create a sensor network. Assuming appropriate parameters, simulate a sensor network with 100 nodes. Find the query propagation time from one end of the network to another end if:
  (a) A flat architecture is used?
  (b) A cluster architecture is used?

P15.14. Using your favorite Web site, find different types of sensors if the idea is to explore the following applications:
  (a) Nuclear plant.
  (b) Underwater project.
  (c) Noise level in a campus.
  (d) Air pollution over an industrial area.
  (e) Maintenance of a large bridge.
  (f) Speeding on a freeway.
  (g) Industrial discharge to a lake or a riverbed.
  (h) Contamination due to an industrial chimney.
  (i) Ozone-level determination in an area.
  (j) Flood-level monitoring.
  (k) Rock-falling (snow-mountain falling) in a mountainous area.
  (l) Underground earth movement determination.
  (m) Movement of ore and manpower in an underground mine.

P15.15. Use your favorite search engine to find out what is meant by the "self-organizing property of sensor networks."

P15.16. You can envision a potential use of wireless technology in having robots with de-centralized decision-making capability. Can you think of at least five applications? What are the limitations and how can you address them? Explain.

P15.17. Can you think of ten important parameters that ought to be sensed in a futuristic automobile? What type of sensors do you need to monitor them constantly?

P15.18. In Problem P15.17, can you comment on the possible use of:
   (a) MANET.
   (b) Sensor network.
   (c) Bluetooth devices.
   (d) HomeRF.
   (e) 802.11-enabled devices.
   (f) WiMax.

P15.19. In an amusement park, wireless devices are used to post the queue length and wait time for each of the rides. Can you think of an infrastructure that could enable patrons to do online interaction? Assume appropriate requirements and a potential solution.

P15.20. For Problem P15.19, you do need some software support. Can you think and outline how you would go about doing this? Explain.

P15.21. For Problem P15.19, you do need some forms of authentication and encryption. What are the most important parameters you would be concerned with? Explain clearly.

P15.22. In a center for mentally challenged people, a decision was made to monitor each individual person using wireless devices. Can you comment on the relative advantages and disadvantages of such an infrastructure?

P15.23. In Problem P15.22, what kind of personal information would you like to maintain in the database and why is it important? Explain clearly.

P15.24. In Problem P15.22, what are the security concerns you would have and how can you address them? Explain with suitable examples.

P15.25. Is it desirable to assign priority to traffic in a MANET? Explain.

P15.26. For Problem P15.4, can you use a normal voltage level for VLSI devices in handling high-priority traffic and a lower voltage level for low-priority traffic? Explain clearly.

P15.27. From your favorite Web site, find the differences between MIMO and smart antennas? What are their relative advantages? Explain clearly.

Erlang B Table

This table represents the relationship among the number of channels, the blocking probability, and the offered load in Erlang. Therefore, given two of these three quantities, the third one can be found using this table.

N*	Blocking Probability							
	0.001	0.002	0.003	0.004	0.005	0.006	0.007	0.008
1	0.0010	0.0020	0.0030	0.0040	0.0050	0.0060	0.0071	0.0081
2	0.0458	0.0653	0.0806	0.0937	0.1054	0.1161	0.1260	0.1353
3	0.1938	0.2487	0.2885	0.3210	0.3490	0.3740	0.3966	0.4176
4	0.4393	0.535	0.6021	0.6557	0.7012	0.7412	0.7773	0.8103
5	0.7621	0.8999	0.9945	1.0692	1.1320	1.1870	1.2362	1.2810
6	1.1459	1.3252	1.4468	1.5421	1.6218	1.6912	1.7531	1.8093
7	1.5786	1.7984	1.9463	2.0614	2.1575	2.2408	2.3149	2.3820
8	2.0513	2.3106	2.4837	2.6181	2.7299	2.8266	2.9125	2.9902
9	2.5575	2.8549	3.0526	3.2057	3.3326	3.4422	3.5395	3.6274
10	3.0920	3.4265	3.6480	3.8190	3.9607	4.0829	4.1911	4.2889
11	3.6511	4.0215	4.2661	4.4545	4.6104	4.7447	4.8637	4.9709
12	4.2314	4.6368	4.9038	5.1092	5.2789	5.4250	5.5543	5.6708
13	4.8306	5.2700	5.5588	5.7807	5.9638	6.1214	6.2607	6.3863
14	5.4464	5.9190	6.2291	6.4670	6.6632	6.8320	6.9811	7.1155
15	6.0772	6.5822	6.9130	7.1665	7.3755	7.5552	7.7139	7.8568
16	6.7215	7.2582	7.6091	7.8780	8.0995	8.2898	8.4579	8.6092
17	7.3781	7.9457	8.3164	8.6003	8.8340	9.0347	9.2119	9.3714
18	8.0459	8.6437	9.0339	9.3324	9.5780	9.7889	9.9751	10.143
19	8.7239	9.3515	9.7606	10.073	10.331	10.552	10.747	10.922
20	9.4115	10.068	10.496	10.823	11.092	11.322	11.526	11.709
21	10.108	10.793	11.239	11.580	11.860	12.100	12.312	12.503
22	10.812	11.525	11.989	12.344	12.635	12.885	13.105	13.303
23	11.524	12.265	12.746	13.114	13.416	13.676	13.904	14.110
24	12.243	13.011	13.510	13.891	14.204	14.472	14.709	14.922
25	12.969	13.763	14.279	14.673	14.997	15.274	15.519	15.739
26	13.701	14.522	15.054	15.461	15.795	16.081	16.334	16.561
27	14.439	15.285	15.835	16.254	16.598	16.893	17.153	17.387
28	15.182	16.054	16.620	17.051	17.406	17.709	17.977	18.218
29	15.930	16.828	17.410	17.853	18.218	18.530	18.805	19.053

N* is the number of channels.

N*	Blocking Probability							
	0.001	0.002	0.003	0.004	0.005	0.006	0.007	0.008
30	16.684	17.606	18.204	18.660	19.034	19.355	19.637	19.891
31	17.442	18.389	19.002	19.470	19.854	20.183	20.473	20.734
32	18.205	19.176	19.805	20.284	20.678	21.015	21.312	21.580
33	18.972	19.966	20.611	21.102	21.505	21.850	22.155	22.429
34	19.743	20.761	21.421	21.923	22.336	22.689	23.001	23.281
35	20.517	21.559	22.234	22.748	23.169	23.531	23.849	24.136
36	21.296	22.361	23.050	23.575	24.006	24.376	24.701	24.994
37	22.078	23.166	23.870	24.406	24.846	25.223	25.556	25.854
38	22.864	23.974	24.692	25.240	25.689	26.074	26.413	26.718
39	23.652	24.785	25.518	26.076	26.534	26.926	27.272	27.583
40	24.444	25.599	26.346	26.915	27.382	27.782	28.134	28.451
41	25.239	26.416	27.177	27.756	28.232	28.640	28.999	29.322
42	26.037	27.235	28.010	28.600	29.085	29.500	29.866	30.194
43	26.837	28.057	28.846	29.447	29.940	30.362	30.734	31.069
44	27.641	28.882	29.684	30.295	30.797	31.227	31.605	31.946
45	28.447	29.708	30.525	31.146	31.656	32.093	32.478	32.824
46	29.255	30.538	31.367	31.999	32.517	32.962	33.353	33.705
47	30.066	31.369	32.212	32.854	33.381	33.832	34.230	34.587
48	30.879	32.203	33.059	33.711	34.246	34.704	35.108	35.471
49	31.694	33.039	33.908	34.570	35.113	35.578	35.988	36.357
50	32.512	33.876	34.759	35.431	35.982	36.454	36.870	37.245
51	33.332	34.716	35.611	36.293	36.852	37.331	37.754	38.134
52	34.153	35.558	36.466	37.157	37.724	38.211	38.639	39.024
53	34.977	36.401	37.322	38.023	38.598	39.091	39.526	39.916
54	35.803	37.247	38.180	38.891	39.474	39.973	40.414	40.810
55	36.631	38.094	39.040	39.760	40.351	40.857	41.303	41.705
56	37.460	38.942	39.901	40.630	41.229	41.742	42.194	42.601
57	38.291	39.793	40.763	41.502	42.109	42.629	43.087	43.499
58	39.124	40.645	41.628	42.376	42.990	43.516	43.980	44.398
59	39.959	41.498	42.493	43.251	43.873	44.406	44.875	45.298
60	40.795	42.353	43.360	44.127	44.757	45.296	45.771	46.199
61	41.633	43.21	44.229	45.005	45.642	46.188	46.669	47.102
62	42.472	44.068	45.099	45.884	46.528	47.081	47.567	48.005
63	43.313	44.927	45.970	46.764	47.416	47.975	48.467	48.910
64	44.156	45.788	46.843	47.646	48.305	48.870	49.368	49.816
65	45.000	46.650	47.716	48.528	49.195	49.766	50.270	50.723
66	45.845	47.513	48.591	49.412	50.086	50.664	51.173	51.631
67	46.692	48.378	49.467	50.297	50.978	51.562	52.077	52.540
68	47.54	49.243	50.345	51.183	51.872	52.462	52.982	53.450
69	48.389	50.110	51.223	52.071	52.766	53.362	53.888	54.361

N*	Blocking Probability							
	0.001	0.002	0.003	0.004	0.005	0.006	0.007	0.008
70	49.239	50.979	52.103	52.959	53.662	54.264	54.795	55.273
71	50.091	51.848	52.984	53.848	54.558	55.166	55.703	56.186
72	50.944	52.718	53.865	54.739	55.455	56.070	56.612	57.099
73	51.799	53.59	54.748	55.630	56.354	56.974	57.522	58.014
74	52.654	54.463	55.632	56.522	57.253	57.880	58.432	58.930
75	53.511	55.337	56.517	57.415	58.153	58.786	59.344	59.846
76	54.369	56.211	57.402	58.310	59.054	59.693	60.256	60.763
77	55.227	57.087	58.289	59.205	59.956	60.601	61.169	61.681
78	56.087	57.964	59.177	60.101	60.859	61.510	62.083	62.600
79	56.948	58.842	60.065	60.998	61.763	62.419	62.998	63.519
80	57.810	59.720	60.955	61.895	62.668	63.330	63.914	64.439
81	58.673	60.600	61.845	62.794	63.573	64.241	64.830	65.360
82	59.537	61.480	62.737	63.693	64.479	65.153	65.747	66.282
83	60.403	62.362	63.629	64.594	65.386	66.065	66.665	67.204
84	61.269	63.244	64.522	65.495	66.294	66.979	67.583	68.128
85	62.135	64.127	65.415	66.396	67.202	67.893	68.503	69.051
86	63.003	65.011	66.310	67.299	68.111	68.808	69.423	69.976
87	63.872	65.897	67.205	68.202	69.021	69.724	70.343	70.901
88	64.742	66.782	68.101	69.106	69.932	70.640	71.264	71.827
89	65.612	67.669	68.998	70.011	70.843	71.557	72.186	72.753
90	66.484	68.556	69.896	70.917	71.755	72.474	73.109	73.680
91	67.356	69.444	70.794	71.823	72.668	73.393	74.032	74.608
92	68.229	70.333	71.693	72.730	73.581	74.311	74.956	75.536
93	69.103	71.222	72.593	73.637	74.495	75.231	75.880	76.465
94	69.978	72.113	73.493	74.545	75.410	76.151	76.805	77.394
95	70.853	73.004	74.394	75.454	76.325	77.072	77.731	78.324
96	71.729	73.896	75.296	76.364	77.241	77.993	78.657	79.255
97	72.606	74.788	76.199	77.274	78.157	78.915	79.584	80.186
98	73.484	75.681	77.102	78.185	79.074	79.837	80.511	81.117
99	74.363	76.575	78.006	79.096	79.992	80.760	81.439	82.050
100	75.242	77.469	78.910	80.008	80.910	81.684	82.367	82.982

N*	Blocking Probability							
	0.009	0.01	0.02	0.03	0.05	0.1	0.2	0.4
1	0.0091	0.0101	0.0204	0.0309	0.0526	0.1111	0.2500	0.6667
2	0.1442	0.1526	0.2235	0.2815	0.3813	0.5954	1.0000	2.0000
3	0.4371	0.4555	0.6022	0.7151	0.8994	1.2708	1.9299	3.4798
4	0.8409	0.8694	1.0923	1.2589	1.5246	2.0454	2.9452	5.0210
5	1.3223	1.3608	1.6571	1.8752	2.2185	2.8811	4.0104	6.5955
6	1.8610	1.9090	2.2759	2.5431	2.9603	3.7584	5.1086	8.1907
7	2.4437	2.5009	2.9354	3.2497	3.7378	4.6662	6.2302	9.7998
8	3.0615	3.1276	3.6271	3.9865	4.5430	5.5971	7.3692	11.419
9	3.7080	3.7825	4.3447	4.7479	5.3702	6.5464	8.5217	13.045
10	4.3784	4.4612	5.0840	5.5294	6.2157	7.5106	9.685	14.677
11	5.0691	5.1599	5.8415	6.3280	7.0764	8.4871	10.857	16.314
12	5.7774	5.8760	6.6147	7.1410	7.9501	9.4740	12.036	17.954
13	6.5011	6.6072	7.4015	7.9667	8.8349	10.470	13.222	19.598
14	7.2382	7.3517	8.2003	8.8035	9.7295	11.473	14.413	21.243
15	7.9874	8.1080	9.0096	9.6500	10.633	12.484	15.608	22.891
16	8.7474	8.8750	9.8284	10.505	11.544	13.500	16.807	24.541
17	9.5171	9.6516	10.656	11.368	12.461	14.522	18.010	26.192
18	10.296	10.437	11.491	12.238	13.385	15.548	19.216	27.844
19	11.082	11.230	12.333	13.115	14.315	16.579	20.424	29.498
20	11.876	12.031	13.182	13.997	15.249	17.613	21.635	31.152
21	12.677	12.838	14.036	14.885	16.189	18.651	22.848	32.808
22	13.484	13.651	14.896	15.778	17.132	19.692	24.064	34.464
23	14.297	14.470	15.761	16.675	18.080	20.737	25.281	36.121
24	15.116	15.295	16.631	17.577	19.031	21.784	26.499	37.779
25	15.939	16.125	17.505	18.483	19.985	22.833	27.720	39.437
26	16.768	16.959	18.383	19.392	20.943	23.885	28.941	41.096
27	17.601	17.797	19.265	20.305	21.904	24.939	30.164	42.755
28	18.438	18.640	20.150	21.221	22.867	25.995	31.388	44.414
29	19.279	19.487	21.039	22.140	23.833	27.053	32.614	46.074
30	20.123	20.337	21.932	23.062	24.802	28.113	33.840	47.735
31	20.972	21.191	22.827	23.987	25.773	29.174	35.067	49.395
32	21.823	22.048	23.725	24.914	26.746	30.237	36.295	51.056
33	22.678	22.909	24.626	25.844	27.721	31.301	37.524	52.718
34	23.536	23.772	25.529	26.776	28.698	32.367	38.754	54.379
35	24.397	24.638	26.435	27.711	29.677	33.434	39.985	56.041
36	25.261	25.507	27.343	28.647	30.657	34.503	41.216	57.703
37	26.127	26.378	28.254	29.585	31.640	35.572	42.448	59.365
38	26.996	27.252	29.166	30.526	32.624	36.643	43.680	61.028
39	27.867	28.129	30.081	31.468	33.609	37.715	44.913	62.690

N*	Blocking Probability							
	0.009	0.01	0.02	0.03	0.05	0.1	0.2	0.4
40	28.741	29.007	30.997	32.412	34.596	38.787	46.147	64.353
41	29.616	29.888	31.916	33.357	35.584	39.861	47.381	66.016
42	30.494	30.771	32.836	34.305	36.574	40.936	48.616	67.679
43	31.374	31.656	33.758	35.253	37.565	42.011	49.851	69.342
44	32.256	32.543	34.682	36.203	38.557	43.088	51.086	71.006
45	33.140	33.432	35.607	37.155	39.550	44.165	52.322	72.669
46	34.026	34.322	36.534	38.108	40.545	45.243	53.559	74.333
47	34.913	35.215	37.462	39.062	41.540	46.322	54.796	75.997
48	35.803	36.109	38.392	40.018	42.537	47.401	56.033	77.660
49	36.694	37.004	39.323	40.975	43.534	48.481	57.270	79.324
50	37.586	37.901	40.255	41.933	44.533	49.562	58.508	80.988
51	38.480	38.800	41.189	42.892	45.533	50.644	59.746	82.652
52	39.376	39.700	42.124	43.852	46.533	51.726	60.985	84.317
53	40.273	40.602	43.060	44.813	47.534	52.808	62.224	85.981
54	41.171	41.505	43.997	45.776	48.536	53.891	63.463	87.645
55	42.071	42.409	44.936	46.739	49.539	54.975	64.702	89.310
56	42.972	43.315	45.875	47.703	50.543	56.059	65.942	90.974
57	43.875	44.222	46.816	48.669	51.548	57.144	67.181	92.639
58	44.778	45.130	47.758	49.635	52.553	58.229	68.421	94.303
59	45.683	46.039	48.700	50.602	53.559	59.315	69.662	95.968
60	46.589	46.950	49.644	51.570	54.566	60.401	70.902	97.633
61	47.497	47.861	50.589	52.539	55.573	61.488	72.143	99.297
62	48.405	48.774	51.534	53.508	56.581	62.575	73.384	100.960
63	49.314	49.688	52.481	54.478	57.590	63.663	74.625	102.630
64	50.225	50.603	53.428	55.450	58.599	64.750	75.866	104.290
65	51.137	51.518	54.376	56.421	59.609	65.839	77.108	105.960
66	52.049	52.435	55.325	57.394	60.619	66.927	78.350	107.620
67	52.963	53.353	56.275	58.367	61.630	68.016	79.592	109.290
68	53.877	54.272	57.226	59.341	62.642	69.106	80.834	110.950
69	54.793	55.191	58.177	60.316	63.654	70.196	82.076	112.620
70	55.709	56.112	59.129	61.291	64.667	71.286	83.318	114.280
71	56.626	57.033	60.082	62.267	65.680	72.376	84.561	115.950
72	57.545	57.956	61.036	63.244	66.694	73.467	85.803	117.610
73	58.464	58.879	61.990	64.221	67.708	74.558	87.046	119.280
74	59.384	59.803	62.945	65.199	68.723	75.649	88.289	120.940
75	60.304	60.728	63.900	66.177	69.738	76.741	89.532	122.610
76	61.226	61.653	64.857	67.156	70.753	77.833	90.776	124.270
77	62.148	62.579	65.814	68.136	71.769	78.925	92.019	125.940
78	63.071	63.506	66.771	69.116	72.786	80.018	93.262	127.610
79	63.995	64.434	67.729	70.096	73.803	81.110	94.506	129.270

N*	Blocking Probability							
	0.009	0.01	0.02	0.03	0.05	0.1	0.2	0.4
80	64.919	65.363	68.688	71.077	74.820	82.203	95.750	130.940
81	65.845	66.292	69.647	72.059	75.838	83.297	96.993	132.600
82	66.771	67.222	70.607	73.041	76.856	84.390	98.237	134.270
83	67.697	68.152	71.568	74.024	77.874	85.484	99.481	135.930
84	68.625	69.084	72.529	75.007	78.893	86.578	100.73	137.60
85	69.553	70.016	73.490	75.990	79.912	87.672	101.97	139.26
86	70.481	70.948	74.452	76.974	80.932	88.767	103.21	140.93
87	71.410	71.881	75.415	77.959	81.952	89.861	104.46	142.60
88	72.340	72.815	76.378	78.944	82.972	90.956	105.70	144.26
89	73.271	73.749	77.342	79.929	83.993	92.051	106.95	145.93
90	74.202	74.684	78.306	80.915	85.014	93.146	108.19	147.59
91	75.134	75.620	79.271	81.901	86.035	94.242	109.44	149.26
92	76.066	76.556	80.236	82.888	87.057	95.338	110.68	150.92
93	76.999	77.493	81.201	83.875	88.079	96.434	111.93	152.59
94	77.932	78.430	82.167	84.862	89.101	97.530	113.17	154.26
95	78.866	79.368	83.134	85.850	90.123	98.626	114.42	155.92
96	79.801	80.306	84.100	86.838	91.146	99.722	115.66	157.59
97	80.736	81.245	85.068	87.826	92.169	100.82	116.91	159.25
98	81.672	82.184	86.035	88.815	93.193	101.92	118.15	160.92
99	82.608	83.124	87.003	89.804	94.216	103.01	119.40	162.59
100	83.545	84.064	87.972	90.794	95.240	104.11	120.64	164.25

Simulation Projects

Simulating one specific aspect of wireless and mobile systems gives a better feel for the technology and the underlying support. The simulation can be done either using a standard simulator such as ns-2, OPNET, QualNeT, or can be done using languages such as Java, C, C++, Visual Basic, Visual C, Visual J, or Matlab. The authors have used most of the following topics for their classes and found them very intriguing for the students if there is enough time and if assigned in the beginning of the course. Each of these projects can be allocated to a group of two to four students, and implementation level depends on the time and grade allocated for the project and the personal enthusiasm of both the students and the instructor. It is better to set up some target dates and designate credits. For example, a 2-page outline describing what is being planned may be due in two weeks. A brief flowchart of each simulation project may be due by mid-term of the course and the demo is required by the last week of the class. The final report could consist of a description of the program in about ten pages and a soft-copy of the simulation with comments. It is very important to emphasize that the working model is required and the instructor is not interested in the piece of code that does not work. Some of the suggested topics are as follows:

1. Blocking probability and forced termination probability in a FDMA system.
2. Blocking probability and forced termination probability in a TDMA system.
3. Blocking probability and forced termination probability in a CDMA system.
4. Handoffs (intra/inter cell, hard or soft) in two-level architectures.
5. Location determination of a MS based on signals received at three BSs.
6. Mobility characterization based on statistical information.
7. Encryption in a wireless environment.
8. Authentication using challenge and response.
9. Routing using home and foreign agents.
10. Encoding of convolutional codes.
11. Encoding of Turbo codes.
12. Changing codes when ARQ is present.
13. Simulation of different ARQ schemes.
14. Near-far power control in a CDMA system.
15. Location determination of a MS based on signals received at three BSs.
16. Illustration of frequency hopping techniques.
17. Channel allocation in multitier hierarchical cellular architecture.
18. Computing efficiency, throughput, etc. for one of the MAC protocols.
19. Denial of service in a wireless system.
20. Intrusion detection in wireless networks.

21. Multicasting in wireless networks.
22. Cluster formation in MANETs.
23. Clustering in sensor MANETs.
24. Multipath routing in a MANET
25. Route selection in MANET based on shortest path and path stability.
26. Effective use of cache for routing in MANETs.
27. Multicasting in MANETs using dominating sets.
28. Ordering grocery using a WAP-enabled PDA.
29. Head selection in a sensor network.
30. Query processing in a sensor network.
31. Use of sensors for finding open spots in a parking lot.
32. Use of sensors for finding noise level or movement in different parts of a university campus.
33. Generation of an XML message from any HTML Web site.
34. Simulation of a Bluetooth piconet.

Acronyms

ABR	Associativity-Based Routing
AC	Access Control
ACK	Acknowledgment
ACL	Asynchronous Connectionless
ACSE	Association Control Service Element
ADPCM	Adaptive Differential Pulse Code Modulation
ADSL	Asymmetric Digital Subscriber Line
AES	Advanced Encryption Standard
AFH	Adaptive Frequency Hopping
AIFS	Arbitration Interframe Space
AIRMAIL	Asymmetric Reliable Mobile Access in Link Layer
AM	Access Manager
AM	Amplitude Modulation
AMPS	Advanced Mobile Phone Service
AMRIS	Ad hoc Multicast Routing Protocol Utilizing Increasing ID Number
ANSI	American National Standards Institute
AODV	Ad hoc On-demand Distance Vector
AODV–BR	Ad hoc On-demand Distance Vector–Backing Routing
AP	Access Point
API	Application Programming Interface
APTEEN	Adaptive Periodic Threshold-Sensitive Energy Efficient Sensor Network Protocol
ARIB	Associate of Radio Industries and Business
ARQ	Automatic Repeat Request
AS	Autonomous System
ASCENT	Adaptive Self-Configuring Sensor Network Technologies
ASE	Applications Service Element
ASF	Advanced Streaming Format
ATM	Asynchronous Transfer Mode
AUC	Authentication Center
BCC	Bellcore Client Company
BCCH	Broadcast Control Channel
BCH	Broadcast Channel
BCH	Bose Chaudhuri Hocquenghem
BCS	Bluetooth Carrier Sense
BER	Bit Error Rate
BGP	Border Gateway Protocol
BIOS	Basic Input/Output System
BOOTP	Bootstrap Protocol
BPSK	Binary Phase Shift Keying
BQ	Broadcast Query
BQ-REPLY	BQ and Await-Reply
BRCA	Bandwidth Reallocation Channel Assignment
BRP	Border Resolution Protocol

BS	Base Station
BSC	Base Station Controller
BSS	BS Subsystem
BT	Bidirectional Tunneling
BTMA	Busy Tone Multiple Access
BTS	Base Transceiver System
BW	Bandwidth
BWG	Bluetooth Wireless Gateways
CAP	CAMEL Application Part
CAP	Contention Access Period
CBRP	Cluster-Based Routing Protocol
CBT	Core Based Trees
CCCH	Common Control Channel
CCIR	Co-Channel Interference Ratio
CCITT	Comité Consultatif International Téléphonique et Télégraphique
C-CoA	Co-located CoA
CDF	Cumulative Distribution Function
CDMA	Code Division Multiple Access
CDPD	Cellular Digital Packet Data
CEPT	Conference European des Post of Telecommunications
CFP	Contention-Free Period
CGSR	Cluster-head Gateway Switch Routing
CH	Cluster Head
CHAMP	Caching and Multipath Routing Protocol
C/I	Carrier-to-Interference
CID	Connection Identifier
CIR	Carrier to Interference Ratio
CLR	Clear Packet
CM	Communication Management
CMA	Constant Modulus Algorithm
CMOS	Complementary Metal Oxide Semiconductor
CMT	Cluster Member Table
CN	Core Network
CoA	Care of Address
CoS	Classes of Services
CP	Complete Partitioning
CP	Control Point
CPCH	Common Packet Channel
CPU	Certral Processing Unit
CRC	Cycle Redundancy Check
CS	Complete Sharing
CS	Circuit Switched
CSMA	Carrier Sense Multiple Access
CSMA/CA	CSMA with Collision Avoidance
CSMA/CD	CSMA with Collision Detection
CSS	Cellular Subscriber Station
CSU	Channel Service Unit
CT	Count Time
CT2	Cordless Telephone
CTCH	Common Traffic Channel
CTS	Clear to Send
CVSD	Continuous Variable Slope Delta

CW	Contention Window
DAG	Directed Acyclic Graph
DARPA	Defense Advanced Research Projects Agency
DC	Direct Current
DCA	Dynamic Channel Allocation
DCC	Digital Color Code
DCCH	Dedicated Control Channel
DCF	Distributed Coordination Function
DCH	Dedicated Channel
DCM	Data Collection Module
DCS	Digital Cellular System
DCTS	Directional CTS
DECT	Digital European Cordless Telecommunications
DES	Data Encryption Standard
DFA	Domain Foreign Agent
DFS	Dynamic Frequency Selection
DFT	Discrete Fourier Transform
DFWMAC	Distributed Foundation Wireless MAC
DGPS	Differential GPS
DHCP	Dynamic Host Configuration Protocol
DL	Downlink
DLL	Data Link Layer
DL-MAP	DL Map
DiffServ	Differentiated Service
DIFS	Distributed Interframe Space
DIUC	Downlink Interval Usage Code
DMSP	Designated Multicast Service Provider
DNS	Domain Name Server
DoD	Department of Defense
DoS	Denial of Service
DQDB	Distributed Queue Dual Bus
DREAM	Distance Routing Effect Algorithm for Mobility
DRP	Dynamic Routing Protocol
DRTS	Directional RTS
DS	Direct Sequence
DSA	Dynamic Slot Assignment
DSB	Double Sideband
DSCH	Downlink Shared Channel
DSC–UWB	Direct Sequence Phase Coded UWB
DSDV	Destination-Sequenced Distance Vector
DSR	Dynamic Source Routing
DSSS	Direct Sequence Spread Spectrum
DTCH	Dedicated Traffic Channel
DTD	Document Type Definition
DVB–C	Digital Video Broadcasting–Cable
DVS	Dynamic Voltage Scaling
EBSN	Explicit Bad State Notification
EDCF	Enhanced Distributed Coordination Function
EHF	Extremely High Frequency
EIR	Equipment Identity Register
ELF	Extremely Low Frequency
ELN	Explicit Loss Notification

ES	Earth Station
ESN	Electronic Serial Number
ETSI	European Telecommunications Standardization Institute
FA	Foreign Agent
FAC	Final Assembly Code
FACCH	Fast Associated Control Channel
FACH	Forward Access Channel
FAUSCH	Fast Uplink Signalling Channel
FBN	Fixed Beam-forming Network
FCA	Fixed Channel Allocation
FCC	Federal Communications Commission
FCCH	Frequency Correction Channel
FCFS	First-Come-First-Service
FCS	Frame Check Sequence
FDD	Frequency Division Duplexing
FDMA	Frequency Division Multiple Access
FEC	Forward Error Correction
FER	Frame Error Rate
FH	Fixed Host
FH	Frequency Hopping
FHSS	Frequency Hopping Spread Spectrum
FIFO	First-In-First-Out
FM	Frequency Modulation
FN	Failure Notification
FOCC	Forward Control Channel
FSK	Frequency Shift Keying
FSR	Fisheye State Routing
FTP	File Transfer Protocol
FVC	Forward Voice Channel
GBN	Go-back-N
GEO	Geostationary Earth Orbit
GFR	Guaranteed Frame Rate
GGSN	Gateway GPRS Support Node
GIS	Geographical Information System
GL	Gateway Link
GLMS	Group and List Management Server
GMSC	Gateway MSC
GMSK	Gaussian Minimum Shift Keying
GPC	Grant Per Connection
GPRS	General Packet Radio Services
GPS	Global Positioning Systems
GPSS	Grant Per SS
GRM	Global Response Module
GSM	Global System for Mobile Communications
GSM	Groupe Speciale Mobile
GTS	Guaranteed Time Slot
GUI	Graphical User Interface
HA	Home Agent
HCA	Hybrid Channel Allocation
HEO	Highly Elliptical Orbit
HF	High Frequency
HLR	Home Location Register

HT	Hard Threshold
HTML	Hyper Text Markup Language
HTTP	Hyper Text Transfer Protocol
IARP	Intrazone Routing Protocol
IBLUES	Interference Aware Bluetooth Dynamic Segmentation
ICMP	Internet Control Message Protocol
ICO	Intermediate Circular Orbit
ID	Identification
iDEN	Integrated Digital Enhanced Network
IDFT	Inverse Discrete Fourier Transform
IDS	Intrusion Detection System
IERP	Interzone Routing Protocol
IETF	International Engineering Task Force
IF	Intermediate Frequency
IFS	Interframe Space
IGMP	Internet Group Management Protocol
IID	Independent Identically Distributed
ILF	Infra Low Frequency
IMEP	Internet MANET Encapsulation Protocol
IMSEI	International MS Equipment Identity
IMSI	International Mobile Subscriber Identity
IMT	International Mobile Telecommunications
IntSev	Internet Uses Integrated Service
IP	Internet Protocol
IPng	Internet Protocol Next Generation
IPR	Intellectual Property Right
IPv4	Internet Protocol version 4
IPv6	Internet Protocol version 6
IR	Infrared
ISDN	Integrated Services Digital Network
ISI	Inter-Symbol Interference
ISL	Inter Satellite Link
ISM	Industrial, Scientific, and Medical
ISMA	Idle Signal Multiple Access
ISO	International Standards Organization
ISP	Internet Service Provider
I-TCP	Indirect TCP
ITU	International Telecommunication Union
ITU-R	International Telecommunications Union–Radio Communications
L2CAP	Logical Link Control and Adaptation Layer Protocol
LA	Location Area
LAI	Location Area Identity
LAN	Local Area Network
LANMAR	Landmark Ad Hoc Routing
LAR	Location Aided Routing
LCC	Least Cluster Change
LEACH	Low-Energy Adaptive Clustering Hierarchy
LEO	Low Earth Orbit
LF	Low Frequency
LGT	Location Guided Tree
LLC	Logical Link Control
LMP	Link Manager Protocol

LOS	Line of Sight
LQ	Localized Query
LRM	Local Response Module
LSA	Link State Advertisement
LSB	Least Significant Bit
LUS	Lookup Service
LZRW	Lempel-Ziv Ross Williams
MAC	Message Authentication Code
MAC	Medium Access Control
MACAW	Multiple Access Collision Avoidance Protocol for Wireless LAN
MADCAP	Multicast Address Dynamic Client Allocation Protocol
MAN	Metropolitan Area Network
MANET	Mobile Ad Hoc Network
MAODV	Multicast Ad Hoc On-Demand Distance Vector
MAP	Mobile Application Part
MBCA	Minimum Blocking Channel Assignment
MBONE	Multicast Backbone
MCC	Mobile Country Code
MCDN	MicroCellular Data Network
MEMS	Micro Electro Mechanical System
MEO	Medium Earth Orbit
MF	Medium Frequency
MFJ	Modified Final Judgement
MFR	MAC Footer
MH	Mobile Host
MHA	Multicast Home Agent
MHR	MAC Header
MIN	Mobile Identification Number
MLME	MAC Layer Management Entity
MM	Mobility Management
MMP	Multicast for Mobility Protocol
MMS	Multimedia Message Service
MNC	Mobile Network Code
MoM	Mobile Multicast
MP3	MPEG-1 Layer 3
MPEG	Moving Picture Expert Group
MPLS	Multiprotocol Label Switching
MPT	Ministry of Posts and Telecommunications
MRL	Message Retransmission List
MROUTER	Multicast-Capable Router
MS	Mobile Station
MSC	Mobile Switching Center
MSIC	Mobile Subscriber Identification Code
MSISDN	Mobile System ISDN
MSR	Mobile Support Router
MSRN	MS Roaming Number
MTP	Message Transfer Part
MTU	Maximum Transmission Unit
MUL	Mobile User Link
NAI	Network Access Identifier
NAK	Negative Acknowledgment
NAVSTAR	Navigation System with Time and Ranging

NDC	National Destination Code
NDP	Neighbor Discovery Protocol
NFS	Network File Server
NMT	Nordic Mobile Telephone
NPDU	Network Protocol Data Unit
NTBMR	Neighbor-Table-Based Multipath Routing
NTT	Nippon Telephone & Telegraph
OAM	Operations, Administration & Maintenance
OCCCH	ODMA Common Control Channel
ODCCH	ODMA Dedicated Control Channel
ODCH	ODMA Dedicated Channel
ODMA	Opportunity Driven Multiple Access
ODTCH	ODMA Dedicated Traffic Channel
OFDM	Orthogonal Frequency Division Multiplexing
OHG	Operators Harmonization Group
OMA	Open Mobile Alliance
ORACH	ODMA Random Access Channel
OSI	Open Systems Interconnection
OSPF	Open Shortest Path First
PA	Paging Area
PACS	Personal Access Communications System
PAN	Personal Area Network
PAR	Peak to Average Ratio
PASC	PCS Access Service for Controllers
PASD	PCS Service for Data
PASE	PCS Access Service for External Service Providers
PASN	PCS Access Service for Networks
PASP	PCS Access Service for Ports
PCC	Power Control Channel
PCCH	Paging Control Channel
PCH	Paging Channel
PCMCIA	Personal Computer Memory Card International Association
PCS	Personal Communication Service
PDA	Personal Digital Assistant
PDAN	Packet Data Access Node
PDC	Personal Digital Cellular
pdf	Probability Density Function
PDU	Protocol Data Unit
PER	Packet Error Rate
PHS	Personal Handyphone System
PHY	Physical Layer
P–K	Pollaczek–Khinchin
PLMN	Public Land Mobile Network
pmf	Probability Mass Function
PN	Pseudo Noise
PoC	PTT over Cellular
PPDU	PHY Protocol Data Unit
PPM	Pulse Position Modulation
PPP	Point-to-Point Protocol
PPR	Pseudo PaRtitioning
PRN	Pseudo Random Number
PRN	Packet Radio Network

PRS	Pseudo Role Switching
PS	Packet Switched
PSAP	Public Safety Answering Point
PSDU	PHY Service Data Unit
PSK	Phase Shift Keying
PSP	PCS Service Provider
PSTN	Public Switched Telephone Network
PTT	Push-to-Talk
PWP	Wireless Provider
QAM	Quadrature Amplitude Modulation
QoS	Quality of Service
QPSK	Quadrature Phase Shift Keying
RACH	Random Access Channel
RANAP	Radio Access Network Application Protocol
RAN	Radio Access Network
RAND	Random Number
RBMoM	Range-Based MoM
RBOC	Regional Bell Operating Companies
RC4	Ron's Code 4 or Rivest's Cipher 4
RCC	Radio Common Carrier
RD	Relative Distance
RD	Route Delete
RDM	Relative Distance Micro-discovery
RDMAR	Relative Distance Micro-discovery Ad Hoc Routing
RECC	Reverse Control Channel
RERR	Route Error
RF	Radio Frequency
RFID	Radio Frequency Identification
RIP	Routing Information Protocol
RLC 10	Radio Link Control
Rlogin	Remote Login
RLS	Recursive Least Squares
RM	Reliable Mobile Multicast
RM	Response Model
RMDP	Reliable Multicast data Distribution Protocol
RN	Route Notification
RNC	Radio Network Controller
RNL	Radio Network Layer
RNS	Radio Network Subsystem
ROSE	Remote Operation Service Element
RP	Radio Port
RPC	Remote Procedure Call
RPCU	Radio Port Control Unit
RR	Radio Resource Management
RRC	Route Reconstruction
RREP	Route Reply
RREQ	Route Request
RS	Retransmission Services
RSA	Ron Rivest, Adi Shamir, Len Adleman
RSC	Recursive Systematic Convolutional
RSVP	Resource Reservation Protocol
RT	Routing Table

RTP	Real-time Transport Protocol
RTR	Ready-to-Receive
RTS	Request to Send
RTSP	Real-Time Streaming Protocol
RVC	Reverse Voice Channel
S/N	Signal to Noise Ratio
SA	Selective Availability
SACCH	Slow Associated Control Channel
SACK	Selective Acknowledgment
SAT	Supervisory Audio Tone
SAW	Surface Acoustic Wave
SAW	Stop-And-Wait
SCADDS	Scalable Coordination Architectures for Deeply Distributed Systems
SCCP	System Connection Control Part
SCH	Synchronization Channel
SCO	Synchronous Connection Oriented
SDCCH	Stand-alone Dedicated Control Channel
SDMA	Space Division Multiple Access
SDP	Service Discovery Protocol
SDU	Service Data Unit
SGSN	Serving GPRS Support Node
SHCCH	Shared Channel Control Channel
SHF	Super High Frequency
SID	System Identification Number
SIFS	Short Interframe Space
SIG	Special Interest Group
SIM	Subscriber Identity Module
SIP	Session Initiation Protocol
SIR	Signal-to-Interference Ratio
SMIL	Synchronized Multimedia Integration Language
SMR	Split Multipath Routing
SMS	Short Message Service
SMTP	Simple Mail Transfer Protocol
SN	Subscriber Number
SNMP	Simple Network Management Protocol
SNR	Signal-to-Noise Ratio
SNR	Serial Number
SP	Service Provider
SPF	Shortest Path First
SPIN	Sensor Protocols for Information via Negotiation
SR	Selective-Repeat
SRP	Static Routing Protocol
SS7	Signalling System 7
SSA	Signal Stability Adaptive
SSB	Single Sideband
SSI	Single System Image
SSR	Signal Stability-based Adaptive Routing
SST	Signal Stability Table
ST	Soft Threshold
SUDC	United States Digital Cellular
SUMR	Satellite User Mapping Register
SV	Space Vehicle

SV	Sensed Value
SVM	Support Vector Machine
SVMDM	Support Vector Machine-based Intrusion Detection Module
SWAP	Shared Wireless Access Protocol
SWID	Switch Identification
SWNO	Switch Number
TAC	Type Approval Code
TC	Transmission Convergence
TC	Traffic Categories
TCAP	Transaction Capabilities Application Part
TCP	Transmission Control Protocol
TDD	Time Division Duplexing
TDM	Time Division Multiplex
TDMA	Time Division Multiple Access
TD–SCDMA	Time Division–Synchronous CDMA
TEEN	Threshold sensitive Energy Efficient sensor Network
TG	Task Group
THF	Tremendrously High Frequency
TKIP	Temporal Key Integrity Protocol
TIA	Telecommunications Industry Association
TLU	Time Last Update
TLV	Threshold Limit Value
TMA	Telephone Modem Access
TMSI	Temporary Mobile Subscriber Identity
TM–UWB	Time-Modulated UWB
TNL	Transport Network Layer
TORA	Temporally Ordered Routing Algorithm
TTL	Time to Live
TULIP	Transport Unaware Link Improvement Protocol
UDP	User Datagram Protocol
UE	User Equipment
UHF	Ultra High Frequency
UL	Uplink
UL–MAP	UL Map
UIUC	Uplink Interval Usage Code
UMTS	Universal Mobile Telecommunication Systems
UNRELDIR	Unreliable Roamer Data Directive INVOKE
URL	Uniform Resource Locator
USCH	Uplink Shared Channel
UTRAN	UMTS Terrestrial RAN
UWB	Ultra-Wideband
UWB–RT	UWB-Radio Technology
VHF	Very High Frequency
VLF	Very Low Frequency
VLR	Visitor Location Register
VLSI	Very Large Scale Integration
VoIP	Voice over IP
WAN	Wide Area Network
WAP	Wireless Access Point
WARC	World Administrative Radio Conference
W–CDMA	Wideband–CDMA
WECA	Ethernet Compatibility Alliance

WEP	Wired Equivalent Privacy
WG	Working Group
WLAN	Wireless LAN
Wi-Fi	Wireless Fidelity
WiMAX	Worldwide Interoperability for Microwave Access
WMAN	Wireless MAN
WML	Wireless Markup Language
WPA	Wi-Fi Protected Access
WPAN	Wireless PAN
WRP	Wireless Routing Protocol
WTCP	Wireless TCP
WWW	World Wide Web
XML	Extensible Markup Language
ZRP	Zone Routing Protocol

Index

1-persistent CSMA 132–133
3GPP 252
3G system 249
7-layer 126
8PSK 165
16QAM 165
64QAM 165
256QAM 165

A

Access channel 244–245
Access control 286
Access grant channel 232
Access manager (AM) 242
Access point (AP) 8, 10–11, 293, 369–370, 372–374, 397, 405
Accidental attack 440
Acknowledgment (ACK) 92, 137–138, 293
ACSE 227
Active attack 440
Adaptive antenna array 425
Adaptive beam 425
Adaptive differential pulse code modulation (ADPCM) 241
Adaptive equalization technique 148
Adaptive frequency hopping (AFH) 432, 435
Adaptive periodic threshold-sensitive energy efficient sensor network protocol (APTEEN) 347, 428
Adaptive self-configuring sensor network topologies (ASCENT) 338
Ad hoc network 11, 22, 304–305, 334–335, 341
Ad hoc on-demand distance vector (AODV) 308
Ad hoc topology 23
Adjacent cell 29, 104
Adjacent channel 31, 145, 152, 154, 169
  interference 154
Adjacent cluster 171, 173

Advanced encryption standard (AES) 417
Advanced mobile phone system (AMPS) 194
Advanced Streaming Format (ASF) 404
A-GPS 280
AIRMAIL 295
ALOHA 127–128, 134
Amplitude 65
Amplitude modulation (AM) 159–160
AMPS 221–225, 227–228, 243
Angular velocity 263
ANSI-41 (American National Standards Institute-41) 251
Antenna 12, 23, 57
AODV 308, 312, 327, 442
AODV-BR 327
Application layer 285, 287
Arbitrary random phenomenon 29
Arbitration interframe space (AIFS) 416
ARIB 250
Arithmetic mean 38
Arrival rate 41
ASCENT 338
Association control 374
Associativity-based routing (ABR) 318
Asymmetric digital subscriber line (ADSL) 375
Asymmetric link 303
Asymmetric traffic 384
Asynchronous connectionless (ACL) 383–384
Asynchronous transfer mode (ATM) 18, 403
  backbone 18
Atmospheric condition 12
Attack 207, 212–213, 215, 305
Attractive force 262
Attribute 340, 347
Audio signal 79

Authentication 11, 19, 190, 193–194, 198, 209–210, 212–213

Authentication authorization and accounting (AAA) 216

Authentication center (AUC) 190, 230

Automatic repeat request (ARQ) 92, 295, 374, 384, 421–422

Autonomous systems (AS) 290

Availability 213

B

Backbone network 190, 198–199, 203

Backoff counter 137

Backoff period 137

Band A 2, 222

Band B 2, 222

Bandwidth 2, 6, 13, 110, 143, 145, 304, 405
  allocation 2
  constrained link 304
  scalability 405

Base station (BS) 7

Base transceiver system (BTS) 19

Basic handoff 238

Basic input/output system (BIOS) 438

Bayes's theorem 36

Bayesian statistics 36

Beacon 193

Beacon kernel table 193

Beacon signal 193–195, 197, 201

Beam 157–158

Bellcore client company (BCC) 241

Bellman-Ford algorithm 291

Bellman-Ford routing algorithm 309

Bessel function 68

Best effort 289, 296, 403

Bi-directional link 303–304

Bidirectional tunneling (BT) 205, 419

Big LEO 267

Binary countdown 125

Binary phase shift keying (BPSK) 161–163, 165

Binary weight 81

Binomial distribution 34

Binomial random variable 34

Birth death process 42

Bit error rate (BER) 74, 79, 165, 196, 271

Bit-map protocol 125

Blocking probability 109, 169, 172, 174–175, 177, 180–182, 184, 198, 221

Block interleaver 90

BlueStar 435

Bluetooth 8, 24, 358, 377–384, 432, 434, 435

Bluetooth carrier sense (BCS) 435

Bluetooth wireless gateways (BWG) 434

Boolean function 208

Bootstrap protocol (BOOTP) 289

Border gateway protocol (BGP) 289–290

Border router 419

Break before make 196

Broadband 2, 4, 22, 149, 405

Broadband CDMA 2

Broadband multimedia service 405

Broadband PCS 2

Broadband radio access networks (BRAN) 418

Broadcast 3, 307

Broadcast channel (BCH) 79, 256
  codes 79

Broadcast control channel (BCCH) 231, 257

Broadcast ID 312

BS 190, 193–196, 198, 212, 231

BSC 190, 192, 196, 229

BS controller (BSC) 18

BS handoff 273

BT 205

BTS 190, 229

Buffer 39

Burst error 73, 90
  rate 73

Busy period 137

Busy tone multiple access (BTMA) 128

C

Caching and multipath routing protocol (CHAMP) 330

Call arrival 33

Call arrival rate 108

Call duration 31

Call holding time 31

Call rate 31

Call setup 271

Capacity 6, 108

Care-of-address (CoA) 201, 419

Carrier frequency 57, 60

Carrier sense multiple access (CSMA) 128, 131, 340–341. *See also* CSMA

Carrier-to-interference ratio (CIR) 154–155, 196

C band 265

cdma2000 2, 6, 250

Cell area 12

Cell sectoring 116–117

Cell shape 103

Cell splitting 116

Cellular concept 102

Cellular digital packet data (CDPD) 194

Cellular operation 2

Cellular structure 18

Cellular system 1–2, 12, 15, 17, 79, 102

Cellular telephone 3, 24

Central hub 370

Centralized master-slave polling scheme 436

Centralized routing 307

Central limit theorem 38

Central moment 32

Centrifugal force 262

Channel acquisition 364

Channel allocation 11, 169–170, 175–176, 179

Channel bandwidth 10

Channel capacity 11

Channel coding 79

Channel decoder 90

Check polynomial 85

Chipping rate 400

Chunnel 279

Circuit-switched (CS) 254

Circular cell 102

Circular interleaver 90

Civilian GPS 274

Classes of services (CoS) 403

Clear packet (CLR) 318

Clear to send (CTS) 138

Closed loop power control 249

Cluster 110

Cluster-based routing protocol (CBRP) 344

Cluster head (CH) 309, 341, 343, 344, 345, 347

Cluster-head gateway switch routing (CGSR) 309

Cluster member table (CMT) 310

Coarse timeout 293

Cochannel 23, 75, 114, 173, 175, 180
  cell 175
  distance 175

Cochannel interference 23, 75, 173, 175, 180

Cochannel interference ratio (CCIR) 115, 175

Code division multiple access (CDMA) 2, 13, 14, 16, 143–144, 148–150, 152, 154, 196, 243, 339–341, 345, 347

Code rate 87

Code word 15, 80

Coexistence mechanism 432

Coherence bandwidth 74

Collaborative coexistence mechanism 432

Collaborative mechanism 432

Collision 125, 135–137
  avoidance 137
  detection 135–136
  free protocol 125

Colocated CoA (C-CoA) 201

Commercial and military environment 8

Common control channel (CCCH) 258

Common packet channel (CPCH) 256

Common traffic channel (CTCH) 258

Common transport channel 256

Communication graph 22

Communication management (CM) 236

Compatibility 6

Complete partitioning (CP) 413

Complete sharing (CS) 413

Complementary metal oxide semiconductor (CMOS) 402

Complex borrowing scheme 172

Composite signals 68

Compressed data 79

Compression 287, 405
  layer 405

Computer network 126

Conditional probability 36

Confidentiality 212–213
Conflict-free protocol 125, 127
Congestion 11, 292, 293, 294
 avoidance 292
 control 293
 window 294
Connection control 287
Connection identifier (CID) 361
Connectionless 11
Connection oriented 11
Constant modulus algorithm (CMA) 425
Contention access period (CAP) 385
Contention-based protocol 125, 127–128
Contention resolution 286
Contention window 137
Continuous random variable 30–31
Continuous variable 29
Continuous variable slope delta (CVSD) 381
Control channel 125, 243, 257
Control message 19
Convolutional codes 79, 87
Cordless telephone 1, 12
Core based trees (CBT) 419
Core network (CN) 253
Correlation 74
COUGAR 342, 428
Count Time (CT) 347–348
Count-to-infinity 311
Coverage area 7, 11–13
Coverage pattern 23
Crisis-management 306
CSMA 131, 134, 137–138, 340–341, 370, 375
 CA with ACK 137–138
 CA with RTS and CTS 138
 with collision avoidance (CSMA/CA) 136–137, 370
 with collision detection (CSMA/CD) 135–137, 375
CT2 241
CT2 TDD 241
Cumulative distribution function (CDF) 30
Cumulative probability distribution (CDF) 68
Cyclic codes 85
Cyclic redundancy check (CRC) 243, 361
Cyclic redundancy code (CRC) 86

D

Data collection module (DCM) 444
Data encryption standard (DES) 207
Data link layer 127, 285–286, 288
Data link layer (DLL) 388
Data polynomial 85
Decentralized ad hoc network 11
Decibels (dB) 66
Decoding 15, 85
Dedicated channel (DCH) 257
Dedicated control channel (DCCH) 258
Dedicated traffic channel (DTCH) 258
Dedicated transport channel 257
Defense Advanced Research Projects Agency (DARPA) 338
Degenerate (or deterministic) distribution 41
Delay functions 73
Delay spread 72
Denial of service (DoS) 215, 305, 440
Depth of fading 70–71
Designated multicast service provider (DMSP) 205
Destination-sequenced distance vector (DSDV) 309
DGPS 278
 beacon receiver 278
Dialog control 287
Diameter 216
differentiated services (DiffServ) 403
Diffraction 59
Digital cellular standard 2
Digital color code (DCC) 225
Digital European cordless telecommunications (DECT) 239, 241, 250
Dijkstra algorithm 290
Direct current (dc) 159
Directed acyclic graph (DAG) 316
Directed diffusion 341
Directional antenna 117, 423
Directional channel locking 172
Directional CTS (DCTS) 426
Directional RTS (DRTS) 426
Directory service 287
Direct sequence (DS) 150

Direct sequence phase coded UWB (DSC–UWB) 399, 400–401

Direct sequence spread spectrum (DSSS) 150, 370

Discrete Fourier transform (DFT) 155

Discrete Markov chain 47

Discrete random variable 29–30

Discrete time 29

Disjoint channel 169

Distance routing effect algorithm for mobility (DREAM) 323

Distance table 311

Distance vector protocol 308

Distance-vector routing protocol 290

Distributed coordination function (DCF) 138

Distributed foundation wireless MAC (DFWMAC) 136

Distributed interframe space (DIFS) 137–138

Distributed queue dual bus (DQDB) 243

Distribution 29

Distribution infrastructure 10

Distribution of waiting time 52

Diversity 11, 270

Diversity coding 326

DL burst profile change request (DBPC-REQ) 364

DL burst profile change response (DBPC-RSP) 364

DL channel descriptor (DCD) 364

DL interval usage code (DIUC) 363

DL-MAP 362

DMSP 205–206

Document type definition (DTD) 439

Domain foreign agent (DFA) 419

Domain name server (DNS) 291

Doppler effect 70–71

Doppler frequency 70–71

Doppler shift 71

Double sideband (DSB) 159

Downlink (DL) 19, 79, 144–145, 222, 265, 361

Downlink shared channel (DSCH) 256

Dual-IP stack 297

Duplex traffic channel 143

DVB-C 165

Dwell time 47, 51

Dynamic channel allocation (DCA) 170, 174–177, 340

Dynamic host configuration protocol (DHCP) 201, 289, 361

Dynamic networking 305

Dynamic routing protocol (DRP) 319

Dynamic source routing (DSR) 313

Dynamic topology 304

Dynamic voltage scaling (DVS) 437

E

E 911 280

Earth station (ES) 21, 226, 261

Eavesdropping 305

Education via the Internet 306

Electromagnetic wave 60

Electronic serial number (ESN) 223, 226

Element 29

Elevation angle 263, 271

Emergency handoff 271

Encoder 87

Encoding 15, 85

Encryption 11, 190, 207–208, 287, 374
    algorithm 374

End-to-end security 442

Energy-constrained operation 305

Energy consumption 195

English channel tunnel 279

Enhanced DCF 416

Envelope 67

Envelope of composite signals 68–69

Ephemeris error 276

Equilibrium state equation 46

Equipment identity register (EIR) 190, 230

Erlang 41, 108, 109, 110
    B formula 109
    C formula 110
    distribution 41

Error control 79, 286–287

Error correction 87

Error polynomial 85

Error-reporting message 289

Error vector 83

Establishment and termination of a connection 286

Ethernet 369

Ethernet card 288

ETSI 250, 403

Event 31

Expected value 31–33

Experiment 29

Explicit bad state notification (EBSN) 294

Explicit loss notification (ELN) 293

Exponential distribution 35

Extensible markup language (XML) 439

Eye opening 412

F

Fabrication 213

Fading 57, 61, 67, 270

    characteristics 67

    duration 70–71

    rate 70–71

    signal 67

Fast associated control channel (FACCH) 232

Fast fading 57, 61–62, 65, 67

    loss 61–62

Fast uplink signaling channel (FAUSCH) 257

Federal Communication Commission (FCC) 2

Feedback shift register 85

File format 403–404

File transfer 287

File transfer protocol (FTP) 291

Final assembly code (FAC) 234

Firewall 215

First central moment 32

First-come-first-service (FCFS) 43

First generation 1, 13

First generation cellular system 13

First-in-first-out (FIFO) 43, 413

First moment 32

Fisheye state routing (FSR) 321

Fixed channel allocation (FCA) 170–171, 174–175

Fixed host 295

Fixed internetwork 23

Fixed microwave 3

Fixed network 23

Fixed time slot 14

Fixed wireless sensor network 338

Flat network 309, 427

Flat routing 341

Flexibility 6

Flexible channel allocation 176–177

Flexmux layer 405

Flooding algorithm 308

Flow control 286–287

Footprint 263, 267, 270, 273

Foreign agent (FA) 200, 201, 203, 205–206, 419

Forward access channel (FACH) 256

Forward channel 19, 79, 144–145

Forward control channel (FOCC) 224

Forward error correction (FEC) 85, 294–295, 384, 421

Forward link 222

Forward traffic channel 245

Forward voice channel (FVC) 225, 226

Frame check sequence (FCS) 389

Frame error rate (FER) 79, 249

Frame synchronization 254

Framing 286

Free space path loss 60

Free space propagation 60

Freeze-TCP 294

Frequency band 13–14, 16, 114, 169, 173, 222

Frequency bandwidth 145

Frequency correction channel (FCCH) 232

Frequency division duplexing (FDD) 144, 147

Frequency division multiple access (FDMA) 13–14, 16, 143–145, 196, 339–340

Frequency division multiplexing (FDM) 230

Frequency domain 144

Frequency hopping 16, 207, 367–368, 381

Frequency hopping (FH) 150–152

Frequency hopping pattern 152

Frequency hopping sequence 16, 152, 207

Frequency hopping spread spectrum (FHSS) 151, 370

Frequency hopping technique 16

Frequency interference 12

Frequency modulation (FM) 160, 221

Frequency reuse 110

Frequency reuse distance 171

Frequency reuse factor 115

Frequency-selective fading 75

Frequency shift 161
Frequency shift keying (FSK) 161, 221
Frequency spectrum 222
Front end component 342
Full dump 309
Full duplex 2

G

Gateway 23, 193, 198, 216, 230
  address 193
  MSC 230
Gaussian 67–68
Gaussian distribution 67, 69
  Gaussian minimum shift keying (GMSK) 372
General distribution (arbitrary distribution) 41
General packet radio services (GPRS) 254
Generating function 49
Generator matrix 81
Generator polynomial 81, 85
Geographical information systems (GIS) 306
Geometric distribution 33–34
Geometric random variable 33–34
Geostationary Earth Orbit (GEO) 261, 267
Geosynchronous 273
Global personal communication network 411
Global positioning system (GPS) 273
Global response module (GRM) 444
Global roaming 251, 306
  capability 306
Global system for mobile communications (GSM) 2, 229
Go-Back-N ARQ (GBN ARQ) 93, 94
Grant per connection (GPC) 363
Grant per SS (GPSS) 363
Graphical user interface (GUI) 342
Gravitational acceleration 263
Ground wave 57
Group and list management server (GLMS) 409
Groupe speciale mobile (GSM) 194, 229
GSM-MAP core network 251
Guaranteed frame rate (GFR) 360
Guaranteed time slot (GTS) 385, 390
Guard band 145

Guard channel 413
Guard frame 230

H

Hadamard matrix 152
Hamming codes 79
Handoff 11, 104
Handoff allocation 176
Handoff area 105
Handoff call 171, 180–181, 183–184, 198
Handoff management 411
Handoff rate 106
Hard-decision algorithm 88
Hard-decision decoding 88
Hard handoff 196, 200
Hard threshold (HT) 346, 347–348
Hardware and software resources 11
Hash function 209–210
Hata 62
Heinrich Hertz 399
Hello message 313, 345
Hexagon 12, 102
Hex-cell cluster 111
Hierarchical clustering 345
Hierarchical network 427
High bandwidth 6
Highly directional 23
Highly elliptical orbit (HEO) 261
High-power transmitter 17
Highway network 305
HiperLAN 372–373, 375
  type 1 372
  type 2 372
HiperLAN2 358
Historical query 337
Holding time 108
HOLD mode 383
Home agent (HA) 200–201, 203, 205, 418
Home location register (HLR) 19, 190, 192, 197–199, 201, 228–230
Home MSC 192, 200
HomeRF 369, 375–376
Hopping sequence 207, 368

Host-based firewall 215
Hybrid channel allocation (HCA) 170, 176
Hyperexponential 41
Hyper text markup language (HTML) 438
Hyper text transfer protocol (HTTP) 404–405

I

I/Q modulation 162
ID 193
Identification code 233
Identity matrix 81
Idle signal multiple access (ISMA) 128
ID number 233
IEEE 802.3 375
IEEE 802.11 24, 136, 369, 432–433, 441–442
IEEE 802.11a 358, 369
IEEE 802.11b 358, 369, 377
IEEE 802.11g 358
IEEE 802.15 377
IEEE 802.15.1 358
IEEE 802.15.2 432
IEEE 802.15.3 358
IEEE 802.15.4 358
IEEE 802.15 coexistence TG2 377
IEEE 802.15 WPAN/Bluetooth TG1 377
IEEE 802.15 WPAN/High Rate TG3 378
IEEE 802.15 WPAN/Low Rate TG4 378
IEEE 802.16 358
IEEE 1394 375
Imbedded Markov chain 47
Immobile sensor 23
Independence 36
Independent identically distributed (IID) 43
Indirect-TCP (I-TCP) 295
Industrial, scientific, and medical (ISM) 3
Industrial system 8
Infinite-bit quantization 90
Infinite-dimensional 48
Information channel 19, 125
Information service 2
Infrared (IR) 375
In-phase (I) 162
Inquiry message 381

Instantaneous amplitude 70
Integrated digital enhanced network (iDEN) 408
Integrated service 411
Integrated services (IntServ) 403
Integrity 213
Intellectual property right (IPR) 251
Interarrival time 39
Interautonomous 290
Inter-BSC handoff 237
Inter-cell handoff 237
Interception 213
Interdomain 290
Interference 12
Interference-aware Bluetooth dynamic segmentation (IBLUES) 433
Interleaver 90
Intermediate circular orbit (ICO) 261
Intermediate node 22
Inter-MSC handoff 238
Internal node 23
International Engineering Task Force (IETF) 205, 297, 419
International mobile subscriber identity (IMSI) 232
International mobile telecommunications 2000 (IMT-2000) 6, 249, 403, 411
International MS equipment identity (IMSEI) 234
International Telecommunication Union (ITU) 2, 257, 411
Internet access 6, 10
Internet backbone 292
Internet control message protocol (ICMP) 289
Internet group management protocol (IGMP) 289, 422
Internet Protocol Version 4 (IPv4) 22, 284, 296–297
Internet Protocol Version 6 (IPv6) 22, 284, 296–297
Internet routing protocol 289
Internet service provider (ISP) 367, 375
Internetworking 284
Interruption 213
Intersatellite handoff 273
Intersymbol interference (ISI) 72, 73
Intersystem handoff 273
Inter-zone routing protocol (IERP) 320
Intra-BSC handoff 237
Intra-BTS handoff 237

Intracell handoff 237
Intra-MSC handoff 237
Intrasatellite handoff 273
Intra-UTRAN handoff 254
Intra-zone routing protocol (IARP) 320
Intrusion detection 442
Intrusion detection mechanism (IDM) 442
Intrusion detection system (IDS) 213
Inverse discrete Fourier transform (IDFT) 155
Ionosphere 57
IP 204, 284, 289
IP address 289
IP broadband network 405
IP connectivity 364
IP network 403
IPng (Internet Protocol next generation) 296
IPR 251
IPv4 22, 284, 296–297
IPv4 header format 296
IPv4-host 296–297
IPv4-router 296–297
IPv4-stack 297
IPv6 22, 284, 296–297
IPv6 header format 297
IPv6-stack 297
IS-41 226–228
IS-95 243
ISDN 375
ISI 155
ISM 278
ISM band 367, 369, 378, 380
ISO 221, 284
Isotropic directed antenna 57
I-TCP 295
ITU-R 249

J

Jamming 16, 150, 207
   effect 16
   transmitter 207
Java virtual machine 429
Joint distribution function 35

K

Ka band 265
Kendall's notation 41
Kronecker delta 134
Ku band 265

L

L'Hôpital's rule 49
LAN 136–137, 369, 371, 403
Landmark ad hoc routing (LANMAR) 321
Land mobile radio channel 61
Laplace transform 51
Large area 21
Large city 63
Large zone 17
Least significant bit (LSB) 224
Level crossing rate 70
Levels of security 11
Limited physical security 305
Linear block codes 79–80
Line of sight (LOS) 21
Link configuration 286
Link-cost table 311
Link manager protocol (LMP) 383
Link-state advertisement (LSA) 290
Link state routing 307
Little LEO 267
Little's law 41, 45
Live streaming 405
Load of a cell 108
Local area network (LAN) 127
Local response module (LRM) 444
Location-aided routing (LAR) 321
Location area identity (LAI) 233
Location areas (LA) 234
   code 234
Location management 411
Location registration 411
Locator 3
Locator transmitter 3
Logical addressing 286
Logical channel 257

Logical link control and adaptation layer protocol (L2CAP)   383

Logical link control (LLC)   388

Log-normal distribution   66

Log-normal fading   65, 271

Log-normal fading or shadowing   65

Long-term fading   61

Long-term variation   65

LOS   59, 261, 268

Low earth orbit (LEO)   261, 267

Low energy adaptive clustering hierarchy (LEACH)   345, 427

Low power   4, 387, 398, 437

   design   437

   pulse   398

   signaling   398

   WPAN   387

M

M/D/1   47

M/G/1   47

M/M/1   43

M/M/S   45

M/M/S/S model   181

Macrocellular   11

Macrodiversity   254

Macrosensor   334

Magnet OS   429

Mail service   287

Make before break   196

Manchester frequency modulation   222

Maritime mobile service   3

Markov chain   42

Markovian   41

Markov model   43, 271

Markov process   42

Markup language   438–439

Masquerade   215, 440

Master   381, 383–384

Maximum throughput   129

Maximum-transmission unit (MTU)   289

MBCA/BRCA   407

Mean rate   39

Mean-square value   70

Mean value   32

Media codecs   403

Medical telemetry   3

Medium access control (MAC)   127, 361, 369, 372, 389

   common part sublayer (MCPS-SAP)   388

   footer (MFR)   389

   frame   389

   header (MHR)   389

   layer management entity (MLME-SAP)   389

   PDU   361

   protocol data unit (MPDU)   389

   service data unit (MSDU)   389

Medium and small city   63

Medium earth orbit (MEO)   261

Memoryless property   40

Merging property   40

Message bit   80

Message retransmission list (MRL)   311

Meta-data   404

Meta-language   439

Metropolitan area network (MAN)   127

Microcell radio   368

Microcellular data network (MCDN)   367

Micro electro mechanical systems (MEMS)   429

Microwave   3

Microwave transmitting tower   12

Military defense   10

Mobile ad hoc networks (MANETs)   22, 23, 303–307, 321

Mobile application part (MAP)   235

Mobile communication network   6

Mobile communication system   2

Mobile host (MH)   22

Mobile identification number (MIN)   223, 224, 226

Mobile infrastructure network   6

Mobile internet protocol (Mobile IP)   200, 205, 419

Mobile mesh networking   305

Mobile military network   305

Mobile multicast (MoM)   205–206, 419

Mobile multimedia service   411

Mobile network code   234

Mobile packet radio networking   305

Mobile phone service   2

Mobile platform 23

Mobile radio 3

Mobile radio channel 61

Mobile radio communication 2

Mobile radio propagation 57

Mobile station (MS) 11, 57, 190, 192–201, 203, 205–206, 212

Mobile support router (MSR) 295

Mobile switching center (MSC) 18, 190, 192–193, 196–201, 226, 229–230

Mobile system ISDN (MSISDN) 233

Mobile user 11

Mobile wireless network 22

Mobility 11

Mobility management (MM) 236, 411

Mobility pattern 11

Modification of data 215

Modulation 155

Moment function 31

Monitoring service 3

Moving picture expert group (MPEG) 372

MPEG-1 404

MPEG-2 404

MPEG-4 404

MPEG-4 format 404

MPEG-4 intermedia 404

MSISDN 233

MS roaming number (MSRN) 234

M-TCP protocol 295

Multicast address dynamic client allocation protocol (MADCAP) 422

Multicast backbone (MBONE) 204, 421

Multicast capable routers (MROUTERS) 204

Multicast delivery tree 289

Multicast for mobility protocol (MMP) 419

Multicast home agent (MHA) 419

Multicasting 11, 204

Multihop 22–23, 303, 305

Multilevel fisheye 321

Multilevel overlapped 11

Multimedia 6

Multimedia messaging service (MMS) 252

Multipath 57, 324

Multipath channel 57

Multipath fading 79

Multipath propagation 57

Multipath propagation channel 57, 61

Multipath routing 326

Multiple access 125, 143

multiple access collision avoidance protocol for wireless LANs (MACAW) 340

Multiple access protocol 127

Multiple access technique 143

Multiple address system 3

Multiple channel 31

Multiple destination 11

Multiple paths 148

Multiple radio access 143

Multiple random variable 35

Multiple servers 45

Multiple subscribers 15

Multiplexing 13

Multipoint 3

Multiprotocol label switching (MPLS) 403

Multitier cellular system 180

N

Nakagami distribution 69

Nakagami fading factor 69

Nakagami-m distribution 69

Narrowband 4, 22

NAVSTAR 274

Near-far problem 149, 152, 154

Nearly orthogonal 391

Near-optimal interleaver 90

Negative acknowledgment (NAK) 92

Neighbor discovery protocol (NDP) 320

Neighbor-table-based multipath routing (NTBMR) 331

Network access identifier (NAI) 373

Network characteristic 8

Network file system (NFS) 291

Network layer 127, 285–286, 289

Network protocol 22

Network protocol data unit (NPDU) 309

Network remote control 405

Network topology 303–304, 307–308, 338

Network virtual terminal 287
Newton's gravitational law 262
Next generation 2
Noiselike spectrum 150
Non-collaborative coexistence algorithm 433
Non-collaborative coexistence mechanism 432
Non-collaborative mechanism 432
Non-iDEN 408
Non-light-of-sight (NLOS) 365
Non-negative 30
Nonoverlapping hexagonal subregion 12
Nonpersistent CSMA 131, 133
Non-prioritized handoff 413
Non real-time 29, 411
    service 411
Nonrepudiation 212–213
Normal distribution 34, 38
Normalized condition 44
Normal random variable 34
NPDU 309
nth central moment 33
nth moment 33
Number of events 39

O

OAM 236
Odd-even interleaver 90
ODMA common control channel (OCCCH) 258
ODMA dedicated channel (ODCH) 257
ODMA dedicated control channel (ODCCH) 258
ODMA dedicated traffic channel (ODTCH) 258
Offered load 128
Offered traffic load 108
Offered traffic rate 132
Okumura curves 62
Omnidirectional 23, 117
    antenna 117
On-demand 308, 341
One-time query 337
Open area 63
Open loop power control 249
Open shortest path first (OSPF) 289, 290

Open systems interconnection (OSI) 22, 127, 221, 227, 284
Operator 2
Operators Harmonization Group (OHG) 251
Opportunity driven multiple access random access channel (ORACH) 256
Optimal interleaver 90
Orbit 261–262
Orbiting shape 261
Originating call 171, 180–181, 184, 198
Orthogonal code 15, 152
Orthogonal frequency division multiplexing (OFDM) 13, 143, 159, 340, 369, 372
Orthogonality 15
Orthogonalization 143
Orthogonal waveform 152
Outer loop power control 254
Overlapping area 110
Overloading of resources 11

P

Packet data access node (PDAN) 254
Packet error rate (PER) 435
Packet radio network (PRN) 127
Packet-switched (PS) 254
Packet transmission time 132
PAGE message 381
PAGE state 381
Paging 4
Paging area (PA) 193, 198, 203–204
Paging channel 232, 244
Paging channel (PCH) 256
Paging control channel (PCCH) 257
Parallel input 87
Parallel output 87
Parity bit 80
Parity check matrix 81
Parity matrix 81
PARK mode 383
Passive attack 215, 440
Password 209
Path loss 60–62
PCMCIA 368
PCS access service for controllers (PASC) 241

PCS access service for external service providers (PASE) 241
PCS access service for networks (PASN) 241
PCS access service for ports (PASP) 241
PCS service for data (PASD) 241
PCS wireless provider (PWP) 241
Peak frequency deviation 160
Peak to average ratio (PAR) 402
Peer-to-peer 11, 22, 303
Permutation 207–208
Persistent query 337
Personal access communications systems (PACS) 239
Personal area network (PAN) 127
Personal communication 4
Personal communications services (PCS) 2, 239
Personal digital assistant (PDA) 17
Personal digital cellular 2
Personal handyphone system (PHS) 250
Personal radio service 4
Phase distribution 67–68
Phase offset 162
Phase shift 161–164
Phase shift keying (PSK) 161–162, 164
Physical address 289
Physical addressing 286
Physical channel 258
Physical layer 288
Physical layer (PHY) 127, 285, 369, 391
    protocol data unit (PPDU) 391
    service data unit (PSDU) 391
Physical topology 286
Piconet 381, 383–384, 432
Piggyback 315
Pilot channel 243–244
Ping pong 105
Point-to-multipoint 361
Point-to-point 23, 379
Point-to-point protocol (PPP) 374
Poisson distribution 33–34
Poisson process 39, 128
Poisson random variable 33
Pollaczek-Khinchin (P-K) 51
    transform equation 51

Polling mode 371
Population 31
Population mean value 31
Position indicating radio beacon 3
Power aware routing 324
Power consumption 11
Power control 154–155, 244, 247
Power control channel (PCC) 243
Power level 114
Power saving algorithm 374
p-persistent CSMA 133
p-persistent parameter 132
PPM 402
Presentation layer 285, 287
Primary route 326
Priority handoff 413
Priority reservation handoff 414
Priority reservation with preemptive priority 414
Private key 210–212
Proactive 308, 339
Probabilistic, statistical, and traffic theories 29
Probability density function (pdf) 29, 30
Probability distribution 29, 32
Probability mass function (pmf) 29, 30
Probability of loss 109
Probability of rejection 109
Propagation delay 131, 136, 147
Propagation loss 61, 432
    law 432
Propagation property 57
Protection ratio 75
Protocol 125
Pseudo noise (PN) 243
Pseudo partitioning (PPR) 436
Pseudorandom sequence 150–152
Pseudo role switching (PRS) 436
Public key 210–212
Public land mobile network (PLMN) 233
Public safety answering point (PSAP) 281
Public switched telephone network (PSTN) 18, 198, 376
Pure ALOHA 127, 128, 130
Push-to-talk (PTT) 408
    over Cellular (PoC) 408

Q

QoS 286, 372, 403
Quadrature (Q) 162
Quadrature amplitude modulation (QAM) 164–165
Quadrature phase shift keying (QPSK) 162–163, 165
Quality indicator 412
Quality of service (QoS) 196
Query proxy 342
Queue 40
Queue length 45, 47, 52
Queuing model 43
Queuing scheme 413
Queuing theory 40
Quick time format 404

R

Radar 4
Radio access network (RAN) 254
Radio access system 8
Radio area 102
Radio Common Carrier (RCC) 2
Radio communication 1, 79
Radio communication resource 102
Radio dispatch 2
Radio frequency (RF) 1, 57, 150, 398
    band 57
Radio frequency/intermediate frequency (RF/IF) 402
Radio link 105
Radio link control (RLC) 255, 361, 374
Radio network controller (RNC) 254
Radio network layer (RNL) 255
Radio network subsystem (RNS) 254
Radio path 59
Radio port control unit (RPCU) 242
Radio port (RP) 242
Radio range 12
Radio resource 11
Radio resource control 374
Radio resource management (RR) 236
Radio spectrum 57, 169
Radio telephone service 4
Radio wave 57, 61
RAKE 149, 243

RANAP 255
Random access channel (RACH) 232, 256
Random access protocol 126, 128
Random interleaver 90
Random projection 445
Random variable 29–31
Range-based MoM (RBMoM) 419
Ranging request (RNG-REQ) 363
Ranging response (RNG-RSP) 363
Rapidly deployable communication 305
Rayleigh distribution 67, 69
Rayleigh fading 271
RC4 371
RDMAR 323
Reactive 308, 339–340
Ready-to-receive (RTR) 426
Real number 29
Real-time 6, 29
Real-time service 411
Real-time streaming protocol (RTSP) 405
Real-time transport protocol (RTP) 405
Reassembly 286–287
Receiving antenna 266
    gain 60
Recovery 213
Recursive least squares (RLS) 425
Recursive systematic convolutional (RSC) 91
Reed-Solomon codes 79
Reed-Solomon GF 365
Reflection 59
Relative distance microdiscovery ad hoc routing (RDMAR) 323
Reliable mobile multicast (RM2) 422
Reliable multicast data distribution protocol (RMDP) 421
Remainder 81
Remote authentication dial-in user service (RADIUS) 216
Remote location 8
Remote login 291
Remote procedure call (RPC) 429
Replay 215
Report time 340
Representation of bits 286

Request to send (RTS) 138
Response model (RM) 442
Response time 307
Retransmission 292
Retransmission server (RS) 422
Reuse cluster 169
Reuse distance 110, 114, 169, 171, 173–175, 178–179
Reuse factor 110
Reverse 143
Reverse channel 19, 144–145
Reverse control channel (RECC) 224–225
Reverse link 222
Reverse traffic channel 247
Reverse voice channel (RVC) 225, 226
Rician distribution 68–69
Rician fading 271
Ricochet 366–367
Rlogin 291
RMDP 421–422
Roaming 411
Roofnet 372
Root mean square (rms) 70
ROSE 228
Rotational frequency 263
Round trip propagation time 133
Route computation 307
Route discovery 311
Route error 315
Route maintenance 312
Route network traffic 307
Router 22
Route reply (RREP) 312
Route request (RREQ) 312
Routing 11, 190, 286, 305–307
Routing information protocol (RIP) 289, 290
Routing loop 309
Routing protocol 307, 309
Routing table (RT) 307, 309, 311, 319
Routing zone 320
RREQ 308
RSA algorithm 211
RTCM SC-104 278
RTP 405

Rural radio telephone 4

S

Sample mean 38
Sample size 38
Satellite beam 21, 263
Satellite communication 6, 79
Satellite system 11, 21, 261
Satellite user mapping register (SUMR) 272
Satellite wave 57
SAW ARQ 93
Scalability 305
Scalable communication paradigm 11
Scalable coordination architectures for deeply distributed systems (SCADDS) 338
Scattering 59
SCH 232
Schedule 347
Scheduling 40
SDCCH 232
Second central moment 32–33
Second generation 1, 6, 14–15, 194
Second generation cellular system 14
Sectoring 172
Sectors 24
Secure system 213
Security 11, 190, 212–213, 440
Security attack 213
Security detection 213
Security prevention 213
Seeking 405
Segmentation 286–287
Selective acknowledgement (SACK) 293
Selective availability (SA) 278
Selective frequency fading 148
Selective-Repeat ARQ (SR ARQ) 93, 96
Self-organizing Bluetooth scatternet 436
Self-organizing capability 305
Semirandom interleaver 90
Sensed value (SV) 346
Sensor database 428
Sensor information technology (SensIT) 338
Sensor network 23, 427

Sensor protocols for information via negotiation (SPIN) 342

Serial number (SNR) 234

Service area 17

Service awareness 294

Service data unit (SDU) 361

Service discovery protocol (SDP) 383

Service-point addressing 287

Service specific convergence sublayer (SSCS) 388

Service time 41

Serving RNS relocation 255

Session initiation protocol (SIP) 405

Session layer 285, 287

SGSN 254

Shadowing 61, 65, 271, 432

Shape of a cell 102

Shared channel control channel (SHCCH) 258

Shift register 85, 87

Shortest path 291

Shortest path first (SPF) 290

Short interframe space (SIFS) 138

Short message 1, 8

Short message service (SMS) 238

Short PN codes 243

Short-range RF communication 379

Short-term fading 61, 67

SIFS 138

Signal amplitude 67

Signal bandwidth 79

Signaling 22

Signaling system 7 (SS7) 242

Signaling traffic 244

Signal level 65

Signal polynomial 85

Signal power 61

Signal processing operation 20

Signal quality 271

Signal stability-based adaptive routing protocol (SSR) 319

Signal stability table (SST) 319

Signal strength 12, 61, 104

Signal-to-interference ratio (SIR) 178–179, 412

Signal-to-noise ratio (SNR) 79, 161

Signal transmission power 79

SIM card 233

Simple borrowing scheme 172

Simple mail transfer protocol (SMTP) 291

Simple network management protocol (SNMP) 361

Simplex 2

Simplex channel 19

Single-hop 128

Single server 47

Single sideband (SSB) 159

Single system image (SSI) 429

Size of a cell 102

Sky wave 57

Slave 381, 383–384

Sleep state 371

Slotted 1-persistent CSMA 133

Slotted ALOHA 130–131

Slotted nonpersistent CSMA 132

Slow associated control channel (SACCH) 232

Slow fading 61, 65

Slow fading loss 61

Slowing down 278

Smart antenna 156, 423

Smearing 72

SNIFF mode 383

Snoop protocol 295

Soft-decision algorithm 88

Soft-decision decoding 88

Soft handoff 196

Soft threshold (ST) 346

Source initiated on-demand protocol 308

Source initiated on-demand routing 311

Source routing 313

Space division multiple access (SDMA) 13, 143, 156, 196, 340

Space vehicle (SV) 275, 346

Space wave 57

Specialized mobile radio (SMR) 409

Spectrum 2, 14, 57

Spectrum drought 403

Spectrum spread 149

  codes 149

Speed of light 60

SPIN 342

Split multipath routing (SMR) 329

Split TCP 292, 295

Splitting property 40

Spoofing 305

Spreading 72

Spread spectrum 14, 150, 154, 367

Spread spectrum modulation 150–151

SSCS 389

ST 347–348

Stale route 307

Stand-alone dedicated control channel (SDCCH) 232

Standard deviation 66–67

Standard normal distribution 34

STANDBY mode 381

Stand-by option 271

State diagram 87

State equation 42

State propagation overhead 307

State transition diagram 43, 45

Static channel allocation 340

Static routing protocol (SRP) 319

Statistical characteristics 66

Steady state probability 42

Stochastic process 47

Stop-and-wait ARQ (SAW ARQ) 93

Store-and-forward method 22

Stub network 23

Sub bands 13

Subscriber 7

Subscriber identity module (SIM) 233

Subsequent handoff 238

Subset 30

Subsystem 7

Suburban area 63

Success probability 34

Supervisory audio tone (SAT) 226

Support vector machine-based intrusion detection module (SVMDM) 444

Surface acoustic wave (SAW) 402

Surface of the earth 21

Sustainability of path 11

SVM 444

Switched beam 425

Switch identification (SWID) 227

Switch number (SWNO) 227

Symbol interval 73

Symbol rate 72

Symmetric link 303, 314

Symmetric traffic 384

Sync channel 244

Synchronization 19, 150, 287

Synchronization channel (SCH) 232

Synchronization of bits 286

Synchronous connection oriented (SCO) 383

Syndrome 83, 85

System identification number (SID) 223, 226–227

T

Table driven protocol 308

Table-driven routing protocol 308

TAC 234

Task group (TG) 377

TCP 284, 290, 403

TCP/UDP 215

TCP over wireless 292

TCP-SACK 293

TEEN 427–428

Telecommunication 2

Tele-geoprocessing 306

Telegraphs 1

Telemedicine 306

Telephone modem access (TMA) 368

Telephony 1

Telnet 291

Temporal key integrity protocol (TKIP) 417

Temporally ordered routing algorithm (TORA) 308, 315

Temporary mobile subscriber identity (TMSI) 235

Terrestrial mobile radio communication 79

Third generation 2, 6, 15

Threshold 105, 347

Throughput 94–97, 132–134, 307

TIA 250

Time division duplexing (TDD) 144, 147

Time division multiple access (TDMA) 2, 13, 14, 16, 143–144, 146–147, 196, 339–341, 345, 347–348

TDMA/FDD 147

TDMA/TDD 147

Time division multiplex (TDM) 254

Time division - synchronous CDMA (TD-SCDMA) 2

Time modulated-UWB (TM-UWB) 399

Time multiplexing 16

Time slot 14, 146

Timestamp 293

Time-to-live (TTL) 331

Time-varying communication path 57

TM-UWB 399–401

TORA 308

Traffic 6, 11, 23, 29

Traffic categories (TC) 416

Traffic channel 19, 125, 143, 170, 243, 258

Traffic channels (TCH) 257

Traffic density 180

Traffic filtering 215

Traffic flow 40

Traffic intensity 44, 177

Traffic load 169

Traffic pattern 29

Traffic theory 39

Transition probability 42

Translation 287

Transmission control protocol/internet protocol (TCP/IP) 22, 284, 288

Transmission convergence (TC) 366

Transmission error 79

Transmission frequency 14

Transmission mode 286

Transmission rate 286

Transmission time 94–95, 97

Transmitter 3

Transmitting antenna 12, 60, 266

gain 60

Transmitting power 79

Transmitting station 12, 102

Transmux layer 405

Transport channel 256

Transport layer 285–286

Transport network layer (TNL) 255

Transport unaware link improvement protocol (TULIP) 294

TREE 128

Tree diagram 88

Trellis diagram 88

Triangulation technique 275

Trivial file transfer protocol (TFTP) 361

Troposphere 57

TULIP 294

Tunneling 297

Turbo code decoder 92

Turbo code encoder 91

Turbo codes 79, 91

Type approval code (TAC) 234

Type of traffic 29

U

UE 256

UHF 67

UL channel descriptor (UCD) 364

UL interval usage code (UIUC) 363, 366

UL-MAP 362

Ultra low power sensor platform 429

Ultra-wideband (UWB) 2, 398

UMTS 250, 253, 411

UMTS terrestrial RAN (UTRAN) 254, 255

Uncoded data vector 81

Unconditional (or a priori) probability 36

Unicasting 11

Unidirectional link 303

Uniform distribution 35, 67

Uniform random variable 35

Uniform resource locator (URL) 421

Unipath 324

Universal mobile telecommunication systems (UMTS) 2

User Datagram Protocol (UDP) 294, 405

Unslotted 1-persistent CSMA 133

Unslotted nonpersistent CSMA 132

Uplink 19, 143–145, 222, 265

Uplink shared channel (USCH) 256

Uplink (UL) 361

Urban area 63
Utilization 11
UWB 399–403
UWB-radio technology (UWB-RT) 402

V

Variable capacity link 304
Variance 31
VHF 67
Video signal 79
Virtual local area network (VLAN) 361
Virtual navigation 306
Visiting MSC 192
Visitor location register (VLR) 19, 190
Viterbi algorithm 87
VLR 192, 197–199, 201, 228, 230
VLSI 281
Voice over IP (VoIP) 360, 403
Voice service 24

W

Waiting line 40
Waiting time 45, 47, 52
Walsh codes 152, 243
Walsh function 152
Walsh set 152
WAN 402
WAP-capable 439
Wavelength 57
W-CDMA 6, 250
Wearable computing 305
Web-enabled phone 18
Weighted function 31
Wideband-CDMA (W-CDMA) 2
WiMAX 11, 358–359
WiMedia 385
WINDOW 128
Window size 65
Wired equivalent privacy (WEP) 417, 441, 442
Wired network 23
Wired sensor 334
Wireless access card 370
Wireless and mobile communications 12

Wireless and mobile networking 24
Wireless and mobile systems 6, 57
Wireless ATM 403
Wireless broadband access 402
Wireless card 18
Wireless communication 1, 10
Wireless communications service 4
Wireless control 8
Wireless data communication 20
Wireless Ethernet compatibility alliance (WECA) 371
(Wireless fidelity (Wi-Fi) 358, 371, 417
   alliance 417
   network 417
   protected access (WPA) 417
Wireless home network 402
Wireless LAN 24
Wireless link 22
Wireless local area networks (WLAN) 358
Wireless MAN 24, 359, 369–370, 378, 402, 432
   hub 370
Wireless MAN-SC 364
Wireless markup language (WML) 439
Wireless metropolitan area networks (WMAN) 24, 358
Wireless mobile channel 57
Wireless mode 8
Wireless multimedia 24
Wireless network 23
Wireless networking 305
Wireless pager 24
Wireless personal area networks (WPAN) 358, 377–378, 401–402, 432
Wireless phone service 7
Wireless routing protocol (WRP) 311
Wireless sensor network 334–335, 337, 341–344
Wireless service 13
Wireless subscriber 13
Wireless system 1, 6
Wireless telephone 17
Wireless topology 22
Wireless universal serial bus (WUSB) 401
Wireless widearea transmission control protocol (WTCP) 293
WMAN 24

WMAN-SC 364

Word error indicator 412

Word vector 82

World administration radio conference (WARC) 250

WTCP 293

WWW (World Wide Web) 403

X

X.509 373

Z

Zero window 294

Zone routing protocol (ZRP) 320